Lecture Notes in Biomathematics

Managing Editor: S. Levin

57

Mathematics in Biology and Medicine

Proceedings of an International Conference held in Bari, Italy, July 18–22, 1983

Edited by
V. Capasso, E. Grosso and S. L. Paveri-Fontana

Springer-Verlag

Editors

V. Capasso
Dipartimento di Matematica, Università di Bari
Palazzo Ateneo, 70121 Bari, Italy

E. Grosso †
Istituto di Igiene, Università di Bari
Policlinico, 70100 Bari, Italy

S. L. Paveri-Fontana
Istituto Universitario Navale
Via Acton 38, 80133 Napoli, Italy

AMS-MOS Subject Classification (1980): 34-XX, 35-XX, 45-XX, 60 G XX, 60 H XX, 60 J XX, 60 K 99, 92-06

ISBN-13: 978-3-540-15200-2 e-ISBN-13: 978-3-642-93287-8
DOI: 10.1007/978-3-642-93287-8

PREFACE

These Proceedings contain a large part of the papers presented at the International Conference "Mathematics in Biology and Medicine" organized by the Dipartimento di Matematica and the Istituto di Igiene at the University of Bari, Italy, July 18-22, 1983.

The main objective of the Conference was to bring together scientists in pure and applied mathematics and scientists in biology and medicine. The purpose was to exchange ideas and discuss the common problems encountered in the formulation, analysis and numerical treatment of mathematical models in the biomedical sciences. Simulation methods and problems of validation of models vs. experimental data were also treated. Surveys on recent mathematical results motivated by biological questions were given. Altogether, we think that a balance between mathematical and biomedical aspects was maintained.

The Scientific Committee consisted of E. Biondi (Milano), P. Colli-Franzone (Udine), I. Galligani (Bologna), M. Iannelli (Trento), G. Koch (Roma), E. Marubini (Milano), C. Matessi (Pavia), M. Primicerio (Firenze), L.M. Ricciardi (Napoli), R. Rinaldi (Milano), A. Zampieri (Roma), and the European Liaison Committee consisted of R.M. Anderson (United Kingdom), N.T.J. Bailey (Switzerland), O. Diekmann (The Netherlands), K. Dietz (F.R. Germany), K.P. Hadeler (F.R. Germany), J.P. Kernevez (France).

The Conference was made possible by support from the following institutions: Istituto Universitario Navale di Napoli, Università di Bari, Consiglio Nazionale delle Ricerche - Comitato Nazionale per le Scienze Biologiche e Mediche, Comitato Nazionale per le Scienze Matematiche, Progetto Finalizzato Medicina Preventiva e Riabilitativa, IRMA Bari -, Regione Puglia, Provincia di Bari, Provincia di Taranto, Comune di Bari, Cassa di Risparmio di Puglia, Istituto S. Paolo di Torino, IBM-Italia.

During five days, 19 general lectures were given and 77 contributed papers were presented, 10 of which being part of a poster session; 23 countries were represented through 148 scientists.

We would like to thank all participants for their enthusiastic cooperation. It is a pleasure to acknowledge the help of the Local Organizing Committee. In particular we wish to thank G. Serio for the key work she has done during all stages of preparation and conduction of the Conference. The help of G. Galeone and L. Maddalena is also gratefully acknowledged. Finally we would like to mention that, as a consequence of the severe space limitation on these Proceedings, the members of the Bari group in Mathematical Biology have kindly decided to submit elsewhere their contributions to the Conference.

It is a great grief for us to mention that on May 10, 1984, during the final stages of preparation of this volume Professor Enea Grosso, our Co-Chairman, left us at the age of 62 years. We wish to remember him for his constant inspiration of new scientific work on the biomedical applications of mathematics. To Professor Grosso this volume is dedicated. We hope that young researchers will maintain his enthusiasm alive in developing and studying mathematical models for biomedical phenomena.

It is our hope that -- with the help of modern methods of mathematical analysis and with the support of the revolutionary computer technology -- it will be possible to solve a growing number of challenging problems in the biomedical sciences, bringing mankind to a better quality of life.

V. Capasso

S.L. Paveri-Fontana

OBITUARY

PROFESSOR ENEA GROSSO

On May 10, 1984, the scientific community suffered a great loss with the sudden death of Professor Enea Grosso at the age of 62. It was a particular blow to the Mathematical Epidemiology Group in Bari, which he had constantly encouraged to analyze open problems related to infectious diseases.

Professor Grosso graduated from the University of Milan, his native city, in Medicine in 1949. He remained there as part of the School of Hygiene. His important contributions to the field of atmospheric pollution, food and milk contamination, and the prevention of tuberculosis and polio gave him a very strong reputation in the Italian scientific community and also abroad, as he also spent much time in foreign universities during that period (he spent one year with Prof. A. Sabin at Cincinnati, and another period at the Institut "L. Pasteur" in Paris).

With the recognition of the importance of his research, in 1962 he was appointed Professor of Hygiene at the University of Camerino and in 1964 was called to the University of Bari, in Italy, and remained there for the rest of his life. After his arrival there, he devoted himself to the study of the endemic diseases of Bari and, more generally, of Puglia (the region containing Bari) and of the neighboring region of Basilicata. He soon became the central figure working on problems affecting the health of these two regions.

Professor Grosso had always stressed the importance of primary prevention in the control of epidemics, and this stimulated him to look for new methods of analyzing epidemic systems. The search for answers to the tragic 1973 cholera epidemic, which spread throughout the Mediterranean area, led him to joint work with mathematicians at the University of Bari (and in particular the two editors of this volume) on the mathematical modelling of fecal-oral diseases. From 1978 until his death he was part of the research unit of Mathematical Epidemiology in the Special Programs "Preventive Medicine" and "Control of Infectious Diseases" of the Italian National Research Council (Consiglio Nazionale delle Ricerche).

(V.C.)

LIST OF CONTRIBUTORS

Ahmad R.: Department of Mathematics, University of Strathclyde, G1 1XH Glasgow Scotland, U.K.

Alberghina L.: Cattedra di Biochimica Comparata, Facoltà di Scienze, Università di Milano, Via Celoria 26, 20133 Milano, Italy.

Alt W.: SFB 123 Universität, Im Neuenheimer Feld 293, D-6900 Heidelberg, F.R.G.

Anderson R.M.: Department of Pure and Applied Biology, Imperial College, London University, London SW7 2BB, U.K.

Aronson D.G.: University of Minnesota, School of Mathematics, 127 Vincent Hall, 206 Church Street S.E., Minneapolis, Minnesota 55455, USA.

Audoly S.: Istituto di Matematica per Ingegneri, Università, Viale Merello, 09100 Cagliari, Italy.

Aulbach B.: Mathematisches Institut der Universität, D-8700 Würzburg, F.R.G.

Banks H.T.: Division of Applied Mathematics, Brown University, Providence, RI 02912, USA

Barbi M.: Istituto di Biofisica, C.N.R., Via S. Lorenzo 26, 56100 Pisa, Italy.

Barni N.: Istituto Internazionale di Genetica e Biofisica, CNR, 80100 Napoli, Italy.

Benham C.J.: Department of Mathematics, University of Kentucky, Lexington, KY, Usa 40506.

Beretta E.: Istituto di Biomatematica, Facoltà di Scienze, Università di Urbino, Via A. Saffi 1, 61029 Urbino, Italy.

Bernardi M.L.: Dipartimento di Matematica, Università di Pavia, Strada Nuova 65, 27100 Pavia, Italy.

Bertuzzi A.: Istituto di Analisi dei Sistemi ed Informatica del CNR, Viale Manzoni 30, 00185 Roma, Italy.

Bohl E.: Universität Konstanz, Fakultät fur Mathematik, D-7750 Konstanz 1, R.F.G.

Bozikov J.: Andrjia Stampar School of Public Health, Medical School, University of Zagreb, Zagreb, Yugoslavia.

Bracht A.: Institut für Physiologische Chemie, Physikalische Biochemie und Zellbiologie der Universität München, Goethestrasse 33, D-8000 München 2 F.R.G.

Braumann C.A.: Universidade de Evora, 7000 Evora, Portugal.

Bruni C.: Istituto di Automatica, Università di Roma "La Sapienza", 00100 Roma, Italy.

Calvi Parisetti C.: Istituto matematico, Università di Parma, Via Università 12 43100 Parma, Italy.

Capelo A.C.: Laboratorio di Analisi Numerica del CNR, Corso Carlo Alberto 5 27100 Pavia, Italy.

Caratozzolo F.: Biophysical and Electronic Engineering Division, Electrical Engineering Department, University of Genoa, Viale Causa 13, 16145 Genova, Italy.

Cascino A.: Istituto Internazionale di Genetica e Biofisica,CNR, 80100 Napoli, Italy.

Celentano F.C.: Dipartimento di Biologia, Sezione di Zoologia e Citologia, Via Celoria 26, 20133 Milano, Italy.

Chelazzi G.: Istituto di Zoologia dell'Universtà di Firenze, Via Romana 17 50100 Firenze, Italy.

Chiabrera A.: Biophysical and Electronic Engineering Division, Electrical Engineering Department, University of Genoa, Viale Causa 13, 16145 Genova, Italy.

Chiang C.L.: University of California, Berkeley, California 94720 USA.

Cipollaro M.: III Servizio Analisi, Seconda Scuola Medica, Università di Napoli, 80100 Napoli, Italy.

Clark C.W.: Department of Mathematics, University of British Columbia, Vancouver V6T 1W5 Canada.

Clerico L.: Istituto di Analisi dei Sistemi ed Informatica del CNR, Viale Manzoni 30 00185 Roma,Italy.

Cobelli C.: Istituto di Elettronica e di Elettrotecnica, Facoltà di Ingegneria, Via Gradenigo 6/a, 35131 Padova, Italy.

Colli Franzone P.: Istituto di Matematica, Informatica e Sistemistica dell'Università di Udine, Via Mantica 1, 33100 Udine, Italy.

Cooke K.: Pomona College, Claremont, California 91711, USA.

Costa A.: Oncologia Sperimentale C, Istituto Nazionale per lo Studio e la Cura dei Tumori, 20100 Milano, Italy.

Cvjetanovic B.: Andrjia Stampar School of Public Health, Medical School, University of Zagreb, Rockfellerova 4, 41000 Zagreb.

Dal Passo R.: Istituto per le Applicazioni del Calcolo "M. Picone", CNR, Viale Policlinico 137, I-00161 Roma, Italy.

D'Angiò L.: Istituto Matematico, Università di Cagliari, Via Ospedale 72, 09100 Cagliari, Italy.

Del Bino C.: Oncologia Sperimentale C, Istituto Nazionale per lo Studio e la Cura dei Tumori, 20100 Milano, Italy.

Del Grosso G.: Dipartimento di Matematica, Università degli Studi di Roma "La Sapienza", Piazzale A. Moro 5, 00185 Roma, Italy.

Delimar N.: Immunoloski Zavod, Zagreb 41000, Yugoslavia.

De Mottoni P.: Istituto di Matematica Applicata, Università, 67100 L'Aquila, Italy.

Deneubourg J.L.: Service de Chimie-Physique II ULB, Campus Plaine, Bruxelles, Belgium.

Dezelic G.: Andrjia Stampar School of Public Health, Medical School, University of Zagreb, Rockfellerova 4, 41000 Zagreb.

Dick G.: Zoologisches Institut der Universitat zu Koln, Weyertal 119, 5000 Koln 41, Bundesrepublik Deutschland.

Di Cola G.: Istituto di Matematica, Università degli Studi di Parma, Via università 12, 43100 Parma, Italy.

Diekmann O.: Mathematisch Centrum, Kruislaan 413, 1098 SJ Amsterdam, The Netherlands.

Dietz K.: Institute of Medical Biometry, Tübingen University, Westbahnhofstrasse 55, 7400 Tübingen 1, F.R.G.

Doedel E.: Computer Science Department, Concordia University, 1455 de Maisonneuve, Blvd. W. Montreal, Quebec H3G 1M8, Canada.

Dvorak I.: Center of Biomathematics, Institute of Physiology, Czech. Acad. Sci., Prague 4,

Eisenfeld J.: Department of Mathematics, University of Texas at Arlington, Arlington, 76019 Texas.

Ferrante L.: Istituto di Medicina Sperimentale e Clinica, Università di Ancona, 60100 Ancona, Italy.

Focardi S.: Istituto Nazionale di Biologia della Selvaggina, Via Stradelli Guelfi 23/a, Ozzano Emilia (Bologna), Italy.

Franzè A.: Istituto Internazionale di Genetica e Biofisica, CNR, 35100 Padova, Italy.

Gandolfi A.: Istituto di Analisi dei Sistemi ed Informatica del CNR, Via Manzoni 30, 00185 Roma, Italy .

Gastaldi L.: Dipartimento di Matematica, Università degli Studi di Pavia, Strada Nuova 65, 27100 Pavia, Italy.

Germani A.: Istituto di Analisi dei Sistemi ed Informatica del CNR, Viale Manzoni 3C 00185 Roma, Italy.

Grattarola M.: Dipartimento di Ingegneria Biofisica ed Elettronica, Via dell'Opera Pia 11a, Università di Genova, 16145 Genova, Italy.

Gripenberg G.: Institute of Mathematics, Helsinki University of Technology, SF-02150 Espoo 15, Finland.

Grossman Z.: School of Mathematical Sciences, Tel Aviv University, Tel Aviv 69978, Israel.

Gyllenberg M.: Helsinki University of Technology, Institute of Mechanics, SF-0215C Espoo 15, Finland.

Hadeler K.P.: Lehrstuhl fur Biomathematik der Universitat Tübingen, Institut fur Biologie II, Auf der Morgenstelle 28, 7400 Tübingen 1, Deutschland.

Henry J.: INRIA, Domaine de Voluceau BP 105 - Rocquencourt, 78153 Le Chesnay Cedex, France.

Holden A.V.: Department of Physiology, University of Leeds, Leeds LS2 9JT, U.K.

Iannelli M.: Istituto Matematico, Facoltà di Scienze, Università degli Studi di Trento, 38050 Povo di Trento (Trento), Italy.

Ishii H.: Department of Mathematics, Chuo University, Tokyo 112, Japan.

Kareiva P.: Division of Biology and Medicine, Brown University, Providence, RI 02912, USA.

Kernevez J.P.: Université de Technologie de Compiègne, Department de Mathematiques. Appliquees et d'Informatique, BP 233, 60206 Compiegne Cedex, France.

Kindlmann P.: Centre of Biomathematics, Institute of Entomology, Na Sadkach 702, 370 05 Ceske Budejovice, Czechoslovakia.

Koch G.: Dipartimento di Matematica, Istituto Matematico "G. Castelnuovo", Piazzale A. Moro 5, 00185 Roma, Italy.

Kosicek M.: Immunoloski Zavod, Zagreb 41000, Yugoslavia.

Kubinova L.: Center of Biomathematics, Institute of Physiology, Czech. Acad. Sci., Prague 4, Videnska 1083, Czechoslovakia.

Lakshmikantham V.: Department of Mathematics, University of Texas at Arlington, Arlington, 76019 Texas, USA.

Lambrecht R.M.: Department of Chemistry, Brookhaven National Laboratory, Upton, New York 11973, USA.

Lamm P.K.: Department of Mathematics, Southern Methodist University, Dallas, Texas 75275, USA.

Lazzari C.: Istituto di Biomatematica, Facoltà di Scienze, Università di Urbino, Via A. Saffi 1, 61029 Urbino, Italy.

Lefevre C.: Université Libre de Bruxelles, Campus Plaine, C.P. 219, B-1050 Bruxelles, Belgique.

Lolli E.: Dipartimento di Matematica, 2° Università degli Studi, Via Orazio Raimondo, 00173 Roma, Italy.

Longini I.M.: Department of Biostatistics, University of Michigan, Ann Arbor, Michigan, 48109, USA.

Macchiato M.: Istituto Internazionale di Genetica e Biofisica, CNR, 80100 Napoli. Italy

Malice M.P.: Université Libre de Bruxelles, Campus Plaine, C.P. 219, B-1050 Bruxelles, Belgique.

Marchetti F.: Dipartimento di Matematica, Università di Roma "La Sapienza" Piazzale A. Moro 5, 00185 Roma, Italy.

Mariani L.: Istituto di Elettrotecnica ed Elettronica dell'Università and LADSEB-CNR, 35131 Padova, Italy.

Matis J.H.: Institute of Statistics, Texas A&M University, College Station, Texas 77843, USA.

Mazzini G.: Istituto di Automatica, Università di Roma "La Sapienza",00185 Roma, Italy.

Messia M.G.: Istituto di Matematica Applicata, Facoltà di Ingegneria, 67040 Poggio di Roio, (L'Aquila), Italy.

Mohammedi A.: Ecole Polytechnique Fédérale de Lausanne, Département de Mathématiques, MA Ecublens, CH-1015 Lausanne, Switzerland.

Monticelli G.: Dipartimento di Biologia, Sezione di Zoologia e Citologia, Via Celoria 26, 20133 Milano, Italy.

Musso E.: Istituto di Fisiologia Generale, Università di Parma, 43100 Parma, Italy.

Pierno G.: Istituto Internazionale di Genetica e Biofisica, CNR, 80100 Napoli, Italy.

Pozio M.A.: Dipartimento di Matematica, 2° Università degli Studi, Via Orazio Raimondo, 00173 Roma, Italy.

Renner H.: Institute of Medical Biometry, Tübingen University, Westbahnhofstrasse 55, 7400 Tübingen 1, F.R.G.

Rescigno A.: Istituto di Clinica e Medicina Sperimentale, Università di Ancona, 60100 Ancona, Italy.

Ricciardi L.M.: Istituto Matematico "R. Caccioppoli", Via Mezzocannone 8, 80134 Napoli, Italy.

Rizzoni M.: Dipartimento di Matematica, 2° Università degli Studi, Via Orazio Raimondo; 00173 Roma, Italy.

Rossi C.: Dipartimento di Matematica, 2° Università degli Studi, Via Orazio Raimondo, 00173 Roma, Italy.

Santi E.: Istituto di Matematica Applicata, Facoltà di Ingegneria, 67040 Poggio di Roio (L'Aquila), Italy.

Saunders P.T.: Department of Mathematics, Queen Elizabeth College, University of London, London SW7 2BB, U.K.

Savageau M.A.: Department of Microbiology and Immunology, The University of Michigan, 6643 Medical Science II, Ann Arbor, MI 48109, USA.

Scarlato V.: Istituto Internazionale di Genetica e Biofisica, CNR, 80100 Napoli, Italy.

Schatzman M.: Centre de Mathématiques Appliquées, Ecole Polytechnique, 91 128 Palaiseau Cedex, France.

Schenzle D.: Department of Medical Biometry, Tübingen University, D-7400 Tübingen, F.R.G.

Scholz R.: Institut für Physiologische Chemie, Physikalische Biochemie und Zellbiologie der Universität München, Goethestrasse 33, D-8000 München 2, F.R.G.

Schwab A.J.: Institut für Physiologische Chemie, Physikalische Biochemie und Zellbiologie der Universität München, Goethestrasse 33, D-8000 München 2, F.R.G.

Shonkwiler R.: School of Mathematics, Georgia Institute of Technology, Atlanta, Georgia 30332 USA.

Silvestrini R.: Istituto di Automatica, Università di Roma "La Sapienza", 00100 Roma, Italy.

Siska J.: General Computing Centre, Czech. Acad. Sci., Prague 8, Pod Vodarenskou vezi 4, Czechoslovakia.

Spirito F.: Istituto di Genetica, Facoltà di Scienze, Università di Roma "La Sapienza" Piazzale A. Moro 5, 00185 Roma, Italy.

Spoljaric B.: Immunoloski Zavod, Zagreb 41000, Yugoslavia.

Spouge J.L.: Los Alamos National Laboratory, Theoretical Biology, T-10, Mail Stop 465 P.O. Box 1663, Los Alamos, New Mexico 87545, USA.

Stilli D.: Istituto di Fisiologia Generale, Università degli Studi di Parma, 43100 Parma, Italy.

Taccardi B.: Istituto di Fisiologia Generale, Università degli Studi di Parma, 43100 Parma, Italy.

Takagi I.: Department of Mathematics, Chuo University, Tokyo 112, Japan.

Takeuchi Y.: Department of Applied Mathematics, Faculty of Engineering,Shizuoka University, Hamamatsu 432, Japan.

Tesei A.: Dipartimento di Matematica, 2° Università degli Studi, Via Orazio Raimondo, 00173 Roma, Italy.

Thieme H.R.: Universität Heidelberg, SFB 123, D-6900 Heidelberg, F.R.G.

Thomaseth K.: Istituto di Elettronica e di Elettrotecnica,Università di Padova, Via Belzoni 7, and LADSEB-CNR, 35100 Padova, Italy.

Toffolo G.: Istituto di Elettronica e di Elettrotecnica, Università di Padova, Via Belzoni 7, and LADSEB-CNR, 35100 Padova, Italy.

Tomarelli G.: Dipartimento di Matematica, Università degli Studi di Pavia, Strada Nuova 65, 27100 Pavia, Italy.

Totaro S.: Istituto di Matematica Applicata "G. Sansone", Via S. Marta 3, 50139 Firenze, Italy.

Tramontano A.: Istituto Internazionale di Genetica e Biofisica, CNR, 80100 Napoli, Italy.

Vauchaussade de Chaumont M.: Université de Bordeaux II, 33076 Bordeaux, France.

Viennot G.: Université de Bordeaux I, 33405 Talence, France.

Vitelli R.: Istituto di Analisi dei Sistemi ed Informatica del CNR, Viale Manzoni 30 00185 Roma, Italy.

Viviani R.: Biophysical and Electronic Engineering Division, Electrical Engineering Dept., University of Genoa, Viale Causa 13, 16145 Genova Italy.

Voit E.O.: Zoologisches Institut der Universität zu Koln, Weyertal 119, 5000 Koln 41, F.R.G.

Wherly T.E.: Institute of Statistics, Texas A&M University, College Station, Texas 77843, USA.

Yang G.L.: Department of Mathematics, University of Maryland, College Park, Maryland 20742, USA.

CONTENTS

PART I. - POPULATION GENETICS AND ECOLOGY

Invited Papers

Contributed Papers

PART III. - RESOURCE MANAGEMENT

Invited Papers

Contributed Papers

PART IV. - PHYSIOLOGY AND MEDICINE

Invited Papers

Contributed Papers

PART I

POPULATION GENETICS AND ECOLOGY

THE ROLE OF DIFFUSION IN MATHEMATICAL POPULATION BIOLOGY: SKELLAM REVISITED

D.G. Aronson
School of Mathematics
University of Minnesota
Minneapolis, MN 55455/USA

R.A. Fisher [F] and the famous troika of Kolmogorov, Petrovskii, and Piscunov [KPP] studied diffusion models in population dynamics as early as 1936. However, the first really systematic attempt at a critical examination of the role of diffusion in population biology was begun by J.G. Skellam in his 1951 paper [S1] and essentially completed in his 1973 paper [S2]. Skellam's work fits my definition of the truly classic, that is, work which is often cited but seldom read. It is my object in this lecture to pay homage to J.G. Skellam and to expound <u>some</u> of his ideas. In particular, I want to discuss how the form of the diffusion operator in a diffusion model is determined by the behaviour patterns of the dispersing animals. My hope is that some members of the "we want the beasties to move so lets slap on a Laplacian" - school of diffusion modeling will be inspired to read Skellam and mend their ways. An elegant exposition of Skellam's ideas (and a great deal more) can be found in Okubo's monograph [O] . My presentation will differ somewhat from both [S2] and [O] because I want to emphasize the underlying discrete process and delay as long as possible the passage to the diffusion approximation.

One of the most frequent objections to diffusion models in population biology is that animals do not do important things purely at random. They search for comfort, food, mates, safety, etc. and have evolved specific mechanisms for accomplishing these tasks efficiently. However, just as in the kinetic theory of gasses, a large population moving according to deterministic laws can still be represented by a diffusion process on a scale which is sufficiently large compared to the scale of the individual steps. On the other hand it is not at all clear that all animals respond to environmental stimuli in a completely deterministic way. Many animals can only perceive stimuli at distances which are small compared to the average dispersal step. For example, insects feeding on a plant often cannot gauge the conditions at neighbouring plants without actually visiting them. These insects will therefore have to decide whether to move or not on the basis of local information and their best response may in effect be the choice of an appropriate probability distribution of dispersal steps.

Let me illustrate these ideas with some examples. I will take the attitude that the underlying discrete process is the basic object of study. The partial differential equations of diffusion theory are obtained by taking the so-called diffusion limit and should be regarded as approximate descriptions of the discrete process. In specifying the discrete process the assumptions about the animal's response to environmental stimuli will determine the transition probabilities. Here the only stimulus which I will consider is population density, but the same method can be applied much more generally.

Consider, for simplicity, a population inhabiting a linear array of loci spread a distance h apart and indexed by $\ell \in \mathbb{Z}$. Let

$$u_\ell^n = \text{population density at locus } \ell h \text{ at time } n\tau .$$

Suppose that at time $t = \tau n$ for each nonnegative integer n each individual in the population evaluates his situation and decides whether or not to move from his present location. To completely describe the process it is necessary to know how these decisions are made and I will do this by specifying transition probabilities. Throughout the discussion which follows I will focus exclusively on dispersal mechanisms and ignore birth and death.

Suppose first that the individuals act exclusively on the basis of local information. Let

$$p_j(u_\ell) = \text{probability of a jump of } jh \text{ units starting from locus } \ell h \text{ given that the population density at } \ell h \text{ is } u_\ell ,$$

where it is assumed that $p_j(\cdot) = 0$ for all sufficiently large $|j|$. In keeping with the assumption that no deaths occur, every individual must go somewhere so that we have the conservation law

$$\sum_{j \in \mathbb{Z}} p_j(u) = 1$$

for all $u \in R^+$. By a standard argument, the population density satisfies the recurrence relation

$$u_\ell^{n+1} = \sum_{j \in \mathbb{Z}} u_{\ell-j}^n p_j(u_{\ell-j}^n) \quad \text{for all } \ell \in \mathbb{Z},\ n \in \mathbb{Z}^+ . \tag{1}$$

To obtain the diffusion approximation, one assumes that $u_\ell^n = u(\ell h, n\tau)$ and $p_j(u) = p(jh,u)$ are quite smooth. Then, by formal Taylor expansions, (1) can be put in the form

$$\begin{aligned} \frac{u_\ell^{n+1} - u_\ell^n}{\tau} &= \frac{u_\ell^{n+1} - u_\ell^n \sum_j p_j(u_\ell^n)}{\tau} \\ &= \frac{h^2}{2\tau} \frac{\partial^2}{\partial x^2} \{u_\ell^n \sum_j j^2 p_j(u_\ell^n)\} - \frac{h^2}{\tau} \{u_\ell^n \frac{1}{h} \sum_j j p_j(u_\ell^n)\} + O(\frac{h^3}{\tau}) . \end{aligned} \tag{2}$$

Let $\tau, h \to 0$ and $\ell, n \to \infty$ in such a way that $n\tau \to t$, $\ell h \to x$, and $\frac{h^2}{\tau} \to \lambda \in R^+$. Assume that

$$\frac{1}{h}\sum_{j\in\mathbb{Z}} j p_j(u_\ell^n) \xrightarrow[C^1]{} \psi(u) \quad \text{and} \quad \sum_{j\in\mathbb{Z}} j^2 p_j(u_\ell^n) \xrightarrow[C^2]{} \varphi(u) .$$

Then in the limit (2) becomes the diffusion equation

$$\frac{\partial u}{\partial t} = \frac{\lambda}{2}\frac{\partial^2}{\partial x^2}\{u\varphi(u)\} - \lambda\frac{\partial}{\partial x}\{u\psi(u)\} . \tag{3}$$

Several aspects of equation (3) should be noted. The second term on the right in (3) is a drift term which is zero if the discrete process is symmetric, that is, if $p_j(\cdot) = p_{-j}(\cdot)$ for all $j \in \mathbb{Z}^+$. The diffusion operator in (3) is not in the usual (Fickian) form since all of the derivatives are on the outside. This is the inevitable consequence of our assumption that the transition probabilities depend only on the local density. In particular, it has nothing to do with the fact that we are working in one space dimension. A similar but more tedious argument in R^d leads to the diffusion operator

$$\Delta\{u\varphi(u)\}$$

where Δ denotes the d-dimensional Laplacian.

To see further how the form of the diffusion operator depends on the structure of the transition probabilities, suppose now that they depend not on local conditions but on the average conditions between the beginning and end of the step. Specifically, let

$p(u_{\ell+\frac{1}{2}})$ = probability of a jump from locus ℓh to $(\ell+1)h$ or a jump from $(\ell+1)h$ to ℓh given that the population density at $(\ell+\frac{1}{2})h$ is $u_{\ell+\frac{1}{2}}$

and

$p_0(u_{\ell+\frac{1}{2}}, u_{\ell-\frac{1}{2}})$ = probability of no jump from locus ℓh given that the population density at $(\ell\pm\frac{1}{2})h$ is $u_{\ell\pm\frac{1}{2}}$
$= 1 - p(u_{\ell+\frac{1}{2}}) - p(u_{\ell-\frac{1}{2}})$.

Here I assume, for simplicity, that $u_{\ell+\frac{1}{2}}$ is defined for all $\ell \in \mathbb{Z}$. In practice it suffices to use $\frac{1}{2}(u_\ell + u_{\ell+1})$. In this case the recurrence relation for u_ℓ^n is

$$u_\ell^{n+1} = u_{\ell-1}^n p(u_{\ell-\frac{1}{2}}^n) + u_\ell^n\{1 - p(u_{\ell-\frac{1}{2}}^n) - p(u_{\ell+\frac{1}{2}}^n)\} + u_{\ell+1}^n p(u_{\ell+\frac{1}{2}}^n) . \tag{4}$$

After formal manipulations with Taylor series, (4) becomes

$$\frac{u_\ell^{n+1} - u_\ell^n}{\tau} = \frac{h^2}{2\tau}\frac{\partial}{\partial x}\{p(u_\ell^n)\frac{\partial u_\ell^n}{\partial x}\} + O(\frac{h^3}{\tau}) \quad .$$

Thus, if $\frac{h^2}{\tau} \to \lambda \in R^+$ the diffusion approximation is

$$\frac{\partial u}{\partial t} = \frac{\lambda}{2}\frac{\partial}{\partial x}\{p(u)\frac{\partial u}{\partial x}\} \qquad (5)$$

which is, of course, the standard Fickian form with diffusivity $p(u)$. In R^d the diffusion operator in (5) would be given by

$$\mathrm{div}\{p(u)\,\mathrm{grad}\, u\} \ .$$

This form of the diffusion operator was derived by continuum arguments by Gurtin and MacCamy in [GMC].

Other interesting examples can be found in [O] and [S2], but the two given here should suffice to make the point that a reasonable diffusion approximation must incorporate in a very direct way the specific assumptions about the behaviour of the dispersing animals. Similar considerations can also be applied to the study of populations consisting of several species which disperse and interact (cf. [A]).

Finally I want to comment briefly on the validity of the diffusion approximation. For this purpose consider again the case in which the transition probabilities depend only on local information and suppose, for simplicity, that the process is drift-free. Thus, the diffusion equation is

$$\frac{\partial u}{\partial t} = \frac{\partial^2}{\partial x^2}\{\Phi(u)\}$$

where $\Phi(u) = u\varphi(u)$. Note that, in general, $\Phi(0) = 0$ so that (6) is a degenerate parabolic equation. However, if

$$\Phi'(u) > 0 \quad \text{for} \quad u > 0 \qquad (7)$$

then equation (6) is analogous to the porous medium equation and can be dealt with by means of existing theory ([A], [GMC]). For example, (7) is satisfied by populations which try to avoid crowds. On the other hand, (7) does not hold in general for aggregating populations. For such populations, (6) is not only degenerate but also involves a negative diffusivity so that the standard initial-boundary value problems are not well-posed. The corresponding problems are, however, well-posed for the underlying discrete process. This suggests that the diffusion approximation is not always a good approximation. Further study is needed in this area.

REFERENCES

[A] D.G. Aronson. "Density-dependent interaction-diffusion systems" in _Dynamics and Modelling of Reactive Systems_ edited by W.E. Stewart et al., Academic Press, 1980.

[GMC] M.E. Gurtin and R.C. MacCamy. "On the diffusion of biological populations", Math Biosci., _33_ (1977), 35-49.

[O] A. Okubo. _Diffusion and Ecological Problems: Mathematical Models._ Biomathematics vol. 10, Springer-Verlag, 1980.

[S1] J.G. Skellam. "Random dispersal in theoretical populations", _Biometrica_, _38_ (1951), 196-218.

[S2] J.G. Skellam. "The formulation and interpretation of mathematical models of diffusionary processes in population biology" in _The Mathematical Theory of the Dynamics of Biological Populations_ edited by M.S. Bartlett et al., Academic Press, 1973.

THE DYNAMICS OF STRUCTURED POPULATIONS: SOME EXAMPLES

Odo Diekmann
Mathematisch Centrum, Kruislaan 413
1098 SJ Amsterdam, the Netherlands

1. INTRODUCTION

In realistic models in population ecology individuals are distinguished from one another according to relevant quantities such as age, weight, amount of toxic substances accumulated in the body etc. (Streifer, 1974). The *state of the individual* (i-state) is given by the values of these quantities, whereas the *state of the population* (p-state) is given by the distribution (or density) function describing the number of individuals within each i-state.

In the course of time each individual passes through a trajectory in the i-state space. The beginning and the end of this trajectory are simply its i-state at birth and death, respectively. In between the trajectory is determined by a differential equation describing the instantaneous rate of change of the i-state (aging, growth or accumulation of toxic substances).

Just as in statistical mechanics the evolution of the p-state is governed by a partial differential equation which describes the consequences of these processes at the individual level for the distribution function. The birth and death processes are described by source and sink terms, respectively (one assumes that the number of individuals in the relevant i-states is so large that one can use a deterministic approximation). The continuous change in the i-state is described by a differential operator (such that the characteristics of the first order p.d.e. are precisely the trajectories in the i-state space).

Thus any model for a population with physiological structure consists of at least the three submodels for birth, "growth" and death. The model specification is a description of:

(i) the chances that an individual with some specific i-state is born or dies;

(ii) the rate of i-state change;

both as a function of the p-state, the environmental state and the p-states of all other populations which interact with the one under consideration.

The aim of structured population dynamics is to derive information about the dynamics of the population from information about the dynamics of the inviduals or vice versa. These models provide links between life history studies on the one hand and measurements of population distributions, as a function of time, on the other. They relate knowledge of physiological processes and behavioural patterns to the development of the population as a whole.

A characteristic feature of biological (as opposed to physical models) is the occurence of *non-local terms* (transformed arguments) in the birth term. Consequently, the analysis of such models poses nontrivial and challenging mathematical problems.

One of the reasons to take the population structure into account is to provide a framework for the detailed modelling of the interaction of a population and its environment (or some other population) on the basis of biological knowledge. So, as a rule, one obtains *nonlinear* equations. Although the long term objective is the study of nonlinear problems, I shall here mainly review some recent work on *linear* equations and only in passing will I comment on the incorporation of density dependence. In particular I shall concentrate on the concept of a *stable distribution*. I hope that the examples presented below will give some feeling for the general ideas underlying models of structured populations and that their mathematical analysis serves, apart from its intrinsic interest, as a finger-exercise for the solution of nonlinear problems.

The present paper is based on work by T. Aldenberg, F.H.D. van Batenburg, H.J.A.M. Heijmans, H.A. Lauwerier, J.A.J. Metz, H. Thieme and the author.

2. GROWTH AND DIVISION

Consider a population of unicellular organisms and assume that the physiological state of an arbitrary individual is determined completely by the value of one quantity, denoted by x and called "size", which obeys a physical conservation law (for example, total mass or the amount of nitrogen atoms in the cell). The cells are subject to the following processes: growth, death (= outflow in a chemostat) and fission. Growth is a deterministic process: the rate of size increase of a cell of size x is described by some function $g(x)$, which we assume to be known and to be strictly positive. Death is a stochastic process: the chance (per capita, per unit of time) that a cell of size x dies is described by a nonnegative function $\mu(x)$, which we assume to be known. Fission is a stochastic process. There are (at least) two ways to describe fission into two identical daughters. Let

(i) $b(x)$ be the rate at which individuals of size x divide;

(ii) $\gamma(x)$ be the chance (per unit of size) that an individual will have size x at the moment of division;

then one can assume that either b or γ is known. Under constant environmental conditions this just amounts to two different ways of representing the statistics (or, bookkeeping). By following a cohort one finds the relations

$$b(x) = g(x)\,\frac{\gamma(x)}{1-\int_a^x \gamma(\xi)\,d\xi}\,; \quad \gamma(x) = \frac{b(x)}{g(x)}\,e^{-\int_a^x \frac{b(\xi)}{g(\xi)}d\xi}, \tag{2.1}$$

where by definition a is the smallest size at which fission can occur (i.e., the smallest point in the support of b and γ). So let us assume for the moment that conditions are constant indeed and return to this point later.

Let $n(t,x)$ be the size distribution function. The balance law for n is (Bell & Anderson (1967), Fredrickson, Ramkrishna & Tsuchiya (1967), Sinko & Streifer (1967, 1971))

$$\frac{\partial n}{\partial t}(t,x) = -\frac{\partial}{\partial x}(g(x)n(t,x)) - \mu(x)n(t,x) - b(x)n(t,x) + 4b(2x)n(t,2x). \tag{2.2}$$

In order to let it fit into our general description one should interpret fission as the "death" of the mother cell and the "birth" of two daughter cells, each of them having half the size of the mother. This equation is supplemented by the boundary condition

$$n(t,\tfrac{1}{2}a) = 0, \tag{2.3}$$

which expresses that no cells are created with a size less than $\frac{1}{2}a$, and the initial condition

$$n(0,x) = n_0(x). \tag{2.4}$$

Let us assume that cells have to divide before they reach a maximal size, which we normalize to be one. This amounts to the assumption that $\int_a^1 \gamma(\xi)d\xi = 1$, or (see (2.1))

$$\lim_{\varepsilon\downarrow 0} \int_a^{1-\varepsilon} b(\xi)d\xi = +\infty. \tag{2.5}$$

As a consequence we have to interpret the term $4b(2x)n(t,2x)$ in (2.2) as zero for $x \geq \frac{1}{2}$.

Now suppose that $a \geq \frac{1}{2}$ (i.e., cells cannot undergo two divisions immediately after each other) and put

$$B(t) = n(t,\tfrac{1}{2}). \tag{2.6}$$

Using elementary integration techniques we find the relation

$$B(t) = \int_{\frac{1}{2}a}^{\frac{1}{2}} K(\xi)B(t+G(\xi)-G(2\xi))d\xi, \tag{2.7}$$

for $t \geq \bar{t} = \max\{G(2\xi)-G(\xi) \mid \frac{1}{2}a \leq \xi \leq \frac{1}{2}\}$, where by definition

$$K(x) = 4\frac{b(2x)}{g(2x)}\exp\left(-\int_x^{2x} \frac{\mu(\xi)+b(\xi)}{g(\xi)}d\xi\right) \tag{2.8}$$

and

$$G(x) = \int_{\frac{1}{2}a}^{x} \frac{d\xi}{g(\xi)} \tag{2.9}$$

(note that $G(x)$ = the time it takes a cell to grow from size $\frac{1}{2}a$ to size x; also note that K is integrable and that $\int_{\frac{1}{2}a}^{\frac{1}{2}} K(\xi)d\xi = 2$ when $\mu \equiv 0$). B depends in some complicated manner on the initial condition for $0 \le t \le \bar{t}$. This reduction to a scalar equation reflects the fact that any cell which takes part in the reproduction process necessarily has to pass the size $x = \frac{1}{2}$, so that we can base our bookkeeping on the traffic of cells at this size. We shall analyse three different cases.

<u>Case (i): $g(2x) < 2g(x)$ for $\frac{1}{2}a \le x \le \frac{1}{2}$.</u>

Under this condition the transformation

$$\eta = G(2\xi) - G(\xi) \tag{2.10}$$

is invertible (note that $\frac{d\eta}{d\xi} = \frac{2}{g(2\xi)} - \frac{1}{g(\xi)}$) and we can write (2.7) as the Volterra convolution integral equation

$$B(t) = \int_{G(a)}^{G(1)-G(\frac{1}{2})} K(\xi(\eta))\frac{d\xi}{d\eta}(\eta)B(t-\eta)d\eta, \tag{2.11}$$

which has a *positive* kernel. Hence (see, for instance, Hoppensteadt (1975) or Diekmann (1980))

$$B(t) \sim C_1 e^{\lambda_d t}, \quad t \to +\infty, \tag{2.12}$$

where λ_d, the *dominant eigenvalue*, is the unique real root of the *characteristic equation*

$$\int_{\frac{1}{2}a}^{\frac{1}{2}} K(\xi)e^{\lambda(G(\xi)-G(2\xi))}d\xi = 1, \tag{2.13}$$

which is obtained by substituting $B(t) = \exp \lambda t$ into (2.7). Note that for $\lambda = 0$ the left-hand side of (2.13) can be interpreted as the expected offspring (measured at $x = \frac{1}{2}$) of an arbitrary expectant mother cell passing size $x = \frac{1}{2}$. So $\lambda_d > 0$ iff this quantity exceeds one. The asymptotic behaviour (2.12) is a consequence of the fact that, due to the positivity of the kernel in (2.11), all other roots of (2.13) satisfy $\text{Re}\,\lambda < \lambda_d$. For the size distribution function one finds, after some more calculations,

$$n(t,x) = e^{\lambda_d t}\{C_2 n_d(x) + o(1)\}, \quad t \to +\infty, \tag{2.14}$$

where

$$n_d(x) = \frac{1}{g(x)} \exp\left(-\int_{\frac{1}{2}a}^{x} \frac{\lambda_d+\mu(\xi)+b(\xi)}{g(\xi)}\, d\xi\right)\rho(x) \tag{2.15}$$

with

$$\rho(x) = \begin{cases} 1 \quad , & \tfrac{1}{2} \le x \le 1 \\ \displaystyle\int_{\frac{1}{2}a}^{x} K(\xi) e^{\lambda_d (G(\xi)-G(2\xi))} d\xi, & \tfrac{1}{2}a \le x \le \tfrac{1}{2}, \end{cases} \tag{2.16}$$

and

$$C_2 = \frac{\int_{\frac{1}{2}a}^{\frac{1}{2}} e^{\lambda_d (G(\xi)-G(\frac{1}{2}))} \{\frac{n_o(\xi)}{g(\xi)} + K(\xi) \int_{\frac{1}{2}}^{2\xi} e^{\lambda_d (G(\eta)-G(2\xi))} \frac{n_o(\eta)}{g(\eta)} d\eta\} d\xi}{\int_{\frac{1}{2}a}^{\frac{1}{2}} e^{\lambda_d (G(\xi)-G(2\xi))} (G(\xi)-G(2\xi)) K(\xi) d\xi} . \tag{2.17}$$

In words this says that the population grows exponentially with exponent λ_d (or decays when $\lambda_d < 0$), while the size distribution converges towards the *stable distribution* $n_d(x)$. The initial distribution n_o manifests itself only in the constant C_2. The infinite dimensional dynamics are asymptotically only one-dimensional!

<u>Case (ii): $g(2x) = 2g(x)$ for $\frac{1}{2}a \le x \le \frac{1}{2}$.</u>
Now $G(2\xi) - G(\xi) = G(a)$, a constant, and (2.7) degenerates into the difference equation

$$B(t) = \int_{\frac{1}{2}a}^{\frac{1}{2}} K(\xi) d\xi \; B(t-G(a)). \tag{2.18}$$

The corresponding characteristic equation is

$$e^{-\lambda G(a)} \int_{\frac{1}{2}a}^{\frac{1}{2}} K(\xi) d\xi = 1, \tag{2.19}$$

and all roots

$$\lambda_k = \frac{1}{G(a)} \{\ell n \int_{\frac{1}{2}a}^{\frac{1}{2}} K(\xi) d\xi + 2k\pi i\}, \; k \in \mathbb{Z}, \tag{2.20}$$

lie on a vertical line. This vertical periodicity of the spectrum corresponds to the fact that the evolution according to (2.18) is given by multiplication and periodic continuation. As another manifestation of the big difference between this case and the former we mention that, although the cone of nonnegative functions is left invariant, the solution does attain the value zero for arbitrary large time, if the initial function attains zero.

The biological reason for this remarkable behaviour should be clear from the following observation (Bell & Anderson, 1967): if two cells with equal size divide some time after each other their respective daughters will again have the same size since in the time interval between the two divisions the second mother grows exactly twice as fast as each of the daughters of the first mother! The relation "equal size" is hereditary. Of course this behaviour hinges upon the assumption that each daughter has exactly half the size of the mother. Heijmans (in preparation 1) shows that, also for the case $g(2x) = 2g(x)$, one obtains a stable distribution if the size of a daughter is related to the size of the mother by a smooth probability distribution.

Case (iii): for some $\beta \in (a,1)$ $g(2x) = 2g(x)$ for $\frac{1}{2}a \le x \le \frac{1}{2}\beta$ and $g(2x) < 2g(x)$ for $\frac{1}{2}\beta < x \le \frac{1}{2}$.

Equation (2.7) can be rewritten as the difference-integral equation

$$B(t) = \int_{\frac{1}{2}a}^{\frac{1}{2}\beta} K(\xi)d\xi\, B(t-G(a)) + \int_{G(\beta)-G(\frac{1}{2}\beta)}^{G(1)-G(\frac{1}{2})} K(\xi(\eta))\frac{d\xi}{d\eta}(\eta)B(t-\eta)d\eta. \tag{2.21}$$

The characteristic equation takes the form

$$e^{-\lambda G(a)} \int_{\frac{1}{2}a}^{\frac{1}{2}\beta} K(\xi)d\xi + \int_{\frac{1}{2}\beta}^{\frac{1}{2}} K(\xi)e^{\lambda(G(\xi)-G(2\xi))}d\xi = 1 \tag{2.22}$$

and the unique real root λ_d is dominant: Re $\lambda < \lambda_d$ for all $\lambda \neq \lambda_d$ which satisfy (2.22). The Laplace transform method may be used to show that the asymptotic behaviour is again given by (2.12) and (2.14)-(2.17). We conjecture that the same results hold in every case in which the complement of $\{x \mid g(2x) = 2g(x)\}$ has positive measure. Finally, we mention that cases with the opposite inequality are mathematically similar but biologically irrelevant.

In Diekmann, Heijmans & Thieme (in preparation) the results reported above are proved in a somewhat different manner (without the restriction $a \ge \frac{1}{2}$). Key ingredients are the theory of semigroups of operators, positivity theory (Krein-Rutman theorem) and compactness arguments. The expansion of the solution into generations (in fact finitely many at each fixed time) turns out to be a very useful tool. In case (i) the semigroup is compact after finite time and this guarantees that the spectrum of the semigroup operators consists of isolated eigenvalues which are related to the spectrum of the infinitesimal generator by the mapping $\lambda \to \exp \lambda t$. In case (iii) this relation remains valid in the region $\{\mu \mid |\mu| \ge e^{\sigma t}\}$ for some $\sigma < \lambda_d$ (the essential spectrum is bounded inside the circle $|\mu| = e^{\sigma t}$). The determination of the eigenvalues of the generator (including the explicit derivation of the characteristic equation in the general case) is presented in Heijmans (preprint, 1982). Extensions of these results to periodic environments (periodic g, μ and b) are in preparation.

When trying to apply these calculations to "real" microbial populations it might be difficult to determine g and b experimentally whereas the measurement of the stable distribution might be relatively easy. Thus one is led to consider the inverse problem: given the left-hand side of (2.15), derive information about μ, b and g. This is discussed is some detail in the pioneering paper of Bell & Anderson (1967).

The present model allows for the incorporation of density dependence (or, more precisely, nutrient limitation) in a natural and biologically justified manner.

Indeed, one can describe the available substrate by a dynamical variable S and specify how the growth rate g depends on S and how, in turn, the consumption influences S. However, since now the growth rate becomes a function of time, it matters how the fission process is described: should one take b or γ independent of time? Or, possibly, still some other function? What is the intrinsic quantity?

If one assumes that (i) γ is time-independent; (ii) the substrate concentration influences the growth rate as a factor; (iii) the death rate μ is independent of x; then one can show that the dynamics is described by the linear problem and a nonlinear, implicit, time-scaling (these assumptions imply that growth and fission scale with the same factor). One finds convergence towards the stable distribution (which does *not* depend on the dilution rate or the inflowing substrate concentration) and an asymptotic dynamical behaviour described by an unstructured total population system of ordinary differential equations. So, under these conditions, the time evolution of the size structure decouples from the nonlinear interaction. In other words: because of the stable distribution it is safe to ignore the size structure (note, however, that these results might still be relevant in view of the inverse problem). We refer to Diekmann, Lauwerier, Aldenberg & Metz (preprint, 1983) and Heijmans (in preparation 1) for the details. We intend to study the case where (ii) is not satisfied (i.e., the basal metabolism is taken into account) in the near future.

This example shows that a population can be stabilized through a density-dependent effect on the growth-rate of individuals only. Another recent example of the same phenomenon is presented in Nisbet & Gurney (1983). Gyllenberg (1983) uses a somewhat different approach.

3. THE FUNCTIONAL RESPONSE DERIVED FROM THE BALANCE OF DIGESTION AND PREY CONSUMPTION

The functional response of a predator is the number of prey eaten (per predator, per unit of time) as a function of the prey density. Holling (1959, 1966) has analysed prey-predator interactions in some detail in order to derive the functional response from the underlying processes. An important realistic assumption is that these processes take place on a much shorter time-scale than the population reproduction so that, effectively, prey and predator densities may be considered to be constant when calculating the functional response. As a consequence, the functional response may be derived from a linear equation and subsequently used as an input into the nonlinear equations for the prey-predator dynamics.

Neglecting handling times and concentrating on prey consumption and digestion, Metz & van Batenburg (preprint, 1983) arrive at the equation

$$\frac{\partial p}{\partial t}(s,t) = -\frac{\partial}{\partial s}(f(s)p(s,t)) - xg(s)p(s,t) + xg(s-w)p(s-w,t) \qquad (3.1)$$

where t = time, s = satiation (i.e., some measure for stomach and gut filling or, in other words, the inverse of hunger), and

$p(s,t)$ = the distribution of predators with respect to satiation at time t,

$f(s)$ = digestion rate (experimentally found to be described by - as for some constant $a > 0$),

w = prey weight (assumed to be a constant; Holling (1966) kept it constant in his experiments),

x = prey density,

$g(s)$ = catching tendency ($xg(s)$ = catching rate).

In the terminology of the introduction, a predator which eats a prey "dies" and at the same time a new predator with satiation $s+w$ is "born". One can, alternatively, interpret $p(s,t)$ as the probability that one given predator will be in the i-state s at time t. The term $xg(s-w)p(s-w,t)$ should be interpreted as zero for $0 \leq s \leq w$. Experimentally one finds a satiation threshold and this is reflected in the assumption

$$g(s) = 0 \quad \text{for} \quad s \geq c \text{ (the predator is full up).}$$

As a consequence one has to supplement (3.1) with the boundary condition

$$p(c+w,t) = 0. \tag{3.2}$$

Heijmans (in preparation 2) shows that one can associate with (3.1)-(3.2) a semigroup of bounded linear operators on a space of functions on the interval [0,c+w]. In fact he first treats the backward equation

$$\frac{\partial q}{\partial t}(s,t) = f(s)\frac{\partial q}{\partial s}(s,t) - xg(s)q(s,t) + xg(s)q(s+w,t) \tag{3.3}$$

on the space C[0,c+w] and then interprets the forward equation (3.1)-(3.2) as the adjoint problem of (3.3) in the space of (normalized) bounded variation functions provided with the weak* topology. Again there exists a dominant real eigenvalue λ_d (of course $\lambda_d = 0$; note that the number of predators $\int_0^{s+w} p(\sigma,t)d\sigma$ is a constant which we take to be 1 with a corresponding nonnegative eigenfunction $p_d(s)$ (also normalized to have integral 1). Moreover, the essential spectrum of the semigroup operators consists of a full circle whose radius is given by $\exp(-xg(0)t)$. Thus one finds the asymptotic behaviour

$$p(s,t) = p_d(s) + o(1), \quad t \to +\infty. \tag{3.4}$$

Or, in other words: under the influence of prey consumption and digestion the satiation structure stabilizes in the course of time. The functional response is now defined to be the function

$$x \to x \int_0^{c+w} g(s)p_d(s)ds$$

(note that this is a nonlinear function since $p_d(s)$ depends on x).

Taking formally the limit $w \to 0$, $x \to \infty$, $\xi = xw$ constant, we find

$$\frac{\partial p}{\partial t}(s,t) = -\frac{\partial}{\partial s}((f(s)+\xi g(s))p(s,t)). \tag{3.5}$$

Prey capture is now conceived as a deterministic process: the predator is constantly slurping prey soup. It appears that (3.5) has a stable Dirac delta distribution $\delta(s-\hat{s})$, where $\hat{s}$ is the unique value for which $f(s) + \xi g(s) = 0$. The functional response $\xi \to \xi g(\hat{s})$ describes the amount of prey soup eaten. In the special case that $g(s) = b(1-\frac{s}{c})$ and $f(s) = -as$ one finds $\hat{s} = bc\xi(ac+b\xi)^{-1}$ and the functional response

$$\xi \to \frac{b\xi}{1+\frac{b}{ac}\xi}. \tag{3.6}$$

Numerical experiments of Metz & van Batenburg (in preparation) indicate that the deterministic limit yields a very good approximation in most cases of practical interest. Finally, we mention that Heijmans (in preparation 2) uses a Trotter-Kato type theorem to justify the limiting procedure.

4. REMARKS

In recent years the study of age-structured population dynamics has flourished (Busenberg & Iannelli (to appear), Cushing (1980) Gurney & Nisbet (1980), Gurtin & MacCamy (1979), Prüss (1981, to appear), Webb (1982, to appear)) and in the mathematical analysis semigroup methods have proved to be useful. The present note calls **attention to two points:**

(i) other-than-age-structures are biologically relevant and mathematically interesting;

(ii) semigroup methods are appropriate in this context as well.

Until now the area of nonlinear structured models is largely unexplored. Some special examples of age-structured interactions have been studied (Auslander, Oster & Huffaker (1974), Cushing & Saleem (1982), Frauenthal (1983)), Gurtin & Levine (1979), Hastings & Wollkind (1982); sometimes, but not always, the analysis is based on a reduction to a system of ordinary differential equations which is possible under certain restrictive assumptions) and first attempts to investigate the effect of a density-dependent individual growth rate (due to competition for food) have been made (Nisbet & Gurney (1983) Diekmann et al. (preprint 1983)). In a very interesting paper Botsford (1981) argues that the combined effects of a density dependent individual growth rate and cannibalism can lead to multiple stable equilibria and catastrophic effects of parameter (fishing pressure, for example) variation. (See May (1977) for a discussion of similar phenomena in unstructured population models.) Botsford indeed finds this behaviour in numerical simulations.

Let me mention some more recent work on structured population models (without any claim of completeness) Lasota (1981) and Brunovský (1983) find stable but also chaotic behaviour in a model for the proliferation of differentiating red blood cells (see Lasota, Mackey & Wazenska-Czyżewska, 1981).

Kooijman & Metz (preprint, 1983) study the effects of toxic chemicals on the population growth rate, given the effects on individuals, in the context of a general model for the age- and food-dependent growth and reproduction of individuals (the model is shown to fit the available data on the development of *Daphnia magna* quite well).

Thieme (1982) presents results on stable distributions which apply to many linear and nonlinear models (for instance in epidemiology).

In my opinion these examples underline the need for a general qualitative and quantitative mathematical theory of nonlinear first order partial differential equations with nonlocal terms. At the moment such a theory seems still far-off, but I hope that it will ultimately arise.

ACKNOWLEDGEMENT

The author has had much benefit from many discussions with J.A.J. Metz.

LITERATURE CITED

Auslander, D.M., Oster, G.F. & Huffaker, C.B. (1974). *Dynamics of interacting populations*, J. Franklin Inst. 297: 345-376.

Bell, G.I. & Anderson, E.C. (1967). *Cell growth and division. I. A mathematical model with applications to cell volume distributions in mammalian suspension cultures*, Biophys. J. 7: 329-351.

Botsford, L.W. (1981). *The effects of increased individual growth rates on depressed population size.* American Naturalist 117: 38-63.

Brunovský, P. (1983). *Notes on chaos in the cell population partial differential equation.* Nonlinear Analysis TMA 7: 167-176.

Busenberg, S. & Iannelli, M. (preprint 1982). *Nonlinear diffusion problems in age-structured population dynamics.*

Cushing, J.M. (1980). *Model stability and instability in age structured populations.* J. Theor. Biol. 86: 709-730.

Cushing, J.M. & Saleem, M. (1982). *A predator-prey model with age structure.* J. Math. Biol. 14: 231-250.

Diekmann, O. (1980). *Volterra integral equations and semigroups of operators.* MC Report TW 197.

Diekmann, O., Heijmans, H.J.A.M. & Thieme, H. (in preparation). *On the stability of the cell size distribution.*

Diekmann, O., Lauwerier, H.A., Aldenberg, T. & Metz, J.A.J. (preprint 1983). *Growth, fission and the stable size distribution.* To appear in J. Math. Biol.

Frauenthal, J.C. (1983). *Some simple models of cannibalism.* Math. Biosc. 63: 87-98.

Fredrickson, A.G., Ramkrishna, D. & Tsuchiya, H.M. (1967). *Statistics and dynamics of procaryotic cell populations.* Math. Biosc. 1: 327-374.

Gurney, W.S.C. & Nisbet, R.M. (1980). *Age - and density - dependent population dynamics in static and variable environments.* Theor. Pop. Biol. 17: 321-344.

Gurtin, M.E. & MacCamy, R.C. (1979). *Some simple models for nonlinear age-dependent population dynamics.* Math. Biosc. 43: 199-211.

Gurtin, M.E. & Levine, D.S. (1979). *On predator-prey interactions with predation dependent on age of prey.* Math. Biosc. 47: 207-219.

Gyllenberg, M. (1982). *Nonlinear age-dependent population dynamics in continuously propagated bacterial cultures.* Math. Biosc. 62: 45-74.

Hastings, A. & Wollkind, D. (1982). *Age structure in predator-prey systems. I. A general model and a specific example.* Theor. Pop. Biol. 21: 44-56.

Heijmans, H.J.A.M. (preprint 1982). *An eigenvalue problem related to cell growth.*

Heijmans, H.J.A.M. (in preparation 1). *On the stable size distribution of populations reproducing by fission into two unequal parts.*

Heijmans, H.J.A.M. (in preparation 2). *Holling's 'hungry mantid' model for the functional response considered as a Markov process.* Part III: *Mathematical elaborations.*

Holling, C.S. (1959). *Some characteristics of simple types of predation and parasitism.* Can. Entomol. 91: 385-398.

Holling, C.S. (1966). *The functional response of invertebrate predators to prey density.* Mem. ent. Soc. Canada 48.

Hoppensteadt (1975). *Mathematical theories of populations: demographics, genetics and epidemics.* SIAM.

Kooijman, S.A.L.M. & Metz, J.A.J. (preprint 1983). *On the dynamics of chemically stressed populations: The deduction of population consequences from effects on indiduals.*

Lasota, A. (1981). *Stable and chaotic solutions of a first order partial differential equation.* Nonlinear Anal., Th. Meth. & Appl. 5: 1181-1193.

Lasota, A., Mackey, M.C. & Ważewska-Czyżewska, M. (1981). *Minimizing therapeutically induced anemia.* J. Math. Biol. 13: 149-158.

May, R.M. (1977). *Thresholds and breakpoints in ecosystems with a multiplicity of stable states.* Nature 269: 471-477.

Metz, J.A.J. & van Batenburg, F.H.D. (preprint 1983). *Holling's 'hungry mantid' model for the functional response considered as a Markov process.* Part 0: *A survey of the main ideas and results.*

Metz, J.A.J. & van Batenburg, F.H.D. (in preparation). *Holling's 'hungry mantid' model for the functional response considered as a Markov process.* Part I: *The full model and the deterministic limit.* Part II: *The case of negligible handling time only.*

Nisbet, R.M. & Gurney, W.S.C. (1983). *The systematic formulation of population models for insects with dynamically varying instar duration.* Theor. Pop. Biol. 23: 114-135.

Prüss, J. (1981). *Equilibrium solutions of age-specific population dynamics of several species.* J. Math. Biol. 11: 65-84.

Prüss, J. (preprint 1982). *Stability analysis for equilibria in age-specific population dynamics.*

Sinko, J.W. & Streifer, W. (1967). *A new model for age-size structure of a population.* Ecology 48: 910-918.

Sinko, J.W. & Streifer, W. (1971). *A model for populations reproducing by fission.* Ecology 52: 330-335.

Streifer, W. (1974). *Realistic models in population ecology.* In: MacFadyen, A. (ed.). Advances in Ecological Research 8: 199-266.

Thieme, H.R. (1982). *Linear and nonlinear renewal theorems.* Habilitationsschrift and preprints. SFB 123, Heidelberg.

Webb, G.F. (1982). *Nonlinear semigroups and functional differential equations.* Lecture Notes Scuola Normale Superiore, Pisa.

Webb, G.F. (in preparation). *Theory of nonlinear age-dependent population dynamics.*

MATHEMATICAL PROBLEMSIN THE DESCRIPTION OF AGE STRUCTURED POPULATIONS

Mimmo Iannelli
Dipartimento di Matematica
Università degli Studi di Trento
38050 POVO (Trento)
Italy

1. Introduction

Mathematical models describing the evolution of age-structured populations have received increasing attention in recent years, both for the biological interest and for the mathematical one. In fact age dependent fertility and mortality rates are among the most basic parameters in the theory of population dynamics and demography; and the mathematical problems arising, are interesting and challenging in their own rights.

Here we try to give an idea of the development of the field following the evolutions of the models since the early linear model (section 2) up to recent non-linear models (section 3) including diffusion (section 4). Our exposition obviously touches only the general features of the problems, not being possible to give, in a few pages, a complete description of the many specific models and specific problems. Thus we will be concerned with the description of a single population and will present some classes of models and some general methods of approach.

Elsewhere in this Conference some other mechanisms and problems will be presented such as models relative to cellular growth, epidemics and interacting species. Also, the bibliography we present is somewhat restricted; the reader will find further references in the cited works; in particular the two monographs by M.E. Gurtin [17] and G. Webb [41] will furnish up-to-date material.

2. The linear model

Here we present the early linear model for the description of the evolution of one single population. The dynamics is assumed to be the simplest one, like in the Malthus model.

The state of the population is described by a non-negative function:

$$u(a,t)\ , \quad a \geq 0,\ t \geq 0 \tag{2.1}$$

which denotes the density in age at time t. Thus the quantities:

$$\int_{a_1}^{a_2} u(a,t)\,da\ , \qquad P(t) = \int_0^{\infty} u(a,t)\,da \tag{2.2}$$

respectively give the number of individuals that at time t have ages in $[a_1,a_2]$ and the total population at time t.

The dynamics is due to birth and death processes depending only on age and is described by means of the two non negative functions :

<u>age-specific fertility</u> $\beta(a)$, $a \geq 0$

<u>age-specific mortality</u> $\mu(a)$, $a \geq 0$

with the following meaning :

$\beta(a)u(a,t)da\,dt$ <u>is the number of new borns occurring in the time interval</u> $[t,t+dt]$, <u>due to individuals of ages in</u> $[a,a+da]$

$\mu(a)u(a,t)da\,dt$ <u>is the number of individuals with ages in</u> $[a,a+da]$ <u>dying in the time interval</u> $[t,t+dt]$

From this we see that the <u>birth rate</u> and the <u>death rate</u> of the total populations are respectively given by:

$$B(t) = \int_0^{\infty} \beta(a)u(a,t)\,da \tag{2.3}$$

$$D(t) = \int_0^{\infty} \mu(a)u(a,t)\,da \tag{2.4}$$

so that:

$$\frac{dP}{dt} = B(t) - D(t) \tag{2.5}$$

With the previous definitions for $\beta(a)$ and $\mu(a)$, the equations that have to be satisfied by u can be easily obtained by setting up a balance of birth and death. Namely the following set of equations can be derived

$$u_t + u_a + \mu(a)u = 0$$

$$u(0,t) = \int_0^\infty \beta(a)u(a,t)\,da \tag{2.6}$$

$$u(a,0) = u_o(a)$$

where $u_o(a)$ denotes the initial age distribution and is supposed to be assigned, together with $\beta(a)$ and $\mu(a)$, in order to determine the evolution of $u(a,t)$.

The analysis of (2.6) is strictly related to the following Volterra integral equation on the birth rate $B(t)$ (see 2.3)

$$B(t) = F(t) + \int_0^t K(t-s)B(s)\,ds \qquad t \geq 0 \tag{2.7}$$

where:

$$F(t) = \int_t^\infty \beta(\sigma)u_o(\sigma-t)\Pi(\sigma) \Big/ \Pi(\sigma-t)\,d\sigma \tag{2.8}$$

$$K(t) = \beta(t)\Pi(t) \tag{2.9}$$

and

$$\Pi(t) = \exp\left(-\int_0^t \mu(\sigma)\,d\sigma\right) \tag{2.10}$$

The relation between (2.7) and (2.6) is stated via the formula:

$$u(a,t) = \begin{cases} u_o(a-t)\Pi(a) \Big/ \Pi(a-t) & \text{if} \quad a < t \\ B(t-a)\Pi(a) & \text{if} \quad a \geq t \end{cases} \tag{2.11}$$

Under some mild assumptions on the functions β and μ, it can be shown that the asymptotic behaviour of $B(t)$ (hence of u via 2.11) is related to the unique real root p^* of the characteristic equation :

$$\hat{K}(\lambda) = \int_0^\infty e^{-\lambda a}\beta(a)\Pi(a)\,da = 1 \tag{2.12}$$

in fact :

$$B(t) = b_o e^{p^*t}(1+0(t)) \tag{2.13}$$

where :

$$b_o \geqq 0 \quad \text{and} \quad 0(t) \to 0 \quad \text{as} \quad t \to +\infty \tag{2.14}$$

and

$$b_o = 0 \text{ if and only if } u_o(a) = 0 \text{ for } a \leq \text{maxsupp } \beta$$

The root p^* is an important parameter in the description of the behaviour of the population. It is related to the <u>net rate of reproduction</u> $R = \int_0^\infty \beta(a)\Pi(a)da$ because we have

$$R \gtreqless 1 \quad \text{if and only if} \quad p^* \gtreqless 0 \tag{2.15}$$

But p^* is also related to the existence of <u>product solution</u>, that is solutions of the form $u(a,t) = A(a)T(t)$. In fact it is easy to prove that such a solution has necessarily the following form

$$u^*(a,t) = p_o e^{p^*t}\Pi(a)e^{-p^*a} \Big/ \int_0^\infty \Pi(a)e^{-p^*a}da \tag{2.16}$$

where p_o is the size of the total population at $t = 0$. In correspondence with such a solution, the total population size is given by:

$$P^*(t) = p_o e^{p^*t} \tag{2.17}$$

That is we have Malthusian growth with growth-rate p^*.

A further insight into the model is given by the use of the two functions:

<u>total population size</u>: $P(t) = \int_0^\infty u(a,t)da$

<u>age-profile</u> : $w(a,t) = u(a,t) / P(t)$

In fact, it turns out that the age profile w satisfies an equation of its own, namely:

$$w_t + w_a + \mu(a)w = -w(a,t)\int_0^\infty [\beta(a)-\mu(a)]\, w(a,t)da$$

$$w(0,t) = \int_0^\infty \beta(a)w(a,t)da; \quad \int_0^\infty w(a,t)da = 1 \tag{2.18}$$

$$w(a,0) = w_o(a) = u_o(a) \Big/ \int_0^\infty u_o(a)da$$

Thus we see that the age profile $w(a,t)$ depends only on the initial profile $w_o(a)$. Moreover $w^*(a) = u^*(a,t) / P^*(t) = e^{-p^*a}\Pi(a) / \int_o^\infty \Pi(a)e^{-p^*a}da$ is a stationary solution of (2.18) and under some reasonable assumptions we also have

$$w(a,t) \xrightarrow{t\to+\infty} w^*(a) \quad \text{in} \quad L'([0,+\infty)) \tag{2.19}$$

$$\int_o^\infty [\beta(a)-\mu(a)]\, w(a,t)da = \lambda(t) \xrightarrow{t\to+\infty} p^* \tag{2.20}$$

Once $w(a,t)$ is determined from (2.18) we can look at the equation satisfied by $P(t)$ and get:

$$dP(t)/dt = \lambda(t)P(t) \; ; \quad P(0) = P_o \tag{2.21}$$

where $\lambda(t)$ is given by (2.20) and therefore depends on $w(a,t)$, that is on w_o. Thus different initial profiles give a different $\lambda(t)$ which can be interpreted as a transient Malthusian coefficient, the limiting equation for (2.21) being:

$$dP/dt = p^*P \tag{2.22}$$

A detailed treatment of (2.6) can be found in Gurtin [17] Hoppensteadt [26], Webb [41], extensive applications of the model in demography are developed in Coale [5] and Keyfitz [27].

3. NON LINEAR MODELS

The limitations of the model described in the previous section are the same as those usually claimed for the Malthus model. Now, because of age structure, a large variety of ways can be envisaged through which this dependence is introduced. In 1974 Gurtin and MacCamy [19] first proposed that β and μ depend on the total population size so that eq. (2.6) is modified by substituting $\beta(a)$ and $\mu(a)$ with ;

$$\beta(a,P(t)) \; , \quad \mu(a,P(t))$$

where $P(t) = \int_o^\infty u(a,t)da.$

A very general model allowing functional dependence on $u(a,t)$ has been recently discussed by G. Webb [41]. Problem (2.6) is generalized into:

$$u_t + u_a + \mu(a,u(\cdot,t))u(a,t) = 0$$
$$u(0,t) = G(u(\cdot,t)) \; , \quad u(a,0) = u_o(a) \tag{3.1}$$

where $G(\cdot) : L'([0,+\infty)) \to [0,+\infty)$ and $\mu(\cdot,\cdot) : [0,+\infty) \times L'([0,+\infty)) \to [0,+\infty)$ are Lipschitz continuous. The approach to (3.1) developed in [41], based on the theory of non-linear semigroups and accretive-operator, provides a general reference scheme for the understanding of basic features of the model 3.1. Other approaches dealing with more specific models allow a more detailed analysis in specific situations. Namely some classes of models allow reduction of (3.1) to equations for which established methods of analysis are available.

A class of models allowing reduction to systems of ordinary differential equations has been introduced by Gurtin and MacCamy in [19], and widely used in a number of papers (see for instance [18], [19], [21], [39]) to analyze various population mechanisms. Here β and μ have special forms such as

$$\beta(a,P) \stackrel{def}{=} e^{-\alpha a}\beta(P) \quad , \quad \mu(a,P) \stackrel{def}{=} \mu(P) \tag{3.2}$$

where $\alpha > 0$ and $\beta(\cdot)$, $\mu(\cdot)$ are non-negative functions of the total population size $P(t) = \int_0^\infty u(a,t)da$.

Cosidering the variables $P(t) = \int_0^\infty u(a,t)da$ and $Q(t) = \int_0^\infty e^{-\alpha a}u(a,t)da$ we get the following system:

$$dP/dt = \beta(P)Q - \mu(P)P$$
$$dQ/dt = [\beta(P) - \alpha - \mu(P)]\,Q \tag{3.3}$$

Many different behaviours of (3.3) are then possible, versus different forms for β and μ, including logistic behaviour ([19]) and existence of periodic solutions ([39]).

More general forms for β can also be considered: for instance $\beta(a,P) \stackrel{def}{=} \sum_{oi}^{M} a^i e^{-\alpha a} \beta_i(P)$ allows reductions to a system of $M+2$ equations.

A different class of models, leading to separate analysis for age structure and total population size, comes from assumptions such as :

$$\beta(a,u(\cdot,t)) \stackrel{def}{=} \beta(a) \quad , \quad \mu(a,u(\cdot,t)) \stackrel{def}{=} \mu(a) + F(S(t)) \tag{3.4}$$

with $S(t) = \int_0^\infty \gamma(a)u(a,t)da$. These assumptions can be interpreted saying that $\beta(a)$ and $\mu(a)$ rule an "intrinsic" birth-death process which is age-dependent, while $F(S(t))$ models an "external" mortality which depends on the weighted population size $S(t)$.

If the age-profile $w(a,t) = u(a,t)/P(t)$ is considered, it turns out that w satisfies the same equations (2.18) of the linear model considered in section 2. Thus the evolution of the age profile is not affected by the external mortality $F(S(t))$. Concerning the total population size $P(t) = \int_0^\infty u(a,t)da$ we have :

$$dP(t)/dt = \lambda(t)P(t) - F(\Gamma(t)P(t))P(t) ; \; P(0) = P_0 = \int_0^\infty u_0(a)da \qquad (3.5)$$

where

$$\lambda(t) = \int_0^\infty [\beta(a)-\mu(a)] \, w(a,t)da \; ; \quad \Gamma(t) = \int_0^\infty \gamma(a)w(a,t)da$$

are defined by means of $w(a,t)$ and characterize the transient behaviour of (3.5), in fact (see (2.19), (2.20))

$$\lambda(t) \to p^* , \quad \Gamma(t) \to \Gamma^* = \int_0^\infty \gamma(a)w^*(a)da$$

and (3.5) has the following limiting equations:

$$dP(t)/dt = p^* P(t) - F(\Gamma^* P(t))P(t) \qquad (3.6)$$

Thus the solution of the problem can be put in separated form:

$$u(a,t) = w(a,t)P(t)$$

and its behaviour analyzed via (2.18) and (3.5), (3.6). These <u>separable models</u> have been considered in [1] - [4] also in connection with the diffusion mechanisms which we will discuss in section 4.

Finally, in a wide class of models it is possible to get <u>reduction to Volterra integral and integro-differential equations</u>, applying the same procedure that in the linear case leads to the renewal equation (2.7). An example of this procedure is the following. Let

$$\beta(a,u(\cdot,t)) \stackrel{\text{def}}{=\!=} \beta(a)\Phi(S(t)) , \qquad \mu(a,u(\cdot,t)) \stackrel{\text{def}}{=\!=} \mu(a) \qquad (3.7)$$

where $S(t) = \int_0^\infty \gamma(a)u(a,t)da$. In this case we again have:

$$u(a,t) = \begin{cases} u_0(a-t)\Pi(a) / \Pi(a-t) & a > t \\ B(t-a)\Pi(a) & a \leq t \end{cases}$$

where $B(t)$ is the birth rate. Then we have

$$B(t) = \int_0^\infty \beta(a,u(\cdot,t))da = \int_0^\infty \beta(a)u(a,t)da \; \Phi\left(\int_0^\infty \gamma(a)u(a,t)da\right) =$$

$$= \left[F_1(t) + \int_0^t K_1(t-a)B(a)da \right] \; \Phi\left(F_2(t) + \int_0^t K_2(t-a)B(a)da\right)$$

where $F_1(t) = \int_t^\infty \beta(a)u_o(a-t)\Pi(a)/\Pi(a-t)da$, $F_2(t) = \int_t^\infty \gamma(a)u_o(a-t)\Pi(a)/\Pi(a-t)da$,

$K_1(a) = \beta(a)\Pi(a)$ and $K_2(a) = \gamma(a)\Pi(a)$. Thus we have a non-linear Volterra equation of the type (* denotes convolution)

$$B(t) = F(t, K_1 * B, K_2 * B)$$

Other constitutive forms for β and μ lead to similar equations on variables other than $B(t)$. This class of models has been considered by several authors giving rise to several interesting mathematical problems (see for instance [6], [7], [8], [10], [14], [24], [25], [28], [35], [36], [37], [38], [40]).

4. AGE STRUCTURE AND DIFFUSION

We now consider a single age-structured population, diffusing in a one dimensional habitat. The state of the population is now described by the non-negative function

$$u(a,t,x) \quad a \geqq 0, \; t \geqq 0, \quad x \in I \tag{4.1}$$

where I is a given interval on the real line. (4.1) denotes the density of the population, with respect to age and space, at time t, so that the integral:

$$\int_{a_1}^{a_2} \int_{x_1}^{x_2} u(a,t,x)dx\,da \qquad x_1, x_2 \in I$$

denotes the number of individuals that at time t live in the interval $[x_1, x_2]$ and have ages in $[a_1, a_2]$. $P(t,x) = \int_0^\infty u(a,t,x)da$ is the total population space-density. Now we have to introduce a diffusion mechanism in addition to the birth-death dynamics previously discussed; this is done by defining the population flux $q(a,t,x)$ which has the following meaning:

$q(a,t,x)da\,dt$ is the number of individuals with ages in $[a,a+da]$ crossing at the point x in the time interval $[t,t+dt]$ (4.2)

Different diffusion mechanism will arise from different constitutive equations for the population flux.

A fairly general model for the evolution of the population is the following

$$u_t + u_a + \mu(a,P)u = - \partial/\partial x\, q(a,t,x)$$
$$u(0,t,x) = \int^{\infty} \beta(a,P)u(a,t,x)\,da \qquad (4.3)$$
$$u(a,o,x) = u_o(a,x)$$

where $a \geq 0$ $t \geq 0$ and $x \in [0,1]$. To this system, spatial boundary conditions have to be added such as either:

$$u(a,t,0) = u(a,t,1) = 0 \qquad (4.4)$$

or

$$q(a,t,0) = q(a,t,1) = 0 \qquad (4.5)$$

describing extremely inhospital boundary and invalicable boundary respectively (see [22]

The introduction of diffusion in age structured population was first discussed by Gurtin [16], in the case of a linear birth and death dynamics. The present non-linear model was introduced in Gurtin-MacCamy [20] , [22]. Different assumptions on $\beta(a,P)$, $\mu(a,P)$ and constitutive laws on $q(a,t,x)$ lead to different classes of mathematical problems which we are going to discuss below.

A fully linear model comes from (4.3) when the following constitutive law:

$$q(a,t,x) \stackrel{\text{def}}{=} - \int_o^{\infty} K(a,a')u_x(a',t,x)\,da' \qquad (4.6)$$

is combined with the linear birth-death dynamics:

$$\beta(a,P) \stackrel{\text{def}}{=} \beta(a) \quad , \quad \mu(a,P) \stackrel{\text{def}}{=} \mu(a) \qquad (4.7)$$

The law (4.6) was proposed in Gurtin [16] and looked at as an extension of the analogous equation derived for the case in which age-structure is neglected.

The first mathematical treatment of this linear case was given by Gopalsamy [13] and Marcati-Serafini [33] with the simplifying assumption $K(a,a') = K_o\, \delta(a-a')$ where $K_o > 0$ and δ is the Dirac function centered at 0. With this and (4.7) the first equation in (4.3) becomes:

$$u_t + u_a + \mu(a)u = K_o\, u_{xx} \qquad (4.8)$$

so that the problem can be approached following the lines of the treatment for the linear case without diffusion. In the general case the problem requires a different mathematical treatment which was performed in Di Blasio-Lamberti [11] with the technical assumption on $K(a,a')$:

$$\int_0^\infty \int_0^\infty K(a,a')u(a)u(a')da\, da' \geq 0 \qquad u \in L^2([0,\infty)) \tag{4.9}$$

The same assumption was later used in other works (Langlàis [29], Di Blasio-Iannelli-Sinestrari [9]) where different approaches to the problem are used and a better understanding of the mathematical properties of the problem is reached.

However, more recently the interest has been pointed toward non linear mechanisms in the field of population diffusion (see [15], [20] and also [34]); non linear constitutive laws for $q(a,t,x)$ have been consequently introduced. They were first considered in Gurtin-MacCamy [20], [22] where two different mechanisms are proposed:

$$q(a,t,x) = -K(a,P(t,x))u_x(a,t,x) \quad \underline{\text{random dispersal}} \tag{4.10}$$

$$q(a,t,x) = -K(a,P(t,x))P_x(t,x)u(a,t,x) \quad \underline{\text{directed dispersal}} \tag{4.11}$$

giving rise to different mathematical problems.
Problem (4.3) with (4.4) (resp. (4.5)) and either (4.10) or (4.11) has been the object of several studies, all concerning the case of $K(a,P)$ independent of age a. First in [22] a wide discussion is provided about the problems arising when $\beta(a,P)$ and $\mu(a,P)$ have special forms allowing reduction to systems of partial differential equations. Namely the following cases are considered:

$$\beta(a,P) \overset{\text{def}}{=} \beta(P)a^i e^{-\alpha a}, \quad \mu(a,P) \overset{\text{def}}{=} \mu(P) \tag{4.12}$$

with $i = 0,1$, $\alpha \geq 0$, combined with $K(a,P) \overset{\text{def}}{=} K_0 > 0$, and some aspects of the resulting systems are pointed out. An interesting feature is that in the <u>directed dispersal</u> case one gets degenerate parabolic equations with the consequent complications from the mathematical point of view. In MacCamy [31] the case

$$\beta(a,P) = \beta_0 e^{-\alpha a}, \quad \mu(a,P) = \mu_0, \quad K(a,P) = 1$$

with $\beta_0 > 0$, $\mu_0 > 0$, $\alpha > 0$, is considered and a complete mathematical treatment for the resulting system:

$$P_t = \beta_0 B - \mu_0 P + (PP_x)_x$$

$$B_t = (\beta_0 - \mu_0 - \alpha)B + (BP_x)_x$$

is provided, in connection with the boundary condition (4.4). The existence of a solution which converges to a stationary solution as $t \to +\infty$ is proved. Another set of assumptions leading to "separated" solutions are the following

$$\beta(a,P) \overset{\text{def}}{=} \beta(a) \quad , \quad \mu(a,P) \overset{\text{def}}{=} \mu(a) + F(P), \tag{4.13}$$

$$q(a,t,x) \overset{\text{def}}{=} K_o(P(t,x))P_x(t,x)u(a,t,x). \tag{4.14}$$

(4.13) are the same as (3.4), thus the model describes <u>directed dispersal</u> combined with a linear "intrinsic" birth-death age dependent dynamics plus an external mortality independent of age. This kind of assumption has been considered in Gurtin-MacCamy [23] in connection with the search of product solutions. In Busenberg-Iannelli [1], [2], [3] it is shown that the analysis of problem (4.3) with (4.13) and (4.14) can be carried through by the use of the following variables:

<u>Eulerian total population size</u> $\quad P(t,x) = \int_o^\infty u(a,t,x)\,da$ (4.15)

<u>Flow characteristic curves</u> $\quad \phi(t,t_o,x)$ defined by the equation (4.16)

$$\phi_t(t,t_o,x) = -K_o(P(t,\phi))P_x(t,\phi) \; ; \; \phi(t_o,t_o,x) = x$$

<u>Lagrangian age profile</u> $\quad w(a,t,x) = u(a,t,\phi(t,0,x))\,/\,P(t,\phi(t,0,x))$ (4.17)

With these definitions $u(a,t,x)$ can be recovered via the formula

$$u(a,t,x) = P(t,x)w(a,t,\phi(0,t,x))$$

The interest of these definitions is in that, for each fixed $x \in [0,1]$, the lagrangian profile $w(a,t,x)$, as a function of a and t, satisfies the equation (2.18). Thus for each fixed x, it evolves intrinsically depending only on β and μ. Once $w(a,t,x)$ is known $P(t,x)$ and ϕ can be determined by :

$$\begin{aligned} P_t &= (K_o(P)P_x)_x + \lambda(t,\phi(0,t,x))P \; ; \quad P(0,x) = P_o(x) \\ \phi_t &= -K_o(P(t,\phi))P_x(t,\phi) \quad , \quad \phi(t_o,t_o,x) = x \end{aligned} \tag{4.18}$$

where :

$$\lambda(t,x) = \int_o^\infty [\beta(a)-\mu(a)]\; w(a,t,x)\,da$$

In [1], [3] existence, regularity and asymptotic behaviour are investigated with different choices of $K_o(P)$. The most interesting problems arise when the first equation in (4.18) is degenerate parabolic (see [3]).

The <u>random dispersal</u> case has been analyzed in Langlais [30] . Here $K(a,P)$ (see (4.10)) is again independent of age; some technical assumptions on β and μ as a function of age allow one to state the existence of weak solutions of the problem in the non degenerate case ($K(P) \geq K_o > 0$).

REFERENCES

[1] Busenberg, S. and Iannelli, M.: A class of nonlinear diffusion problems in age-dependent population dynamics, J. Nonlinear Analysis T.M.A., vol. 7 n°5 (1983) 501-529.

[2] Busenberg, S. and Iannelli, M.: Nonlinear diffusion problems in age-structured population dynamics, to appear in Proceedings of the Meeting "Autumn Course on Mathematical Ecology" Trieste 1982.

[3] Busenberg, S. and Iannelli, M.: A degenerate nonlinear diffusion problem in age-structured population dynamics, J. Nonlinear Analysis T.M.A. Vol. 7 n°12 (1983), 1411-1429.

[4] Busenberg, S. and Iannelli, M.: Separable models in age dependent population dynamics. In preparation.

[5] Coale, A.J.: The Growth and Structure of Human Populations, Princeton University Press, Princeton, 1972.

[6] Cushing, J.M.: Volterra integrodifferential equations in population dynamics, Mathematics of Biology, Centro Internazionale Matematico Estivo, Napoli, 1979, 81-148.

[7] Cushing, J.M. and Saleem, M.: A predator-prey model with age structure, J. Math. Biol. 14 (1982) 231-250.

[8] Di Blasio, G.: Nonlinear age-dependent population growth with history-dependent birth rate, Math. Biosci. 46 (1979), 279-291.

[9] Di Blasio, G., Iannelli, M. and Sinestrari, E.: An abstract partial differential equation with a boundary condition of renewal type, Boll. U.M.I. Serie V, Vol. XVIII, - C, n°1 (1981), 260-274.

[10] Di Blasio, G., Iannelli, M., and Sinestrari, E.: Approach to equilibrium in age structured populations with an increasing recruitment process, J. Math. Biol. 13 (1982), 371-382.

[11] Di Blasio, G., Lamberti, L.: An initial-boundary value problem for age-dependent population diffusion, SIAM J. Appl. Math. 35 (1978), 593-615.

[12] Diekmann, O.: The stable size distribution: An example in structured population dynamics, Mathematisch Centrum Report TW 231, Amsterdam.

[13] Gopalsamy, K.: On the asymptotic age distributions in dispersive population, Math. Biosci. 31 (1976), 191-205.

[14] Gripenberg, G.: On a nonlinear integral equation modelling an epidemic in an age-structured population . J. für die reine und angewandte Math. 341 (1983) 54-67.

[15] Gurney, W.S.C. and Nisbet, R.M.: The regulation of inhomogeneous populations. J. Theor. Biol. 52 (1975) 441-457.

[16] Gurtin, M.E.: A system of equations for age-dependent population diffusion, J. Theoret. Biol. 40 (1973), 389-392.

[17] Gurtin, M.E., The Mathematical Theory of Age-Structured Populations, to appear.

[18] Gurtin, M.E. and Levine, D.S.: On predator-prey interaction with predation dependent on age of prey, Math. Biosci. 47 (1979), 207-219.

[19] Gurtin, M.E. and MacCamy, R.C.: Nonlinear age-dependent population dynamics, Arch. Rat. Mech. Anal. 54 (1974), 281-300.

[20] Gurtin, M.E. and MacCamy, R.C.: On the diffusion of biological population, Math. Biosci. 38 (1977), 35-49.

[21] Gurtin, M.E. and MacCamy, R.C.: Some simple models for nonlinear age-dependent population dynamics, Math. Biosci. 43 (1979), 199-211.

[22] Gurtin, M.E. and MacCamy, R.C.: Diffusion models for age structured populations Math. Biosc. 54 (1981) 49-59.

[23] Gurtin, M.E. and MacCamy, R.C.: Product solutions and asymptotic behavior for age-dependent, dispersing populations, Math. Biosc. 62 (1982) 157-167.

[24] Gyllenberg,M.: Nonlinear age-dependent population dynamics in continuously propagated bacterial cultures, Math. Biosc. 62 (1982), 45-74.

[25] Gyllenberg, M.: Stability of a nonlinear age-dependent population model containing a control variable, to appear.

[26] Hoppensteadt, F.: Mathematical Theories of Populations: Demographics Genetics, and Epidemics, SIAM Reg. Conf. Series in Appl. Math. 1975.

[27] Keyfitz, N.: Introduction to the Mathematics of Population, Addison-Wesley, Reading, 1968.

[28] Lamberti, L. and Vernole, P.: Existence and asymptotic behavior of solutions of an age structured population model, Boll. UMI, Anal. Funz. Appl., Serie Vol. XVIII-C (1981).

[29] Langlais, M.: On a linear age dependent population diffusion model, Quarterly of Applied Mathematics 40 (1983) 447-460.

[30] Langlais, M.: A nonlinear problem in age dependent population diffusion, to appear in SIAM J. Math. Anal.

[31] MacCamy, R.C.: A population model with nonlinear diffusion, J. Differential Equations 39 (1981), 52-72.

[32] MacCamy, R.C.: Simple population models with diffusion, Comp. Math. Appl. 8 (1982).

[33] Marcati,P. and Serafini, R.: Asymptotic behavior in age dependent population dynamics with spatial spread, Boll. Un. Mat. Ital. 16-B (1979), 734-753.

[34] Okubo, A.: Diffusion and Ecological Problems: Mathematical Models. Springer Verlag. Biomathematics. Vol. 10 (1980).

[35] Rorres, C.: Stability of an age specific population with density dependent fertility, Theoret. Population Biol. 10 (1976), 26-46.

[36] Rorres, C.: A nonlinear model of population growth in which fertility is dependent on birth rate, SIAM J. Appl. Math. 37 (1979), 423-432.

[37] Sinestrari, E.: Nonlinear age-dependent population growth, J. Math. Biol. 128 (1980), 1-15.

[38] Swick, K.E.: A nonlinear age-dependent model of single species population dynamics, SIAM J. Appl. Math. 22 (1977), 484-498.

[39] Swick, K.E.: Periodic solutions of a nonlinear age-dependent model of single species population dynamics, SIAM J. Math. Anal. 11 (1980), 901-910.

[40] Thieme, H.R.: Renewal theorems for linear periodic Volterra integral equations, to appear.

[41] Webb, G.F.: Theory of non-linear Age-Dependent Population dynamics, to appear.

MODELS FOR MUTUAL ATTRACTION AND AGGREGATION OF MOTILE INDIVIDUALS

Wolfgang Alt, SFB 123 Universität
Im Neuenheimer Feld 293
D-6900 HEIDELBERG, West Germany

Let $u = u(t,x)$ be the density distribution of individuals over $x \in \mathbb{R}$ and $w = w(t,x)$ their mean flux. Then without birth and death the simple conservation law holds

(1) $$\partial_t u + \partial_x w = 0 \quad .$$

Modelling dispersion by Ficks law would result in

(2) $$w = -\mu_0(u) \cdot \partial_x u \quad , \qquad \mu_0(u) \geq 0$$

and (1) would be the usual diffusion equation. In contrast, modelling aggregation would require the opposite sign of μ_0 leading to an ill-posed problem for (1) in general.

A well known idea to override this difficulty originates in the Ginzburg-Landau energy ansatz and gives a fourth order parabolic equation with flux

(3) $$w = \chi_0(u) \cdot \partial_x u + \mu_0 \cdot \partial_x^3 u \quad ;$$

compare [3], for example, where aggregation in a morphogenetic context is described. It can be shown, that under suitable hypotheses on the "attractivity" coefficient $\chi_0(u)$ the equation (1),(3) has locally stable aggregation patterns (joint work with Hans Engler, in preparation).

1. Direct mutual attraction

A different approach was used in an ecological model by MIMURA and YAMAGUTI [8], who included the first integral of u instead of its third derivative, namely

(4) $$w = \chi_0(u) \cdot G[u] - \mu_0(u) \cdot \partial_x u$$

with degenerate motility , for instance

$$\mu_0(u) = \chi_0(u) = u \quad ,$$

and the functional

(5) $$G[u] = \bar{u} - 2\int_{-\infty}^{x} u = \int_{\mathbb{R}} k(y)\, u(x+y)\, dy \quad ,$$

where $\bar{u} = \int_{\mathbb{R}} u$ is the total population size and the convolution kernel is

(6) $\qquad k(y) = \text{sign } y$.

This models the simple rule that individuals are attracted towards the region of higher density. More generally, mutual attraction (in a mean field limit) may be modelled by a flux expression

(7) $$w(x) = \int_{\mathbb{R}} h(u(x),y)\, u(x+y)\, dy$$

with an odd kernel $h(u,\bullet)$. It is easy to see, for example, that for the kernel

(8) $$h_\varepsilon(u,y) = 9\frac{\mu_o}{\varepsilon^3}\begin{cases} 1/\varepsilon - 1/y & : \ o < y \le 2\varepsilon + \varepsilon^3 \delta(u) \\ 0 & : \ \text{otherwise} \end{cases}$$

with $\qquad \delta(u) = \frac{1}{18\mu_o} \cdot \chi_0(u)$

the expression (7) for w coincides with (3) modulo small terms of order ε^2 .

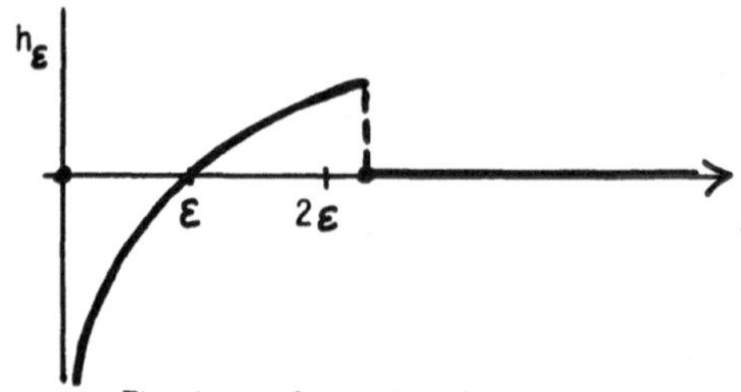

The kernel $h_\varepsilon(u,\bullet)$ in (8).

This means that the Ginzburg-Landau flux (3) represents the localized limit of an integral kernel model with small range repulsion and longer range attraction, whose influence domain is sharply bounded. The last property is not important, since smooth kernels h_ε with the same limiting behavior can be constructed.

However, if the "essential support" of the kernel h_ε does not shrink to $\{0\}$ as $\varepsilon \searrow o$, the resulting flux expression is no longer local. Take, for example,

(9) $$h_\varepsilon(u,y) = \min \left\{ \frac{6}{\varepsilon^2}\mu_0(u)\,(y/\varepsilon - 1) \ , \ \chi_0(u) \right\} ;$$

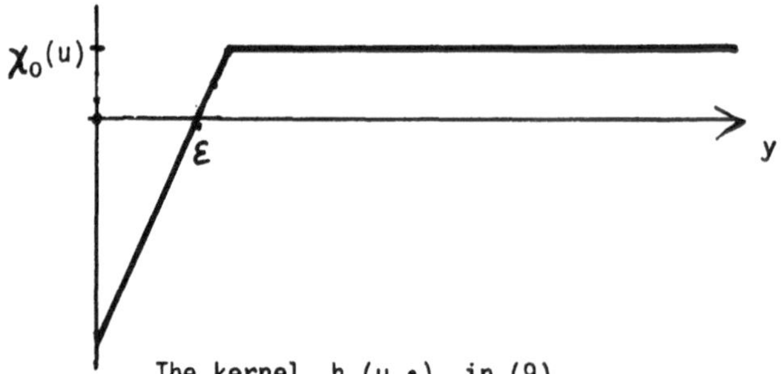

The kernel $h_\varepsilon(u,\bullet)$ in (9).

then the expression (7) for w is equivalent to (4) modulo $O(\varepsilon^2)$ - terms with G as in (5). This means that the Mimura-Yamaguti model is the limit case of a model with short range repulsion together with a finite range attraction. This is obvious from the formulation of (4).

2. Indirect mutual attraction

An individual can be attracted by diffusing mediators produced from other individuals. A classical example is the chemotaxis - system [4, 6] for the cell density u and the mediator concentration ϱ, for instance

$$\left.\begin{aligned} &(10) \qquad \partial_t u = \partial_x\{\mu_0(u)\cdot\partial_x u - \chi_0(u)\cdot\partial_x\varrho\} \\ &(11) \qquad \varepsilon\cdot\partial_t\varrho = \partial_x^2\varrho + 2\cdot u \end{aligned}\right\} \text{ on } [0,1]$$

$$(12) \qquad \left.\begin{aligned} \partial_x u &= 0 \\ \partial_x\varrho &= (-1)^i\delta\cdot\varrho \end{aligned}\right\} \text{ for } x = i = 0, 1 .$$

Solving the equations for ϱ in the limit case $\varepsilon \searrow 0$ and $\delta \searrow 0$ provides a pseudo-steady-state expression for the gradient

$$(13) \qquad \partial_x\varrho = \int_0^1 u - 2\int_0^x u$$

which is exactly the functional (5) for the finite interval $[0,1]$. This means that the above chemotaxis system, in the limit of fast diffusion and production of the mediator, is identical with the Mimura-Yamaguti equations (1),(4),(5). These then lead to stable aggregation phenomena, as shown in [8].

If the dilution (12) of the mediator at the boundaries is replaced by a uniform degradation in the interior, such that the total amount of mediator is preserved, then we can set $\delta = 0$ in (12) and replace (11) by

$$(14) \qquad \varepsilon\cdot\partial_t\varrho = \partial_x^2\varrho + u - \bar{u} .$$

In the same limit $\varepsilon \searrow 0$ we get the pseudo-steady-state gradient

$$(15) \qquad \partial_x\varrho = \int_0^x (\bar{u} - u)$$

and

$$(16) \qquad w = \chi_0(u)\int_0^x(\bar{u} - u) - \mu_0(u)\partial_x u$$

a slight variation of the Mimura-Yamaguti model (4),(5).

3. Hyperbolic equations

In cases where the random motility of individuals is small compared to the attractive flux, we might assume

$$\mu_0 \approx 0 .$$

The special assumption $\chi_0(u) = u$ in model (16) then gives

$$w = u \cdot \int_0^x (\bar{u} - u) . \tag{17}$$

Together with the conservation law (1) for u this system can be regarded as one prototype of a simple "hyperbolic" aggregation model. In fact, the same flux (17) was obtained in a model for a highly viscous actomyosin system with mutual attraction between the actin filaments, see [2, 5].

Introducing the mean velocity

$$v = \int_0^x (\bar{u} - u) \tag{18}$$

with the properties $v = 0$ at $x = 0,1$ and

$$\partial_t v + v \cdot \partial_x v = \lambda \cdot v \quad , \quad \lambda = \bar{u} ,$$

we can solve this problem using the method of characteristics, compare also [2] : Each constant steady state $u \equiv u_\infty , v \equiv 0$ is unstable. A symmetric initial perturbation $u_0(x) = u_0(1-x)$, for example, with exactly one maximum leads to a shock formation in finite time

$$T_0 = \frac{1}{\lambda} \log \left(1 + \frac{1}{u_0(\frac{1}{2})/\lambda - 1} \right) .$$

For $t > T_0$ the density

$$u(t,\cdot) = \tilde{u}(t,\cdot) + M(t) \cdot \delta_{1/2}$$

is the sum of a smooth term $\tilde{u}$ and a Dirac distribution on the shock line $\{x = \frac{1}{2}\}$ with mass $M(t)$. From $M(T_0) = 0$ this mass grows and converges to

$$M_\infty = \overline{u_0} = \lambda$$

as $t \to \infty$, while the smooth part $\tilde{u}(t,x)$ converges to zero, uniformly for all points $x \neq 1/2$, see the figure below. (In the general case of a perturbation with several maxima several shock lines can appear at different times, which then for larger time converge to each other or towards one of the boundary points.)

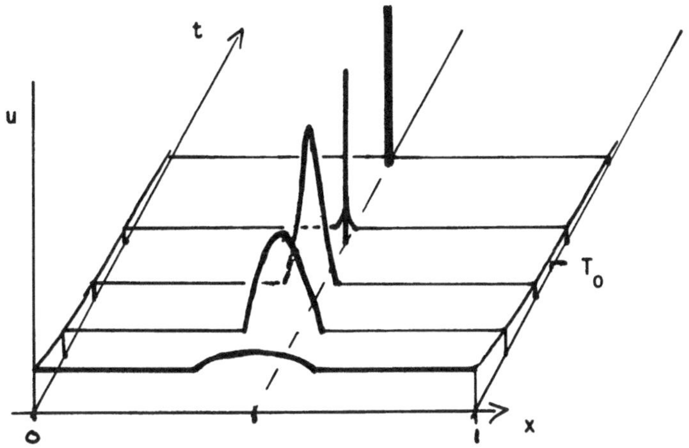

Solution of problem (1),(17) for a symmetric initial perturbation u_0 .

This means that the total population mass λ of the initial distribution u_0 collapses into a sharp aggregation center. With nonvanishing motility μ_0 the same phenomenon occurs, but the aggregative distribution is smeared out, compare [4, 8] , for example.

4. Degenerate parabolic equations

The choice $\mu_0(u) = \chi_0(u) = u$ in (1),(4) means that motility and attractivity degenerate for low densities $u \gtrsim 0$. This assumption is essential for the proofs in [8]. On the other hand, many locomoting cells as myxobacteria, for example, [7], have a normal motility for low concentrations, but as the cells aggregate the high density u prohibits any movement. In these situations a degeneration

(19) $$\mu_0(u) \sim [a - u]_+$$

seems to be more realistic, compare also [9] . If moreover

(20) $$\chi_0(u) \sim u \cdot \mu_0(u) \;,$$

then the resulting degenerate parabolic equation (1),(16) has been investigated in [1]. Under certain conditions on the parameters we can establish locally stable aggregation patterns of the following form

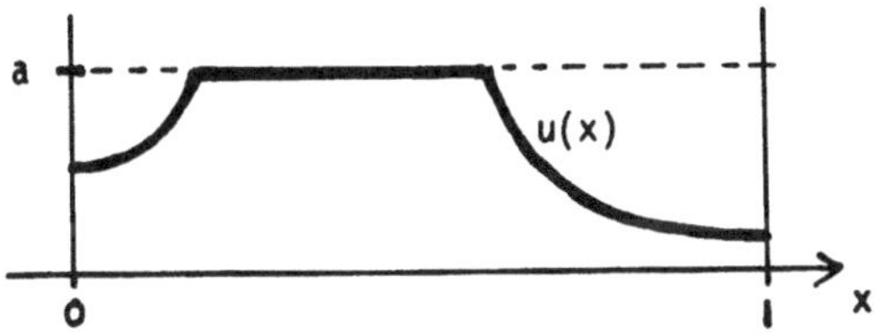

Steady state solution of (1),(17) with no flux boundary condition and with the hypotheses (19),(20).

This result is proven by constructing certain Lyapunov functions for the density u and by applying a comparison principle for the velocity v defined in (18).

It should be emphasized that both techniques cannot be used for the analysis of the strongly coupled parabolic system (10)-(12) with $\varepsilon > o$. Also, comparison techniques for the forth order parabolic equation (1),(3) do not seem to work. From this we conclude, that the mathematical description of "aggregation" can be performed in an easier and sufficient way by scalar diffusion equations which include drift functionals as (5) or (15). Moreover, these kindsof functionals are able to model direct or indirect attraction between individuals over non-local ranges. This seems to occur in many ecological situations.

References

1. ALT, W.: Degenerate diffusion equations with drift functionals modelling aggregation (preprint)
2. ALT, W.; DEMBO, M.: A contraction disassembly model for intracellular actinggels. Proc. Equadiff Conf. Würzburg, August 1982 (to appear in Springer Lect. Notes in Math.)
3. COHEN, D.S.; MURRAY, J.D.: A generalized diffusion model for growth and dispersal in a population. J. Math. Biol. 12, 237-249 (1981)
4. CHILDRESS, S.; PERCUS, J.K.: Nonlinear aspects of chemotaxis. Math. Biosciences 56, 217-237 (1981)
5. DEMBO, M.; HARLOW, F.H.; ALT, W.: The biophysics of cell surface motility. In: Cell surface dynamics: Concepts and models (eds. Perelson, deLisi,Weigel) Marcel Dekker, New York (to appear 1983)
6. KELLER, E.F.; SEGEL, L.A.: Models for chemotaxis. J. Theor. Biol. 30, 225-234 (1971)
7. LAUFFENBURGER, D.A.: Chemotaxis and cell aggregation models in microbial ecology and inflammation. Workshop on Patterns in Space and Time, Heidelberg 1983 (to appear in Springer Lect. Notes in Math.)
8. MIMURA, M.; YAMAGUTI, M.: Pattern formation in interacting and diffusing systems in population biology. (preprint 1981)
9. De MOTTONI, P.: Stabilization properties for nonlinear degenerate parabolic equations with cut-off diffusivity. In: Systems of nonlinear partial differential equations (ed. J.M.Ball) NATO ASI-Series C 111. Reidel, Dordrecht 1983

A Problem on Invariant Manifolds arising in Population Genetics

Bernd Aulbach*
Mathematisches Institut
der Universität
D-8700 Würzburg
West Germany

1. Introduction

In the Fisher - Wright - Haldane model from population genetics (see Crow and Kimura [6], Edwards [7], Hadeler [9], [10]) the mean fitness of a population of diploid individuals increases as time evolves and remains constant only when the population is in equilibrium. In mathematical terms this means that the mean fitness is a global Ljapunov function, which ensures that any solution of the underlying differential equation or difference equation, respectively, approaches the set E of stationary solutions. Both in the continuous time and the discrete time version of the Fisher-Wright-Haldane model the set E generally is the union of linear submanifolds of the state space (see Hughes and Seneta [15] and Aulbach and Hadeler [5]) and so the following problem arises. IF A SOLUTION APPROACHES A CONTINUUM C OF STATIONARY SOLUTIONS, DOES IT THEN CONVERGE TO A POINT ON C OR MAY IT IGNORE THE STATIONARY FLOW ON C BY CREEPING ALONG C FOR ALL FUTURE TIME? As far as the concrete equations of the Fisher-Wright-Haldane model are concerned this question has found partial answers by Feller [8], an der Heiden [12], and Aulbach and Hadeler [5]. Meanwhile the general answer has been given by Losert and Akin [17]. They have shown by means of certain local Ljapunov functions that any solution of the corresponding differential equation or difference equation, respectively, converges to a stationary solution as time goes to infinity. In this paper we want to approach the above problem, which also arises in other model equations, without any reference to the particular form of the underlying equation.

* Sponsored by the Volkswagenwerk foundation.

2. The continuous time case

We consider autonomous differential equations

$$\dot{x} = f(x), \tag{1}$$

where $f: \mathbb{R}^n \to \mathbb{R}^n$ is at least three times continuously differentiable, and we suppose that (1) admits an m-dimensional $(0 \leq m \leq n)$ differentiable manifold M of stationary solutions. Our main result for this situation is as follows.

THEOREM 1: Let $x(t)$ be any solution of (1) and denote the ω-limit set of $x(t)$ by Ω. Suppose that

(1) there exists a point $x^* \in \Omega$,

(2) there exists an $\mathbb{R}^n$- neighborhood U of x^* such that $\Omega \cap U \subset M$,

(3) $n - m$ eigenvalues of the Jacobian of f at x^* have real parts different from 0.

Then $\lim_{t\to\infty} x(t) = x^*$.

REMARKS:

1. It was essentially a result of this type that was applied in Aulbach and Hadeler [5] to the Fisher-Wright-Haldane model. The outcome was a positive answer to the above convergence problem; however, the solution behavior near some exceptional points was not covered.

2. Theorem 1 can be viewed as a (not quite trivial) consequence of Hirsch, Pugh, Shub [14, Theorem 4.1]. A proof based on purely classical analysis is given in [3] where it is also shown by means of examples that the theorem becomes false if either one of the three hypotheses is dropped. Remarks on the history of the convergence problem and more references can be found in [1],[2],[3],[4].

3. The discrete time case

We consider autonomous difference equations

$$x(k+1) = f(x(k)), \qquad (2)$$

where f is a C^3-mapping of $\mathbb{R}^n$ into itself. The discrete time k ranges in the set $\mathbb{N}_0$ of nonnegative integers. We suppose that there exists an m-dimensional $(0 \leq m \leq n)$ C^3-manifold M of fixed points of f. In this case we get the following result.

THEOREM 2: Let x(k) be any solution of (2) for $k \geq 0$ with ω-limit set Ω (= set of accumulation points of the sequence $x(k), k \in \mathbb{N}_0$). Suppose that

(1) there exists a point $x^* \in \Omega$,

(2) there exists an $\mathbb{R}^n$-neighborhood U of x^* such that $\Omega \cap U \subset M$,

(3) n - m eigenvalues of the Jacobian of f at x^* have moduli different from 1,

(4) x(k) is bounded as $k \to \infty$,

(5) each ω-limit point of x(k) is a stationary solution of equation (2).

Then $\lim_{k \to \infty} x(k) = x^*$.

Remarks:

1. Unlike the case of differential equations the solution paths of difference equations are not connected. This deficiency causes problems in our situation which can be overcome by means of the hypotheses (4) and (5) of Theorem 2, whose counterparts were not neccessary in the continuous time case. Those hypotheses guarantee that $\lim_{k \to \infty} [x(k+1)-x(k)] = 0$, a relation which is sufficient for our purposes. The proof hereof is the same as that in Hadeler [10,p.101] for a similar situation.

2. As in the continuous time case this rather general theorem can be applied to the Fisher-Wright-Haldane model. The results one obtains then are the discrete analogues of those in Aulbach and Hadeler [5] for differential equations. We do not state them here because they are contained in the paper [17] by Losert and Akin.

3. The validity of Theorem 2 is known in case f is supposed to be a global diffeomorphism. This follows from Hirsch, Pugh, Shub [14, Theorem 4.1]. In this paper we show that the theorem remains true if f is supposed to be only a differentiable map which means that the machinery of differential topology is not available. We want to emphasize

that in the Fisher-Wright-Haldane model just this non-diffeomorphism case occurs. This is because one of the eigenvalues of the Jacobian of f at any fixed point of f is always 0.

4. Although our convergence problem is a global one the key idea of the proof is to use a suitable local normal form of the map f near x*. In the diffeomorphism case the map f is topologically conjugate to its linearization around x*. Since the concept of topological conjugacy is not available if the linearization of f around x* is singular, we derive a certain quasilinear normal form (see (6) below) which allows us to solve the nonlinear global convergence problem by means of an essentially linear and local analysis.

4. Proof of Theorem 2

Rather than considering solutions of the difference equation (2) we temporarily study the action of the map f on points of $\mathbb{R}^n$. Using a local coordinate chart (φ, U_φ) of M near x* we may represent f in Euclidean (u,v,w)-coordinates as a C^3-map $g = \varphi f \varphi^{-1}$ which is defined on an open neighborhood of the origin in $\mathbb{R}^n$. We arrange the local coordinates such that the set $\varphi(U_\varphi)$ of g-fixed points is, in the relative topology, an open neighborhood of the origin in the subspace of $\mathbb{R}^n$ with vanishing u- and v-coordinates. Thus g assumes the form

$$g\begin{pmatrix} u \\ v \\ w \end{pmatrix} = \begin{pmatrix} A^- u + r_1(u,v,w) \\ A^+ v + r_2(u,v,w) \\ w + r_3(u,v,w) \end{pmatrix} \qquad (3)$$

where $u \in \mathbb{R}^{n_-}$, $v \in \mathbb{R}^{n_+}$, $w \in \mathbb{R}^m$, $n_- + n_+ + m = n$. Here A^-, A^+ are constant matrices whose eigenvalues have moduli less than 1 and greater than 1, respectively. The functions $r_i(u,v,w)$, i=1,2,3, are C^2-nonlinearities of order $o(\|(u,v,w)\|)$ as $(u,v,w) \to 0$. Furthermore, the location of the manifold of fixed points implies that the functions $r_i(u,v,w)$ vanish for $u = 0, v = 0$ and $\|w\|$ suitably small. This in particular means that near $(u,v,w) = (0,0,0)$ the map g may be represented in the quasilinear form (see Hartman [11,V.Lemma 3.1])

$$g\begin{pmatrix}u\\v\\w\end{pmatrix} = \begin{pmatrix}A^-u + A_1(u,v,w)u + B_1(u,v,w)v\\ A^+v + A_2(u,v,w)u + B_2(u,v,w)v\\ w + A_3(u,v,w)u + B_3(u,v,w)v\end{pmatrix} \tag{4}$$

where A_i, B_i, $i=1,2,3$, are C^1-matrices which vanish at $(u,v,w) = (0,0,0)$. The key trick of this proof is to get rid of the matrix $A_2(u,v,w)$ in the representation (4) in order to obtain a certain kind of decoupling. The main tool we are going to use in this endeavor is a result on invariant manifolds due to Kirchgraber (see Kirchgraber and Stiefel [16,12.Satz 1, Satz 2]). We note in passing that we cannot use the standard references (Hartman [11,IX.Theorem 5.1] or Hirsch, Pugh and Shub [14,Theorem 4.1]) since they require a nonsingular linearization whereas our matrix A^- is allowed to have zero eigenvalues. Kirchgraber's result says that the map (4) has a locally invariant manifold, sometimes called center-stable manifold, with representation $v = s(u,w)$ near $(u,w) = (0,0)$. A crucial property of this manifold is that it contains all fixed points of f near $(u,v,w) = (0,0,0)$, i.e. $s(0,w) \equiv 0$ for $\|w\|$ suitably small. The above mentioned trick can be accomplished by means of the coordinate change

$$\bar{u} = u, \quad \bar{v} = v - s(u,w), \quad \bar{w} = w \tag{5}$$

whose effect, in geometrical terms, is that the center-stable manifold gets a local representation $\bar{v} = 0$ while the representation for the manifold of fixed points is kept to be $\bar{u} = 0$, $\bar{v} = 0$. It is a tedious though straightforward calculation to show that in the $(\bar{u},\bar{v},\bar{w})$-coordinate system our map g has the desired form

$$g\begin{pmatrix}\bar{u}\\\bar{v}\\\bar{w}\end{pmatrix} = \begin{pmatrix}[A^- + \bar{A}_1(\bar{u},\bar{v},\bar{w})]\bar{u} + \bar{B}_1(\bar{u},\bar{v},\bar{w})\bar{v}\\ [A^+ + \bar{B}_2(\bar{u},\bar{v},\bar{w})]\bar{v}\\ \bar{w} + \bar{A}_3(\bar{u},\bar{v},\bar{w})\bar{u} + \bar{B}_3(\bar{u},\bar{v},\bar{w})\bar{v}\end{pmatrix} \tag{6}$$

where $\bar{A}_1, \bar{A}_3, \bar{B}_1, \bar{B}_2, \bar{B}_3$ are continuous matrices which vanish at $(\bar{u},\bar{v},\bar{w}) = (0,0,0)$. This map now exhibits the kind of decoupling which allows us to complete the proof. Before doing this we restate the assumptions of Theorem 2 in the new coordinates. There exists a solution $(\bar{u}(k),\bar{v}(k),\bar{w}(k))$ of the difference equation whose right-hand side is the function (6) and a ball B_σ of radius σ around $(\bar{u},\bar{v},\bar{w}) = (0,0,0) \in \mathbb{R}^n$ such that

(I) $\lim_{j\to\infty}(\bar{u}(k_j),\bar{v}(k_j),\bar{w}(k_j)) = (0,0,0)$ where k_j is a subsequence of the positive integers with $k_j \to \infty$ as $j \to \infty$, (7)

(II) each ω-limit point $(\bar{u}_\infty,\bar{v}_\infty,\bar{w}_\infty)$ of $(\bar{u}(k),\bar{v}(k),\bar{w}(k))$ in B_σ satisfies $\bar{u}_\infty = 0$, $\bar{v}_\infty = 0$, (8)

(III) $\lim_{k\to\infty}[(\bar{u}(k+1),\bar{v}(k+1),\bar{w}(k+1) - (\bar{u}(k),\bar{v}(k),\bar{w}(k)] = (0,0,0)$. (9)

For (II) we refer to the remark succeeding Theorem 2. Assumption (3) of this theorem is being reflected in the spectral properties of the linearization of g around (0,0,0). These properties particularly imply that there exist constants $\gamma \geq 1$ and $0 < \delta < 1$ such that $\|(A^-)^{k-i}\| \leq \gamma\, \delta^{k-i}$ for all $k \geq i \geq 0$ and $\|(A^+)^{k-i}\| \leq \gamma\, \delta^{i-k}$ for all $i \geq k \geq 0$. Here δ is a number which is greater than the moduli of all eigenvalues of both A^- and $(A^+)^{-1}$.

In the remainder of the proof we return to the difference equation point of view. A crucial role is played by the following lemma which is the discrete analogue of [3,Lemma 2].

LEMMA 1: For A^-, A^+, γ, δ as above consider the class of linear difference equations of the form

$$\begin{aligned} u(k+1) &= [A^- + A_1(k)]u(k) + B_1(k)v(k) \\ v(k+1) &= [A^+ + B_2(k)]v(k) \\ w(k+1) &= w(k) + A_3(k)u(k) + B_3(k)v(k) \end{aligned} \tag{10}$$

where the matrices $A_1(k), A_3(k), B_1(k), B_2(k), B_3(k)$ are defined and bounded (in norm) above by $(1-\delta)/2\gamma$ on a set $J_0 := \{k_0, k_0+1, \ldots, K_0-1\}$ of consecutive integers.
Then any solution $(u(k), v(k), w(k))$ of (10) on J_0 satisfies the estimate

$$\|w(K_0)\| \leq \|w(k_0)\| + \|u(k_0)\| + 2\|v(K_0)\|. \tag{11}$$

Remark: The crucial point of this lemma is that the estimate (11) is valid uniformly for the whole class of equations (10) as long as the boundedness conditions for the matrices $A_1(k), A_3(k), B_1(k), B_2(k), B_3(k)$ are satisfied. In particular, the "length" of J_0 has no influence on the coefficients occurring in the estimate (11). The proof (see [4,Appendix B]) of this lemma goes parallel to the continuous time situation that has been dealt with in [3].

With Lemma 1 at hand we are able to complete the proof of Theorem 2. Our aim is to show that the sequence $(\bar{u}(k), \bar{v}(k), \bar{w}(k))$ converges to (0,0,0) as $k \to \infty$. To this end we suppose to the contrary that this sequence does not converge to (0,0,0) as $k \to \infty$. More explicitly this can be stated in the following way: There exists a positive constant $\rho \leq \sigma$ and a sequence of pairwise disjoint sets $T_j := \{k_j, k_j+1, \ldots, K_j\}$, $k_j+1 \leq K_j$, $j \in \mathbb{N}_0$, of consecutive integers such that $(\bar{u}(k_j), \bar{v}(k_j), \bar{w}(k_j)) \in B_\rho$ for all $k \in \bigcup_{j \in \mathbb{N}_0} T_j$,

$$(\bar{u}(K_j), \bar{v}(K_j), \bar{w}(K_j)) \in B_\rho \setminus B_{\rho/2} \quad \text{for all } j \in \mathbb{N}_0. \tag{12}$$

Without loss of generality the constant ρ may be chosen so small that

$$\| \bar{A}_i(\bar{u}(k),\bar{v}(k),\bar{w}(k)) \| \leq \frac{1-\delta}{2\gamma} \text{ for all } k \in \bigcup_{j\in\mathbb{N}_0} T_j, \; i=1,3,$$

$$\| \bar{B}_\iota(\bar{u}(k),\bar{v}(k),\bar{w}(k)) \| \leq \frac{1-\delta}{2\gamma} \text{ for all } k \in \bigcup_{j\in\mathbb{N}_0} T_j, \; \iota=1,2,3.$$

Thus, for each $j \in \mathbb{N}_0$, we may consider $(\bar{u}(k),\bar{v}(k),\bar{w}(k))$ as solution of the difference equation

$$\begin{aligned}
\bar{u}(k+1) &= [A^- + \bar{A}_1(\bar{u}(k),\bar{v}(k),\bar{w}(k))]\bar{u}(k) + \bar{B}_1(\bar{u}(k),\bar{v}(k),\bar{w}(k))\bar{v}(k)\\
\bar{v}(k+1) &= [A^+ + \bar{B}_2(\bar{u}(k),\bar{v}(k),\bar{w}(k))]\bar{v}(k)\\
\bar{w}(k+1) &= \bar{w}(k) + \bar{A}_3(\bar{u}(k),\bar{v}(k),\bar{w}(k))\bar{u}(k) + \bar{B}_3(\bar{u}(k),\bar{v}(k),\bar{w}(k))\bar{v}(k)
\end{aligned}$$

on T_j. Now we may apply Lemma 1 providing the estimate

$$\|\bar{w}(K_j)\| \leq \|\bar{w}(k_j)\| + \|\bar{u}(k_j)\| + 2\|\bar{v}(K_j)\| \text{ for each } j \in \mathbb{N}_0. \quad (13)$$

Because of (12) the sequence $(\bar{u}(K_j),\bar{v}(K_j),\bar{w}(K_j))$, $j \in \mathbb{N}_0$, is bounded and therefore has a convergent subsequence $(\bar{u}(K_{j_i}),\bar{v}(K_{j_i}),\bar{w}(K_{j_i}))$ with limit $(\bar{u}_\infty,\bar{v}_\infty,\bar{w}_\infty)$, say. Employing (8) we get

$$\begin{aligned}
\lim_{i\to\infty} \bar{u}(K_{j_i}) &= \bar{u}_\infty = 0,\\
\lim_{i\to\infty} \bar{v}(K_{j_i}) &= \bar{v}_\infty = 0,
\end{aligned} \quad (14)$$

and with (12)

$$\| \lim_{i\to\infty} \bar{w}(K_{j_i}) \| = \|\bar{w}_\infty\| \geq \frac{\rho}{2} > 0. \quad (15)$$

On the other hand, combining (7), (13) and (14) leads to

$$\lim_{i\to\infty} \|\bar{w}(K_{j_i})\| = \|\bar{w}_\infty\| = 0$$

which contradicts (15). Hence Theorem 2 is proved.

References

1. B.Aulbach, Behavior of Solutions near Manifolds of Periodic Solutions. J. Differential Equations 39 (1981), 345 - 377.
2. B.Aulbach, Invariant Manifolds with Asymptotic Phase. Nonlinear Analysis TMA 6 (1982), 817 - 827.
3. B.Aulbach, Approach to Hyperbolic Manifolds of Stationary Solutions, in "Equadiff 82", Lecture Notes in Mathematics, Springer, to appear.
4. B.Aulbach, Continuous and Discrete Dynamics near Manifolds of Equilibria. Preprint University of Würzburg.
5. B.Aulbach and K.P.Hadeler, Convergence to Equilibrium in the Classical Model of Population Genetics. J.Math.Anal.Appl., to appear.
6. J.F.Crow and M.Kimura, An Introduction to Population Genetics Theory. Harper and Row, New York 1970.
7. A.W.F.Edwards, Foundations of Mathematical Genetics. Cambridge University Press, London 1977.
8. W.Feller, A Geometric Analysis of Fitness in Triply Allelic Systems. Math. Biosciences 5 (1974), 19 - 38.
9. K.P.Hadeler, Selektionsmodelle in der Populationsgenetik, in "Methoden und Verfahren der Mathematischen Physik 9", 136 - 160, Bibliographisches Institut, Mannheim 1973.
10. K.P.Hadeler, Mathematik für Biologen. Springer, Berlin 1974.
11. P.Hartman, Ordinary Differential Equations. Wiley & Sons, New York 1964.
12. U. an der Heiden, On Manifolds of Equilibria in the Selection Model for Multiple Alleles. J.Math.Biol. 1 (1975), 321 - 330.
13. M.Hirsch, J.Palis, C.Pugh and M.Shub, Neighborhoods of Hyperbolic Sets. Inv.Math. 9 (1970), 121 - 134.
14. M.W.Hirsch, C.C.Pugh and M.Shub, Invariant Manifolds. Lecture Notes in Mathematics 583, Sringer, Berlin 1977.
15. P.J.Hughes and E.Seneta, Selection Equilibria in a Multiallele Single-Locus Setting. Heredity 35 (1975), 185 - 194.
16. U.Kirchgraber and E.Stiefel, Methoden der analytischen Störungsrechnung und ihre Anwendungen. Teubner, Stuttgart 1978.
17. V.Losert and E.Akin, Dynamics of Games and Genes, Discrete versus Continuous Time. Preprint.

ON A MATHEMATICAL PROBLEM ARISING IN CELL POPULATION BIOLOGY (*)

Marco Luigi BERNARDI
Dipartimento di Matematica-Università di Pavia-I-27100 PAVIA (Italy)
Antonio Candido CAPELO
Istituto di Analisi Numerica del C.N.R. Pavia (Italy)

ABSTRACT: We present some results concerning an initial and boundary value problem for a system of first order linear partial differential equations. This problem comes from modelling the evolution of a homogeneous cell population, where the cells may replicate and are submitted to the action of some mutagenic agent.

1. THE MATHEMATICAL MODEL .

Our aim is here to model the evolution of a homogeneous cell population. First of all, we suppose that the cells undergo a replication process, such that, at any time t, there are cells (say $U_r(t)$) in a "resting phase" and the other cells (say $U_c(t)$) are somewhere along the replication cycle. We assume that the cells may die both in the resting phase and in the replication cycle, the cells in the resting phase may, at any time, enter into the replication cycle, and that every cell arriving at the end of the replication cycle (mitosis) divides into two "daughter cells" which enter into the resting phase.

We specify the "position" reached by a cell in its replication cycle by the variable ω ($0 \leqslant \omega \leqslant T$, where T is the physiological replication time) to be called the "maturity".

The cells in our population are submitted to the action of some mutagenic agent, which damages them. We measure the damage of a cell by the variable x ($0 \leqslant x \leqslant 1$; x can be interpreted as the "fraction" of some kind of "targets" (being in the cell) which are made inert by the administered mutagenic agent). Moreover, we suppose that the damage is reversible (thanks to the action of a "repairing mechanism"). Denoting by s(t) the instantaneous intensity of

---

(*) This work was partially supported by the Istituto di Analisi Numerica of the C.N.R. in Pavia (Italy) and the G.N.A.F.A. of the C.N.R. (Italy).

administration of the agent to a cell, we assume that the dynamics of the damage in this cell is governed by the following equation:

$$(1)\qquad \frac{dx}{dt} = a(1-x)s(t) - bx \ .$$

The coefficients a and b in (1) represent respectively the "sensitivity" to the mutagenic agent and the "efficiency" of the repairing mechanism. We suppose that $a = a_c(\omega)$ and $b = b_c(\omega)$ for the cells being along the cycle, and $a = a_r$ and $b = b_r$ for the cells in the resting phase. Moreover, we assume that, for undamaged cells, the "speed" in the replication process is constant and equal to 1 (that is, $d\omega/dt = 1$), and that the damage changes the speed according to

$$(2)\qquad \frac{d\omega}{dt} = v(x) \qquad (\text{where } 0\leq v(x)\leq 1).$$

We consider a homogeneous cell population: we suppose that, for every cell, the replication process and the dynamics of the damage (due to the administration of some mutagenic agent) are governed by the above assumptions. Of course, the cells in the resting phase [resp. in the replication cycle] are distributed according to their current damage x [resp. x and maturity ω]. We represent these distributions by means of the density functions $u_r(x,t)$ and $u_c(\omega,x,t)$ ($t\geq 0$; $0\leq x\leq 1$; $0\leq \omega\leq T$), such that:

$$(3)\qquad U_r(t) = \int_0^1 u_r(x,t)dx; \quad U_c(t) = \int_0^T d\omega \int_0^1 u_c(\omega,x,t)dx.$$

Our assumptions then imply (see [2]) that $u_r(x,t)$ and $u_c(\omega,x,t)$ satisfy the following equations:

$$(4)\qquad \begin{cases} \frac{\partial}{\partial t} u_c(\omega,x,t) + \frac{\partial}{\partial x}\left[(a_c(\omega)(1-x)s(t)-b_c(\omega)x)u_c(\omega,x,t)\right] + \\[2mm] + v(x)\frac{\partial}{\partial \omega} u_c(\omega,x,t) = -\mu_c(\omega,x,t)u_c(\omega,x,t), \quad 0<x<1, \ 0<\omega<T, \ t>0, \end{cases}$$

$$(5)\qquad \begin{cases} \frac{\partial}{\partial t} u_r(x,t) + \frac{\partial}{\partial x}\left[(a_r(1-x)s(t)-b_rx)u_r(x,t)\right] = -(\mu_r(x,t)+ \\[2mm] + \nu_r(x,t))u_r(x,t) + 2v(x)\, u_c(T,x,t), \quad 0<x<1, \ t>0, \end{cases}$$

($\mu_c(\omega,x,t)$ and $\mu_r(x,t)$ are "death coefficients"; $\nu_r(x,t)$ represents the "tendency" of the cells to enter into the cycle), together with the boundary condition (see [2])

(6) $\quad v(x)u_c(0,x,t) = \nu_r(x,t)u_r(x,t),\ 0\leqslant x\leqslant 1,\ t\geqslant 0.$

We add to (4) (5) and (6) the following initial conditions:

(7) $\quad u_c(\omega,x,0) = g_c(\omega,x),\ u_r(x,0) = g_r(x), \quad 0\leqslant\omega\leqslant T,\ 0\leqslant x\leqslant 1.$

2. STATEMENT OF THE MATHEMATICAL PROBLEM AND FIRST RESULTS.

We are led by the considerations from the previous section to consider the following

Problem 1. Given the non-negative functions a_c, b_c, a_r, b_r, μ_c, μ_r, s, v, ν_r, g_c, g_r, find u_c and u_r ($0\leqslant\omega\leqslant T$; $0\leqslant x\leqslant 1$; $t\geqslant 0$) verifying (4), (5), (6) and (7), provided that the following compatibility condition holds:

(8) $\quad \nu_r(0,x)g_r(x) = v(x)g_c(0,x),\ 0\leqslant x\leqslant 1.$

The mathematical treatment of Problem 1 is still an open question. We consider here a very particular case only:

Problem 2. As Problem 1, but $a_c = a_r \equiv a$, $b_c = b_r \equiv b$, $\mu_c = \mu_r \equiv \mu$, s, v and $\nu_r \equiv \nu$ are positive constants.

It is clear that in dealing with Problem 2 we are decoupling the population dynamics and the damage done by the mutagenic agent. Note that, in this special case, if we want to calculate the total number of cells $U(t) = U_r(t) + U_c(t)$, we can proceed direct, by observing that the pair $(U_r(t), U_c(t))$ verifies a suitable linear delay differential system (see [3]). Starting from this system, we can obtain that

(9) $\quad k_1U(0) \leqslant \exp(\lambda t)U(t) \leqslant k_2U(0), \quad t \geqslant 0,$

where the positive constants k_1 and k_2 and the real constant λ (in particular, the sign of λ) depend only on T, v, μ, ν.
U(t) is an easily observable quantity. However, we are here mainly interested in studying how s influences the densities u_r and u_c. (Remark that the following procedures apply also, "mutatis mutandis", to the case where s is not constant; we take here s as a

constant, for sake of brevity).

We observe at once that Problem 2 can be reduced, of course, to the following Problem 3, where we set for sake of brevity

(10) $d \equiv as+b;\ h \equiv \frac{as}{as+b};\ u(\omega,x,t) \equiv u_c(\omega,x,t);\ g(\omega,x) \equiv g_c(\omega,x)$.

Problem 3. Given the same data as in Problem 2, find $u(\omega,x,t)$ ($0 \leqslant \omega \leqslant T$; $0 \leqslant x \leqslant 1$; $t \geqslant 0$) such that:

$$(11)\quad \begin{cases} \frac{\partial u}{\partial t}(\omega,x,t) + d(h-x)\frac{\partial u}{\partial x}(\omega,x,t) + v\frac{\partial u}{\partial \omega}(\omega,x,t) = \\ = (d-\mu)u(\omega,x,t),\ 0<\omega<T,\ 0<x<1,\ t>0; \end{cases}$$

$$(12)\quad u(\omega,x,0) = g(\omega,x),\ 0 \leqslant \omega \leqslant T,\ 0 \leqslant x \leqslant 1;$$

$$(13)\quad \frac{\partial u}{\partial t}(0,x,t)+d(h-x)\frac{\partial u}{\partial x}(0,x,t)=(d-\mu-\nu)u(0,x,t)+2\nu u(T,x,t),\ 0<x<1,t>0.$$

It is clear that in order to study Problem 3 one has first to analyse carefully the characteristic base curves (c.b.c.) of equations (11) and (13). It is also convenient, for this analysis, to think (11) (resp. (13)) as to be verified for $(\omega,x,t) \in \mathbb{R}^3$ (resp. $(x,t) \in \mathbb{R}^2$). The c.b.c. of (11) are immediately evaluated; the one passing through $(\omega_0,x_0,t_0) \in \mathbb{R}^3$ is given by

$$(14)\quad \omega = v(t-t_0) + \omega_0;\ x = h+(x_0-h)\ e^{-d(t-t_0)},\ t \in \mathbb{R}^1.$$

Of course, the c.b.c. of (13) are obtained by projecting the ones of (11) on any plane which is orthogonal to the ω-axis.

With this information, we are able to determine a unique solution $u(\omega,x,t)$ for Problem 3 not in the whole Q, but only in the subset V of Q, where:

$$(15)\quad \begin{cases} Q \equiv [0,T]\times[0,1]\times[0,+\infty[; \\ V \equiv \{(\omega,x,t) \in \mathbb{R}^3 \mid 0 \leqslant \omega \leqslant T;\ t \geqslant 0;\ h(1-e^{-dt}) \leqslant x \leqslant h+(1-h)e^{-dt}\} \end{cases}$$

(define also: $\tilde{O} \equiv \{(x,t) \in \mathbb{R}^2 \mid (0,x,t) \in V\}$).

(See Figure 1 for a qualitative picture of V).

We can obtain $u(\omega,x,t)$ in V by a step-by-step procedure. In the first step, we determine $u(\omega,x,t)$ in V_0 (see Figure 1), starting from the initial condition (12) and following the c.b.c. of (11)

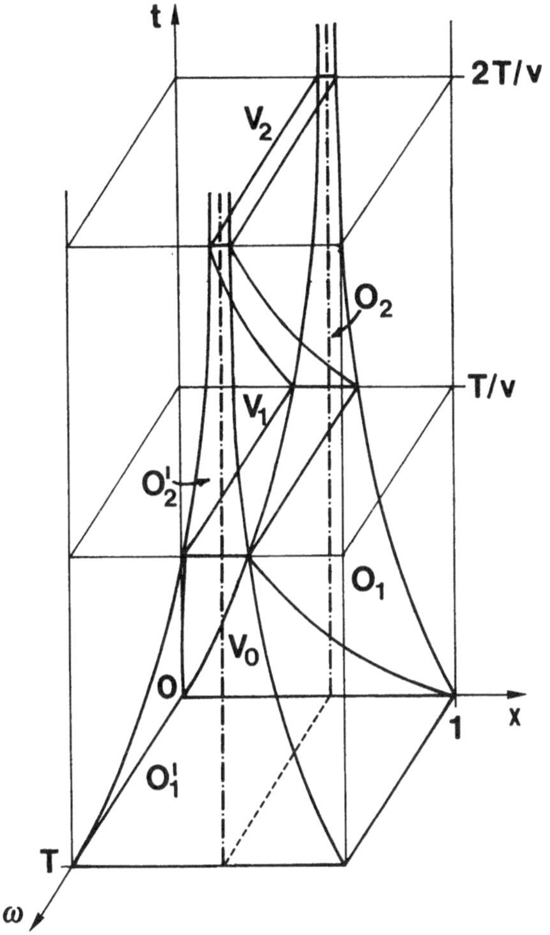

$V_0 \equiv \{(\omega,x,t) \in V \mid 0 \leqslant t \leqslant \frac{\omega}{v}\}$;

$V_{n+1} \equiv \{(\omega,x,t) \in V \mid \frac{1}{v}(\omega+nT) \leqslant t \leqslant \frac{1}{v}(\omega+(n+1)T)\}$, $\quad n \geqslant 0$;

$O_n \equiv V_n \cap \{\text{plane } \omega=0\}$, $\quad n \geqslant 1$;

$O'_{n+1} \equiv V_n \cap \{\text{plane } \omega=T\}$, $\quad n \geqslant 0$.

Figure 1

which pass through $\{(\omega,x,0) \in \mathbb{R}^3 \mid 0 \leqslant \omega \leqslant T \; ; \; 0 \leqslant x \leqslant 1\}$. In the second step, we determine $u(0,x,t)$ in O_1 (always see Figure 1), starting from (12) (with $\omega=0$) and following the c.b.c. of (13) which pass through $\{(x,0) \in \mathbb{R}^2 \mid 0 \leqslant x \leqslant 1\}$ (we just know $u(T,x,t)$ in O'_1 from the first step). In the third step, we determine $u(\omega,x,t)$ in V_1, starting from the "initial datum" $u(0,x,t)$ in O_1 (obtained in the second step) and following the c.b.c. of (11) which pass through O_1. Then, the procedure continues in an obvious way. (For more details, we refer to [3]).

Now, let us remark (see the above procedure) that condition (12) does not influence the values of $u(\omega,x,t)$ in $Q\setminus V$. Indeed, there exist infinitely many functions which verify (11) in the interior of $Q\setminus V$ and also (13) on (the suitable part of) the boundary of $Q\setminus V$. Among these functions, only $u\equiv 0$ in $Q\setminus V$ is suitable for the original biological problem.

In the next section, we present another way of solving Problem 3: we turn our problem into that of the study of a suitable delay differential equation (of course, because of the previous conclusions, it suffices to work directly in V rather than in Q).

3. REDUCTION OF THE PROBLEM TO A DELAY DIFFERENTIAL EQUATION.

Consider all the c.b.c. of (11) which pass through some point in V, and concentrate on the set consisting of the points where these c.b.c. intersect the plane $t=0$: the half strip $\Omega \equiv \{(\omega,x,0)\in R^3 | -\infty<\omega\leqslant T;\ 0\leqslant x\leqslant 1\}$. Now, considering the initial condition (12), suppose that $g(\omega,x)$ is "sufficiently smooth" and can be extended "in a smooth way" to some $G(\omega,x)$, $(\omega,x)\in\Omega$. Then, taking $G(\omega,x)$ as initial datum and following the c.b.c. of (11) which pass through Ω, we determine the corresponding solution $u(\omega,x,t)$ of (11): in particular, this $u(\omega,x,t)$ is well defined in V. We require that this $u(\omega,x,t)$ has to satisfy (13) for all $(x,t)\in\tilde{O}$. Then (as some calculations show), this implies that $G(\omega,x)$ must verify

$$(16)\qquad \frac{\partial G}{\partial \omega}(\omega,x) = \frac{\nu}{v}\left[G(\omega,x)-2G(\omega+T,x)\right],\ -\infty<\omega<0,\ 0\leqslant x\leqslant 1;$$

moreover, of course,

$$(17)\qquad G(\omega,x) = g(\omega,x),\ 0\leqslant\omega\leqslant T,\ 0\leqslant x\leqslant 1.$$

Equation (16) is a very simple linear delay differential equation. We can use some well known results (see, e.g., [1] and [5]) to obtain that equations (16) and (17) have a unique solution. Taking the so determined $G(\omega,x)$, $(\omega,x)\in\Omega$, as initial datum, considering the corresponding solution of (11), we obtain the unique solution u of Problem 3 in V.

Let us finish by stating our result in a precise form regarding the regularity as well (for more details and for other results, see [3]). To do this, we have to introduce some additional notation. Keeping Figure 1 in mind and considering V_0, V, and denoting by intA the interior of some set A, we define:

(18) $\quad C^1_{\#}(V) \equiv C^1(\text{int } V_0 \cup \text{int}(V \setminus V_0))$.

Theorem 1. Suppose that $g(\omega,x) \in C^1([0,T]\times[0,1])$. Then, there exists a unique $u(\omega,x,t) \in C^0(V) \cap C^1_{\#}(V)$, with $u(0,x,t) \in C^1(\text{int}\tilde{O})$, which verifies: (12) in $[0,T]\times[0,1]$; (11) in $\text{int } V_0 \cup \text{int}(V\setminus V_0)$; (13) in $\text{int}(\tilde{O})$. If g verifies moreover:

(19) $$\frac{\partial G}{\partial \omega}(0+,x) = \frac{\nu}{V}\,[g(0,x)-2g(T,x)], \text{ for } 0<x<1,$$

then we have further that: $u(\omega,x,t) \in C^1(\text{int}V)$; u verifies (11) in the whole int V.

For various problems (in particular, for some arising in the theory of electrical circuits) a similar procedure, i.e. the reduction to a delay problem, was developed in [4], although in a different form.

REFERENCES

[1] R. Bellman and K.L. Cooke, Differential-Difference Equations, Mathematics in Science and Engineering, vol. 6, Academic Press, New York, London (1963).

[2] M.L. Bernardi, A.C. Capelo and P. Periti, A mathematical model for the evolution of cell populations under the action of mutagenic agents, submitted to Math. Biosciences.

[3] M.L. Bernardi and A.C. Capelo, On a differential problem coming from cell biology, to appear.

[4] K.L. Cooke and D.W. Krumme, Differential-difference equations and nonlinear initial-boundary value problems for linear hyperbolic partial differential equations, J. Math. Anal. Appl., 24, 372-387 (1968).

[5] J. Hale, Theory of Functional Differential Equations, Applied Mathematical Sciences, vol. 3, Springer, New York, Heidelberg, Berlin (1977).

THE EXTERNAL MEMORY OF INTERTIDAL MOLLUSCS: A THEORETICAL STUDY OF TRAIL-FOLLOWING

S. Focardi
Istituto Nazionale di Biologia della Selvaggina
via Stradelli Guelfi 23/a, Ozzano Emilia, Italy

J.L. Deneubourg
Service de Chimie-Physique II, Campus Plaine, U.L.B.
Bruxelles, Belgium

G. Chelazzi
Istituto di Zoologia dell'Università
via Romana 17, Firenze, Italy

Introduction

The secretion of a slimy mucus from the foot, besides being mechanically essential for the locomotion of many molluscs, seems to constitute a system of external storage of information used for orientation towards biologically significant goals such as hiding places, sexual partners or prey. Field observations and laboratory experiments have shown that many gastropod species are able to interpret chemical information contained in their mucous trails. In some cases the trail-following mechanism can involve the recognition of "personal" trails and the detection of trail polarization (Wells & Buckley, 1972; Cook & Cook, 1975).

The trail-following behaviour can also be considered as a mechanism for information exchange between members of a population when it is used for congregating into common shelters or mating aggregates. For some littoral species, at least, it seems reasonable to exclude that the periodical onshore clustering could be based on independent and concordant responses of the different animals to the same environmental cues such as light, gravity or micromorphology of the shore.

The traditional ethological approach to this problem, consisting in the experimental manipulation of the trail, can be partially flanked by theoretical studies on the validity of different hypotheses concerning the orientation mechanisms. The purpose of such mathematical models is to link the outputs of the two levels of empirical analysis : the "micro-analysis" of the physio-ethological aspects of orientation and the "macro-analysis" of such bulk phenomena as the aggregation.

The biological phenomenon : clusterization and collective homing

A large number of littoral species exhibit a daily rhythm characterized by a foraging and a resting phase during which animals stay frequently in a sheltered place. Numerous species show fidelity to their sheltered place : after each foraging phase the animal comes back to its home and stays in it during succesive resting phases. The home is personal or collective (cluster). Homing and clustering are common stress reducing behaviours present in many intertidal molluscs species (Newell, 1979; Underwood, 1979). We discuss here the dynamic of the cluster formation in protected places (We consider a hole containing i animals as being equivalent to a cluster containing i animals) Despite the differences in the specific behaviour, this phenomenon may be described as in Fig. 1 (left). During the non-active periods, the population of snails present on the shore (P) can be divided into a scattered (X) and a protected population (PP) inhabiting a number of the H_t equidimensional protected sites (holes).

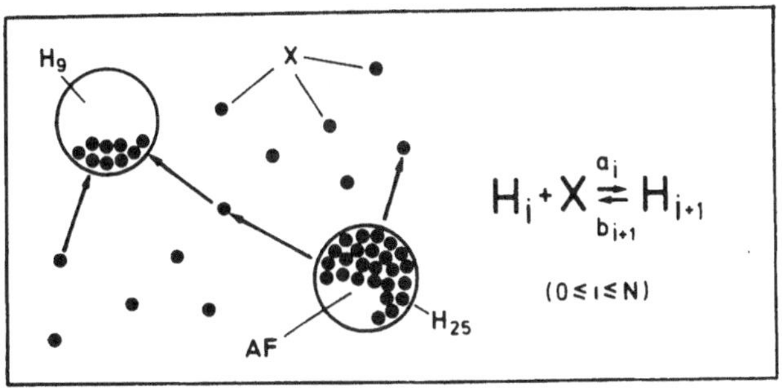

Figure 1. Left: schematic representation of the spatial distribution of snails on a rocky shore. Two holes sheltering clusters and some scattered animals (X) are shown. Right: symbolic representation of the mechanism determining the variation of cluster dimension.

The total area of the H_t holes present on a given surface constitutes the protected surface (PS). PS, H_t and P are constant. Under natural conditions the clusters disaggregate during each activity phase and the scattered and protected snails melt into a common group of feeding animals. The clusters are re-formed at the end of the feeding period. During each aggregation-disaggregation cycle a dynamic exchange of animals between the two populations PP and X can be observed (Fig. 1, right). These kinetics affect H_i, the number of holes containing i snails. n is the maximum number of animals which one hole can shelter a_i ($0 \leq i \leq n-1$) and b_i ($1 \leq i \leq n$) are the rate-constants of the process.

Model and Results

The time evolution equations for $H_0, \ldots, H_n$ are:

$$\dot{H}_0 = -a_0 X H_0 + b_1 H_1$$

(1) $$\dot{H}_i = a_{i-1} X H_{i-1} - b_i H_i - a_i X H_i + b_{i+1} H_{i+1}$$

$$\dot{H}_n = a_{n-1} X H_{n-1} - b_n H_n$$

$$i = 1, \ldots, n-1$$

with $X = P - PP$; $PP = \sum_1^n i H_i$; $H_t = \sum_0^n H_i$

System (1) admits an unique and stable steady-state (Nicolis & Prigogine, 1977). The distribution of cluster size at the steady-state is:

(2) $$H_i = H_t U_i / (\sum_0^n U_j) \qquad i = 0, \ldots, n$$

with $$U_i = X^i (\prod_0^{i-1} a_j / \prod_1^i b_j) \quad i = 1, \ldots, n \quad \text{and } U_0 = 1$$

Two other functions are used : the protected fraction (PF= PP/P) and the turn-over TO between X and PP which is at the steady state $\sum_1^n b_i H_i$

We shall now examine two cases of orientation and communication.

Individual orientation mechanismns.

a_i, the rate-constant at which the unprotected snails become members of a cluster containing i animals, is assumed to be proportional to the fraction of free places in the hole (1.- i/n) and to the frequence at which an animal of "type X" crosses a hole of "type H_i". In the individual orientation, this contact is by chance with a frequency $f_r = (L/S)(D+d)Q$ (Curio, 1976), where L is the length of the homeward branch of the feeding excursion, Q is the probability of stopping when a hole is found and S is the surface. The holes and the animals are assumed to be circular with respective diameters D and d ($D = n^{0.5} d$).

b_i, the rate-constant of transformation of a cluster i to i-1 by loss of one animal, is proportional to the number of animals inside the hole (i) and to the rate at which one animal escapes from the hole (which is the inverse of the mean time of an individual presence in a hole). In the individual strategy each animal acts independently thus the individual escape-rate, g, is independant of the hole composition ($b_i = g.i$).

In this case the hole-population distribution (2) may be recognized as a binomial distribution :

(3) $$H_i / H_t = n! / (i! \; (n - i)!) \; p^i q^{n-i}$$

where: $p = f_r X / (n g + f_r X)$; $q = n g / (n g + f_r X)$

And the turn-over is proportional to the protected population :

(4) TO = g PP or TO / PP = g

Trail-following behaviour.

As in the indivudal strategy

(5) a_i =(contact frequency)(1. - i/n)

but now the contact frequency is the summation of two processes : the random meeting which is the same as in the individual strategy (f_r) and the meeting induced by trail-following : a member of the unprotected population, X ,crosses a trail laided by a member of a cluster during its feeding excursion, follows this trail to the hole and stops. This process is assumed to be proportional to the number of trails i.e. to the number of animals in the cluster (i) and to a constant $f_c = L^2C/S$ where C is the probability of following the trail to the cluster. L^2 results from the length of the excursion made by the scattered animal and by the trail-maker. These two lengths are equal.

The individual escape-rate in this collective strategy depends on the hole-composition : a member of a cluster loosing its way can be trapped by the web of trails surrounding its hole. To simulate this trapping effect the individual rate-constant is given by g/(1 + e i), where e measures the trail's trapping force (b_i = g i/(1 + e i)).

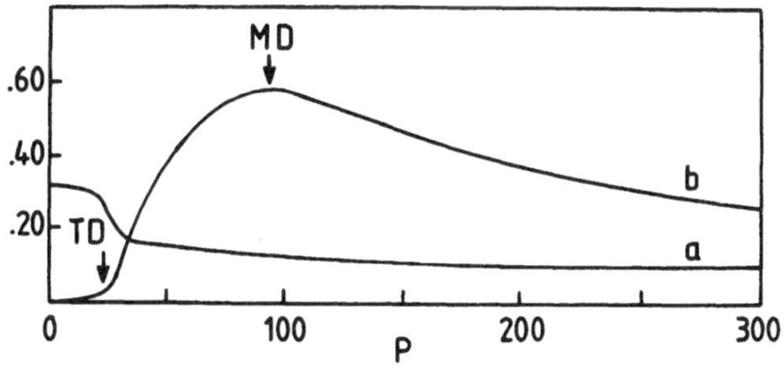

Figure 2. Variation of TO / PP (a) and PF (b) as a function of P. N = 50, f_r= 8.071 10^{-4} , f_c = .01, g = .3333, e = .05.

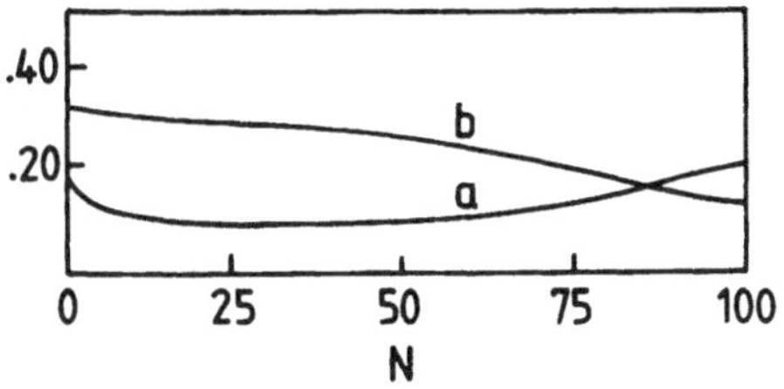

Figure 3. TO / PP (a) and PF (b) as a function of N. P = 25. The other parameter values as in Fig. 2.

The variation of TO / PP and PF as function of the density of population (P) is shown in Fig. 2 (a and b, respectively). Line b shows two values of P deserving special attention: firstly TD, the threshold value producing an abrupt rise in the PF and secondly MD, the density giving the maximum value of PF.

In Fig. 3 TO / PP (a) and PF (b) are plotted as a function of the hole dimension (n). PF exhibits a minimum value if the efficency of the individual orientation is high (f_r large). When n is small the individual orientation is efficient and when n is large the cooperative behaviour becomes efficient.

Under the hypothesis of collective (trail-following) orientation the cluster dimension distribution given with increasing values of n is clearly non-binomial (Fig. 4), showing a marked bimodality (Fig. 4, b) or a strong kurtosis (Fig. 4, c).

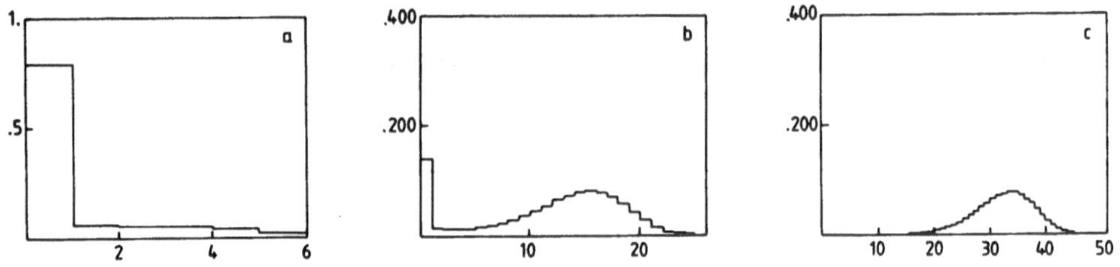

Figure 4. Distribution of cluster dimension for n = 6 (a), 25 (b), 50 (c). PS = 60, P= 75 and the other parameter values are as in Fig. 2.

Discussion

The model constitutes a useful tool for the interpretation of field data. Two instances of output have been analyzed : the cluster size distribution and the turn-over, and comparison with empirical results can yield information on the clustering dynamics. For example, a preliminary study shows that the ratio TO/PP in Nerita textilis population decreases when the protected population increases as forecast by the model in the cooperative orientation.

More generally, the model shows that small differences in morphological (body size), etho-physiological (excursions length) and ecological (population density) parameters may produce abrupt changes in the snail population, producing a shift from clustering to non-clustering behaviour as they move from areas with small holes to a part of the shore characterized by large holes (Vannini & Chelazzi, 1978).

Different orientation mechanisms are selected under different shore conformations in the congeneric and sympatric species N. plicata and textilis (Focardi, Deneubourg & Chelazzi, in prep.) or between Acmae digitalis and A. scabra (Willoughby, 1973).

Another interesting result of the model concerns the social organization of a species. When dealing with animal aggregation it is important to have quantitative information on the level of the interaction between the members of the group, in order to judge the sociability level attained. In fact aggregations may result by individual response to environmental cues, as in the isopods or by information exchange as in many intertidal gastropods. In this respect the trail-following deserves a special interest because it is a kind of long-range system of communication widespread among molluscs and presenting the important feature of making a collective goal-directed orientation possible.

We would like to thank Prof. Nicolis and Prigogine and Dr. Goss for their suggestions. This study was supported by the NATO grant 267.81.

References

Cook, S.B. & Cook, C. "Directionality in the trail-following response of the pulmonate limpet Siphonaria alternata" (1975) Mar. Behav. Physiol. 3: 147-155.

Curio, E. The ethology of predation, Springer-Verlag (1976).

Newell, R.C. Biology of intertidal animals. Marine Ecological Surveys, Faversham, Kent (1979).

Nicolis, G. & Prigogine, I. Self Organization in Non-equilibrium Systems, Wiley-Interscience (1977).

Underwood, A.J. "The ecology of intertidal gastropods" (1979) Adv. Mar. Biol. 16: 111-210.

Vannini, M. & Chelazzi, G. "Field observations on the rhythmic behaviour of Nerita textilis (Gastropoda : Prosobranchia)" (1978) Mar. Biol. 45: 113-121.

Wells, M.J. & Buckley, S.K.L. "Snails and trails" (1972) Anim. Behav. 20: 345-355.

Willoughby, J.M. " A field study of the clustering and movement behaviour of the limpet Acmea digitalis " (1973) The Veliger 15: 223-230

STATIONARY DISTRIBUTIONS FOR POPULATIONS SUBJECT TO RANDOM CATASTROPHES

Gustaf Gripenberg
Institute of Mathematics
Helsinki University of Technology
SF-02150 Espoo 15, Finland

1. The Model

Let $X(t)$ denote the size of a certain population at time $t \geq 0$. It is assumed that at times when no catastrophes occur, the population grows according to the differential equation

$$X'(t) = \alpha(X(t)). \tag{1}$$

The hazard function for the occurrence of a catastrophe is $\beta(X(t))$, i.e.

$$\Pr\{\text{ no catastrophe occurs in the interval } (T_1,T_2)\} = \exp(-\int_{T_2}^{T_1} \beta(X(s))ds). \tag{2}$$

If a catastrophe or downward jump takes place at time T, then it is assumed that

$$\Pr\{ X(T) \leq y \mid X(T-) = x \} = h(x,y), \tag{3}$$

where h is a given function. The problem will be to study the distribution of $X(t)$, especially as $t \to \infty$.

In [4] - [7] the time to extinction of a population described by a model of this kind is studied and in [3] the convergence of the distribution of $X(t)$ towards a stationary one is established. In [1] and [2] the existence of a stationary distribution for some closely related models is considered.

2. A Theoretical Result

In [3] the following theorem is proved:

THEOREM. *Assume that* $I = [0,x_0)$, $I_0 = (0,x_0)$, $0 < x_0 \leq \infty$, $\sigma > 0$ *and that*

(4) *the function* α *is continuous on* I, *positive on* I_0 *and* $\int_0^y \alpha(s)^{-1}ds = +\infty$ *for* $y \in I_0$ *if* $\alpha(0) = 0$,

(5) *the function* β *is continuous and nonnegative on* I,

(6) *the function* $\gamma(x) \overset{def}{=} \beta(x)/\alpha(x)$, $x \in I$, *satisfies* $\int_0^y \gamma(s)ds < \infty$ *and* $\int_y^{x_0} \gamma(s)ds = +\infty$, $y \in I_0$,

(7) *the function* h *is nonnegative on* I×I, *nonincreasing in its first, nondecreasing and right-continuous in its second argument,* $h(x,y) = 1$ *if* $x \leq y$ *and* $\lim_{y \to x-} h(x,y) = 1$ *if* $x \in I_0$,

(8) $\liminf_{y \to x_0} \int_y^{x_0} \gamma(s)\exp(-\sigma\int_y^s \gamma(u)du)\ h(s,y)ds > 1,$

(9) $\int_0^y \alpha(x)^{-1} \int_I \gamma(s)\exp(-\int_0^s \gamma(u)du)\ h(s,x)ds\ dx < \infty,\ y \in I_0,$

(10) $\int_y^{x_0} \alpha(s)^{-1}\exp(-\sigma\int_y^s \gamma(u)du)ds < \infty,\ y \in I_0,$

If

(11) $$G_0(y) \equiv 1,\ G_{n+1}(y) = \lim_{x \to x_0-} h(x,y) + \\ + \int_I \int_x^{x_0} \gamma(s)\exp(-\int_x^s \gamma(u)du)(h(x,y) - h(s,y))ds\ \gamma(x)G_n(x)dx, \\ y \in I,\ n \geq 0,$$

then

(12) $G(y) \overset{def}{=} \lim_{n\to\infty} G_n(y) = \inf_{n\geq 0} G_n(y),\ y \in I$ *exists*

and there is a positive constant c *such that if*

(13) $$f(x) = c\alpha(x)^{-1}(G(x) - \exp(-\int_0^x \gamma(u)du)G(0) - \\ - \int_0^x \gamma(s)\exp(-\int_s^x \gamma(u)du)\ G(s)ds),\ x \in I_0,$$

then

(14) f *is nonnegative on* I_0 *and* $\int_I f(s)ds = 1$

and for every $x \in I$ *such that* $x > 0$ *if* $\alpha(0) = 0$

(15) $$\lim_{t\to\infty} \Pr\{\ X(t) \leq y \mid X(0) = x\ \} = \int_0^y f(s)ds = \\ = \int_I f(s)\Pr\{\ X(\tau) \leq y \mid X(0) = s\ \}ds,\ \tau > 0,\ y \in I_0.$$

The statements (14) and (15) imply that f is the density function of a stationary attractive probability distribution for the process X(t) and from (11) - (13) one sees that this density function can be found in a constructive way. The assumption that h(x,y) is nonincreasing in its first argument is by (3) quite reasonable and implies that the sequence G_n is nonincreasing. The assumption (8) is used to prove that $G \not\equiv 0$ and the assumptions (9) and (10) are needed when one shows that f is integrable. It follows from (4) that the differential equation (1) has unique solutions.

3. A Numerical Method

In this section we will only consider cases where

$$\lim_{y\to x_0-} \lim_{x\to x_0-} h(x,y) = 1.$$

This assumption is quite natural in the case when $x_o < \infty$ and we define $h(x_o,y) = \lim_{x\to x_o-} h(x,y)$. We will not use the iteration procedure given in the Theorem, but we observe instead that the function G satisfies the equation

$$(16)\qquad G(y) = h(x_o,y) + \int_I \int_x^{x_o} \gamma(s)\exp(-\int_x^s \gamma(u)du)(h(x,y) - h(s,y))ds \times \times \gamma(x)G(x)dx,\ y \in I.$$

Next we note that if we define

$$(17)\qquad g(y) = \int_{[0,y]} \exp(-\int_x^y \gamma(u)du)\ dG(x),\ y \in I,$$

then $G(y) = g(y) + \int_0^y \gamma(x)g(x)dx$, $y \in I$, and therefore we can show, since $h(x,y) = 1$ if $x \leq y$, that g satisfies the equation

$$(18)\qquad g(y) = \int_y^{x_o} g(x)\gamma(x)h(x,y)dx,\ y \in I.$$

Moreover, the density function f is given in the form

$$(19)\qquad f(x) = c\alpha(x)^{-1}g(x),\ x \in I_o.$$

A discrete version of equation (16) can, for example, be of the form

$$(20)\qquad G_j = \sum_{k=1}^n \sum_{i=k}^n c_i(\Pi_{m=k}^j (1 + c_m))^{-1}(h_{k,j} - h_{i,j})c_kG_k,\quad j = 0, 1, \dots, n,$$

where $0 = s_0 < s_1 < \dots < s_n = x_o$, $h_{i,j} = h(s_i,s_j)$, $c_0 = 0$, $c_{j+1} = \int_{s_j}^{s_{j+1}} \gamma(u)du$ and $G_j \approx G(s_j)$. Note that $c_n = \infty$, so that $1/(1 + c_n) = 0$ and $c_n/(1 + c_n) = 1$. With these conventions we see that if we in analogy with (17) define

$$(21)\qquad g_j = \sum_{k=1}^j (G_k - G_{k-1})(\Pi_{m=k}^j (1 + c_m))^{-1},\ j = 0, 1, \dots, n,$$

$(G_{-1} = 0)$, then we have the following discrete version of (18):

$$g_j = \sum_{i=j+1}^n g_ic_ih_{i,j},\ j = 0, 1, \dots, n-1.$$

The sum on the right hand side in this equation is not, however, well defined, since $g_n = 0$ and $c_n = \infty$, but an inspection of the definition (21) shows that we can rewrite this equation as

$$(22)\qquad g_j = \sum_{i=j+1}^{n-1} g_ic_ih_{i,j} + h_{n,j}(g_{n-1} + G_n - G_{n-1}),\quad j = 0, 1, \dots, n-1.$$

Thus we see that if $h_{n,n-1} < 1$, then we choose $G_n - G_{n-1}$ to be some positive number, (e.g. 1), and from (22) we then have $g_{n-1} = h_{n,n-1}(G_n - G_{n-1})/(1 - h_{n,n-1})$. If $h_{n,n-1} = 1$, then $G_n - G_{n-1} = 0$ and we choose g_{n-1} to be some positive number, (e.g. 1). The remaining numbers g_j, $j = 0, 1, \dots, n-2$, can now be solved recursively from (22). Now we let $f_j^* = g_j/\alpha(s_j)$, $j = 1, 2, \dots, n-1$, and determine f_0^* and f_n^* by linear extrapolation, (but we recall that these numbers should be nonnegative). Finally, the desired approximations f_j of $f(s_j)$ are given by $f_j = cf_j^*$ where $c = (\sum_{j=0}^{n-1} (s_{j+1} - s_j)(f_{j+1}^* + f_j^*)/2)^{-1}$.

The approximations that we obtain using the method described above seem to be reasonably accurate as can be checked by comparing the numerical results with the analytical expressions that one can find in some simple cases, see e.g. [3, Sect. 4].

4. Numerical Results

In all the examples that we will consider here we take

$$x_o = 1,\ \alpha(x) = x(1-x),\ \beta(x) = \beta_o x \text{ and } h(x,y) = H(y/x),$$

where

$$H(t) = \begin{cases} 0 \text{ if } t < \rho_1, \\ (t-\rho_1)/(\rho_2-\rho_1) \text{ if } \rho_1 < t < \rho_2, \\ 1 \text{ if } t \geq \rho_2. \end{cases}$$

This assumption concerning h means that the fraction of the population that is left after a catastrophe is uniformly distributed on the interval $[\rho_1, \rho_2]$. In the numerical algorithm we take $n = 500$ and $s_j = j/n$. The graphs of the density functions f that are given below for various values of β_o, ρ_1 and ρ_2 have been obtained by linear interpolation from the approximate values f_j.

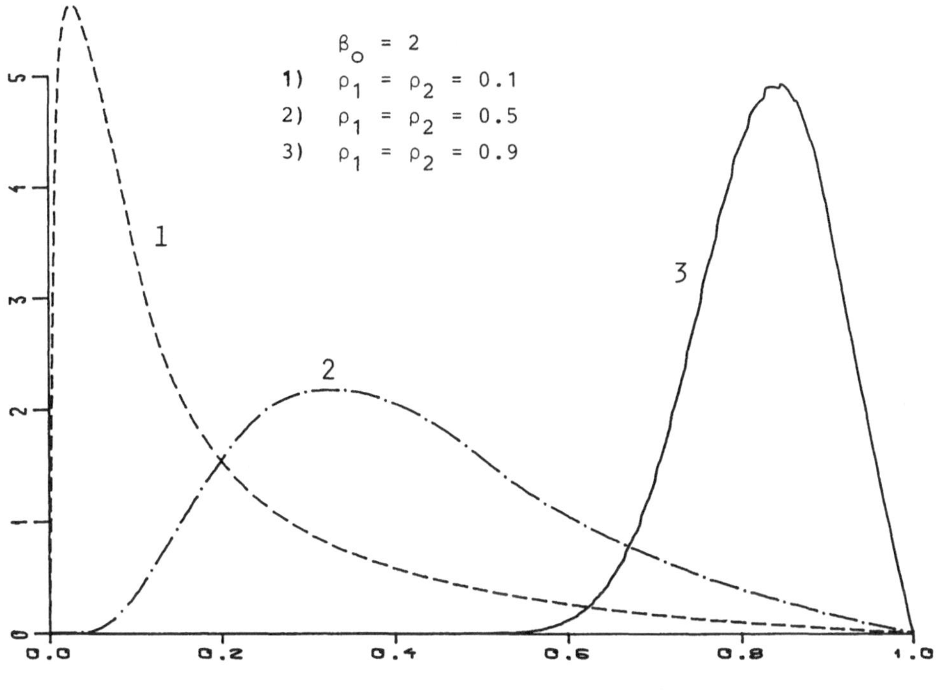

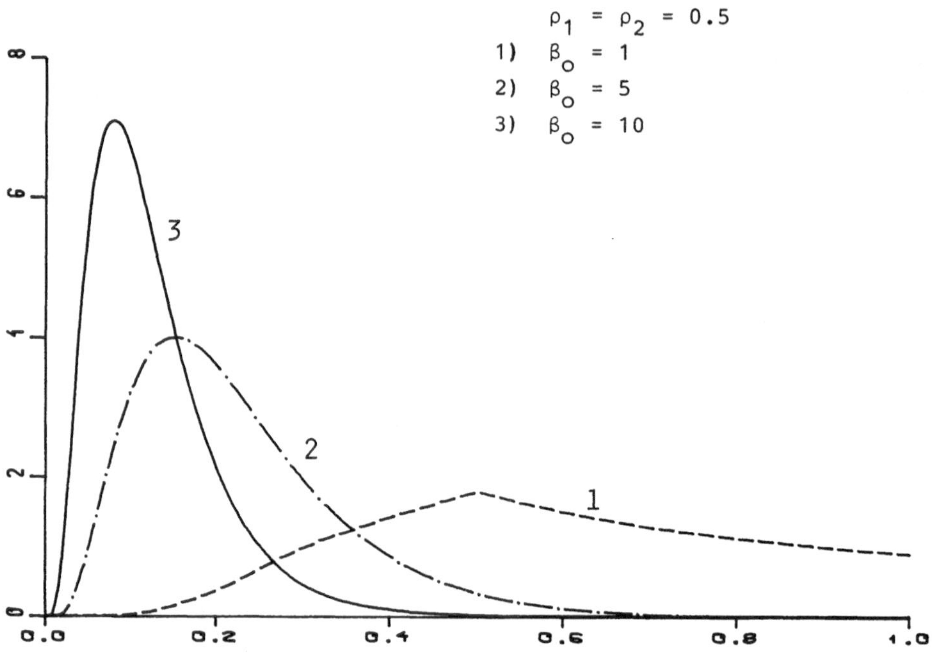
ρ1 = ρ2 = 0.5
1) βo = 1
2) βo = 5
3) βo = 10
3
2
1
0
2
4
6
8
0.0
0.2
0.4
0.6
0.8
1.0

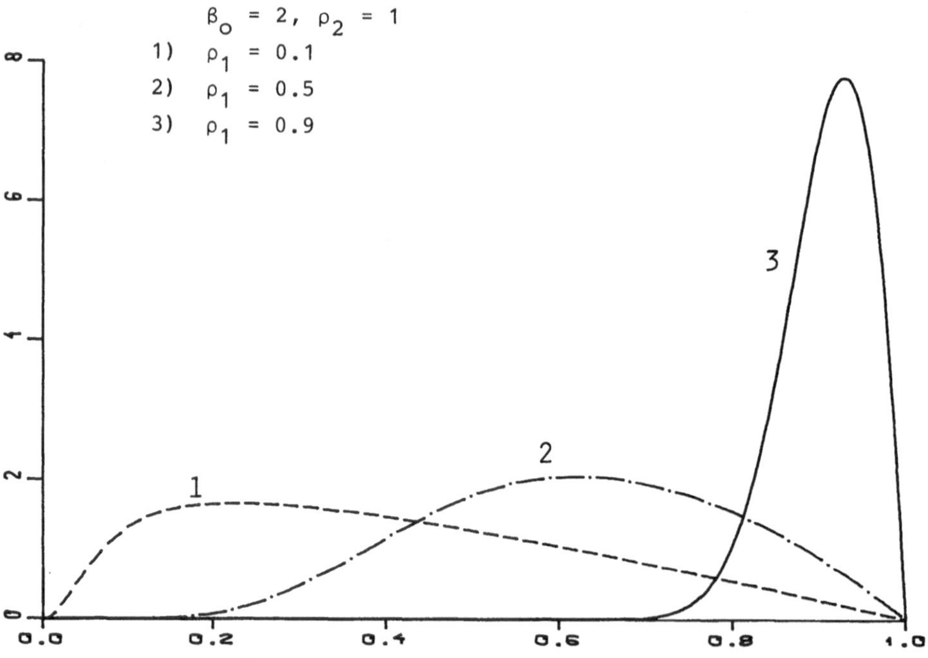
βo = 2, ρ2 = 1
1) ρ1 = 0.1
2) ρ1 = 0.5
3) ρ1 = 0.9
3
2
1
0
2
4
6
8
0.0
0.2
0.4
0.6
0.8
1.0

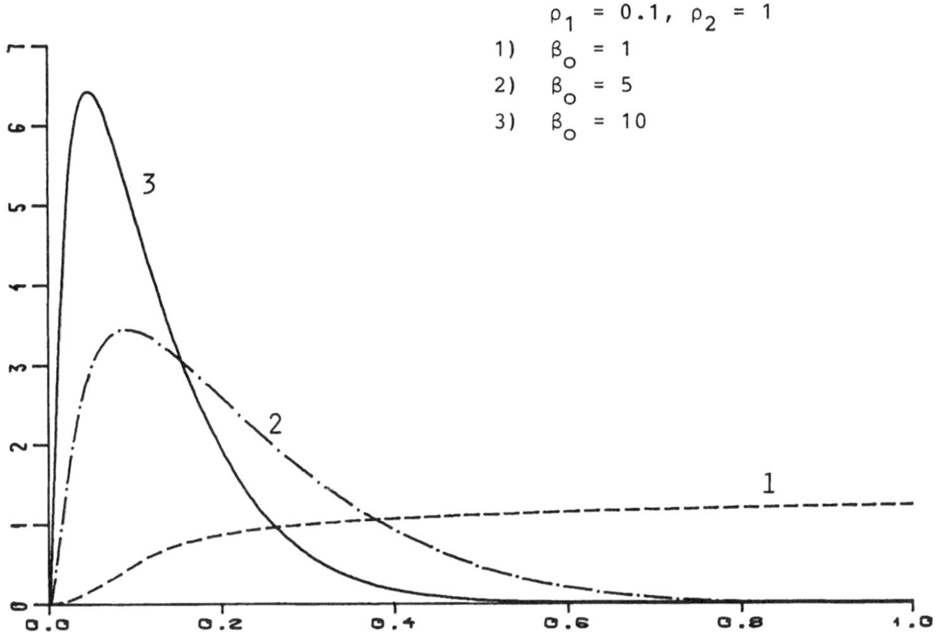

5. REFERENCES

1. P.J. Brockwell, J.M. Gani and S.I. Resnick, Birth, immigration and catastrophe processes, Adv. in Appl. Probab. 14 (1982), 709 - 731.
2. P.J. Brockwell, J.M. Gani and S.I. Resnick, Catastrophe processes with continuous state space, Austral. J. Statist., to appear.
3. G. Gripenberg, A stationary distribution for the growth of a population subject to random catastrophes, J. Math. Biol., to appear.
4. F.B. Hanson and H.C. Tuckwell, Persistence times of populations with large random fluctuations, Theoret. Population Biol. 14 (1978), 46 - 61.
5. F.B. Hanson and H.C. Tuckwell, Logistic growth with random density independent disasters, Theoret. Population Biol. 19 (1981), 1 - 18.
6. A.G. Pakes, A.C. Trajstman and P.J. Brockwell, A stochastic model for a replicating population subject to mass emigration due to population pressure, Math. Biosci. 45 (1979), 137 - 157.
7. A.C. Trajstman, A bounded growth population subjected to emigrations due to population pressure, J. Appl. Probab. 18 (1981), 571 - 582.

A NONLINEAR DIFFUSION EQUATION IN PHYTOPLANKTON DYNAMICS WITH SELF-SHADING EFFECT

H. Ishii
Department of Mathematics
Chuo University
Tokyo 112, Japan

and

I. Takagi
Mathematical Institute
Tohoku University
Sendai 980, Japan

1. Introduction

We consider the nonlinear diffusion equation

(1.1) $$p_t = p_{xx} - \omega p_x - \lambda p + f(I(p))p \qquad \text{in} \quad (0,\ell) \times (0,\infty),$$

where

$$I(p) \equiv I(p)(x,t) = \int_0^x (\kappa + p(\xi,t))d\xi.$$

Here $\kappa \geq 0$, $\lambda > 0$, $\omega \geq 0$, and $0 < \ell \leq \infty$ are given constants, $f: [0,\infty) \to (0,\infty)$ is a decreasing Lipschitz function such that $\lim_{r\to\infty} f(r) = 0$, and $p: [0,\ell] \times [0,\infty) \to [0,\infty)$ is the unknown. Imposed are the boundary condition

(1.2) $$p_x(0,t) = \omega p(0,t),$$

(1.3) $$\begin{cases} p_x(\ell,t) = \omega p(\ell,t) & \text{if} \quad \ell < \infty, \\ \lim_{x\to\infty} p(x,t) = 0 & \text{if} \quad \ell = \infty \end{cases}$$

for all $t > 0$ and the initial condition

(1.4) $$p(x,0) = p_0(x) \qquad \text{for} \quad x \in (0,\ell).$$

This is a mathematical model describing phytoplankton population dynamics with self-shading effect. The function p represents the vertical distribution of phytoplankton in the water. The variable x represents the depth in the water, $x = 0$ at the surface, and t represents the time. The constants ω and λ are, respectively, the sinking velocity and the loss rate of phytoplankton. The function $f(I(p))$ represents the production rate of phytoplankton which depends on light intensity. κ is the light absorption rate of water.

The purpose here is to study the asymptotic behaviour of the solution of (1.1)-(1.4) as $t \to \infty$. The case $\ell = \infty$, $\kappa = 0$ has been investigated by Shigesada-Okubo [2]. The most interesting characteristic of the equation (1.1) is that the nonlinear term $f(I(p))p$ is not local, i.e., $f(I(p))$ depends on p through its integral over $[0,x]$. We will state our recent results on this subject and give an outline of the proof of one of them.

2. The case of infinite depth

We assume $\ell = \infty$, $\omega > 0$ and

$$(2.1) \qquad p_0 \in C[0,\infty) \cap L^\infty(0,\infty) \cap L^1(0,\infty), \qquad p_0 \geq 0.$$

Let ϕ be the (unique) solution of the Cauchy problem

$$(2.2) \qquad \begin{cases} \phi'' - \omega\phi' - \lambda\phi + f(\kappa x)\phi = 0 & \text{in } (0,\infty), \\ \phi(0) = 1, \qquad \phi'(0) = \omega. \end{cases}$$

Then there are two cases as to the behaviour of ϕ:

(A) $\phi(x) > 0$ for all $x \in [0,\infty)$.

(B) $\phi(x_0) = 0$ for some $x_0 \in (0,\infty)$.

Theorem 1. If (A) is satisfied, then $p(x) \equiv 0$ is the unique nonnegative solution of the stationary problem

$$(2.3) \qquad \begin{cases} p'' - \omega p' - \lambda p + f(I(p))p = 0 & \text{in } (0,\infty), \\ p'(0) = \omega p(0), \qquad \lim_{x\to\infty} p(x) = 0 \end{cases}$$

and is globally stable, i.e., the solution $p(x,t)$ of (1.1)-(1.4) tends to zero uniformly on any compact subset of $[0,\infty)$ as $t \to \infty$. If (B) is satisfied, then (2.3) has a unique positive solution p^* and it is globally stable, i.e., the solution p of (1.1)-(1.4) converges to p^* uniformly on any compact subset of $[0,\infty)$ as $t \to \infty$ unless $p_0 \equiv 0$.

Next we want to describe the conditions (A) and (B) in terms of κ, λ, ω. Set

$$\mathcal{B} = \{(\lambda,\omega) \in (0,\infty)^2;\ \frac{\omega^2}{4} + \lambda - f(0) > 0\}.$$

Theorem 2. If $\kappa \geq 0$ and $(\lambda,\omega) \notin \mathcal{B}$, then (A) holds. There exists a continuous function $\kappa_c : \mathcal{B} \to (0,\infty)$ such that (A) holds for $\kappa \geq \kappa_c$ and (B) holds for $0 \leq \kappa \leq \kappa_c$ provided $(\lambda,\omega) \in \mathcal{B}$.

We refer to Ishii-Takagi [1] for a proof of the above theorems.

3. The case of finite depth

Let us assume $\ell < \infty$ and

$$p_0 \in C[0,\ell], \qquad p_0 \geq 0. \tag{3.1}$$

Consider the eigenvalue problem

$$\begin{cases} \phi'' - \omega\phi' - \lambda\phi + f(\kappa x)\phi = \mu\phi & \text{in } (0,\ell), \\ \phi'(0) = \omega\phi(0), & \phi'(\ell) = \omega\phi(\ell), \end{cases} \tag{3.2}$$

and let μ_1 be the largest eigenvalue of this problem. Note that all the eigenvalues of (3.2) are real, simple and bounded from above, since (3.2) is equivalent to a Sturm-Liouville eigenvalue problem. Note also that the problem (3.2) is associated with the linearization at $p \equiv 0$ of the stationary problem

$$\begin{cases} p'' - \omega p' - \lambda p + f(I(p))p = 0 & \text{in } (0,\ell), \\ p'(0) = \omega p(0), & p'(\ell) = \omega p(\ell). \end{cases} \tag{3.3}$$

Theorem 3. If $\mu_1 \leq 0$, then $p \equiv 0$ is the unique nonnegative solution of (3.3) and is globally stable. If $\mu_1 > 0$, then (3.3) has a unique positive solution and it is globally stable.

We have the following characterization of the sign of μ_1 in terms of κ, λ, ω.

Theorem 4. If $\lambda \geq f(0)$, then $\mu_1 \leq 0$. If $0 < \lambda < f(0)$, then there exists a continuous function $\kappa_c = \kappa_c(\lambda,\omega) > 0$ such that $\mu_1 > 0$ for $0 < \kappa < \kappa_c$ and $\mu_1 \leq 0$ for $\kappa \geq \kappa_c$.

4. Proof of Theorem 3

We will give here a rough outline of the proof of Theorem 3. We need the following lemmas.

Lemma 5. Let $0 < b < \infty$ and $h, k \in C[0,b]$. Assume $h \geq k$ on $[0,b]$. If $u, v \in C^1[0,b]$ satisfy

$$u' = -u^2 + \omega u + h(x), \qquad v' = -v^2 + \omega v + k(x) \qquad \text{in} \quad (0,b),$$

and $u(0) \geq v(0)$, then $u \geq v$ on $[0,b]$. In addition, if $h \not\equiv k$ then $u(b) > v(b)$.

Lemma 6. Let b, h, k be as in Lemma 5. Let u,v satisfy, respectively,

$$u'' - \omega u' - h(x)u = 0 \qquad \text{in} \quad (0,b), \qquad u'(0) = \omega u(0),$$

and

$$v'' - \omega v' - k(x)v = 0 \qquad \text{in} \quad (0,b), \qquad v'(0) = \omega v(0).$$

If $u(0) \geq v(0)$ and $v > 0$ on $[0,b]$, then $u \geq v$ on $[0,b]$.

Outline of proof. Note that $y = u'/u$ satisfies $y' = -y^2 + \omega y + h(x)$ if $u(x) \not\equiv 0$. Apply Lemma 5 to y and $z = v'/v$ to get $u'/u \geq v'/v$ on $[0,b]$, from which it follows that $u \geq v$ on $[0,b]$. Q.E.D.

Lemma 7. The solution p of (1.1)-(1.4) satisfies

$$\sup_{t \geq 0} \int_0^\ell p(x,t)dx < \infty, \tag{4.1}$$

$$\sup_{t \geq 1} \left\{ \int_0^\ell p_x(x,t)^2dx + \int_t^{t+1}\int_0^\ell p_t(x,s)^2dxds \right\} < \infty . \tag{4.2}$$

Outline of proof. We write $F(r) = \int_0^r f(\sigma)d\sigma$ for $r \geq 0$ and $v(x,t) = \int_0^x p(\xi,t)d\xi$. Note that $v_t \leq v_{xx} - \omega v_x - \lambda v + F(v)$ (see [1]). Since $v_{xx} - \omega v_x = 0$ at $x = \ell$, $y = \int_0^\ell p(\xi,t)d\xi$ satisfies $y' \leq -\lambda y + F(y)$ for $t > 0$. From this (4.1) follows. Setting $\chi(x) = \ell - x$, by standard calculations we have

$$\frac{d}{dt}\int_0^\ell v^2\chi^2dx \leq -\int_0^\ell v_x^2\chi^2dx + C,$$

from which we deduce

$$\sup_{t \geq 0} \int_t^{t+1}\int_0^{\ell/2} p(x,s)^2dxds < \infty.$$

Similar calculations for $v(x,t) = \int_x^\ell p(\xi,t)d\xi$ and $\chi(x) = x$ show

$$\sup_{t \geq 0} \int_t^{t+1}\int_{\ell/2}^{\ell} p(x,s)^2dxds < \infty.$$

Now inequalities

$$\frac{d}{dt}\int_0^\ell p^2 dx \le -\int_0^\ell p_x^2\, dx + C\int_0^\ell p^2\, dx,$$

$$\int_0^\ell p_t^2\, dx + \frac{d}{dt}\int_0^\ell p_x^2\, dx \le C\{\frac{d}{dt}(p(\ell,t)^2 - p(0,t)^2) + \int_0^\ell (p^2 + p_x^2)\, dx\},$$

and

$$p(\ell,t)^2 - p(0,t)^2 \le \int_0^\ell (p^2 + p_x^2)\, dx$$

together with the above observations prove (4.2). Q.E.D.

Lemma 8. Let p and $\tilde{p}$ be solutions of (1.1)-(1.3). If

$$\int_0^x p(\xi,0)\, d\xi \ge \int_0^x \tilde{p}(\xi,0)\, d\xi \qquad \text{for } x \in [0,\ell],$$

then

$$\int_0^x p(\xi,t)\, d\xi \ge \int_0^x \tilde{p}(\xi,t)\, d\xi \qquad \text{for } (x,t) \in [0,\ell] \times [0,\infty).$$

The proof of [1, Lemma 5.1] can be used to prove this lemma with minor changes.

Now we turn to

Proof of Theorem 3. The nonexistence of a positive solution of (3.3) in the case $\mu_1 \le 0$ and its uniqueness in the case $\mu_1 > 0$ are easy consequences of Lemmas 5 and 6. We assume $\mu_1 > 0$ and show the existence of a positive solution of (3.3). For $\alpha > 0$ let p_α be the solution of

$$\begin{cases} p'' - \omega p' - \lambda p + f(I(p))p = 0 & \text{for } x > 0, \\ p(0) = \alpha, \qquad p'(0) = \alpha\omega. \end{cases}$$

We first show that $p_\alpha > 0$ on $[0,\ell]$ and, setting $q_\alpha = p'_\alpha/p_\alpha$, $q_\alpha(\ell) > \omega$ for large α. Let ϕ be the solution of

$$\begin{cases} \phi'' - \omega\phi' - \lambda\phi + f(\kappa x)\phi = 0 & \text{in } (0,\ell), \\ \phi(0) = 1, \qquad \phi'(0) = \omega, \end{cases}$$

and choose $b \in (0,\ell]$ so that $\phi(x) \ge \frac{1}{2}$ in $[0,b]$. By Lemma 6, $\alpha\phi \le p_\alpha$ on $[0,b]$. Hence, for any $0 < \varepsilon < b$, $f(I(p_\alpha)) \to 0$ uniformly on $[\varepsilon,b]$ as $\alpha \to \infty$, and so $q'_\alpha = -q_\alpha^2 + \omega q_\alpha + \lambda - f(I(p_\alpha)) \ge -q_\alpha^2 + \omega q_\alpha + \frac{\lambda}{2}$ on $[\varepsilon,b]$ for large α. With the help of Lemma 5 this implies that $q_\alpha(x) > \omega$ in a neighborhood of $[0,b]$ and thus $q_\alpha > \omega$

on $[0,\ell]$ for large α. We now define $\alpha_0 = \inf\{\alpha > 0;\ p_\alpha > 0$ on $[0,\ell]$ and $q_\alpha(\ell) > \omega\}$. If $\alpha_0 = 0$, then $p_\alpha \downarrow 0$ as $\alpha \downarrow 0$. Compare q_α with the solution ψ of $\psi' = -\psi^2 + \omega\psi + \lambda + \mu_1 - f(\kappa x)$ and $\psi(0) = \omega$ to obtain a contradiction, $\omega = \psi(\ell) > q_\alpha(\ell) > \omega$ for small α. Thus $\alpha_0 > 0$. Since $p_\alpha(x) \downarrow p_{\alpha_0}(x)$ as $\alpha \downarrow \alpha_0$, $p_{\alpha_0} \geq 0$ on $[0,\ell]$. The fact that p_{α_0} solves a linear second order ODE implies $p_{\alpha_0} > 0$ on $[0,\ell)$. Moreover, the fact $q_{\alpha_0}(\ell) \geq \omega$ ensures that $p_{\alpha_0} > 0$ on $[0,\ell]$. It is now easy to see that $q_{\alpha_0}(\ell) = \omega$. Thus we have proved the existence of a positive solution of (3.3).

Assume $\mu_1 \leq 0$, and let p be the solution of (1.1)-(1.4). Let ψ be the solution of $\psi' = -\psi^2 + \omega\psi + \lambda + \mu_1 - f(\kappa x)$ in $(0,\ell)$ and $\psi(0) = \omega$. By direct computations

$$\frac{1}{2}\frac{d}{dt}\int_0^\ell p^2 e^{-\omega x}\,dx + \int_0^\ell (p_x - \psi p)^2 e^{-\omega x} dx$$
$$+ \int_0^\ell (f(\kappa x) - f(I(p)) - \mu_1)p^2 e^{-\omega x}\,dx \leq 0 \qquad \text{for} \quad t > 0.$$

From this and Lemma 7, we may conclude that $p(x,t) \to 0$ uniformly on $[0,\ell]$ as $t \to \infty$.

The global stability of the unique positive solution of (3.3) in the case $\mu_1 > 0$ follows by an argument similar to [1] with the aid of Lemmas 7 and 8. Q.E.D.

Acknowledgement. The first author was supported in part by the (Italian) CNR.

References

1. H. Ishii and I. Takagi, Global stability of stationary solutions to a nonlinear diffusion equation in phytoplankton dynamics. J. Math. Biol. 16, 1-24 (1982)

2. N. Shigesada and A. Okubo, Analysis of the self-shading effect on algal vertical distribution in natural waters. J. Math. Biol. 12, 311-326 (1981)

A MODEL OF APHID POPULATION WITH AGE STRUCTURE

Pavel Kindlmann
Centre of Biomathematics, Institute of Entomology
Na sádkách 702, 370 05 České Budějovice
Czechoslovakia

Aphids are among the most conspicuous and important pests in the greenhouses and in the fields. As such, they have been the subject of intensive studies and various attempts have been made to develop models of their and their predators' population dynamics (see, e.g., Barlow, 1982; Barlow and Dixon, 1980 and references therein).

In our experiments (Okrouhlá, 1983), the population dynamics of the species Aphis fabae was studied. The development of the aphid population was observed at $22\pm1^{\circ}C$ in cages containing 10 bean plants infested by 250 apterous viviparous females of Aphis fabae at the beginning of their reproductive period.

Before their premature death the females laid at most 1/3 of their reproductive potential at an average daily fecundity rate of approximately 4 nymphs per female per day. The total number of individuals in the population rapidly increased at the beginning, reaching its maximum on the 5th day. There was a mild decrease between the 5th-7th day and a rapid decrease in the number of individuals between days 7-11 as a consequence of high mortality of the aphids (Fig. 1).

The excessive population density of Aphis fabae, exceeding the carrying capacity of the cage, resulted in a rapid deterioration of environment through continuous production of honeydew manifested in the development of alate morphs, emigration of adults as well as developing aphids (by leaving the plants), their subsequent death and the gradual withering of the plants caused by excessive sucking by the aphids.

In order to elucidate the rapid decrease in the number of individuals, we have developed the following simple model: Let $x(t)$ be the total number of individuals at time t. The change in the population number, $x'(t)$, is then equal to the difference of the birth rate and the mortality rate. For the first approximation we have chosen the simplest possible form for the birth rate: $rx(t)$, where r is a constant. The mortality rate was very low in the initial phase of the experiment, but then increased dramatically. With regard to the bionomy of our species we have attributed this rapid increase of the mortality rate to the deterioration of the environment by the honeydew excretions of aphids. Honeydew forms a weak cover on the leaf surface preventing the individuals from further sucking and movement and so causing starvation. Evidently, the area covered by the excretions at time t is proportional to the integral $\int_0^t x(s)\,ds$, which is called integral or cumulative density (Barlow and Dixon, 1980). Following Barlow and Dixon (1980), we have assumed linear dependence between integral density and mortality rate:

$$x' = rx - \frac{x}{C}\int_0^t x(s)ds. \tag{1}$$

Such a model is very simple, indeed, and does not describe precisely the actual mechanisms which occured in our cages. But owing to its simplicity it is analytically tractable.

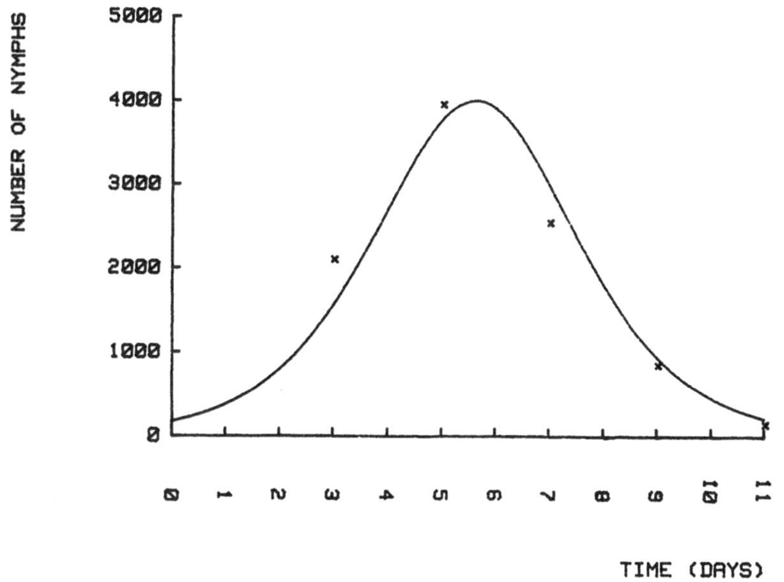

Fig. 1. Total numbers of nymphs of Aphis fabae. Crosses indicate empirical data, curve is solution (10) of eq. (1) with $C = 10^4$, $K = .8$, $K' = -4.5$.

A slight modification gives us

$$(rx - x')\,\frac{C}{x} = \int_0^t x(s)\ ds. \tag{2}$$

Let $x(t) \neq 0$ for all $t \geq 0$, let x be at least twice differentiable. After differentiation we have:

$$x''x - (x')^2 + \frac{x^3}{C} = 0, \tag{3}$$

which must be solved subject to the initial conditions $x(0) = x_0$, $x'(0) = r$. The substitution $x'(t) = p(x)$ now shows us that

$$pp'x - p^2 + \frac{x^3}{C} = 0. \tag{4}$$

We divide (4) by $x^2 p$ and set $q(x) = p(x)/x$. We obtain:

$$q' + \frac{1}{Cq} = 0. \tag{5}$$

Separation of variables gives us

$$q = \pm \sqrt{K - \frac{2x}{C}} \,. \tag{6}$$

Back substitution shows

$$x' = p = \pm x \sqrt{K - \frac{2x}{C}} \,. \tag{7}$$

Again by separation of variables, we find:

$$\int \frac{dx}{x \sqrt{K - \frac{2x}{C}}} = \pm t + K''. \tag{8}$$

The integral on the left-hand side of (8) is equal to

$$K^{-1} \left(\ln\left|\sqrt{K - \frac{2x}{C}} - \sqrt{K}\right| - \ln\left|\sqrt{K - \frac{2x}{C}} + \sqrt{K}\right| \right). \tag{9}$$

From (8) and (9) it follows that

$$x(t) = \frac{2CK \exp(tK + K')}{(1 + \exp(tK + K'))^2}, \tag{10}$$

where $K' = KK''$. The explicit expressions for K and K' may be obtained from (10) by setting $t = 0$ and using initial conditions. The function $x(t)$ attains its maximum at $t = -K'/K$ and it holds $\lim_{t \to \infty} x(t) = 0$ (see Fig. 1).

The qualitative behaviour of the trajectory (10) is very similar to that of real population of Aphis fabae. But in our experiments there was no evidence for the assumption that the birth rate is dependent on

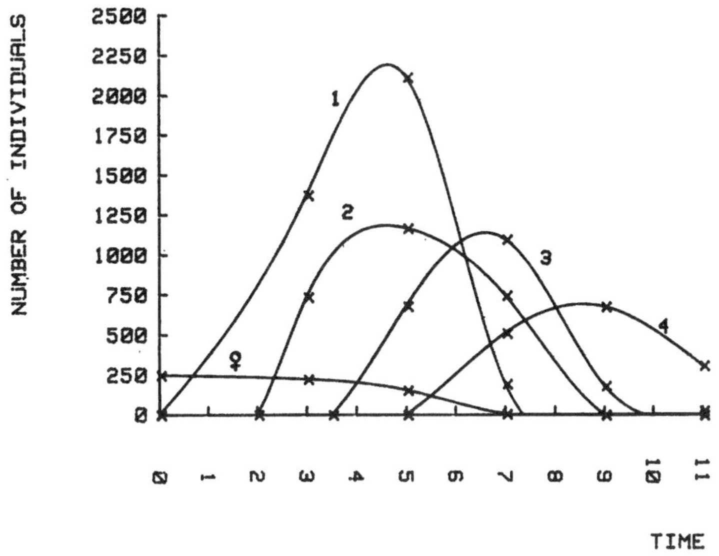

Fig. 2. Spline estimate of the number of individuals in each age class. Crosses are taken from empirical data. The numbers by the curves indicate number of instar, ♀ are females.

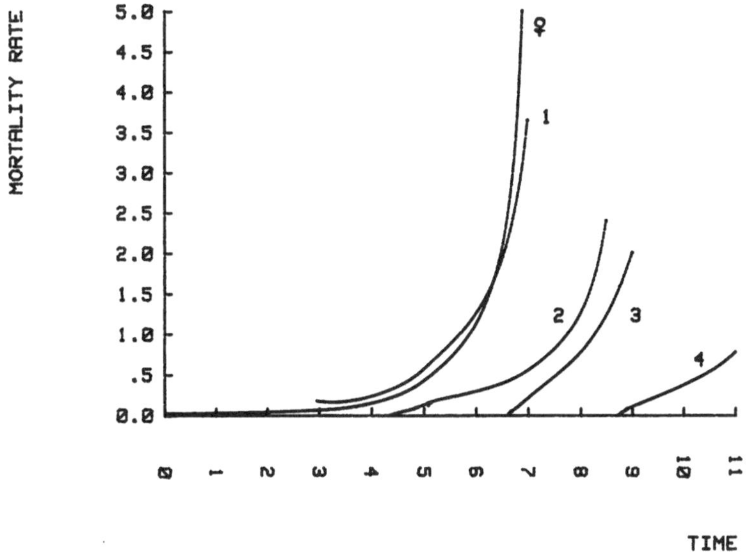

Fig. 3. The estimate of the actual mortality rate based on model (11).

the population density and that the dependence of the mortality rate on integral biomass is really linear. Moreover, age structure was completely ignored in the first model. Therefore, we have tried to estimate the actual mortality rate from our data.

Each individual in the population goes through four juvenile developmental stages (instars) and one adult stage during its life. Therefore, we have divided our population into five age classes and assumed that within each of them all individuals can sensibly be regarded as functionally identical (that is all having the same per capita vital rates). We have used a modification of the model of Gurney, Nisbet and Lawton (1983):

$$
\begin{aligned}
N_i'(t) &= R_i(t) - R_i(t - s_i)P_i(t) - d_i(t)N_i(t),\\
P_i'(t) &= P_i(t)\cdot(d_i(t - s_i) - d_i(t)),\\
R_i(t) &= \begin{cases} \sum_{l=1}^{Q} b_l N_l(t), & i = 1,\\ R_{i-1}(t - s_{i-1})P_{i-1}(t), & i = 2, 3, \ldots, Q, \end{cases}
\end{aligned}
\tag{11}
$$

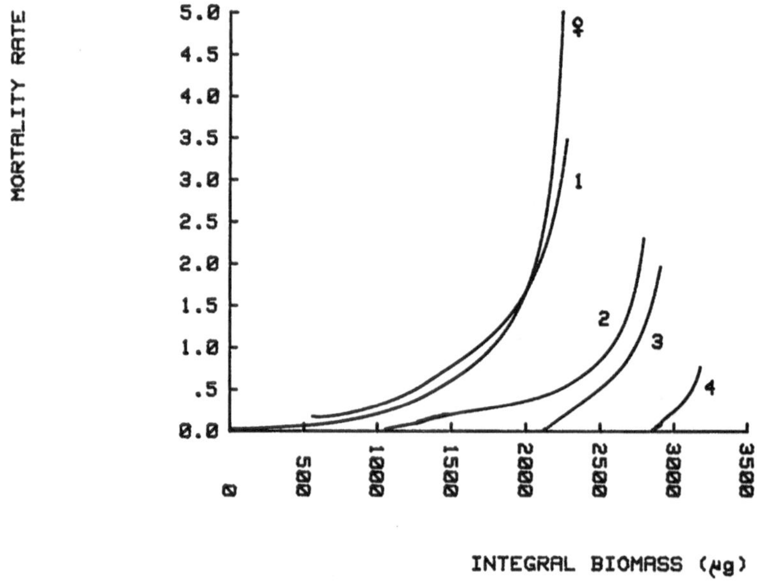

Fig. 4. The dependence of the mortality rate on integral biomass for individual age classes.

where Q is the number of age classes, $N_i(t)$ is the number of individuals of age class i at time t, $R_i(t)$ is the recruitment into age class i at time t, $P_i(t)$ is the proportion of individuals recruited into age class i at time t - s_i who survive to be recruited into age class i + 1 at time t, $d_i(t)$ is the mortality rate of age class i at time t, b_1 is the fecundity of age class 1 and s_i the duration of i-th age class. From (11), the d_i's and P_i's may easily be expressed as functions of the N_i's, N_i''s and R_i's, i.e. of functions which may be estimated from empirical data by means of spline functions (Fig. 2 - N_i's and N_i''s), or directly measured (R_i's).

The estimated dependence of the mortality rates of individual age groups on time is shown in Fig. 3. It is conspicuously nonlinear and tends to infinity in some finite time in each group. It is a consequence of the total extinction of corresponding instars.

The hypothesis about the linear dependence of the mortality rate on integral biomass is tested in Fig. 4, which shows that this dependence may be assumed to be linear only within limited intervals of integral biomass.

On the basis of the second model, some implications for the biological control of aphids may be derived. They will be published elsewhere.

REFERENCES

Barlow, N.D. (1982) Modelling aphid populations. New Zealand J. Ecol. 4:52-55.

Barlow, N.D. and Dixon, A.F.G. (1980). Simulation of lime aphid population dynamics. Ctre Agr. Publ. and Document., Wageningen.

Gurney, W.S.C., Nisbet, R.M. and Lawton, J.H. (1982) The systematic formulation of tractable single species population models incorporating age structure. J. Anim. Ecol. 52:479-495.

Okrouhlá, M. (1983) Potravně ekologická studie slunéčka Cheilomenes sulphurea. PhD Thesis, Charles University, Prague.

ON TWO-POPULATION MODELS WITH SWITCHING DEPLETION

M.G. Messia, P. de Mottoni, E. Santi
Istituto di Matematica Applicata, Fac. Ingegneria
I - 67040 Poggio di Roio (L'Aquila), Italy

1. Generalities

We consider two differential systems, suggested by two different ecological models, describing two mutually interacting populations subject to depletion effects (due to harvesting or to predation), cf. [1,2], which share certain "switching" features.

The first model deals with two mutually competing populations subject to harvesting in such a way that only the most abundant species is depleted - (which may be caused by a predator which develops a search image depending on the size of either population), cf. [5].

The second model describes two populations which do not interact with each other in the absence of predation. The latter models a situation studied in the field by Contoli [3,4], where basically two groups of micromammalia, rodentia and insectivora (the latter being represented by chiroptera and soricidae) are predated by barn owls in such a way that on the one hand they are predated in a completely unselective way, on the other one, predation depends in a very definite way on the presence of just one population (the rodentia): indeed, the scarceness of rodentia cannot be compensated by a quick growth of the insectivora (which are essentially territorial), so that, the overall food supply being not sufficient, the predator abandons the territory, and predation (of both populations) ceases.

This paper is devoted to a concise presentation of the results obtained for such models, which are described in detail in the papers [6] and [7], to which the reader is referred for more details and for the proofs.

2. The first model: switching harvesting

As model for the above described situation we consider

$$(2.1)\qquad \begin{cases} \dot{u} = u(1 - b_1 v - \varepsilon_1 u) - \dfrac{\gamma}{2}\left[\dfrac{u-v}{u+v}\right]_+ \\ \dot{v} = \beta v\,(1 - b_2 u - \varepsilon_2 v) - \dfrac{\gamma}{2}\left[\dfrac{v-u}{u+v}\right]_+ \end{cases}$$

where $[x]_+ := \max\ \{0,x\}$, $\varepsilon_i < b_i/2$ and $\gamma \in (0,\ \min\{\frac{1}{2\varepsilon_1}, \frac{\beta}{2\varepsilon_2}\})$.

This express the fact that:

-1) in the absence of depletion the two species interact via the usual Volterra-Lotka competition dynamics, and competitive exclusion prevails;

-2) the depletion is unaffected by the "absolute" levels of the two populations, depending only on their ratio;

-3) depletion affects either of the two species at a time.

Under the above assumptions we prove that:

-2a) initial positive data give rise to solutions which are well-defined and positive for all future times, and satisfy an a-priori bound.

-2b) A full description of the stationary solutions can be given, along with their stability character. The relevant parameters, in this connexion, turn out to be γb_i (i=1,2); indeed, different ranges of values of γb_i classify different structures of stationary solutions.

-2c) Of particular interest are the periodic solutions. We show that if γb_i exceeds a certain threshold, periodic solutions bifurcate which are asymptotically stable. The detailed conditions characterizing the above situation are illustrated in [7].

One of the outcomes of our investigations is that given two populations in a competitive exclusion regime, a harvesting effort directed towards the most abundant population at a time may prevent extinction of both species and, under appropriate conditions on its strength, may give rise to a permanently oscillating regime.

3. The second model: unselective predation

As a model for the situation outlined above we take:

$$(3.1) \qquad \begin{cases} \dot{u} = u(p_1(1- \frac{u}{p_2}) - H(p_5, p_6; u/v)) \\ \dot{v} = v(p_3(1- \frac{u}{p_4}) - H(p_5, p_6; u/v)) \end{cases}$$

where $\xi \mapsto H(p_5, p_6; \xi)$ is a continuous function having the graph:

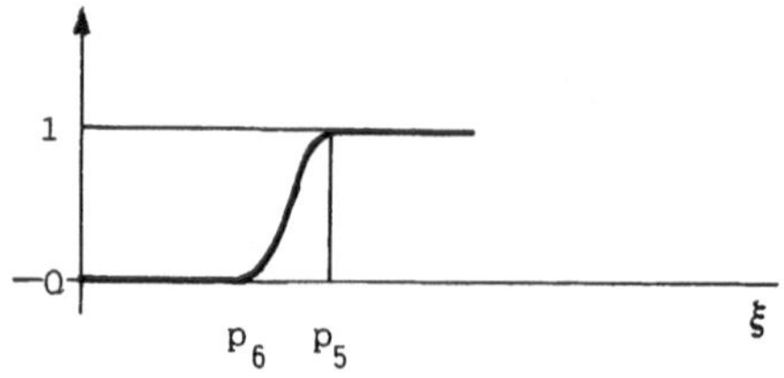

and p_1, $p_3 > \gamma$.

Because of the definition of H, u represents the prey on which predation depends, (namely the rodentia, in the field situation examined above).

The following features are incorporated into the model:

-1) in the absence of predation ($\gamma = 0$), the two populations evolve independently of each other, and follow a logistic growth.

-2) The amount of either species which is predated in the unit time is zero if u is less numerous than v by a factor p_6 , and proportional to the respective abundance if u is more numerous than v by a factor p_5 , $p_5 > p_6$.

-3) The condition on p_1 , p_3 ensures that predation is never so strong as to doom either species to extinction.

The main results are:

-3a) the solutions having non-negative initial values are well-defined and non negative for all future times, independently of the specific form of H;

-3b) The stationary points $(0, p_4)$, $(p_2 \cdot \frac{p_1 - \gamma}{p_1}, 0)$, already present in the absence of predation (the second is the counterpart of $(p_2, 0)$) preserve their saddle-point character. Instead, the stable node (p_2, p_4), which is a global attractor in the absence of predation, may split in as much as three stationary points, one of them stable, and, of the remaining two, if any, one stable, another unstable.

The necessary condition for the occurrence of these three stationary solutions is $p_3 < p_1$; this is also sufficient under some additional conditions.

-3c) We prove that there is only one stationary point which is a stable vortex and, when it does, no other stationary points can exist in the positive orthant.

A necessary condition for the existence of a stable vortex is $p_1 < p_3$, that is when the net growth rate of the "preferred" prey does not exceed that of the other one.

Our analysis puts into evidence that only three cases can prevail in the long run:

-1) evolution, without oscillations, towards a unique stationary state, much as in the case of absence of predation;

-2) a bi-stable regime: the solutions may evolve to either of the two stable states; thus a threshold effect shows up;

-3) damped oscillations approaching a single rest point. The occurrence of the above situations is completely characterized in terms of the system's parameters, and a central rôle is played by the relative size of the carrying capacities.

The fact that in the field situation, the necessary condition for the onset of damped oscillations ($p_1 < p_3$) is never satisfied (for the rodentia have a larger net growth rate than the insectivora) suggests that the observed oscillations might be induced by seasonal variations.

REFERENCES

[1] **F. Brauer, A.C. Soudack**: "Coexistence Properties of Some Predator-Prey Systems Under Constant Rate Harvesting and Stocking" J. Math. Biol. 12 (1), pp.101-114 (1981).

[2] **G.J. Butler, P. Waltman**: "Bifurcation from a limit cycle in a Two Predator-One Prey ecosystem modeled on a Chemostat" J. Math. Biol. 12, pp.295-310, (1981).

[3] **L. Contoli**: "Micromammals and environment in central Italy: data from Tyto alba (scop.) pellets" Boll. Zool. n. 42, pp.223-229, (1975).

[4] **L. Contoli**: "Mammiferi del tolfetano-cerite" Quad. n. 227 Accad. Naz. Lincei, pp.191-226, (1977).

[5] **S.A. Levin, L.A. Segel**: "Models of the Influence of Predation on Aspect Diversity in Prey Population" J. Math. Biol. 14 (3), pp.253-284, (1982).

[6] **M.G. Messia, P. de Mottoni, E. Santi**: "On two population subject to Predation of Non-selective type" in press on Advances in Modelling and Simulation.

[7] **M.G. Messia, P. de Mottoni, E. Santi**: "On a Competitive System subject to switching predation" in press on Ecological Modelling.

ON SOME NONLINEAR DIFFUSION MODELS IN POPULATION DYNAMICS

Maria Assunta Pozio
Dipartimento di Matematica, Università di Trento
I-38050 Povo, Trento (Italy)

Alberto Tesei
Dipartimento di Matematica, II Università di Roma
Via O. Raimondo, I-00173 Roma (Italy)

Reaction-diffusion models in population dynamics have been widely investigated in recent years [6]. In particular, quasilinear models have been recently suggested [11] to deal with situations where diffusivity depends on crowding (see also [7]). In the case of two interacting species such models have the general form

$$(1)\qquad \begin{aligned} u_t &= \Delta y(u,v) + u\, f(x,u,v) \\ v_t &= \Delta j(u,v) + v\, g(x,u,v) \end{aligned}$$

the solution satisfying suitable initial and boundary conditions in a bounded domain (here t is the time variable, x the space variable, Δ the diffusion operator, $u=u(t,x)$, $v=v(t,x)$ are population densities, see [6,7,11]).

Monotonicity methods can be used to investigate the qualitative properties of (1) in the semilinear case (formally when $y=au$, $j=bv$, $a,b>0$), in particular when the system is quasimonotone. Such methods do not apply to general quasilinear systems, due to strong coupling between the equations. However they prove to be useful in two particular cases of (1) :

a) $y=y(u)$, $j=j(v)$, [3,8] ;

b) either $y=0$ or $j=0$, [2] .

We discuss here a competition model with diagonal diffusion, namely of type a) (see [10] for the linear diffusion case). Even in the simple case where f and g are linear functions of u and v , the nonlinearities $y(u)$, $j(v)$ allow a more realistic modelling of the diffusivity. As we will see here , several natural situations arise in this case, which could not be obtained with a linear diffusion model (namely with linear $y(u)$ and $j(v)$).

The results we give here can be proved by monotonicity methods analogously to the results in [8] for the predator-prey case. Such methods are used in [3] for symbiotic interactions and rely on previous results proved in [1,4] for a single equation. A competition model of type (1), with non-diagonal diffusion (case (b)) is investigated in [2] by linearized stability methods.

Here we deal with the following competition model :

$$
(2)\quad
\begin{cases}
u_t = \Delta y(u) + u\,[a_1(x) - u - b_1 v] , & \\
v_t = \Delta j(v) + v\,[a_2(x) - v - b_2 u] , & \text{in } (0,+\infty)\times D , \\
u = v = 0 , \quad \text{in } (0,+\infty)\times\partial D , & \\
u = u_0 , \quad v = v_0 , \quad \text{in } \{0\}\times D ; &
\end{cases}
$$

where D is a bounded connected open set in $\mathbb{R}^n$ with smooth boundary . Let us assume the following

i) $y,j : \bar{\mathbb{R}}_+ \rightarrow \bar{\mathbb{R}}_+$ are continuously differentiable and increasing

$y(0)=j(0)=y'(0)=j'(0)=0$, and $y(s)$ is convex in a neighbourhood of $s=0$.

ii) $a_1,a_2 : D \rightarrow \mathbb{R}$ are Hölder continuous functions which assume positive values at least in some subsets of D; $b_1,b_2 > 0$.

iii) u_0,v_0 are continuous nonnegative functions (the initial data).

Only nonnegative solutions of (2) will be considered. By a steady state solution of (2) we mean a time independent solution, that is a solution of the associated elliptic problem.

In the following we only give a qualitative description of some results concerning problem (2). Conditions under which such results hold are given in [8]. Our aim is to describe some phenomena which have a biological meaning and do not arise, neither in the linear diffusion case ($y(u)=au$, $j(v)=bv$), nor in the absence of diffusion ($y=0, j=0$) . Indeed, if $y=0$ and $j=0$, then $u(t,x)$ and $v(t,x)$ evolve at each point $x \in D$, independently from the population densities in the surroundings and the populations coexist or survive according to pointwise relations involving $a_1(x)$, $a_2(x)$, b_1, b_2 . On the other hand, if $y(u) = au$, $j(v) = bv$ $(a,b>0)$, initial data with nonidentically vanishing components instantaneously diffuse and the solution has both positive components in D for any $t>0$. Each component of any steady state solution is either positive in D or vanishes identically. More realistic phenomena are described by problem (2) in the hypotheses i),ii),iii) . Infact the following general results hold :

1) (Single species) In the absence of one species, say $v=0$, the other may survive at least in the subregions of D where $a_1(x) > 0$. In some cases it may survive in the surroundings of such subregions, depending on the crowding influence on diffusivity (i.e., on the nonlinearity $y(u)$). In other words, nontrivial steady state solutions $(u,0)$ exist, such that $u(x)>0$ whenever $a_1(x)>0$. Moreover, such solutions are not unique in general : steady state solutions $(u_1,0),(u_2,0)$ with $u_1>u_2$ in D , or noncomparable steady state solutions ($u_1(x) \gtrless u_2(x)$, depending on $x \in D$)

may exist . This is to say that the same region may be inhabited at different population densities, or different (nonoverlapping or only partially overlapping) subregions may be inhabited, depending on the initial population density. If two disjoint subregions of D exist, where $a_1(x) > 0$, and only one of them is inhabited at time $t=0$, then the other may stay empty or get inhabited, depending on the nonlinearity $y(u)$ and on the initial population density u_0 . There always exists a maximal steady state solution, say $(u_m,0)$, such that for any steady state $(u,0)$, we have $u_m \gg u$. Analogously $(0,v_m)$ exists (the proof of these results relies on some theorems in [1,4, 5,8,9]) .

2) (Coexistence) If $a_1(x)-b_1v_m(x) > 0$ for some $x \in D$ and $a_2(x)-b_2u_m(x) > 0$ for some $x \in D$, then both species survive in D . It may happen that the two species inhabit the same region, or only partially overlapping regions, or different regions (segregation). The last two phenomena are not possible in the semilinear diffusion case . Under the previous conditions neither species may become extinct if it inhabits the whole region D at time $t=0$.

3) (Coexistence and exinction) In some cases , if the population v is sufficiently large at time $t=0$ and the population u is sufficiently low, u becomes extinct and v tends to the density distribution v_m ; if the initial data are positive on suitable regions and v is not too large, then the populations coexist . This situation occurs if, for instance:

j) $a_2(x)-b_2u_m(x) > 0$ for some $x \in D$,

jj) $a_1(x)-b_1v_m(x) \leq 0$ for all $x \in D$, but $a_1(x)-b_1v(x) > 0$ for some $x \in D$ and some steady state solution $(0,v)$.

It should be pointed out that under conditions j),jj), both the coexistence steady state solutions and and the extinction steady state $(0,v_m)$ are stable in some sense (see [8]) .

REFERENCES

1. D.G.Aronson, M.G.Crandall, L.A.Peletier : Stabilization of solutions of a degenerate nonlinear diffusion problem, Nonlinear Anal.:TMA 6(1982),1001,1022.

2. D.G.Aronson,A.Tesei,H.Weinberger : On a simple density dependent diffusion system,

3. R.Dal Passo, P.de Mottoni :On existence, uniqueness and attractivity of stationary solutions to semilinear parabolic problems,these Proceedings.

4. P.de Mottoni,A.Schiaffino,A.Tesei : Attractivity properties of nonnegative solutions for a class of nonlinear degenerate parabolic problems, Ann. Mat.Pura Appl.(1984).

5. J.I.Diaz,J.Hernandez : On the existence of a free boundary for a class of reaction-diffusion systems, SIAM J.Math.Anal., to appear.

6. P.C.Fife : Mathematical aspects of reacting and diffusing systems, Lecture Notes in Biomathematics 28 (Berlin: Springer, 1979).

7. T.Namba : Density-dependent dispersal and spatial distribution of a population, J.Theor.Biol. 86 (1980), 351,363.

8. M.A.Pozio,A.Tesei : Attractivity results for a class of quasilinear parabolic problems, preprint 1983.

9. M.Schatzman : Stationary solutions and asymptotic behaviour of a quasilinear degenerate parabolic equation , preprint (1981).

10.A.Schiaffino,A.Tesei : Competition systems with Dirichlet boundary conditions, J.Math.Biol. 15 (1982), 93,105.

11.N.Shigesada,K.Kawasaki,E.Teramoto : Spatial segregation of interacting species, J.Theor.Biol. 79 (1979),83,99.

REDUCTION OF THE GENE FLOW OF A "NEUTRAL" LINKED GENE DUE TO THE PARTIAL STERILITY OF HETEROZYGOTES FOR A CHROMOSOME MUTATION

Spirito Franco*, Rizzoni Marco**, Lolli Elena, Rossi Carla***

In order to understand the problem of "chromosomal speciation" (White, 1978) the clarification and quantification of the processes and mechanisms involved as well as theoretical models are required in explaining the onset, fixation, spreading and accumulation of chromosome mutations and their role in reproductive isolation.

Models of the onset and the fixation of chromosome mutations which cause partial sterility of the heterozygotes have been already successfully demonstrated (Bengtsson, 1980; Bengtsson and Bodmer, 1976; Hedrick, 1981; Lande, 1979).

The role of both stochastic and deterministic factors has been evaluated and meiotic drive, inbreeding and genetic drift have been shown to be the most important factors in establishing "new" chromosome forms.

The other steps in the process have so far been looked into very little (Lande, 1979, for the spreading phase).

The role of partial sterility of the heterozygotes by chromosome mutations as a mechanism of reproductive isolation is focused on this paper: we have tried to quantify the efficiency of a single chromosome mutation, with partially sterile heterozygotes, in reducing gene flow, for a neutral linked locus, by the use of a deterministic model with two populations, each initially monomorphic for one of two alternative forms (the normal and the mutated one) of a chromosome.

We have studied the behaviour of:

1) the difference between the frequencies of the mutated chromosome in the two populations $(p_1^{(n)} - p_2^{(n)})$;
2) the normalized difference between the frequencies of one of the two neutral alleles in the two populations $(P_1^{(n)} - P_2^{(n)})$;
3) the normalized gametic phase disequilibrium in the overall system of the two populations $(D_c^{(n)})$

* Department of Genetics and Molecular biology, Faculty of Sciences, I^ University of Rome "La Sapienza".
** Department of Biology, Faculty of Sciences, II University of Rome "Tor Vergata".
*** Department of Mathematics, Faculty of Sciences, II University of Rome "Tor Vergata"

as functions of three parameters:

a) m = the symmetrical and reciprocal migration rate between the two populations;

b) s = the partial sterility of the heterozygotes for the chromosome mutation;

c) R = the recombination frequency between the chromosome mutation and the neutral linked locus.

For this purpose recurrence equations are defined which give the values of the three variables as functions of their values at the previous generation and of the parameters m, s, R:

$$p_1^{(n)} - p_2^{(n)} = [(1-2m)/(1-2sp_1^{(n-1)}p_2^{(n-1)})]\,(p_1^{(n-1)} - p_2^{(n-1)}) \qquad [1]$$

$$P_1^{(n)} - P_2^{(n)} = \frac{(1-2m)\left[(P_1^{(n-1)} - P_2^{(n-1)})\,(1-s/2) + 2sD_c^{(n-1)}(p_1^{(n-1)} - p_2^{(n-1)})\right]}{1-2sp_1^{(n-1)}p_2^{(n-1)}} \qquad [2]$$

$$D_c^{(n)} = \frac{(1/2)D_c^{(n-1)}a(s,R) + (1/2)(p_1^{(n-1)} - p_2^{(n-1)})\,(P_1^{(n-1)} - P_2^{(n-1)})\,b(s,R)}{1-2sp_1^{(n-1)}p_2^{(n-1)}} \qquad [3]$$

where:

$a(s,R) = 1+(1-s)\,(1-2R)$ and $b(s,R) = 1-(1-s)\,(1-2R)$.

Analyzing these equations it turns out that, for $s < 1$, $0 < m \leq 0.5$ and $R > 0$:

1) $\lim_{n\to\infty} p_1^{(n)} - p_2^{(n)} = \sqrt{1-4m/s}$ for $m/s < 1/4$

$\lim_{n\to\infty} p_1^{(n)} - p_2^{(n)} = 0$ for $m/s \geq 1/4$;

2) the equilibrium values of $P_1^{(n)} - P_2^{(n)}$ and $D_c^{(n)}$ are always equal to 0.

To study the speed of the process in approaching equilibria, several simulations were performed for the following values of the parameters:

$m = (1/4)^1, (1/4)^2, \ldots, (1/4)^7$; $s = 0, 0.05, 0.15, 0.5$; $R = 0.02, 0.1, 0.5$.

A good approximation of both $P_1^{(n)} - P_2^{(n)}$ and $D_c^{(n)}$ is given for low values of m and s by the components of the vector $X^{(n)}$ which is the linear image of $X^{(n-1)}$, when $m/s < 1/4$, by the matrix:

$$B(m,s,R) = 1/2 \begin{pmatrix} 2-s & s\sqrt{1-4m/s} \\ \dfrac{b(s,R)\sqrt{1-4m/s}}{1-2m} & a(s,R)/(1-2m) \end{pmatrix}$$

the maximum eigenvalue of which is $\lambda(m,s,R)$:

$$\lambda(m,s,R) = \left[(1-m)(2-s) - R(1-s) + \sqrt{T(m,s,R)}\right]/2(1-2m)$$

where:

$$T(m,s,R) = T_0(m,s) + 2RT_1(m,s) + R^2\,T_2(s)$$

and:

$$T_0(m,s) = (ms-s+2m)^2; \quad T_1(m,s) = 8m^2-8sm^2+5ms+ms^2-s^2-6m+s; \quad T_2(s) = (1-s)^2 \;.$$

On the basis of $\lambda(m,s,R)$ the following approximated equations can be written:

$$P_1^{(n)} - P_2^{(n)} \simeq (1-2m)\left[\lambda(m,s,R)\right]^{n-1} \;; \quad D_c^{(n)} \simeq \left[\lambda(m,s,R)\right]^{n-1}$$

for $s \neq 0$; on the other hand, when $s = 0$ the following equations are obtained:

$$p_1^{(n)} - p_2^{(n)} = P_1^{(n)} - P_2^{(n)} = (1-2m)^{(n-1)}$$

$$D_c^{(n)} = \left[1-Q(R,m)\right](1-R)^{(n-1)} + Q(R,m)(1-2m)^{2(n-1)}$$

Where:

$$Q(R,m) = R(1-2m)^2/\left[(1-2m)^2 - (1-R)\right] .$$

Resulting in:

$$Q(R,m) < 0 \quad \text{for} \quad m > (1-\sqrt{1-R})/2 \quad \text{and} \quad Q(R,m) > 0 \quad \text{for} \quad m < (1-\sqrt{1-R})/2$$

while, for $m \ll 1/4$ $Q(R,m) \simeq 1$ and we can write: $D_c^{(n)} \simeq (1-2m)^{2(n-1)}$.

By comparison, when $s = 0$ the functions $p_1^{(n)} - p_2^{(n)}$ and $P_1^{(n)} - P_2^{(n)}$ show speeds which are approximately 1/2 of the speed of $D_c^{(n)}$, for $m \ll 1/4$. This result is confirmed by numerical tests, too.

In order to quantify the reduction of the gene flow we calculated the number of generations necessary to reach values of $P_1^{(n)} - P_2^{(n)}$ and $D_c^{(n)}$ equal to 1/10 , 1/100 of the initial value and two parameters were introduced:

1) the ratio between the number of generations to reach a given value for $s \neq 0$ and that necessary to reach the same value for $s = 0$ (TR) ;
2) the ratio between the migration rate for $s = 0$ and that for a given $s \neq 0$ necessary to reach in the same number of generations the same value (MRE) .

When $m/s < 1/4$ accurate approximations for TR and MRE can be analytically calculated and a result independent of the particular value of the functions be reached:

$$TR = \log(1-2m)/\log\left[\lambda(m,s,R)\right] \qquad \text{for} \quad P_1^{(n)} - P_2^{(n)}$$

and:

$$TR = 2\log(1-2m)/\log(\lambda(m,s,R)) \quad \text{for} \quad D_c^{(n)} \quad \text{as } Q(R,m) \simeq 1$$

$$MRE = (1-\lambda(m,s,R))/2m \quad \text{for} \quad P_1^{(n)} - P_2^{(n)}$$

and:

$$MRE = (1-\sqrt{\lambda}(m,s,R))/2m \quad \text{for} \quad D_c^{(n)} .$$

Resulting in:

$$\lim_{m\to 0} MRE\,(P_1^{(n)} - P_2^{(n)}) = \lim_{m\to 0} (1-\lambda(m,s,R))/2m = R(1-s)/(s+R(1-s)) = G(s,R)$$

and consequently:

$$\lim_{m\to 0} MRE\,(D_c^{(n)}) = (1/2)G(s,R); \quad \lim_{m\to 0} TR(P_1^{(n)} - P_2^{(n)}) = 1/G(s,R); \quad \lim_{m\to 0} TR(D_c^{(n)}) = 2/G(s,R).$$

The value of $G(s,R)$ for $R=0.5$ becomes $(1-s)/(1+s)$ which is identical to Bengtsson's estimation for the reduction of gene flow (Bengtsson, 1974).

On the basis of both analytical and empirical data we can say that TR is always an increasing function (MRE decreasing) of s.

Concerning the other parameters we note that two patterns are shown for TR :

1) for high migration rate (near panmixy);
2) for low migration rate.

The high migration pattern is analytically described in panmixy $(m=0.5)$ for $D_c^{(n)}$:

$$TR(pan.) = \log(1-R)/(\log(1+(1-s)(1-2R)) - \log 2(1-s/2))$$

Empirical data show that TR values are lowered by increasing m for both $P_1^{(n)}-P_2^{(n)}$ and $D_c^{(n)}$. In addition, the value of TR tends to TR(pan.) for $D_c^{(n)}$, while for $P_1^{(n)} - P_2^{(n)}$ the. TR value tends to 1.

The low migration pattern can be defined for the MRE function too and is analytically described by $G(s,R)$ and related functions.

The TR values for any s and the differences in TR values between any pair of s values are higher in a low than in a high migration pattern and with lower R.

In the transition from the high to the low migration pattern of TR, the latter is approximated first:

1) by $P_1^{(n)} - P_2^{(n)}$ than by $D_c^{(n)}$;
2) for higher s values;
3) for higher R values (only for $D_c^{(n)}$).

Therefore the efficiency of a chromosome mutation with a given s in reducing gene flow increases with decreasing m; this increase is stronger for higher s values and lower R values; thus the interaction between high heterozygote sterility and low recombination frequences can produce a drastic reduction in gene flow.

However, as $D_c^{(n)}$ and its decrease is the most interesting index of the reduction in gene flow and as the low migration pattern and the low s values $(0 < s < 0.3)$ are the conditions which are thought to operate in chromosomal speciation, it turns out that the partial sterility of the heterozygotes for a chromosome mutation has a limited effect in reducing gene flow in natural populations, whereas this sterility can keep apart the two karyomorphs when $m/s < 1/4$ in our model, which does not foresee perturbations. (see Karlin and McGregor, 1972).

The reduction of the gene flow is greater for the lower R, but only in the case of high values of s there is a strong reduction in the gene flow for high R values.

We can conclude that, in this range of values of s and m, the linkage between the chromosome mutation and the neutral locus increases the efficiency in reducing gene flow as a consequence of the partial sterility of the heterozygotes by a chromosome mutation only for a small portion of the genome. Results which were similar qualitatively but not quantitatively were obtained by Barton (1979) using a more general model and being based on a continuous population structure.

So the general conclusion of a limited effect of a single chromosome mutation in reducing gene flow (Spirito et al., to appear) is confirmed.

REFERENCES

1) Barton N.H. (1979) "Gene flow past a cline" Heredity 43, 333-339;
2) Bengtsson B.O. (1974) "Karyotype evolution in vivo and in vitro". Ph. D. Thesis, Chapt. 3, pp. 38-45, Oxford;
3) Bengtsson B.O. (1980) "Rates of karyotype evolution in placental mammals" Hereditas 92, 37-47;
4) Bengtsson B.O. and Bodmer W.F. (1976) "On the increase of chromosome mutation under random mating" Theor. Pop. Biol. 9, 260-281;
5) Hedrick P.V. (1981) "The establishment of chromosomal variants" Evolution 35, 322-332;
6) Karlin S. and McGregor (1972) "Application of method of small parameters to multi--niche population genetic models" Theor. Pop. Biol. 3, 186-209;
7) Lande R. (1979) "Effective deme size during long term evolution estimated from rates of chromosomal rearrangements" Evolution 33, 234-251;
8) Spirito F., Rossi C., Rizzoni M. (1983) "Reduction of gene flow due to the partial sterility of heterozygotes for a chromosome mutation. 1) Studies on a "neutral" gene not linked to the chromosome mutation in a two population model" Evolution 37, 785-797;
9) White M.J.D. (1978) "Modes of speciation" W.H. Freeman & Co. San Francisco.

PREDATOR-MEDIATED COEXISTENCE OF COMPETING SPECIES IN A VOLTERRA MODEL

Yasuhiro Takeuchi
Department of Applied Mathematics, Faculty of Engineering,
Shizuoka University, Hamamatsu 432, JAPAN

ABSTRACT

We examine the possibility of predator-mediated coexistence of all species with model ecosystems of Volterra type. We are concerned with dynamics of a biological community of three competing species to which one predator is added. Stability at equilibria increases when a predator is included. Oscillatory coexistence in limit cycles or in chaotic motions is also possible with a predator. Stable coexistence is enhanced if a predator prefers a dominant competitor. Even when coexistence of species in the subcommunities is impossible, four species can coexist.

1. INTRODUCTION

A lot of experiments have shown that the complexity-stability relation is an important problem in population biology. Paine (1966) discovered that predation pressure on multi-species interacting populations could stabilize a system, although one particular level in isolation was unstable due to competition. Watt (1968) observed that the complexity at one trophic level, which might promote stability of the level, caused instability of the other level or total system. Lubchenco (1978) showed in the herbivores-plants system that there was unimodal relationship between prey density and consumer density, when the consumer preferred a competitively dominant species.

Stability of prey-predator Volterra models has also been discussed intensively (e.g., for two-prey, one-predator system, Parrish and Saila (1970), Cramer and May (1972), Fujii (1977), Vance (1978) and Hsu (1981); for two-prey, two-predator system, May (1971) and Armstrong (1982)). Applying Lyapunov's direct method and Hopf bifurcation theory, Takeuchi and Adachi (1983a) analyzed the stabilizing effect of predation with two-prey, one-predator Volterra model. They showed that predation pressure makes stable persistence possible even when two competitors can not coexist in the subcommunity. Three patterns of coexistence were presented by them mathematically; (1) coexistence at equilibrium state, (2) stable oscillatory coexistence with a small amplitude, (3) spiral chaos of Vance (1978). They also showed that this predator-mediated coexistence can be attained by close relationship between preferences of a predator and competitive ability of prey as suggested by Lubchenco (1978) and further that the patterns of coexistence relate intimately to strength of competitive interactions between prey. Takeuchi and Adachi (1983b) also discussed the case where one more predator is included in two-prey, one-predator system. It is shown that the coexistence of four species is possible at equilibrium states and in limit cycles or in chaotic motions.

For three-prey, one-predator system, Roughgarden and Feldman (1975) proved that predation pressure permits increased niche overlap among prey species. They gave only necessary conditions for species coexistence.

In this paper, we consider the role of predation also with three-prey, one-predator system and give sufficient conditions for species coexistence. It is shown that predation pressure on three competitors enlarges possibility of all species coexisting in the community. We also discuss the relationship between the predator-mediated coexistence and competitive ability of prey or diet preferences of a predator.

2. MODELS AND DEFINITION

The models discussed in this paper are described by the following systems of differential equations :

Three-Prey, One-Predator System

$$\frac{d}{dt}\begin{pmatrix} x_1(t) \\ x_2(t) \\ x_3(t) \\ y(t) \end{pmatrix} = \begin{pmatrix} x_1(t)(\ 1 - x_1(t) - \alpha x_2(t) - \varepsilon_1 y(t)) \\ x_2(t)(\ 1 - \beta x_1(t) - x_2(t) - \beta x_3(t) - \varepsilon_2 y(t)) \\ x_3(t)(\ 1 - \alpha x_2(t) - x_3(t) - \varepsilon_3 y(t)) \\ y(t)(\ -1 + d\varepsilon_1 x_1(t) + d\varepsilon_2 x_2(t) + d\varepsilon_3 x_3(t)) \end{pmatrix}, \qquad (1)$$

Three-Species-Competing System

$$\frac{d}{dt}\begin{pmatrix} x_1(t) \\ x_2(t) \\ x_3(t) \end{pmatrix} = \begin{pmatrix} x_1(t)(\ 1 - x_1(t) - \alpha x_2(t)) \\ x_2(t)(\ 1 - \beta x_1(t) - x_2(t) - \beta x_3(t)) \\ x_3(t)(\ 1 - \alpha x_2(t) - x_3(t)) \end{pmatrix}, \qquad (2)$$

Two-Prey, One-Predator System

$$\frac{d}{dt}\begin{pmatrix} x_1(t) \\ x_2(t) \\ y(t) \end{pmatrix} = \begin{pmatrix} x_1(t)(\ 1 - x_1(t) - \alpha x_2(t) - \varepsilon_1 y(t)) \\ x_2(t)(\ 1 - \beta x_1(t) - x_2(t) - \varepsilon_2 y(t)) \\ y(t)(\ -1 + d\varepsilon_1 x_1(t) + d\varepsilon_2 x_2(t)) \end{pmatrix}. \qquad (3)$$

Here $x_i(t)$ $(i = 1, 2, 3)$ or $y(t)$ are population sizes of prey or predator respectively, positive parameters α and β represent competitive effects between prey, positive parameters ε_i $(i = 1, 2, 3)$ are coefficients of decreases of prey due to predation and $d>0$ is a transformation rate of a predator. System (1) contains a particular subsystem (2), two species (x_1 and x_3) having no interspecific relationship and the other one competing identically (we will use β) with them. This structure of the subsystem is obviously an oversimplified one, but it seems to be difficult mathematically to handle a more general system.

Let the possible equilibria of (1) be denoted by (E_{++++}), (E_{+++0}) etc. Here (E_{+++0}), for example, denotes an equilibrium where three prey remain positive and a predator is extinct. For systems (2) and (3), notations are defined similarly.

DEFINITION A nonnegative equilibrium x^* of the systems is called globally stable if

(i) x^* is locally stable, that is, if for any $\varepsilon > 0$, there exists a $\delta(\varepsilon)$ such that $|x^0 - x^*| < \delta(\varepsilon)$ and $x(t) \in R^n_I$, then $|x(t) - x^*| < \varepsilon$ for $t \geq 0$, and

(ii) every solution converges to x^* as $t \to +\infty$, if $x^0 \in R^n_I$.

Here x^0 is an initial state, $x(t)$ is the solution of the system such that $x(0) = x^0$ and R^n_I is the set such as $\{x | x_i \geq 0$ for $i \in I$ and $x_j > 0$ for $j \in J\}$, where $x^*_i = 0$ for $i \in I$, $x^*_j > 0$ for $j \in J$ and n is the dimension of the system.

3. THE MAIN RESULTS

For system (2) of three-competing species, the following theorem is obtained.

THEOREM 1 *Let us consider system* (2) *satisfying* $\alpha \neq 1$ *or* $\beta \neq 1/2$.

(i) (E_{+++}) *is globally stable if and only if* $\alpha < 1$ *and* $\beta < 1/2$.

(ii) (E_{+0+}) (*or* (E_{0+0})) *is globally stable if and only if* $\alpha \leq 1$ *and* $\beta \geq 1/2$ (*or* $\alpha \geq 1$ *and* $\beta \leq 1/2$).

(iii) *When* $\alpha > 1$ *and* $\beta > 1/2$, (E_{+0+}) *and* (E_{0+0}) *are locally stable.*

For more general three-species competing system, Strobeck (1973) proved that condition of Theorem 1 (i) are necessary and sufficient for (E_{+++}) to be *locally* stable. Theorem 1 implies that the condition ensures the *global* stability for (E_{+++}). Theorem 1 shows that no periodic motion exists, and that three-species coexistence is possible only when interspecific competition is rather weak.

THEOREM 2 A. *Let us consider system* (1) *satisfying* $\varepsilon_1 = \varepsilon_3 \neq \varepsilon_2$.

(i) *Suppose that* $\alpha + \beta < 2^{1/2}$. *If* $(E_{++++}) = (x^*_1, x^*_2, x^*_3, y^*)$ *is nonnegative, then* (E_{++++}) *is globally stable. If* (E_{++++}) *is not nonnegative, then one of the other equilibria is globally stable.*

(ii) *Suppose that* $\alpha\beta < 1/2$ *but* $\alpha + \beta \geq 2^{1/2}$. *Then* (E_{++++}) *can be positive and locally stable for some* ε_i $(i = 1, 2, 3)$.

(iii) *Suppose that* $\alpha\beta \geq 1/2$. *In the parametric regions* H_1 (*or* H_2) *of Figure* 1, *there exists at least one Hopf bifurcation value* ε^*_1 (*or* ε^*_2) *for any* ε_2 (*or* ε_1) *fixed.*

B. *Let us consider system* (1) *satisfying* $\varepsilon_1 = \varepsilon_2 = \varepsilon_3$. *Then* (E_{++++}) *can be globally or locally stable if and only if* $\alpha < 1$ *and* $\beta < 1/2$. *There exists no Hopf bifurcation for any* α, β *and* ε_i $(i = 1, 2, 3)$.

Theorems 1 and 2B show that the possibility of species coexistence is not enlarged by the inclusion of a predator with identical diet preference in prey. Theorems 1 and 2A imply that predation pressure enlarges the possibility of species coexistence at equilibrium or in persistent oscillatory motions (Figure 2). Further, Figure 1 (b), (e) and (g) (or (c), (f) and (h)) show that the predator-mediated coexistence needs relationship $\varepsilon_1 > \varepsilon_2$ (or $\varepsilon_1 < \varepsilon_2$). That is, the predator always prefers a competitively dominant prey when stable coexistence is realized.

Next let us consider the case where prey x_3 is added to two-prey, one-predator system (3). The parametric regions hatched by vertical lines in Fig. 1(b), (c), (e)

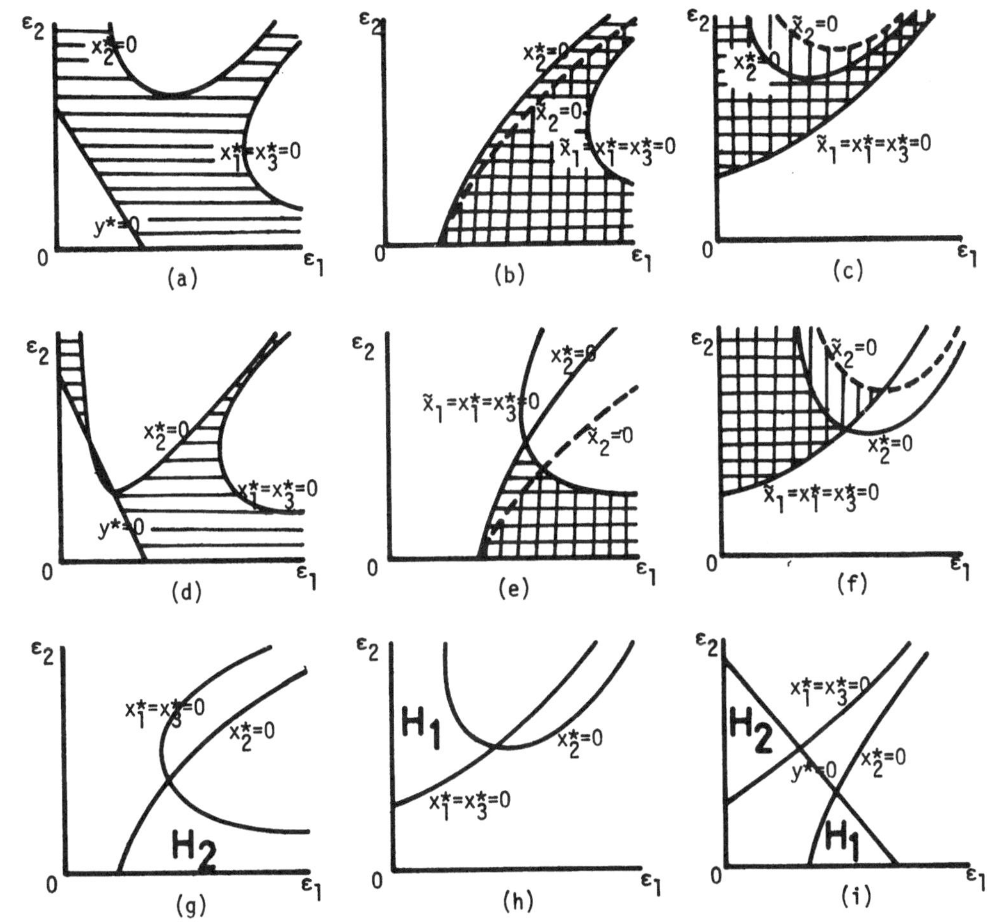

Figure 1 These figures show parametric regions ensuring the existence of globally or locally stable equilibrium and Hopf bifurcation for systems (1) and (3) satisfying $\varepsilon_1 = \varepsilon_3 \neq \varepsilon_2$. Here $(E_{++++}) = (x_1^*, x_2^*, x_3^*, y^*)$ for (1) and $(E_{+++}) = (\tilde{x}_1, \tilde{x}_2, y)$ for (3). $\alpha + \beta < \sqrt{2}$ for (a), (b) and (c). $\alpha + \beta \geqq \sqrt{2}$ and $\alpha\beta < 1/2$ for (d), (e) and (f). $\alpha\beta \geqq 1/2$ for (g), (h) and (i). $\alpha < 1$ and $\beta < 1/2$ for (a) and (d). $\alpha \leqq 1$ and $\beta \geqq 1/2$ for (b), (e) and (g). $\alpha \geqq 1$ and $\beta \leqq 1/2$ for (c), (f) and (h). $\alpha > 1$ and $\beta > 1/2$ for (i). (E_{++++}) is positive and globally (or locally) stable in the region hatched by horizontal lines in each figure (a), (b) and (c) (or (d), (e) and (f)). In the regions hatched by vertical lines of (b), (c), (e) and (f), (E_{+++}) is positive and stable. In the regions H_1 (or H_2) of (g), (h) and (i), there exists at least one Hopf bifurcation parameter ε_1^* (or ε_2^*) for any ε_2 (or ε_1) fixed. See Theorem 2. Examples of oscillatory orbits are given in Figure 2.

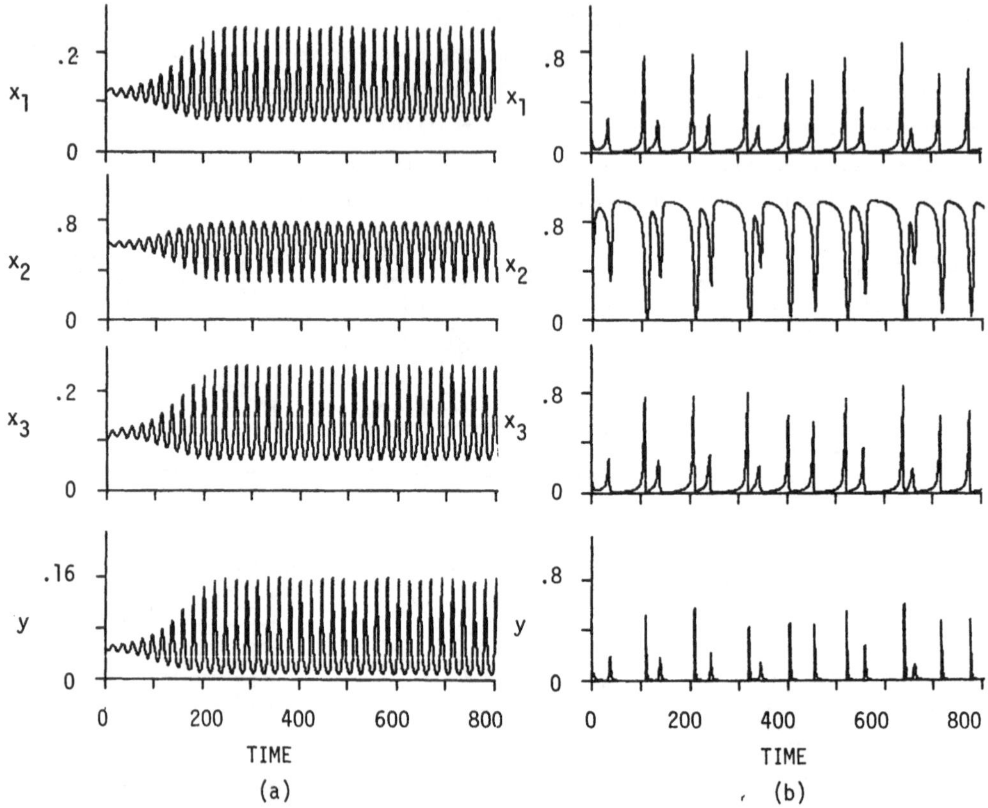

Figure 2 Each figure depicts one-dimensional x_1, x_2, x_3 and y population changes with respect to time of system (1). Each trajectory starts from near (E_{++++}) for $\varepsilon_2 = 1$, $\alpha = 1$, $\beta = 1.5$ and $d = 0.5$. Figure 1(g) corresponds to this case. Hopf bifurcation parameter ε_1^* is about 5.5.
(a); a limit cycle of Hopf type at $\varepsilon_1 = 6$.
(b); a chaotic motion at $\varepsilon_1 = 9$ with a larger amplitude.

and (f) ensure the stability of positive three-species equilibrium of (3). For the cases of Fig. 1(b) and (e), an additional prey is superior to prey x_2 and the predator prefers the additional one. The areas ensuring the coexistence of four species (i.e., areas hatched by horizontal lines) are enlarged by the inclusion of such a dominant prey. On the other hand, for the cases of Fig. 1(c) and (f), the predator prefers resident dominant competitor x_2 to an additional inferior x_3. The possibility of coexistence decreases in these cases by the inclusion of prey.

4. CONCLUSION

Possibility of predator-mediated coexistence in three-competitor Volterra model is discussed. Global and local stable coexistence at equilibrium increases when a predator is included. Oscillatory coexistence becomes possible with the predator. Predator-mediated coexistence is enhanced if the predator prefers a dominant competitor.

The addition of a prey to two-prey, one-predator system increases possibility of stable coexistence if the additional prey is a dominant competitor and the predator prefers the prey.

APPENDIX *OUTLINE OF PROOFS OF THEOREMS*

Theorem 1. For 3 × 3 system matrix C satisfying $\alpha\beta < 1/2$ of (2), there exists a positive definite diagonal matrix W such that $WC + C^tW$ is positive definite. (E_{+++}) is positive if $\alpha < 1$ and $\beta < 1/2$. Applying the method given by Takeuchi and Adachi (1983a), Theorem 1 can be proved similarly to one of two-prey system.
Theorem 2. For 4 × 4 system matrix A of (1), $WA + A^tW$ is nonnegative definite if we choose $W = \mathrm{diag}(1, 1, 1, 1/d)$ and $\alpha + \beta < \sqrt{2}$. Applying Goh's theorem (1978), (i) can be proved. Fig. 1(d), (e) and (f) show that there exist parametric regions ensuring (E_{++++}) positive if $\alpha\beta < 1/2$. The characteristic polynomial of $-E_4A$ is

$$P(\lambda)=(\lambda+x_1^*)[\lambda^3+(x_1^*+x_2^*)\lambda^2+\{(1-2\alpha\beta)x_1^*x_2^*+d(2\varepsilon_1^2x_1^*+\varepsilon_2^2x_2^*)y^*\}\lambda+d(2\varepsilon_1^2+\varepsilon_2^2-2(\alpha+\beta)\varepsilon_1\varepsilon_2)x_1^*x_2^*y^*]$$

where $E_4 = \mathrm{diag}(x_1^*, x_2^*, x_3^*, y^*)$. All the real parts of the roots of $P(\lambda)$ are negative if $\alpha\beta < 1/2$ and (E_{++++}) is positive. This proves (ii). Applying Hopf bifurcation theory (Marsden and McCraken (1976)) and the method used for two-prey, one-predator system (Takeuchi and Adachi (1983a)), (iii) can be proved. For $\varepsilon_1 = \varepsilon_2 = \varepsilon_3$, (E_{++++}) is positive and $\det(-E_4A) > 0$ if and only if $\alpha < 1$ and $\beta < 1/2$. $\det(-E_4A) > 0$ is necessary for Hopf bifurcation. Hence, no Hopf bifurcation occurs.

REFERENCES

Armstrong, R.A.: The effects of connectivity of community stability. Amer. Nat. 120, 391-402 (1982)

Cramer, N.F., May, R.M.: Interspecific competition, predation and species diversity : A comment. J. Theor. Biology 34, 289-293 (1972)

Fujii, K.: Complexity-stability relationship of two-prey-one-predatór species system model : Local and global stability. J. Theor, Biology 69, 613-623 (1977)

Goh, B.S.: Sector stability of a complex ecosystem model. Math. Biosci. 40, 157-166 (1978)

Hsu, S.B.: Predator-mediated coexistence and extinction. Math. Biosci. 54, 231-248 (1981)

Lubchenco, J.: Plant species diversity in a marine intertidal community : Importance of herbivore food preference and algal competitive abilities. Amer. Nat. 112, 23-39 (1978)

Marsden, J.E., McCraken, M.: The Hopf bifurcation and its application. 131-135, Springer-Verlag, New York 1976

May, R.M.: Stability in multispecies community models. Math. Biosci. 12, 59-79 (1971)

Paine, R.T.: Food web complexity and species diversity. Amer. Nat. 100, 65-75 (1966)

Parrish, J.D., Saila, S.B.: Interspecific competition, predation and species diversity. J. Theor. Biology 27, 207-220 (1970)

Roughgarden, J., Feldman, M.: Species packing and predation pressure. Ecology 56, 489-492 (1975)

Strobeck, C.: N species competition. Ecology 54, 650-654 (1973)

Takeuchi, Y., Adachi, N.: Existence and bifurcation of stable equilibrium in two-prey, one-predator communities. Bull. Math. Biology (In Press) (1983a)

Takeuchi, Y., Adachi, N.: Oscillations in prey-predator Volterra models : in H.I. Freedman and C. Strobeck (editor), Population Biology, Lect. Notes in Biomath. 52, 320-326 (1983b)

Vance, R.R.: Predation and resource partitioning in one predator-two prey model communities. Amer. Nat. 112, 797-813 (1978)

Watt, K.E.F.: Ecology and resource management. McGraw-Hill, New York 1968.

ON A NONLINEAR PROBLEM ARISING FROM INTERACTION OF ALGAE WITH LIGHT

SILVIA TOTARO
Istituto di Matematica Applicata
"G. Sansone "

Via S. Marta, 3 - 50139 Firenze (Italy)

1. Introduction

We examine a mathematical model arising from the study of the interaction of algae with light. In fact, let us consider a population of algae suspended in water in sea or in artificial tanks. If $n=n(z;t)$ measures the suspected concentration of the algae that, at time t are suspended in water at depth z ($-\infty<-b\leq z\leq 0$, b is the maximum depth of water), we conjecture the following evolution for n

$$\frac{\partial n(z;t)}{\partial t} = -\frac{1}{\tau} n(z;t)+n(z,t)\alpha\left(\int_{\nu_1}^{\nu_2} d\nu' \int_{-1}^{+1} d\mu' I(z,\nu',\mu';t)-\phi(n(z;t))\right) \qquad (1a)$$

In the first term of right hand side of (1a), $\frac{1}{\tau}$ represents the death rate of algae, whereas the second term of (1a) represents the growth of the algae concentration due to the absorption of light. This growth is thought to be proportional to the product of algae-concentration with a known function of total radiant energy flux (see equation (1b)). The third term in (1a) is a removal term. The radiant energy satisfies the Boltzmann equation [1].

$$\frac{\partial I(z,\nu,\mu;t)}{\partial t} = -c\mu \frac{\partial I}{\partial z}(z,\nu,\mu;t)-c\Sigma(\nu)I(z,\nu,\mu;t)-cn(z;t)\sigma(\nu)I(z,\nu,\mu,t)+$$

$$+ c\int_{\nu_1}^{\nu_2} d\nu' \int_{1}^{1} d\mu' \frac{\nu}{\nu}\Sigma_{sc}(\nu',\nu,\mu',\mu)I(z,\nu',\mu';t) +$$

$$+ cn(z;t)\int_{\nu_1}^{\nu_2} d\nu' \int_{1}^{1} d\mu' \frac{\nu}{\nu}\sigma_{sc}(\nu',\nu,\mu',\mu)I(z,\nu',\mu';t) \qquad (1b)$$

Equation (1b) is a balance equation for $I=I(z,\nu,\mu;t)$ which measures the specific intensity of radiation (here $\nu\epsilon(\nu_1,\nu_2)$ is the photon frequency $\mu\epsilon(-1,+1)$ the angle variable).
The first term in the right of (1b) is the streaming term. The other terms show that light can be scattered and absorbed by algae and water.
The initial and boundary conditions of (1a) and (1b) are the following

$$n(z;0) = n_o(z) \qquad z\epsilon[-b,0] \qquad (1c)$$

$$I(z,\nu,\mu,0) = I_o(z,\nu,\mu) \qquad z\epsilon[-b,0], \quad \nu\epsilon[\nu_1,\nu_2],\mu\epsilon[-1+1] \qquad (1d)$$

$$I(0,\nu,\mu,t) = \Phi(\nu,\mu,t) \qquad \nu\epsilon[\nu_1,\nu_2],\mu\epsilon[-1,0), \quad t\geq 0 \qquad (1e)$$

$$I(-b,\nu,\mu;t)\equiv 0 \qquad \nu\epsilon[\nu_1,\nu_2],\mu\epsilon(0,+1] \quad t\geq 0 \qquad (1f)$$

In (1e) $\Phi(\nu,\mu;t)$ represents the specific intensity of radiation at water level.
We can rewrite (1a) to (1f) as a Cauchy problem with homogeneous boundary conditions by putting

$$\tilde{I}(z,\nu,\mu;t) = I(z,\nu,\mu;t) - \eta(z,\nu,\mu;t) \quad z\epsilon[-b,0],\ \nu\epsilon[\nu_1,\nu_2],\ \mu\epsilon[-1,+1] \quad t\geq 0 \qquad (2)$$

where $\eta=\eta(z,\nu,\mu;t)$ is a quite regular function (see (A5), (A6), (A7), in the following) such that

$$\eta(0,\nu,\mu;t) = \Phi(\nu,\mu;t) \qquad \nu\epsilon[\nu_1,\nu_2] \qquad \mu\epsilon[-1,0) \qquad t\geq 0 \tag{3a}$$

$$\eta(-b,\nu,\mu;t) = 0 \qquad \nu\epsilon[\nu_1,\nu_2] \qquad \mu\epsilon(0,+1] \qquad t\geq 0 \tag{3b}$$

2. ABSTRACT FORMULATION OF THE PROBLEM

This section contains the list of the spaces, the operators and the assumptions we shall use in the abstract formulation of the problem.
We introduce the real Banach space

$$X = X_1 \times X_2 \quad , \|f\| = \|f_1\|_1 + \delta\|f_2\|_2 \qquad \delta>0 \text{ dimensional constant} \tag{4}$$

where

$$X_1 = L^1[-b,0] \, , \qquad \|f_1\|_1 = \int_{-b}^{0} dz|f_1(z)|$$

$$X_2 = L^1[-b,0]\times[\nu_1,\nu_2]\times[-1,+1], \quad \|f_2\|_2 = \int_{-b}^{0} dz \int_{\nu_1}^{\nu_2} d\nu \int_{-1}^{+1} d\mu|f_2(z,\nu,\mu)|$$

and

$$X_\infty = X_{1\infty}\times X_{2\infty} \, , \ \|f\|_\infty = \|f_1\|_{2\infty} + \delta_\infty\|f_2\|_{2\infty} \, , \ \delta_\infty>0 \text{ dimensional constant} \tag{5}$$

where

$$X_{1\infty} = L^\infty[-b,0]. \qquad \|f_1\|_{1\infty} = \text{ess sup } \{|f_1(z)| \ ; \ z\epsilon[-b,0]\}$$

$$X_{2\infty} = L^\infty[-b,0]\times[\nu_1,\nu_2]\times[-1,+1], \|f_2\|_{2\infty} = \text{ess sup}\{|f_2(z,\nu,\mu)|;z\epsilon[-b,0],\nu\epsilon[\nu_1,\nu_2],\mu\epsilon[-1,+1]\}$$

Obviously, if $f_1\epsilon X_1$, then $f_1\epsilon X_2$

$$\|f_1\|_2 \leq 2\bar{\nu}\|f_1\|_1 \quad , \quad \bar{\nu} = \nu_2-\nu_1$$

We use X^+, X_1^+, X_2^+ to denote the positive cones in X, X_1, X_2 respectively.
We also use the following closed subset of X and X^+:

$$S(M_1,M_2) = S_1(M_1)\times S_2(M_2) \tag{6}$$

where M , M are given positive constants and

$$S_i(M_i) = \{f_i\epsilon X_i : f_i\epsilon X_{1\infty} \ ; \ \|f_i\|_{1\infty} \leq M_i\} \qquad i=1,2$$

$$S^+(M_1,M_2) = S_1^+(M_1)\times S_2^+(M_2) = \{S_1(M_1)\cap X_1^+\} \times \{S_2(M_2)\cap X_2^+\} \tag{7}$$

We define the following operators

$$J_1 f_2 = \int_{\nu_1}^{\nu_2} d\nu' \int_{-1}^{+1} d\mu' \, f_2(z,\nu',\mu') \tag{8}$$

$$D(J_1) = X_2, \ R(J_1)\subset X_1 \quad ;$$

$$J_2 f_2 = \int_{\nu_1}^{\nu_2} d\nu' \int_{-1}^{+1} d\mu' \, \frac{\nu}{\nu'} \Sigma_{sc}(\nu',\nu,\mu',\mu) f_2(z,\nu',\mu') \tag{9}$$

$D(J_2) = X_2$ (see A4) , $R(J_3)\subset X_2$;

$$J_3 f_2 = \int_{\nu_1}^{\nu_2} d\nu' \int_{1}^{1} d\mu' \frac{\nu}{\nu'} \sigma_{sc}(\nu',\nu,\mu'\mu) \quad f_2(z,\nu',\mu') \tag{10}$$

$D(J_3) = X_2$ (see A4) , $R(J_3) \subset X_2$;

$$\alpha(J_1 f_2) = \alpha\left(\int_{\nu_1}^{\nu_2} d\nu' \int_{1}^{1} d\mu' f_2(z,\nu',\mu')\right) \quad \text{(see (A9), (A10)} \tag{11}$$

$D(\alpha) = R(J_1)$, $R(\alpha) \subset X_1$;

$$A_1 f_1 = -\frac{1}{\tau} f_1 \tag{12}$$

$D(A_1) = X_1$, $R(A_1) = X_1$;

$$A_2 f_2 = -c\mu \frac{\partial f_2}{\partial z} \tag{13}$$

$D(A_2) = \{f_2\ X_2 \colon c\mu \frac{\partial f_2}{\partial z} \in X_2, f_2(0,\nu,\mu) = 0 \quad \nu\in[\nu_1,\nu_2] \quad \mu\in[-1,0),$

$f_2(-b,\nu,\mu) = 0 \quad \nu\in[\nu_1,\nu_2], \mu\in(0,+1]\}$, $R(A_2) \subset X_2$

We list the assumptions that we shall use in the following

(A1) $\Sigma(\nu) \geq 0$ almost everywhere , $\Sigma(\nu)\ L^{\infty}[\nu_1,\nu_2]$, ess sup $\Sigma(\nu) = \bar{\Sigma}$

(A2) $\sigma(\nu) \geq 0$ almost everywhere , $\sigma(\nu)\ L^{\infty}[\nu_1,\nu_2]$, ess sup $\sigma(\nu) = \bar{\sigma}$

(A3) $\Sigma_{sc}(\nu',\nu,\mu',\mu) \geq 0$ almost everywhere, $\frac{\nu}{\nu}\Sigma_{sc}(\nu',\nu,\mu',\mu)\ L^{\infty}[\nu_1,\nu_2]x[\nu_1,\nu_2]x$ $x[-1,+1]x[-1,+1]$, ess sup $\frac{\nu}{\nu'} \sigma_{sc}(\nu',\nu,\mu';\mu) = \bar{\Sigma}_{sc}$

(A4) $\sigma_{sc}(\nu',\nu,\mu',\mu) \geq 0$ almost everywhere, $\frac{\nu}{\nu'} \sigma_{sc}(\nu',\nu,\mu',\mu) \in L^{\infty}[\nu_1,\nu_2]x[\nu_1,\nu_2]\ x$ $x[-1,+1]x[-1,+1]$, ess sup $\frac{\nu}{\nu'} \sigma_{sc}(\nu',\nu,\mu',\mu') = \bar{\sigma}_{sc}$

(A5) $\eta(t) = \eta(\cdot,\cdot,\cdot;t) \in X_2$ for each $t \geq 0$ and the strong derivative $\frac{d\eta}{dt} \in X_2$ exists for each $t \geq 0$.

(A6) $\eta(t)$ has a generalized derivative $\frac{\partial\eta}{\partial z} \in X \quad \forall t \geq 0$

(A7) $\eta(t) \in X_{2\infty}$ $\forall t \geq 0$ and there exists $k > 0$ such that $\|\eta(t)\|_{2\infty} \leq k \quad \forall t \geq 0$

(A8) the removal term $\phi = \phi(n)$ is a function such that $\phi \colon X_1 \to X_1$
$\phi(f_1) = \bar{\phi} f_1 \quad f_1 \in X_1 \quad \bar{\phi} \in X_1^+ \cap X_{1\infty}$

(A9) $\alpha \colon R(J_1) \to X_1$ is such that for every $f_1, g_1 \in R(J_1)$
$\|\alpha(f_1) - \alpha(g_1)\|_1 \leq \tilde{\alpha} \| f_1 - g_1 \|_1$

(A10) there exists an $\bar{\alpha} > 0$ such that for every $f_1 \in R(J_1)$
$\|\alpha(f_1)\|_{1\infty} \leq \bar{\alpha}$

Hence if we let

$$u(t) = \begin{pmatrix} u_1(t) \\ u_2(t) \end{pmatrix} = \begin{pmatrix} n(\cdot;t) \\ \tilde{I}(\cdot,\cdot,\cdot;t) \end{pmatrix} \tag{14}$$

$$u_o = \begin{pmatrix} n_o \\ I_o - \eta(\cdot,\cdot,\cdot;0) \end{pmatrix} \tag{15}$$

$$\psi(t) = \begin{pmatrix} 0 \\ \psi_2(t) \end{pmatrix} \quad , \ \psi_2(t) = -c\mu\frac{\partial\eta}{\partial z} - c\Sigma\eta(t) - \frac{d\eta}{dt} + cJ_2\eta \tag{16}$$

$$B = \begin{pmatrix} A & 0 \\ 0 & A_2 - c\bar{\Sigma} \end{pmatrix} \quad \begin{matrix} D(B) = X_1 x D(A_2) \subset X \\ R(B) = X \end{matrix} \tag{17}$$

$$F(f) = \begin{pmatrix} g_1 \\ g_2 \end{pmatrix} \quad \begin{matrix} D(F) = \{f\epsilon X \ , \ F(f)\epsilon X\} \\ R(F) \subset X \end{matrix}$$

$$g_1 = \alpha(J_1(f_2+\eta)f_1 - \phi(f_1) \tag{18a}$$

$$g_2 = c(\bar{\Sigma}-\Sigma)f_2 - cf_1\sigma f_2 - cf_1\sigma\eta + cJ_2f_2 + cf_1J_3f_2 + cf_1J_3\eta \ . \tag{18b}$$

System (1a) to (1f) becomes:

$$\frac{du}{dt} = Bu(t) + F(u(t) + \psi(t) \quad t>0$$

$$u(0) = u_o$$

3. PRELIMINARY RESULTS

By using definitions (8) (9) (10) (12) (13) ,assumptions (A1) to (A10) and results of semigroups theory [5],[6] we are able to prove the following lemmas.
Notations are those of [5]:

Lemma 1: "If (A3), (A4) hold:

(i) $J_1 \epsilon \mathcal{B}(X_2,X_1)$, $||J_1|| \leq 1$

(ii) $J_2 \epsilon \mathcal{B}(X_2)$, $||J_2|| \leq 2\bar{\nu}\ \bar{\Sigma}_{sc}$

(iii) $J_3 \epsilon \mathcal{B}(X_2)$, $||J_3|| \leq 2\bar{\nu}\ \bar{\sigma}_{sc}$

(iv) if $f_2 \epsilon X_2^+$ then $J_1 f_2 \epsilon X_1^+$ and $J_i f_2 \epsilon X_2^+$ i=2,3

(v) if $f_2 \epsilon S_2(M_2)$, then $J_1 f_2 \epsilon S_1(2\bar{\nu}M_2)$, $J_2 f_2 \epsilon S_2(2\bar{\nu}\ \bar{\Sigma}_{sc}M_2)$ and $J_3 f_2 \epsilon S_2(2\bar{\nu}\ \bar{\sigma}_{sc}M_2)$"

Lemma 2:

"(i) $B \epsilon G(1,-h;X)$ where $h = \min\ (\frac{1}{\tau}, c\bar{\Sigma}) > 0$

(ii) if $f \epsilon S(M_1,M_2)$, then $Z(t)f = \exp(tB)f \epsilon S(M_1,M_2)$ $\forall t \geq 0$

(iii) if $f \epsilon X^+$, then $Z(t)f \epsilon X^+$ $\forall t \geq 0$ "

Lemma 3: "If (A1) to (A10) hold, then:

(i) $D(F)\supset S(M_1,M_2)$

(ii) $F(f)\in S(h_1,h_2)$ for every $f\in S(M_1,M_2)$

where

$$h_1=(\bar{\alpha}+\|\Phi\|_{1\infty})M_1$$

$$h_2=c\{\bar{\Sigma}\,M_2+M_1\,\bar{\sigma}M_2+M_1\bar{\sigma}k+2\bar{\nu}\,\bar{\Sigma}_{sc}\,M_2+2\bar{\nu}M_1M_2\bar{\sigma}_{sc}+2\bar{\nu}\,\bar{\sigma}_{sc}M_1k\}$$

(iii) $\|F(f)-F(g)\|\le\gamma\|f-g\|\qquad\forall f,g\in S(M_1,M_2)$

where

$$\gamma=\bar{\alpha}+\frac{\tilde{\alpha}M_1}{\delta}+c\bar{\Sigma}-cM_1\bar{\sigma}+2\bar{\nu}c\bar{\Sigma}_{sc}+cM_1 2\bar{\nu}\,\bar{\sigma}_{sc}+\|\bar{\phi}\|_{1\infty}+2\bar{\nu}\delta\bar{\sigma}M_2c+2\bar{\nu}\delta\bar{\sigma}kc+4\bar{\nu}^2\delta M_2\bar{\sigma}_{sc}c+4\bar{\nu}^2\delta\bar{\sigma}_{sc}kc\text{"}$$

To prove the existence of a solution of (19) with physical meaning we choose the auxiliary function $\eta(t)$ in the following way:

$$\eta(z,\nu,\mu;t)=\begin{cases}\Phi(\nu,\mu;t-\frac{z}{c\mu})\exp[-(\Sigma(\nu)+M_1\bar{\sigma})\frac{z}{\mu}], & z\in[-b,0],\nu\in[\nu_1,\nu_2],\mu\in[-1,0),t\ge 0\\ 0 & z\in[-b,0],\nu\in[\nu_1,\nu_2],\mu\in(0,1)\quad t\ge 0\end{cases}\qquad(20)$$

By using definition (20), we can prove the following lemma

Lemma 4: "If $\eta(t)$ is chosen as in (20), $\Phi(t)=\Phi(\cdot,\cdot;t)\in X_{2\infty}$ $\forall t\ge 0$, $\frac{\partial\phi}{\partial t}$ exists $\forall t\ge 0$ and $H_2=c(2\bar{\nu}\,\bar{\Sigma}_{sc}+M_1\bar{\sigma})\|\Phi\|_{2\infty}$, then

(i) (A5), (A6) hold

(ii) $\psi_2(t)=cJ_2\eta(t)+cM_1\bar{\sigma}\eta(t)$, $\|\psi(t)\|_\infty\le\delta_\infty H_2$ $\forall t\ge 0$"

4. LOCAL MILD POSITIVE SOLUTION

Since $u(t)$ represents a couple of physical quantity, we are interested in finding a solution of (19) which belongs to X^+ for every $t\ge 0$.
So, first we prove that the mild solution $u(t)$ of (19) exists for every $t\in[0,\bar{t}]$ and then we show that $u(t)\in X^+$, $t\in[0,\bar{t}]$.
Let $Y=C([0,\bar{t}];X)$, where $\bar{t}$ will be chosen later on, $\|w;Y\|=\max\{\|w(t)\|,\ t\in[0,\bar{t}]\}$ and consider the following closed subset of Y.

$$\Delta(M_1,M_2)=\{w\in Y\ ;\ w(t)\in S(M_1,M_2)\quad\forall t\in[0,\bar{t}]\}\qquad(21)$$

If we put

$$Q(u)(t)=Z(t)u_o+\int_0^t Z(t-s)[F(u(s))+\psi(s)]ds,\qquad D(Q)\subset T\quad,\quad R(Q)\subset Y\qquad(22)$$

the nonlinear Volterra equation corresponding to (19) becomes

$$u=Qu\qquad(23)$$

By using the results of lemmas 2,3,4 the following lemma can be proved

Lemma 5: "If u_o $S(m_1,m_2)$ where m_1,m_2 are given positive constants, (A1) to (A4), (A8) (A9), (A10) hold and the auxiliary function $\eta(t)$ is chosen as in (20), then

(i) $Q(w)\in\Delta(M_1c(\bar{t}),M_2c(\bar{t}))\quad\forall w\in\Delta(M_1,M_2)$

(ii) $\|Q(w)-Q(w');Y\|\le c(\bar{t})\|w-w';Y\|\quad\forall w,w'\in\Delta(M_1,M_2)$

where

$$c(\bar{t})=r+(\bar{h}+\frac{H_2}{M_2}+\gamma)\bar{t}\quad,\quad r=\mathrm{Max}\left\{\frac{m_1}{M_1},\frac{m_2}{M_2}\right\},\ \bar{h}=\max\left\{\frac{h_1}{M_1},\frac{h_2}{M_2}\right\}\text{"}\qquad(24)$$

So, by using contraction mapping theorem, the following theorem holds:

Theorem 1: "Let the assumptions of lemma 5 hold, let $\bar{t}>0$ be chosen suitably small and let $M_i>m_i$, $i=1,2$. Then equation (23) has a unique solution $u\in\Delta(M_1,M_2)$ $Y=C([0,\bar{t}], X)$".

Note that theorem 1 still holds if we choose instead of $\eta(t)$ defined by (20) another $\bar{\eta}(t)\neq\eta(t)$ that satisfies (3a),(3b),(A5),(A6),(A7) and such that an estimate like (21) holds.

Theorem 2: "Let $\eta(t)$ be defined by (20), $\Phi(t)\in X_2^+\cap X_{2\infty}$, (A1) to (A10) hold, ---- $\alpha:R(J_1)\to X_1$ maps X^+ into itself and $u_o\in S^+(m_1,m_2)$. Then if $M_i>m_i$ $i=1,2.$, system (19) has a unique mild solution $u=u(t)$ such that $u(t)\in S^+(M_1,M_2)$ $\forall t\in[0,t]$".

Proof: It is enough to rewrite (19) so that (23) becomes $u=Q^+u$ where Q^+ acts as Q but $D(Q^+)\subset Y^+=C([0,t],X^+)$, $R(Q^+)\subset Y^+$.

5. GLOBAL SOLUTION

Let $u_o\in S^+(m_1,m_2)$ and (A8) (A10) hold. Fix $T>0$ $(T>\bar{t})$. Then

$$\|u_1(t)\|_{1\infty}\leq dm_1 \qquad \forall t\in[0,\bar{t}] \tag{25}$$

where

$$d=\begin{cases} 1 & \text{if } \frac{1}{\tau}\geq\bar{\alpha} \\ \exp\left(-\frac{1}{\tau}+\bar{\alpha}\right)T & \text{if } \frac{1}{\tau}<\bar{\alpha}\end{cases} \tag{26}$$

Now by using (26), we can prove that

$$\|u_2(t)\|_{2\infty}\leq\rho_2 \qquad \forall t\in[0,t] \tag{27}$$

where

$$\rho_2=\left(m_2+\frac{HK}{c\bar{\Sigma}(H-c\bar{\Sigma})}\right)\exp[(H-c\bar{\Sigma})T] \tag{28}$$

with

$$H=c\bar{\Sigma}+cdm_1 2\bar{\nu}\,\bar{\sigma}_{sc}+2\bar{\nu}c\bar{\Sigma}_{sc}\ ;\ K=c(2\bar{\nu}dm_1\bar{\sigma}_{sc}+2\bar{\nu}\,\bar{\Sigma}_{sc}+M_1\bar{\sigma})\,\|\Phi\|_{2\infty}$$

Theorem 3: "Let the assumptions of theorem 2 hold and let $T>0$ be fixed. Then if $M_1>dm_1$ and $M_2>\rho_2$ (23) has a unique solution $u=u(t)$, $t\in[0,T]$. Moreover, $u(t)\in S^+(M_1,M_2)$, $t\in[0,T]$".

REFERENCES

1. POMRANING G.C., The Equation of Radiation Hydrodynamics, Pergamon Press Oxford (1973)
2. DAVISON B., Neutron Transport Theory, Oxford Clarendon Press (1958)
3. MARTIN Jr., R.M., Nonlinear Operators and Differential Equations in Banach Spaces, J.Wiley, New York, (1976)
4. MORTON R., A Model for Light-Scattering by Algae in Water, Math. Biosc. 40, 195-204 (1978)
5. KATO T., Perturbation Theory for Linear Operators, Springer-Verlag, New York (1966)
6. BELLENI-MORANTE A., Applied Semigroups and Evolution Equations, Oxford University Press, Oxford (1979)
7. Atti del Convegno su: Prospettive della Coltura della Spirulina in Italia, Firenze 20-21 Novembre 1980.

PART II

EPIDEMICS

CONGENITAL RUBELLA SYNDROME (C.R.S.): MODELS OF DISEASE CONTROL BY VACCINATION.

by

Roy M. Anderson
Department of Pure and Applied Biology,
Imperial College, London University,
LONDON SW7 2BB

INTRODUCTION

An important consequence of mass immunization programmes is their tendency to increase the average age at which an individual typically acquires an infection over that which pertained before vaccination. As emphasized in a number of recent papers, such an increase in the average age at infection may, under certain circumstances, result in an increased incidence of infection among older age classes of the community than was the case before the implementation of immunization (Knox, 1980; Dietz, 1981; Cvjetanovic, Grab and Dixon, 1982; Hethcote, 1983; Anderson and May, 1983a). This observation is a cause for concern if the effects of infection are typically more severe among adults than among children.

Congenital rubella syndrome (C.R.S.) is a good example of this phenomenon. The syndrome occurs among 20-50% of infants born to women who had acquired inapparent or apparent rubella infection during the first trimester of pregnancy. Rubella is a common directly-transmitted viral infection of children and adolescents throughout the world. Mass immunization against rubella may be of benefit to the community in diminishing the overall incidence of infection but may be disadvantageous to women in their childbearing years as a consequence of its impact on the typical age at which people acquire the infection.

The considerations outlined above have led to the adoption of different vaccination policies by different countries. The policy currently adopted in the U.S.A., seeks to immunize large numbers of children at a pre-school age, with the aim of reducing the overall rate at which the infection circulates within the community and hence the incidence of cases in women of childbearing age. The policy currently adopted in the U.K. seeks to encourage the acquisition of immunity via natural infection during childhood by vaccinating only those individuals who will become at risk (i.e. girls) just before they enter the childbearing age classes. Currently, in the U.K. there is some discussion of the relative merits of the U.S.A. and the U.K. policies, and of the advantages of a 'two stage policy' in which boys and girls are immunized at a pre-school age in addition to the vaccination of girls around 10 to 15 years of age.

This paper builds on recent work on the development of mathematical models of recurrent epidemic phenomena (see Anderson and May, 1983), to examine the effects of a two stage vaccination programme on the incidence of C.R.S.

MATHEMATICAL MODEL

The analysis is based upon a compartmental deterministic model with age structure. The properties of this model are only briefly sketched in this paper and the reader is referred to work by Hoppensteadt (1975), Dietz (1976) and Anderson and May (1983a) for further details.

The numbers of people who are susceptible, infected but not infectious, infectious and immune are denoted by the variables $X(a,t)$, $H(a,t)$, $Y(a,t)$ and $Z(a,t)$ respectively. It is assumed that the age-dependent mortality rate, $\mu(a)$, is a step function such that $\mu(a) = 0$ for $a < L$ and $\mu(a) = \infty$ for $a \geqslant L$ where L = life expectancy (roughly 70 to 75 years in most developed countries at present). Individuals are assumed to leave the latent class (H) to join the infectious class (Y) at a rate σ, and to leave the infectious class (Y) to join the immune class (Z) at a rate γ ($1/\sigma$ is the latent period and $1/\gamma$ is the infectious period). Susceptibles of age a are assumed to acquire the infection at an age independent per capita rate $\lambda(\bar{Y}(t))$; λ is the so-called 'force of infection', which at time t is some function of the total number of infectious people within the community, $\bar{Y}(t)$ (where $\bar{Y}(t) = \int_0^L Y(a,t)da$).

These assumptions generate the basic model:

$$\frac{\partial X(a,t)}{\partial t} + \frac{\partial X(a,t)}{\partial a} = -\left[\mu(a) + \lambda(\bar{Y}(t))\right] X(a,t) \tag{1}$$

$$\frac{\partial H(a,t)}{\partial t} + \frac{\partial H(a,t)}{\partial a} = \lambda(\bar{Y}(t))\, X(a,t) - \left[\mu(a) + \sigma\right] H(a,t) \tag{2}$$

$$\frac{\partial Y(a,t)}{\partial t} + \frac{\partial Y(a,t)}{\partial a} = \sigma H(a,t) - \left[\mu(a) + \gamma\right] Y(a,t) \tag{3}$$

$$\frac{\partial Z(a,t)}{\partial t} + \frac{\partial Z(a,t)}{\partial a} = \gamma Y(a,t) - \mu(a) Z(a,t) \tag{4}$$

It is assumed that the total population size N is constant such that the net death rate exactly balances the net birth rate. With a life expectancy of L years, this implies that the number of new born individuals, X_0, entering the population each year is simply $X_0 = N/L$. The boundary conditions are as follows: $X(0,t) = X_0$, $H(0,t) = Y(0,t) = Z(0,t) = 0$ for all t. The initial conditions are determined by specifying $X(a,0)$, $H(a,0)$, $Y(a,0)$ and $Z(a,0)$ for all a at time $t = 0$. The force of infection λ is assumed to obey the 'mass action principle' such that

$$\lambda(t) = \beta \int_0^L Y(a,t)\, da \tag{5}$$

where β is a 'transmission coefficient'.

At equilibrium the fraction of all individuals in age class a that are susceptible, $x(a)$, is simply

$$x(a) = \exp(-\lambda a) \tag{6}$$

while the proportion susceptible within the total community, $\hat{x}$ is:

$$\hat{x} = \left[1 - \exp(-\lambda L)\right] / \lambda L \tag{7}$$

Under the assumption that λ is independent of age, and where $f(a)$ is the population age distribution, the average age at infection, A, is

$$A = \int_0^\infty a\lambda e^{-\lambda a} f(a)\,da \Big/ \int_0^\infty e^{-\lambda a} f(a)\,da \;=\; \lambda^{-1}\left[1 - \lambda L e^{-\lambda L}/(1 - e^{-\lambda L})\right] \tag{8}$$

For L large in relation to λ (usually the case for viral infections in developed countries) then eqn. (8) simplifies to

$$A \simeq 1/\lambda \tag{9}$$

The basic reproductive rate of the infection, R_0 (defined as the number of secondary cases generated by one primary case in a susceptible population; see Dietz (1976)) is approximately related to the fraction susceptible, $\hat{x}$, (given γL and σL large, and λ independent of age) where

$$R_0 \simeq 1/\hat{x} \tag{10}$$

A TWO STAGE VACCINATION PROGRAMME

A) Equilibrium results

Consider a two stage vaccination programme in which a proportion p' of boys and girls are vaccinated at age b and a proportion p of girls are vaccinated at age c where $c > b$. The force of infection at equilibrium after vaccination is defined as λ'. The proportions susceptible at age a, $x(a)$, are given by

$$x(a) = \exp(-\lambda' a); \qquad a < b \tag{11}$$

$$x(a) = (1-p')\exp(-\lambda' a); \qquad c > a > b \tag{12}$$

$$x(a) = (1-p/2)(1-p')\exp(-\lambda' a); \qquad a > c \tag{13}$$

assuming a 1:1 sex ratio of boys to girls. The fraction susceptible in the total population, $\hat{x}$, is given by

$$\hat{x} = \left[1 - p'\exp(-\lambda' b) - \frac{p}{2}(1 - p')\exp(-\lambda' c) - (1 - p')\left(1 - \frac{p}{2}\right)\exp(-\lambda' L)\right] \Big/ L\lambda' \tag{14}$$

The basic reproductive rate of the infection in the vaccinated community, $\hat{R}_0$, is again related to the fraction susceptible ($\hat{R}_0 \simeq 1/\hat{x}$) hence

$$\hat{R}_0 \simeq \lambda' L \Big/ \left[1 - p'\exp(-\lambda' b) - \frac{p}{2}(1-p')\exp(-\lambda' c) - (1-p')\left(1-\frac{p}{2}\right)\exp(-\lambda' L)\right] \tag{15}$$

However, given the assumption that λ is age-independent the fractions susceptible in the vaccinated and unvaccinated communities are identical in value (this is a consequence of the 'mass action' assumption of infection spread; see Dietz (1976) and Anderson and May (1983a)) and hence $\hat{R}_0 = R_0$. As the infection moves to the point of eradication, for sufficient levels of vaccination coverage, λ' tends to zero and thus

$$(1-p')\left(1-\frac{p}{2}\right) + \frac{p'b}{L} + \frac{p}{2}(1-p')\frac{c}{L} \rightarrow \frac{1}{R_0} \tag{16}$$

This criterion may be expressed in terms of the level of coverage of boys and girls at age b (the value of p') where

$$p' > \frac{\left[1 - \frac{1}{R_0} - \frac{p}{2}\left(1 - \frac{c}{L}\right)\right]}{\left[1 - \frac{b}{L} - \frac{p}{2}\left(1 - \frac{c}{L}\right)\right]} \tag{17}$$

Conversely, it may be expressed in terms of the level of coverage of girls at age c (the value of p) where

$$p > \frac{\left[1 - \frac{1}{R_o} - p'(1 - \frac{b}{L})\right]}{\left[(1 - p')(1 - \frac{c}{L}) / 2\right]} \tag{18}$$

In the case of congenital rubella we are interested in the number of cases of infection that occur in the childbearing age classes. Specifically, our interest lies in the impact of different levels of vaccination coverage (the values of p and p') on the incidence of disease in these high risk age classes. A convenient way to record this information is to define a ratio $w(a_1,a_2)$ which denotes the number of cases of infection arising in the female age classes a_1 to a_2 after vaccination, divided by the number of cases arising in the same age classes before vaccination. With respect to the equilibrium state, the ratio $w(a_1,a_2)$ is defined as follows:

$$w(a_1,a_2) = \frac{\int_{a_1}^{a_2} \lambda' \; x^*(a)da}{\int_{a_1}^{a_2} \lambda \; x^*(a)da} \tag{19}$$

where $x^*(a)$ denotes the proportion of females susceptible at age a. From eqns. (11) to (13) the ratio may be expressed as follows:

$$w(a_1,a_2) = (1-p)(1-p') \frac{\left[\exp(-\lambda' a_1) - \exp(-\lambda' a_2)\right]}{\left[\exp(-\lambda \; a_1) - \exp(-\lambda \; a_2)\right]} \tag{20}$$

For rubella in the U.K., the average age at infection, A, prior to vaccination was approximately 9 years of age (see Anderson and May, 1983). The values of λ, $\hat{x}$ and R_o may therefore be calculated from eqns. (9), (7) and (10) respectively. It is thus possible, using eqns. (5) and (20), to calculate the value of the ratio $w(a_1,a_2)$ for various levels of vaccination coverage within a two-stage policy. The age range of interest is the childbearing age classes which are approximately 16 years of age (a_1) to 40 years of age (a_2). Figure 1 records the value of $w(a_1,a_2)$ for a vaccination programme in which boys and girls are vaccinated at age 2 years (b) and girls are vaccinated at age 12 years (c).

A number of interesting points are illustrated in Fig. 1. (1) Eradication of rubella is predicted to occur when boys and girls are vaccinated at age 2 years, at a coverage level of $p' > 0.9$ (2) Eradication of C.R.S. is predicted to occur if girls, and girls only, are vaccinated at age 12 years only if all the girls in the population are vaccinated ($p = 1.0$). This is a consequence of the fact that the average age at vaccination (12 years) is in excess of the average age at infection prior to vaccination (9 years) and as a result of only vaccinating girls (50% of the total population). (3) Vaccination of boys and girls at a pre-school age may result in an

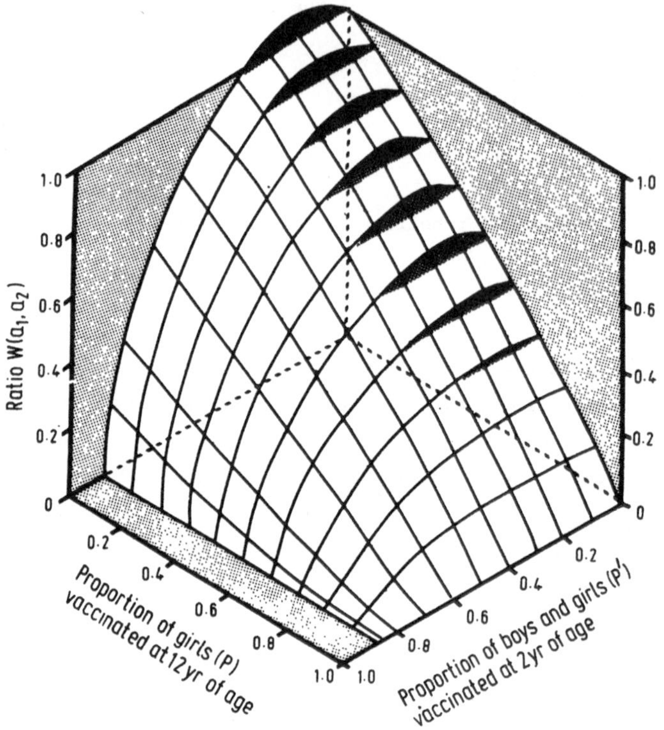

Figure 1: The ratio $w(a_1,a_2)$ of the number of rubella cases arising in the age range 16-40 years after vaccination, divided by the number in the same age range before control, plotted as a function of the proportion (p') of boys and girls vaccinated at 2 years of age and the proportion of girls vaccinated at 12 years of age. In the darkly shaded regions vaccination of boys and girls at a pre-school age increases the value of the ratio over that pertaining when girls and girls only are vaccinated at 12 years of age ($\lambda = 0.1111\ yr^{-1}$, $L = 75$ yr).

increase in the incidence of rubella in the at risk-age classes for low levels of vaccination coverage (the ratio $w(a_1,a_2)$ exceeds unity in value). This is a consequence of the tendency of vaccination to reduce the force of infection ($\lambda' < \lambda$) and hence to raise the average age at infection over that pertaining prior to control. (4) Little benefit accrues from vaccinating boys and girls at 2 years of age in addition to vaccinating girls at 12 years of age unless the level of coverage of the pre-school children is very high. (5) Similarly, the U.S.A. policy ($p' > 0$, $p = 0$) is only predicted to be more effective than the U.K. policy ($p' = 0$, $p > 0$) when coverage levels are in excess of approximately 84% of the pre-school children.

The principal conclusion to emerge from this analysis is that there is little benefit to be gained from changing the current U.K. policy to a two-stage vaccination programme unless very high levels of vaccination can be attained in the pre-school children. Past history of vaccination in Great Britain argues that such levels may be difficult to achieve (see Editorial, Lancet December 10th, 1983).

B) Numerical solutions

Numerical procedures can be employed to generate time-dependent solutions of the model defined by eqns. (1) to (4). A series of simulations are presented in Fig.2, representing a one stage U.S.A. policy (graph (a)), a one stage U.K. policy (graph (b)) and a two-stage policy (graph (c)) (the methods employed in the numerical solution of eqns. (1) to (4) are detailed in Anderson and May (1983a)). The levels of vaccination were chosen to mirror current trends in the U.S.A. and U.K. in graphs (a) and (b). In graph (c) it was assumed that the rubella vaccine would be administered as a combined measles-rubella injection at around 2 years of age with further rubella vaccination for girls between 12 to 13 years of age. The current acceptance rate for measles vaccine in the U.K. is around 50% of each cohort of children.

Note the dramatic differences in the predicted impacts of the three policies. The severe perturbation induced by the U.S.A. policy tends to dramatically increase the average age at infection, lengthen the interepidemic period (from around 4 years to 12 years) and induced the ratio $w(a_1,a_2)$ to rise significantly above unity in value during an epidemic year. Between epidemics the suppression in incidence of rubella in 16-40 year women is, however, very substantial (graph (a)). The U.K. policy has a more gradual impact; the average age at infection decreases slightly (since the average age at vaccination (12 years) is greater than the average age at infection prior to control (9 years)) and the interepidemic period remains at around 4 to 5 years (the natural cycle of rubella in the U.K. prior to immunization). Most importantly, however, the policy does not induce large amplitude oscillations in $w(a_1a_2)$, but tends to gradually suppress the ratio (through time) to a value significantly below unity (graph (b)). The two stage policy has an intermediary effect, with medium amplitude fluctuations and a slight increase in both the average age at infection and the interepidemic period. Note, however, that at a coverage level of 50% of the two year old children, the 2 stage programme is not predicted to be more advantageous than the U.K. policy (compare graphs (c) and (b)).

The overall conclusion suggested by these numerical results is that a two stage policy has advantages over the one stage U.S. policy in the sense of suppressing large amplitude fluctuation in disease incidence, but is unlikely to improve on the one stage U.K. policy unless vaccine acceptance rates attain a very high level (see Fig. 1 for equilibrium predictions). Thus in the U.K. if acceptance rates amongst young children can not be improved over that pertaining for the measles vaccine at present (50%), there is little advantage to be gained from instituting a two stage policy.

Figure 2

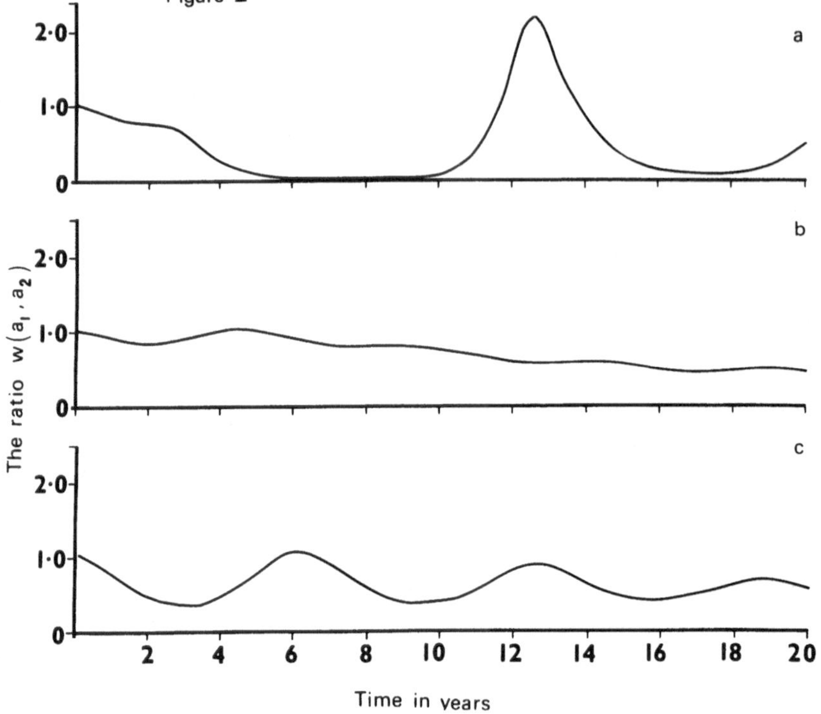

Figure 2: Numerical solutions of eqns. (1) to (4) showing temporal changes in the ratio $w(a_1, a_2)$ over a twenty year period, for three different vaccination strategies. The following parameter values were employed: $a_1 = 16.0$ years; $a_2 = 40.0$ years; $\sigma = 34.8$ year^{-1}; $\gamma = 31.8\ \text{year}^{-1}$; $\lambda = 0.1111\ \text{year}^{-1}$; N(total population size) $= 5 \times 10^7$; $\mu(a) = 0$ for $a \leq 75$ years, $\mu(a) = \infty$ for $a > 75$ years. The starting conditions were set at the equilibrium values for the system and the boundary conditions are as defined in the main text. Graph (a) The U.S.A. policy where boys and girls are vaccinated at a pre-school age. The level of vaccination coverage was set at 80% and the age at vaccination was between 2-3 years. Graph (b) The U.K. policy where girls, and girls only, are vaccinated at between 12-13 years of age. The level of vaccination coverage was set at 80%. Graph (c) A two stage policy in which 50% of boys and girls are vaccinated at between 2-3 years of age and 80% of girls are vaccinated between 12-13 years of age. In all three simulations, maternal antibody was assumed to provide protection to infection in new born infants for a period of 0.25 years.

DISCUSSION

The analysis presented in this paper is based upon a model which incorporates the assumptions that the force of infection, λ, is independent of age. This is probably inaccurate since serological data, and case notification records, suggest that λ changes with age, going from a low value in very young children to a high value in the 5-10 year old age class and then back to a low value thereafter (Anderson and May, 1983b). These effects are probably a consequence of changes in behaviour with age. Such patterns raise important questions concerning the structure of models of recurrent epidemic behaviour. An interesting and important discussion of this problem is presented by Schenzle in this volume (see Schenzle (1984)). A major problem arises in the interpretation of the value of λ for a given age class. For example, is such a value primarily a function of contacts between susceptibles and infecteds within the age class or is it a consequence of both within and between age group contacts? Ideally, we need quantitative information on "who mixes with whom" within a community. Such data, however, is likely to be extremely difficult to obtain in practice. In qualitative terms, the impact of such effects on our conclusions concerning the U.S.A. rubella vaccination policy are likely to be more important when compared with the influence of similar factors in the U.K. policy. This is because vaccination at an early age (well before the average age at infection in the unvaccinated community) acts to increase the average age at infection and to therefore shift individuals into age classes with different rates of exposure to infection. The level of vaccination coverage for eradication, calculated from models with a constant λ, may therefore either overestimate the critical coverage for eradication (if λ changes in value from high in the young age classes to low in the older age classes) or underestimate the critical coverage (if λ changes in value from low in the young age classes to high in the older age classes). Future analytical and numerical studies of recurrent epidemic phenomena must clearly focus on the significance of observed age related changes in the force of infection to the design of vaccination policies (Schenzle, this volume). There is also a concomitant need for better serological data to facilitate the estimation of age-related changes in transmission.

ACKNOWLEDGEMENTS

I am indebted to Robert May for much help and invaluable advice on the research presented in this paper.

REFERENCES

Anderson, R.M. and May, R.M. (1983a). Vaccination against rubella and measles: quantitative investigations of different policies. Journal of Hygiene, 90, 259-325.

Anderson, R.M. and May, R.M. (1983b). Two-stage vaccination programme against rubella. Lancet 17th December 1983, 1416-17.

Cvjetanovic, B., Grab, B. and Dixon, H. (1982). Epidemiological models of poliomyelitis and measles and their application in the planning of immunization programmes. Bulletin of the World Health Organization 60, 405-22.

Dietz, K. (1976). The incidence of infectious diseases under the influence of seasonal fluctuations. Lecture Notes in Biomathematics 11, 1-15.

Dietz, K. (1981). The evaluation of rubella vaccination strategies. In Mathematical Theory of the Dynamics of Biological Populations vol. II (ed. R.W. Hiorns and D. Cooke), pp. 81-98, London, Academic Press.

Hethcote, H. (1983). Measles and rubella in the United States. American Journal of Epidemiology. 117, 2-21.

Hoppensteadt, F.C. (1974). An age-dependent epidemic model. Journal of the Franklin Institute 297, 325-33.

Knox, E.G. (1980). Strategy for rubella vaccination. International Journal for Epidemiology 9, 13-23.

Schenzle, D. (1984).Control of virus transmission in age-structured populations. Mathematics in Biology and Medicine. (V. Capasso et al. Eds.). Lecture Notes in Biomathematics, Springer-Verlag, to appear.

A SIMULATION MODEL FOR THE CONTROL OF HELMINTH DISEASES BY CHEMOTHERAPY

K. Dietz and H. Renner
Institute of Medical Biometry
Tübingen University
7400 Tübingen/FRG

1. INTRODUCTION

If an individual is successfully vaccinated he or she is protected against the corresponding infection during the duration of the induced immunity. The vaccination of a certain proportion in a population is of obvious benefit to those covered. But there is also some effect to the rest of the community: the risk of acquiring an infection is reduced. For certain kinds of infections this also results in a reduction of the incidence of disease. Some infections can only lead to disease in higher age groups, like rubella which may cause congenital malformations if the mother gets infected during early pregnancy. In this case a reduction in infection risk results in an increase of age at infection. This may lead to an increase of the total incidence of disease in the population as has been pointed out by Knox (1980), Dietz (1981) and Anderson & May (1983). This example shows that an evaluation of a vaccination strategy has to take into account the consequences on the *disease* incidence in the total population.

Similarly, the chemotherapy of an infected individual is not only of benefit to him/her by reducing the duration of the infection but also to the community by reducing the infection risk. On the other hand chemotherapy will not prevent reinfection such that a new susceptible is created, and the total incidence of new infections and/or new cases of disease may increase for certain coverage.

In the following we shall quantify these effects both with the help of a simple analytical model and of a simulation model which incorporates the basic elements of the epidemiology of schistosomiasis. The analytical model takes only into account presence or absence of an infection. It is necessarily oversimplified but has the advantage that it allows one to derive explicit expressions for relevant variables such that the effect of certain parameter changes can be explored over a wide range. The simulation model can take into account many realistic aspects of the epidemiology and the control of schistosomiasis but due to its complexity only runs for selected parameter sets can be performed. The analytical model can help in the design and the interpretation of simulation runs.

2. THE POPULATION EFFECT OF CHEMOTHERAPY

The simplest epidemiological model for a parasitic infection is due to Ross (1911) who proposed it in the context of malaria control. Individuals in the population are either susceptible or infected. There is no latent period, i.e. infected individuals are immediately infectious. Let y denote the proportion of infected individuals in the population. The contact rate and the recovery rate is denoted by β and γ, respectively. Then the transmission is described by the following equation:

$$\frac{dy}{dt} = \beta y(1-y)-\gamma y, \tag{1}$$

with the solution

$$y(t) = Y/\left[1+\{[Y/y(0)]-1\}e^{-(\beta-\gamma)t}\right], \tag{2}$$

where

$$Y = 1-\gamma/\beta = 1-1/R_0$$

is the equilibrium proportion for the basic reproductive rate $R_0 > 1$.

Chemotherapy transfers an infected individual to the susceptible state. We consider two schemes for the application of chemotherapy in the population:

"Random" : Each infected individual is treated after an exponential waiting time of average $T = \delta^{-1}$ after becoming infected.

"Scheduled": A certain proportion c of all infected individuals are treated simultaneously in so-called mass drug administrations (MDAs) which take place at fixed intervals of length T.

Random chemotherapy can easily be incorporated into Eq.(1):

$$\frac{dy}{dt} = \beta y(1-y)-(\gamma+\delta)y. \tag{3}$$

The new equilibrium proportion of infected individuals, i.e. the endemic level, is given by

$$Y(n_r) = 1-(1+n_r)/R_0, \tag{4}$$

where $n_r = \delta/\gamma$ is the average number of treatments which one infected individual would receive during one infectious period if treatments were continued at the rate δ. Another interpretation is simply the ratio of the length of the natural infectious period to the waiting time for treatment. If the endemic level is to be reduced to zero by chemotherapy alone, i.e. without reductions in the contact rate β, then n_r has to exceed R_0-1:

$$n_r^* \geq R_0-1. \tag{5}$$

If $\gamma^{-1} = D$ denotes the average infectious period, then Eq.(5) implies

$$T \leq D/(R_0-1). \tag{6}$$

This scheme assumes that *all* infectives are treated sooner or later. If the basic reproductive rate R_0 is large then eradication can only be achieved by short waiting times for treatment.

The equilibrium prevalence $Y(n_r)$ can be written as the product of inci-dence and duration of an infection. The incidence is the product of the incidence rate and the proportion of susceptibles:

Prevalence = {Incidence Rate}{Proportion Susceptible}{Duration}

$$1-(1+n_r)/R_0 = \{\beta[1-(1+n_r)/R_0]\}\{(1+n_r)/R_0\}\{1/[\gamma(1+n_r)]\}.$$

As the rate of chemotherapy expressed by n_r increases, the incidence rate decreases, the proportion susceptibles increases and the duration of the infection decreases. The total incidence

$$\beta[1-(1+n_r)/R_0][(1+n_r)/R_0]$$

increases for small values of n_r for $R_0 > 2$ and then decreases again to zero for $n_r = R_0-1$. In other words, only for $R_0 < 2$ will any level of chemotherapy reduce the total incidence.

The case of scheduled treatment has already been dealt with by Dietz (1975). Let $\bar{Y}(n_f)$ denote the average proportion of infectives if n_f MDAs are applied during one infectious period ($n_f = (\gamma T)^{-1}$) and the average is taken over one interval between two consecutive MDAs. Dietz (1975) derives the following formula:

$$\bar{Y}(n_f) = \max\{0, Y[1-(\ln(1-c)^{-n_f})/(R_0-1)]\}. \tag{7}$$

In order to reduce the endemic level to zero the following inequality has to be satisfied

$$n_f \geq (R_0-1)/\ln(1-c)^{-1}. \tag{8}$$

This inequality can also be used to specify a lower bound for the coverage c:

$$c \geq 1-\exp[-(R_0-1)/n_f]. \tag{9}$$

This formula agrees with Eq.(12) in Anderson and May (1982) if one assumes that treatment is fully effective ($h = 1$). The parameter A in the cited paper is the average life length of a parasite expressed in units of the treatment interval, i.e. it corresponds precisely to our varriable n_f.

The paper by Anderson and May (1982) also studies the effect of a single round of MDA. They assume that the number of parasites per host has a negative binomial distribution. A host with j parasites is treated with probability

$$g_j = a[1-(1-\alpha)\hat{z}^j], \tag{10}$$

where a, α and $\hat{z}$ are constants less than one. They show that the average proportion to be treated in the total population for a given reduction in the average parasite load decreases with the clumping parameter k of the negative binomial distribution, i.e. with increasing clustering of parasites. This approach however does not allow one to assess the long-term effect of repeated MDAs: the parasite distribution after a single MDA with selective treatment is no longer negative binomial. In order to explore the effect of different treatment strategies we therefore apply the simulation approach since it allows one to incorporate many realistic complications, such as heterogeneous and age-dependent exposure and the distinction between infection and disease.

3. A SIMULATION MODEL FOR SCHISTOSOMIASIS

The quantitative epidemiology of schistosomiasis has recently been reviewed by Barbour (1982).

The human host population is described by an immigration-death process. Newbornes enter the population at the rate ν_1. In order to achieve realistic age-distributions, individuals have a survival time with a gamma distribution. This is achieved by assuming K fictitious states through which an individual must pass before dying. The life expectancy is assumed to be μ_1^{-1} = 50 years. The average sojourn time in one state is therefore $1/(\mu K)$. Individuals differ with respect to their contact rates with snails. Strictly speaking one would have to distinguish exposure and contamination rates. But these two parameters are probably highly correlated such that they can be assumed to be equal as done by Barbour (1978). At the time of entering the population we associate with each individual i the exposure parameter b_i which is taken from an exponential distribution with mean one. This parameter determines the life time average contact rate of an individual. During his/her life an individual may undergo changes in the contact rate. Especially children in school age have more water contact than other age groups. For computational reasons we make the contact rates state dependent rather than age dependent. Hence for each of the K = 10 states we specify factors c_k,

$k = 1,\ldots,K$ which are to be multiplied with the contact rate b_i of an individual if he/she is in state k. We set

$$c_{(i)} = c_k, \text{ if individual } i \text{ is in state } k.$$

Also for the snail population we assume an immigration death process with immigration rate ρ_2 and death rate μ_2. The age distribution of susceptible snails is exponential. Snails may have however other death rates μ_3, μ_4 while they are latent (infected but not yet infectious) and infectious, respectively. In the snail we do not take into account superinfections. In the human host the number of immature and mature worms are explicitly incorporated into the model. The immature worms mature at a constant rate δ, and the mature worms die at the rate μ_5. The complications with respect to the pairing of worms in one host are ignored. The rate of acquiring new immature worms is proportional to the number of infectious snails. The model takes into account concomitant immunity by assuming that the presence of mature worms decreases the chance of new immature worms to enter the host. The production of eggs per mature worm is assumed to decrease as the number of mature worms increases.

The effect of chemotherapy in an individual is the reduction of the worm load. Immature and mature worms will survive chemotherapy with probability p_1 and p_2, respectively. The number of immature and mature worms after chemotherapy is described by independent binomial distributions $B(s_i;p_1)$ and $B(w_i;p_2)$ where s_i and w_i denote the number of immature and mature worms in individual i, respectively. The probability for an individual to be included in an MDA may depend on age, on egg output, on his/her water contact rate or on the presence of disease. The model assumes that individuals carrying more than a critical number of mature worms are sick. As soon as the worm load is reduced below the critical level either by natural death of the worms or by chemotherapy the individual is considered healthy again, i.e. chronic sequelae are ignored.

The simulation program is written in FORTRAN. The sojourn time in a given state s of the model has an exponential distribution with parameter $\sum_r q_{sr}$, where q_{sr} is the transition rate from state s into state r. Let u be a random number with a uniform distribution in the unit interval. Then the corresponding sojourn time is determined by

$$T = (-\ln u)/\sum_r q_{sr}. \tag{11}$$

A second random number from the unit interval determines which of the transitions takes place. The transition q_{sr_0} occurs with the probability $q_{sr_0}/\sum_r q_{sr}$. At each step the model allows only few types of transitions, like "an individual enters the population", "a latent snail dies", "a mature worm dies", but due to the average size of the human population of about 500 individuals, the number of possible transitions is in the order of several hundred. The average sojourn time in one state is about six hours. In order to simulate one year it takes in the average eight seconds on a UNIVAC 1100/80.

4. SOME SIMULATION EXAMPLES

Table 1 specifies all the input parameters which were used to compare three types of yearly MDA schemes: treatment is given a) to everybody; b) to all individuals whose age is between five and twenty years; c) to the sick individuals only, i.e. for whom the number of mature worms is more than nine. The MDAs began after the transmission had reached an equilibrium situation. The average worm load per individual is 8.6 with a variance of 64.0. The prevalence of the infection with mature worms is 95 % and the prevalence of disease is 32 %. The highest worm load is 60. If one tries to fit a negative binomial distribution to the worm distribution by the method of moments one gets $\hat{k} = 1.1$, i.e. nearly a geometric distribution with $k = 1$. For this distribution the expected number of individuals without worms is 43.8 corresponding to 8.7 %. The observed number is however only 24, i.e. 4.8 %. This shows that the negative binomial distribution does not provide a satisfactory fit.

The age-specific worm load is given in Table 2 in addition to the age-specific prevalence and morbidity. The prevalence of infection rises quickly to nearly 100 %. The average worm load has its peak in the age-group 10-15 years and the peak prevalence of morbidity is in the age-group 15-20 years. We are dealing here with a highly endemic situation, and we expect that control measures would have to be very efficient in order to achieve eradication. In fact the contact rate β_1 would have to be reduced by the factor 0.0015. This would correspond to a basic reproductive rate in the order of six to seven hundred. Whether there exist endemic levels with such high rates is at present unknown.

TABLE 1

TRANSITION RATES OF SCHISTOSOMIASIS MODEL

Event	Rate
Individual enters population	$\nu_1=10$/year
Individual matures or dies	$\mu_1 KI$; $\mu_1=0.02$/year $K=10$; I=no. of individuals
Susceptible snail enters	$\nu_2=500$/year
Susceptible snail dies	$\mu_2 Q$; $\mu_2=1$/year Q=no. of susceptible snails
Susceptible snail is infected	$\left[\beta_2 \sum_{i=1}^{I} b_i c_{(i)} w_i/(1+d_2 w_i)\right]Q$ $\beta_2=0.04; E(b_i)=V(b_i)=1; d_2=4$ $c_1=c_2=0.05; c_3=c_4=1.0; c_5=..=c_{10}=0.1$; w_i=no. of mature worms in individual i
Latent snail dies	$\mu_3 R$; $\mu_3=1$/year R=no. of latent snails
Latent snail becomes infectious	$\delta_2 R$; $\delta_2=7$/year
Infectious snail dies	$\mu_4 S$; $\mu_4=12$/year S=no. of infectious snails
New immature worm enters one individual	$\beta_1 S \sum_{i=1}^{I} b_i c_{(i)}/(1+d_1 w_i)$ $\beta_1=5$; $d_1=1$
Immature worm matures	$\delta_1 \sum_{i=1}^{I} s_i$; $\delta_1=1$/year s_i=no. of immature worms in individual i
Mature worm dies	$\mu_5 \sum_{i=1}^{I} w_i$; $\mu_5=0.2$/year

TABLE 2

AGE-SPECIFIC WORM LOAD AND PREVALENCE OF INFECTION AND MORBIDITY

Age	No. of ind.	Average worm load	Prevalence infection (%)	morbidity (%)
0- 5-	36	3.2	72.2	5.6
5-10-	55	8.4	96.4	25.4
10-15-	50	12.3	98.0	48.0
15-20-	63	11.7	98.4	54.0
20-25-	42	11.7	100.0	47.6
25-30-	43	8.9	93.0	37.2
30-40-	85	7.8	94.1	27.1
40-50-	66	7.0	100.0	25.8
50-60-	29	5.9	96.6	13.8
60+	32	7.2	96.9	25.0
	501	8.64	95.2	32.3

Simulation models as the one proposed in this paper open a way to bring models in closer agreement with observations.

The equilibrium snail population of about 270 snails contains 7.8 % infectious snails which agrees with observed orders of magnitude.

The incidence of new infections is 15/year in a population of about 500. From the absolute prevalence we estimate the average duration of one infection to be 32 years. Similarly, the average duration of one disease episode is about 4 years.

We now turn to the results of simulating yearly MDAs. Table 3 summarizes the results of nine consecutive MDAs. If everybody is treated (scheme a) the number of worms (immature and mature) is reduced to zero immediately after an MDA. But the presence of infectious snails provides a risk for new cases. The number of diseased individuals rises up to 30 in the last month before the next MDA. The yearly average prevalence of cases is 15. Thus even the yearly treatment of all individuals is not sufficient to eradicate the infection.

The third scheme (c) concentrates treatment on sick individuals only. After each MDA the number of cases is zero but rises again during the

TABLE 3

COMPARISON OF THREE MDA SCHEMES

	baseline	a)	b)	c)
no. of treatments/year	0	500	147	90
yearly average no. of cases (prevalence)	160	15	100	43
yearly incidence of new cases	48	47	66	90
yearly incidence of new infections	15	350	130	118

a) MDA covers everybody every year
b) MDA covers age group 5-20 years every year
c) MDA covers all sick individuals (with more than nine mature worms)

year up to about 90 yielding a yearly average of 3 cases. The yearly incidence of new cases is doubled. The number of treatments per individual is given in Table 4. The expected values are calculated under the assumption of a binomial distribution, i.e. that every individual has the same chance of being treated. There are more individuals simulated who received either no treatment or more than three treatments than expected. There are even 8 individuals who received all nine treatments. Thus there is a strong tendency for concentrating treatments to a relatively small proportion of the population: 66 % of the treatments are given to 18 % of the population.

The scheme (b) covers only the age-group 5-20 years because this corresponds to the years of highest water-contact-rates. From a practical point of view this part of the population is most easily accessible. The yearly average prevalence of cases is however only reduced to 100. All three schemes raise considerably the incidence of new infections and, to a lesser extent, also the incidence of new cases. Table 3 provides the basis for a cost-effectiveness-analysis provided one is able to attach weights to the various variables.

TABLE 4

NUMBER OF TREATMENTS/INDIVIDUAL

No. of treatments	No. of individuals simulated	No. of individuals expected
0	258	89.8
1	69	167.9
2	36	139.5
3	40	67.6
4	30	21.1
5	21	4.4
6	9	0.6
7	12	0.1
8	8	0.0
9	8	0.0

5. DISCUSSION

The present paper only sketches briefly the structure of the simulation model. A sensitivity analysis which explores the effect of varying some of the key parameters will be presented in detail elsewhere. In particular, mean and variance of the contact rates will be modified. Also the parameters d_1 and d_2 deserve detailed study since they determine the strength of density-dependent regulation in the human host.

The model can easily be extended to include further refinements, like differential mortality due to worm load. A further extension would be the simultaneous simulation of the transmission of the two schistosome species *S. haematobium* and *S. mansoni*. This aspect is important since there may be some interference between the two infections and some therapeutic agents are effective against both parasites.

6. ACKNOWLEDGEMENT

The present work was partly supported by the Deutsche Forschungsgemeinschaft.

REFERENCES

Anderson, R.M. & May, R.M. (1982). Population dynamics of human helminth infections: control by chemotherapy. *Nature* 297, 557-563.

Anderson, R.M. & May, R.M. (1983). Vaccination against rubella and measles:quantitative investigations of different policies. *J. Hyg. Camb.* 90, 259-325.

Barbour, A.D. (1978). Macdonald's model and the transmission of bilharzia. *Transactions of the Royal Society of Tropical Medicine and Hygiene* 72, 6-15.

Barbour, A.D. (1982). Schistosomiasis. In *Population Biology of Infectious Diseases: Theory and Applications*(ed. R.M. Anderson), 180-208. London: Chapman & Hall.

Dietz, K. (1975). Models for parasitic disease control. *Bull. Int. Statist. Inst.* 46, Book 1, 531-544.

Dietz, K. (1981). The evaluation of rubella vaccination strategies. In *The Mathematical Theory of the Dynamics of Biological Populations*, vol. II (ed. R.W. Hiorns and D. Cooke), 81-98. London: Academic Press.

Knox, E.G. (1980). Strategy for rubella vaccination. *International Journal for Epidemiology* 9, 13-23.

Ross, R. (1911). *The prevention of malaria* (2nd ed.). London: Murray.

MODELS FOR ENDEMIC DISEASES

K. P. Hadeler, Tübingen

The classical epidemic models derived from the Kermack-McKendrick model consider the prevalence of the disease only. The host population is partitioned into classes of susceptibles S, infectious I, and recovered R. The transition between these classes is described by ordinary differential equations. The following equations take into account the recruitment of new individuals into the class of susceptibles by birth and a death rate depending on the class. The equations read

$$\begin{aligned} \dot{S} &= -\beta SI + \gamma R - \delta_1 S + (\delta_1 S + \delta_2 I + \delta_3 R) \\ \dot{I} &= \beta SI - \alpha I - \delta_2 I \\ \dot{R} &= \alpha I - \gamma R - \delta_3 R \end{aligned} \tag{1}$$

Of course these equations are independent of δ_1, and $P = S + I + R$ is an invariant. Thus the equations simplify to

$$\begin{aligned} \dot{S} &= -\beta SI + (\gamma + \delta_3)\ (P - S) - (\gamma - \delta_2 + \delta_3)\ I \\ \dot{I} &= \beta SI - (\alpha + \delta_2)\ I \end{aligned} \tag{2}$$

The "natural" parameters of the system are β, $\gamma + \delta_3$, $\alpha + \delta_2$, $\alpha + \gamma + \delta_3$. In the following we consider only the endemic case $\gamma + \delta_3 > 0$.
The state space of the system (2) is the triangle $T = \{S \geq 0,\ I \geq 0,\ S + I \leq P\}$ which is positively invariant with respect to the flow.
There are two stationary states

$$\begin{aligned} (S_o, I_o) &= (P, 0) \\ (S_1, I_1) &= \left(\frac{\alpha + \delta_2}{\beta}, \quad \frac{\gamma + \delta_3}{\alpha + \gamma + \delta_3} \quad (P - \frac{\alpha + \delta_2}{\beta})\right) \end{aligned} \tag{3}$$

The threshold condition is $P > S_1 = (\alpha + \delta_2)/\beta$.
If $S_1 < P$ then the point (S_1, I_1) is located in the interior of T, and it is the only stable stationary point. If $S_1 > P$ then (S_1, I_1) is not in the half-plane $I > 0$, and (S_o, I_o) is stable.
Applying Dulac's criterion with the weight function 1/I and the

Poincaré - Bendixson theory one sees that for $S_1 < P$ all trajectories in T except (S_o, I_o) converge to (S_1, P_1), for $S_1 > P$ all trajectories in T converge to (S_o, I_o).

It can be observed, that for a certain range of parameters the system (2) admits a Lyapunov function. Assume that in equations (2) an increase of the number of infectious always leads to a decrease of susceptibles, i. e. assume $\gamma - \delta_2 + \delta_3 > 0$.
Then for

$$V(S,I) = S - \frac{\alpha+\gamma+\delta_3}{\beta} \log \left(\frac{\gamma-\delta_2+\delta_3}{\beta} + S\right) + I - I_1 \log I \tag{4}$$

holds

$$\frac{dV}{dt} = - \frac{\beta(\gamma+\delta_3)}{\alpha+\gamma+\delta_3} \cdot \frac{\gamma-\delta_2+\delta_3+\beta P}{\gamma-\delta_2+\delta_3+\beta S} \cdot (S-S_1)^2 \tag{5}$$

This Lyapunov function has been communicated to the author by P. de Mottoni for the special case $\delta_2 = \delta_3 = 0$, $\gamma = \alpha$.
Consider the number of infectious at the stable equilibrium as a function of the contact rate β. For small β one has I = 0. At $\beta_o = (\alpha+\delta_2)/P$ occurs a bifurcation with an exchange of stability, then $I(\beta) = I_1 \leq I_{max} = (\gamma+\delta_3)P/(\alpha+\gamma+\delta_3)$.

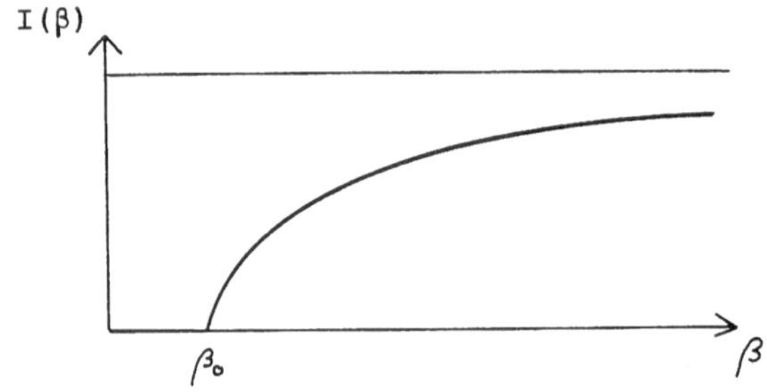

We shall see that a similar picture arises in the model for discrete infections to be derived now.
The classical prevalence model is inappropriate for helminthic infections such as onchocerciasis or other filariosis diseases. In such diseases the host acquires a finite, typically small number of parasites at different times, usually through the mediation of vectors such as insects. Several authors (see e.g. [1]) have designed models for such diseases. A model for such a disease should take account of the discrete structure, the transmission by vectors and the age structure of the hosts. The model should be derived from basic assumptions on the interaction of hosts and parasites rather than from mass action

equations. In [2] such a model has been proposed. Let t denote chronological time and let a denote the age of the host. Let $\mu(a)$ be the age-dependent mortality of hosts in the absence of parasites, and $b(a)$ the birth rate of hosts of age a. Let σ be the death rate of parasites within the host, and ρ their birth rate. In most helminthic infections parasites do not directly multiply within the host, then $\rho = 0$. Let α be the differential mortality inferred on the host by one parasite. It is assumed that parasites do not interact, thus the mortality of hosts of age a carrying r parasites is $\mu(a) + \alpha r$. The action of vectors is described as follows: It is assumed that the rate of parasite acquisition φ is a nonlinear function of the average parasite load $\bar{w}$,

$$\varphi = \beta f(\bar{w}) . \tag{6}$$

Thus the host population is governed by a Lotka model, the parasites satisfy a birth and death process with killing (see [8]), and both populations are coupled by the interaction law (6).

This model is cast into mathematical form in the following way: Let $n(t,a,r)$ be the density of hosts of age a carrying r parasites at time t. These functions then satisfy a system of infinitely many partial differential equations. For $r = 0,1,2,\ldots$

$$\begin{aligned} \frac{\partial n(t,a,r)}{\partial t} + \frac{\partial n(t,a,r)}{\partial a} = & -[\varphi(t) + \mu(a) + (\alpha+\sigma+\rho)r]n(t,a,r) \\ & +[\varphi(t) + \rho(r-1)]n(t,a,r-1) \\ & +\sigma(r+1)n(t,a,r+1) \end{aligned} \tag{7}$$

where the convention $n(t,a,-1) \equiv 0$ is used. Then the average parasite load is

$$\bar{w}(t) = \int_0^\infty \sum_{r=0}^\infty r n(t,a,r)\,da \Big/ \int_0^\infty \sum_{r=0}^\infty n(t,a,r)\,da . \tag{8}$$

With assistance of the generating function

$$u(t,a,z) = \sum_{r=0}^\infty n(t,a,r) z^r \tag{9}$$

one can condense this system into a single partial differential equation

$$u_t + u_a + g(z)u_z - [\varphi(t)(z-1) - \mu(a)]\,u = 0 \tag{10}$$

where

$$g(z) = (\alpha+\sigma+\rho)\,z - \sigma - \rho z^2 . \tag{11}$$

The expression for the average parasite load assumes the form

$$\bar{w}(t) = \int_0^\infty u_z(t,a,1)\,da \Big/ \int_0^\infty u(t,a,1)\,da \tag{12}$$

The equation (10) has to be considered with the acquisition law (12) (6) and appropriate initial conditions

$$u(0,a,z) = u_0(a,z) \tag{13}$$

and boundary conditions. At the boundary $a = 0$ one can assume either a prescribed birth rate

$$u(t,0,z) = N(t) \tag{14}$$

or a Lotka birth law

$$u(t,0,z) = \int_0^\infty b(a)u(t,a,\omega)\,da \tag{15}$$

Here a parameter $\omega \in [0,1]$ is introduced for the variable z which describes the effect of the parasite load on the individual fertility in the form of a geometric law: For $\omega = 1$ there is no effect, in the extreme case $\omega = 0$ only non-infected reproduce. To describe vertical transmission one could let ω depend on z. However this is inappropriate for helminthic infections.

In [2], [6], [7] the following approach towards the solution of problem (10), (6), (13), (14) has been chosen: Assume, for a moment, that the function φ were known. Then (10),(13),(14) is a linear partial differential equation with initial and boundary conditions. This problem can be solved by the method of characteristics. This solution, which of course depends on φ, is then introduced into (12) and then into (6). The result is a nonlinear Volterra equation of a peculiar type

$$\varphi(t) = \beta f(\bar{w}(t)) \tag{16}$$

$$\bar{w}(t) = \mathcal{N}(t)/\mathfrak{D}(t) \tag{17}$$

$$\mathcal{N}(t) = \int_0^t e^{A\varphi - M(a)} NB\varphi\,da + e^{C\varphi} \int_t^\infty e^{M(a-t)-M(a)} [u_{0z}G_z + u_0 D\varphi]\,da$$

$$\mathfrak{D}(t) = \int_0^t e^{A\varphi - M(a)} N\,da + e^{C\varphi} \int_t^\infty e^{M(a-t)-M(a)} u_0\,da$$

Here the integral operators $(A\varphi)(t,a) = \int_{t-a}^t [G(s-t,1)-1]\varphi(s)\,ds$, $(B\varphi)(t,a) = \int_{t-a}^t G_z(s-t,1)\varphi(s)\,ds$, $(C\varphi)(t) = \int_0^t [G(s-t,1)-1]\varphi(s)\,ds$, $(D\varphi)(t) = \int_0^t G_z(s-t,1)\varphi(s)\,ds$, are formed with the solution operator $G(t,z_0)$ of the initial value problem $\dot{z} = g(z)$, $z(0) = z_0$, and $M(a) = \int_0^a \mu(s)\,ds$.

In [2], [6], [7] it has been shown that the integral equation and thus the original problem (10), (6), (13), (14) have a global solution under the following conditions: The function $f: [0,\infty) \to [0,\infty)$ is continuously differentiable, $f(0) = 0$, $f'(0) = 1$, $f(u) > 0$ for $u > 0$, and sublinear, i. e. $f(u) \le f_0 u$ for some f_0. The death rate μ satisfies $\int_0^\infty \exp(-M(a))da < \infty$. Finally the initial datum u_0 is a generating function and satisfies

$$0 < \int_0^\infty u_0(a,1)da < \infty$$

$$\sup \left\{ \frac{u_{oz}(a,1)}{u_0(a,1)} : u_0(a,1) > 0 \right\} < \infty \;.$$

In a similar way the initial boundary value problem for the Lotka condition (15) can be reduced to a pair of coupled integral equations for the parasite acquisition rate φ and the birth rate $u(t,0,z)$ (see [5]).

If the number of newborn hosts N in (14) is constant in t then one can expect stationary solutions. Such solutions correspond to constant acquisition rates φ, where φ is obtained from an equation

$$\varphi = \beta f(W(\varphi)\varphi) \tag{18}$$

where

$$\begin{aligned} W(\varphi) &= I_1(\varphi)/I_0(\varphi) \\ I_1(\varphi) &= \int_0^\infty e^{-Q(a)\varphi - M(a)} q(a)da \\ I_0(\varphi) &= \int_0^\infty e^{-Q(a)\varphi - M(a)} da \end{aligned} \tag{19}$$

For $\omega = 0$ the functions q, Q are given by

$$\begin{aligned} q(a) &= \frac{1}{\varkappa}(1-e^{-\varkappa a}), \qquad \varkappa = \alpha + \sigma \\ Q(a) &= \frac{\alpha}{\varkappa}\left(a - \frac{1}{\varkappa}(1-e^{-\varkappa a})\right) \end{aligned} \tag{20}$$

For the general case see [4].

Since in equation (18) one can solve for β, the existence of a branch of nontrivial stationary solutions $\varphi > 0$ is obvious. Because of $f'(0) = 1$, this branch starts at

$$\beta_0 = \int_0^\infty e^{-M(a)}da \Big/ \int_0^\infty e^{-M(a)} q(a)da \;. \tag{20}$$

In [2] the following has been shown: If f does not grow too fast, i. e. $f(u)/u^2 \to 0$ for $u \to \infty$ then the branch exists for all $\beta > \beta_o$. If f is concave in the sense of Krasnoselskij, i. e. $f'(u) > 0$, $(f(u)/u)' \leq 0$ then the branch is monotone. Then for each $\beta > \beta_o$ there is exactly one solution $\varphi > 0$. Of course the constant β can be interpreted as a contact rate. Thus the bifurcation phenomenon is very similar to that in the classical endemic model.

The asymptotic behavior of solutions is not completely explored. In [2], [3] it has been shown that the zero solution $\varphi = 0$ of the integral equation (1), corresponding to a population without parasites in (10), is asymptotically stable for $\beta < \beta_o$. This solution $\varphi = 0$ loses its stability for $\beta > \beta_o$.

Recently examples have been given [6], where for monotone f and μ = const., $\rho = 0$ the branch of nontrivial solutions is not monotone. At $\varphi = 0$, $\beta = \beta_o$ the branch may bifurcate backwards. Also the branch may have double bends. Both cases lead to hysteresis phenomena in the following sense: An increase of the contact rate β may lead to a stabilization of the endemic on a higher level where it remains even after β is decreased. In a subsequent paper it will be shown that the endemic states changes stability exactly at the turning points of the bifurcation diagram. The endemic state is stable where $d\varphi/d\beta > 0$ and unstable where $d\varphi/d\beta < 0$.

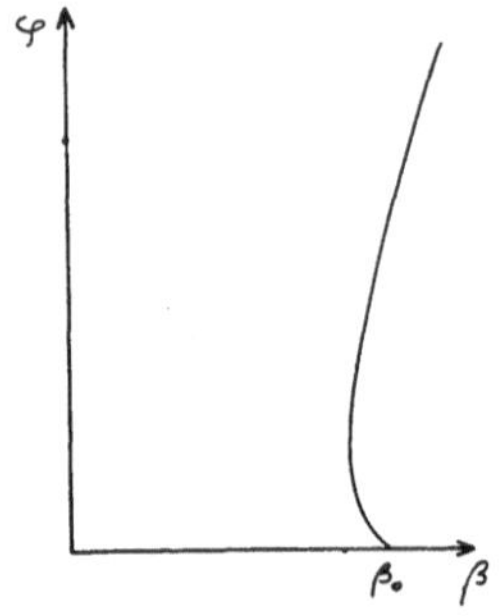

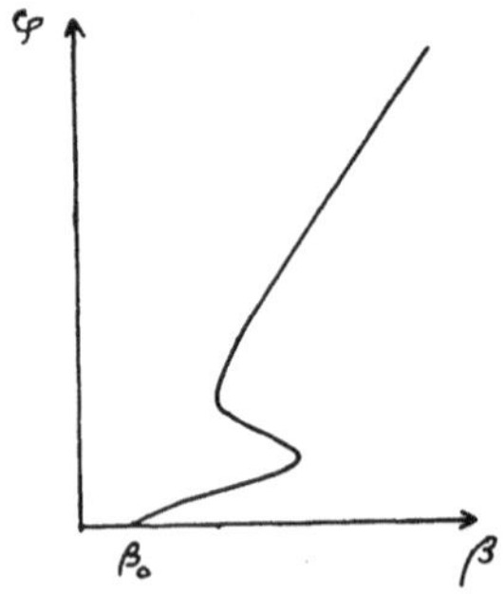

As in the classical Lotka model for populations with age structure in the case of a Lotka birth law (15) one does not expect stationary solutions but rather persistent solutions, i. e. stationary age distributions with exponential growth or decay of the population size

$$u(t,a,z) = N \exp \left[\int_0^a [G(s,z) - 1] ds \varphi - M(a) \right]$$

where the exponent φ is obtained from two coupled equations for φ and λ, which read (for $\omega = 1$, $\rho = 0$)

$$\varphi = \beta f\left(\frac{\int_0^\infty e^{-Q(a)\varphi - M(a) - \lambda a} q(a)\,da}{\int_0^\infty e^{-Q(a)\varphi - M(a) - \lambda a} da}\right) \tag{22}$$

$$\int_0^\infty b(a) e^{-Q(a)\varphi - M(a) - \lambda a} da = 1 \ .$$

Also in the case of a Lotka birth law, there is a branch of nontrivial solutions, i. e. of persistent solutions with positive parasite load (see [4]).

In the present model the parameters are constant in time. In natural populations, however, the parameters fluctuate statistically or vary in a regular manner. Although periodicity in birth rate is not pronounced in humans, it is important in vertebrate host populations. Numerical simulation of (10), (11), (13), (14) or (16) show that with a periodic birth rate $N(t)$ the solution approaches a periodic solution with the same period. The assumption of a time-depending contact rate $\beta = \beta(t)$ is appropriate also for human hosts, reflecting annual periodicity in climate, mobility of vectors, activity in agriculture etc. Again, computer simulations show that every solution of the problem with constant N and periodic β rather quickly approaches a periodic solution with the same period. For given $\beta = \beta(t)$, the periodic solutions are given by an equation

$$\varphi(t) = \beta(t) f(\bar{w}(t)) \tag{23}$$

$$\bar{w}(t) = \frac{\int_0^\infty \exp\left\{\int_{-a}^0 \Gamma(s)\varphi(t+s)\,ds - M(a)\right\} \int_{-a}^0 \Gamma_z(s)\varphi(t+s)\,ds\,da}{\int_0^\infty \exp\left\{\int_{-a}^0 \Gamma(s)\varphi(t+s)\,ds - M(a)\right\} da} \tag{24}$$

where Γ, Γ_z are functions closely related to the solution operator of the Riccati equation (11) (see forthcoming paper with K. Dietz).

A general assumption of the present model says that parasites act on the host independently. This hypothesis is somewhat unrealistic. However, a more general death rate $\mu(a) + \alpha(r)$, where now α is some function, leads to further mathematical problems and possibly less qualitative insight. If the function $\alpha(r)$ is a quadratic polynomial then instead of (10) one obtains a diffusion equation for the generating function, for general birth and death rates one has to work in sequence spaces. The discussion of this approach is deferred to a future publication.

References

[1] May, R.M., Anderson, R.M., Population dynamics of infectious diseases II, Nature 280, 455-461 (1979).

[2] Hadeler, K.P., Dietz, K., Nonlinear hyperbolic partial differential equations for the dynamics of parasite populations, Computers and Math. with Appl. 9, 415-430 (1983).

[3] Hadeler, K.P., An integral equation for helminthic infections: Stability of the non-infected population, Proc.Vth Int. Conf. on Trends in Theory and Practice of Nonl. Diff. Equ. 1982 (Ed. V. Lakshmikantham), Marcel Dekker 1984.

[4] Hadeler, K.P., Integral equations for infections with discrete parasites: Hosts with Lotka birth law. In: Autumn Course in Mathematical Ecology, Trieste 1982 (Ed. S. Levin, T. Hallam).

[5] Hadeler, K.P., Dietz, K., An integral equation for helminthic infections: Global existence of solutions. In: Conference Proceedings "Recent Trend in Mathematics", Reinhardsbrunn 1982, Teubner Texte zur Mathematik 50, Teubner Verlag Leipzig 1982/83.

[6] Hadeler, K.P., Hysteresis in a model for parasitic infection. Proceedings "Numerical methods for bifurcation problems", Dortmund 1983 (Eds. H.D. Mittelmann, T. Küpper) Birkhäuser Verlag.

[7] Hadeler, K.P., Dietz, K., Population dynamics of killing parasites which reproduce in the host. J. Math. Biol., submitted 1983.

[8] Karlin, S., Tavaré, S., Linear birth and death processes with killing., J. App. Prob. 19, 477-487 (1982).

Lehrstuhl für Biomathematik
Universität Tübingen
Auf der Morgenstelle 28
7400 Tübingen
West Germany

STOCHASTIC EPIDEMICS AS POINT PROCESSES

Grace L. Yang

Department of Mathematics
University of Maryland
College Park MD 20742 USA

1. Introduction

This article deals with modeling an epidemic by utilizing the information of the times of occurrences and removals of the infectives, through a bivariate counting process. This approach incorporates the epidemic features into the model more flexibly than the Markov chain approach. It also provides a convenient framework for statistical analysis. After a brief discussion of the general epidemic model (a Markov Chain) we will present a point process model for epidemics and discuss some statistical inference problems.

Consider a finite homogeneous population of size $N+1$. The development of an epidemic of case by case transmissions in the population is often described in terms of the changes in numbers of susceptibles (S_t), infectives (I_t) and immunes (or removals, R_t) as a function of time t. At any time t, these three groups of people comprise the entire population, so that, $N+1 = S_t + I_t + R_t$.

The general epidemic model discussed in Bailey (1975) treats the epidemics $\{(S_t, I_t),\ t \geq 0\}$ as a two-dimensional Markov chain. The possible transitions at any time t are of the forms

$$(S_t, I_t) \rightarrow (S_t - 1, I_t + 1) \quad \text{and} \quad (S_t, I_t) \rightarrow (S_t, I_t - 1) \qquad (1)$$

with respective transition rates $\beta S_t I_t$ and γI_t, and the states $(S_t, 0)$, for $0 \leq S_t \leq N$ are the absorbing states. Model (1), known as the SIR model, is a stochastic version of the Kermack-McKendrick model. In terms of the epidemic phenomenon, Model (1) is very simple, yet it is mathematically complicated. For instance, so far there are no completely satisfactory ways of analyzing the expected values of S_t, I_t and R_t, the distribution of the duration of the epidemic. Many of the available results are of an asymptotic nature, such as letting $t \rightarrow \infty$. One important result is the existence of a threshold (Whittle, 1955 and other references in Bailey, 1975). That is, the distribution of $N+1-S_\infty$ (the final size of the epidemic) is asymptotically U-shaped or J-shaped depending on whether the relative removal rate $\rho = \gamma/\beta \leq N$ or $> N$.

Furthermore, if we then let $N \to \infty$, the probability of ultimate extinction of the infective (I_t) is one if $\rho \geq N$ and is (ρ/N) if $\rho < N$. Both the exact and the approximate (as $N \to \infty$) distribution of the final size of the epidemic have been derived (Bailey 1953, Nagaev and Startsev 1970, Von Bahr and Martin-Löf 1980). These distributions depend only on ρ, rather than on β and γ separately. Thus, one cannot use this distribution to estimate β and γ separately (p. 114, Bailey 1975).

2. Modelling

Suppose the epidemic is being monitored starting at some time, say t_0, then a first event takes place changing an individual from a susceptible to an infective state. We set $t_0 = 0$, so that initially

$$S_0 = N, \quad I_0 = 1 \quad \text{and} \quad R_0 = 0. \tag{2}$$

For easy interpretation, we introduce positive random variables U_i and $U_i + V_i$, for $i = 0, \ldots, N$, U_i and $U_i + V_i$ defined on a fixed probability space (Ω, A, P). Here U_i represents the time of infection, and $U_i + V_i$ the time of removal of the ith individual if he ever gets infected. For every $t \geq 0$, let

$$\begin{aligned} M_{i1}(t) &= I_{[U_i \leq t]}, \\ M_{i2}(t) &= I_{[U_i + V_i \leq t]}, \quad \text{and} \\ C_i(t) &= I_{[U_i \leq t < U_i + V_i]} \end{aligned} \tag{3}$$

where $0 \leq i \leq N$ and I_A is the indicator of the set A. For notational convenience, we set $M_{01}(t) \equiv 0$. At any time t, the status of the ith individual can be described by $M_{i1}(t)$, $M_{i2}(t)$ and $C_i(t)$, where

$$\begin{aligned} M_{i1}(t) &= 1 \quad \text{if the } i^{th} \text{ individual is infected} \\ &\ 0 \quad \text{if he is still a susceptible} \\ M_{i2}(t) &= 1 \quad \text{if he is removed by the time } t \\ C_i(t) &= 1 \quad \text{if he is an infective in circulation (before removal).} \end{aligned} \tag{4}$$

At time t the basic data consists of

$$\min(V_0,t),\ I_{[V_0\le t]} \text{ and}$$

$$\min(U_i,t),\ I_{[U_i\le t]},\ \min(U_i+V_i,t) \text{ and } I_{[U_i+V_i\le t]}, \tag{5}$$

for $i=1,\dots,N.$

Let A_s be the σ-field generated by the collection of random variables in (5) for all $t\in[0,s]$. The collection $M = \{M_{i\ell}(t),\ i=0,\dots,N,\ \ell=1,2,\ t\ge 0\}$ forms a multivariate counting process with respect to $\{A_t,\ t\ge 0\}$. By the construction in (3), the sample paths of the component processes are right continuous with initial values 0 and have at most a single jump of size +1. We further assume that no two component processes can jump at the same time. We refer the reader to Aalen (1977, 1978) and Jacobsen (1982) for rigorous analysis of counting processes.

For the counting processes M in (3), we have $X_0=0$ and the sum

$$X_t = M_{02}(t) + \sum_{i=1}^{N}\sum_{\ell=1}^{2} M_{i\ell}(t), \qquad t\ge 0 \tag{6}$$

is again a counting process. Instead of using (S_t,I_t), the epidemic may be described by the number of infectives and number of removals in $(0,t]$, $(M_1(t),M_2(t))$, where

$$M_1(t) = \sum_{i=1}^{N} M_{i1}(t) \quad \text{and} \quad M_2(t) = \sum_{i=0}^{N} M_{i2}(t). \tag{7}$$

For $n\ge 1$, define T_n as

$$\begin{aligned} T_n &= \{\inf t>0 : X_t = n\} \\ &= \infty, \quad \text{if } X_t<n, \text{ for all } t>0. \end{aligned} \tag{8}$$

The process $X=\{X_t,\ t\ge 0\}$ makes a jump whenever some component of M jumps. Thus T_n coincides either with U_i or U_i+V_i for some i. To determine the exact correspondence of T_n and U_i or U_i+V_i, we need to specify the jump type $J_n(i,\ell)$, where

$$J_n(i,\ell) = \begin{cases} 1 & \text{if the jump at } T_n \text{ occurs in } M_{i\ell} \\ 0 & \text{otherwise.} \end{cases} \tag{9}$$

Since no two components processes jump at the same time, we have $\sum_i \sum_j J_n(i,\ell) = 1.$

The sequence $\{(T_n,J_n(i,\ell)),\ n\geq 1,\ i=0,\ldots,N,\ \ell=1,2\}$ completely determines $M=\{M_{i\ell}(t),\ t\geq 0,\ i=0,\ldots,N,\ \ell=1,2\}$. The epidemic model is to be constructed based on $\{(T_n,J_n(i,\ell))\}$. We will make the following assumptions. Assumption 1 introduces a contagion function f to describe the disease. Assumption 2 relates the contagion function f to the intensities of the counting process M and the distributions of $\{(T_n,J_n(i,\ell)\}$.

Assumption 1. There is given a contagion function f of the disease under consideration. The function $f(x,y)$ is defined on $R^+\times R^+\to R^+ = [0,\infty)$, and it is jointly measurable and nonnegative. For every fixed x, $f(x,y)$ is right continuous in y for $y\geq x$, $f=0$ if $y<x$, and integrable in y.

The contagion function f can be interpreted as follows. Suppose the ith individual becomes infective at time U_i. The function $f(U_i,t)$ is the amount of pathogenic material produced by this individual at time t. A general shape of $f(U_i,t)$ is given in Figure 1 below.

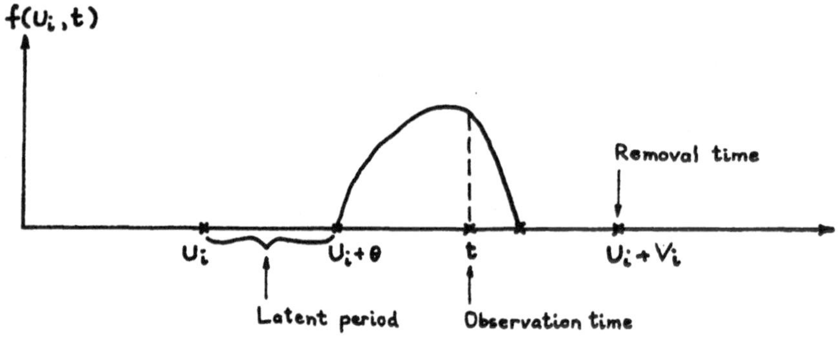

Figure 1. An example of contagion function $f(U_i,t)$.

The contagion function f can be evaluated at any t based only on the information available on U_i (prior to t). An individual can only contribute pathogenic material while still in circulation as indicated by $C_i(t) = I_{[U_i\leq t<U_i+V_i]}$ in (3). Thus, the actual amount he contributes at time t is given by

$$f(U_i,t)C_i(t). \tag{10}$$

Let $0 = U_0 < U_{(1)} < U_{(2)} \cdots < U_{(M_1(t))} \leq t$ be the ordered occurrence times of the infective cases in $(0,t]$. Then the sum $G(t)$, defined by

$$G(t) = \sum_{i=0}^{M_1(t)} f(U_{(i)},t)C_{(i)}(t) = \sum_{i=0}^{N} f(U_i,t)C_i(t) \tag{11}$$

may be interpreted as a measure of pathogenicity in the environment at time t.

Let us consider some simple examples of f under Assumption 1.

Example 1. If there is no removal, we set $V_i = \infty$, for $0 \leq i \leq N$. Let

$$f(U_i,t) = \begin{cases} 1 & \text{for } U_i \leq t \\ 0 & \text{otherwise.} \end{cases} \tag{12}$$

The sum in (11) reduces to $M_1(t)+1$, the number of infectives at time t including the case at time 0.

Example 2. Let f be defined as in (12) and let V_i be a finite random variable. The sum in (11) is then equal to

$$I_t = \sum_{i=0}^{N} C_i(t), \tag{13}$$

the number of infectives in circulation at time t.

Example 3. Let θ be a positive constant representing the length of the latent period and V_i be a finite random variable. Let

$$f(U_i,t) = \begin{cases} 1 & \text{for } U_i + \theta \leq t \\ 0 & \text{otherwise.} \end{cases} \tag{14}$$

Then the sum $G(t)$ in (11) represents the number of infectious individuals at time t. It is different from the I_t in (13) which is the number of infectives at time t.

The contagion function f was studied in Yang (1968, 1972) in which θ and V are treated as nonrandom constants. Here we generalize it to allow for a random removal time. We further note that a deterministic model involving age dependent infectivity rate was considered by Kermack and McKendrick (1927).

The next assumption is concerned with the hazard for susceptibles and the rate of removals of the infectives. We shall put them in the framework of the counting processes. Thus we consider the modelling of the conditional distribution of T_n and $J_n(i,\ell)$ (see (9) and (8)) given the past history. Specifically, let

$$h_n = \{t_k, j_k(i,\ell),\ k=1,\dots,n;\ i=0,\dots,N,\ \ell=1,2\}. \tag{15}$$

The set h_n consists of a particular realization of jump times and jump types in the first n jumps of the process X given in (6). We set $t_0 = 0$ and $h_0 = \{t_0\}$. For $t > t_n$, we let

$$\begin{aligned}
G(t|h_n) &= \beta \sum_{i=0}^{N} f(u_i,t)C_i(t_n),\\
\mu_{i1}(t|h_n) &= \beta G(t|h_n)(1 - M_{i1}(t_n)),\\
\mu_1(t|h_n) &= \sum_{i=1}^{N} \mu_{i1}(t/h_n) = \beta G(t|h_n)(N - M_1(t_n)),\\
\mu_{i2}(t|h_n) &= \gamma C_i(t_n)\\
\mu_2(t|h_n) &= \gamma \sum_{i=0}^{N} C_i(t_n), \quad \text{and}\\
\mu(t|h_n) &= \mu_1(t|h_n) + \mu_2(t|h_n),
\end{aligned} \tag{16}$$

where parameters $\beta > 0$, $\gamma \geq 0$, $C_i(t)$ and $M_{i1}(t)$ are defined in (3).

Assumption 2. The conditional distribution of T_n given any arbitrarily fixed set h_n is given by

$$P[T_{n+1} > t \mid h_n] = \exp\{-\int_{t_n}^{t} (\mu_1(\tau|h_n) + \mu_2(\tau|h_n))\,d\tau\} \tag{17}$$

for $t > t_n$, $n = 1,2,\dots$, and on the event $[t_n < \infty]$. The jump probability for $J_{n+1}(i,\ell)$ is given by a multinomial distribution,

$$\begin{aligned}
&P[J_{n+1}(i,\ell) = j_{n+1}(i,\ell),\ \text{for } i=0,\dots,N,\ \ell=1,2 \mid h_n, t_{n+1}]\\
&= \prod_{i=0}^{N} [\beta G(t_{n+1}^-|h_n)(1-M_{i1}(t_n))]^{j_{n+1}(i,1)} [\gamma C_i(t_n)]^{j_{n+1}(i,2)} / \mu(t_{n+1}^-|h_n), \quad n \geq 0
\end{aligned} \tag{18}$$

where $j_{n+1}(i,\ell) = 0$ or 1 subject to $\sum_{i=0}^{N} \sum_{\ell=1}^{2} j_{n+1}(i,\ell) = 1$, and $J_n(0,1) = 0\ \forall\ n$, $J_0(i,\ell) = 0$.

In the epidemic model, the parameters β and γ are interpreted as infection and removal rates respectively. For simplicity, we let

$$\Lambda_1(t) = \mu_1(t|h_{X_t}), \qquad \Lambda_2(t) = \mu_2(t|h_{X_t}),$$
$$\Lambda(t) = \Lambda_1(t) + \Lambda_2(t). \tag{19}$$

Recall that X_t is the point process which counts the number of jumps in $(0,t]$. The functions $\Lambda_1(t^-)$ and $\Lambda_2(t^-)$ are the infection and removal intensities of the epidemic as described by $M_1(t)$ and $M_2(t)$.

The function $G(t|h_n)$ in (18) and implicitly in (17) is defined in terms of the u_i's and the v_i's. Clearly, it can be expressed in terms of the t_n's and $j_n(i,\ell)$'s, since given h_n, the u_i and u_i+v_i can be identified with the corresponding t_n and $j_n(i,\ell)$. However, it is awkward to make this correspondence explicit.

3. The Likelihood and Estimation of Parameters

Let the observation interval be $[0,t]$. It follows from (18) and (19) that the joint density of $T_1,\ldots,T_{X_t}$, $J_1(i,\ell),\ldots,J_{X_t}(i,\ell)$ and X_t, for $X_t > 0$, is given by

$$\prod_{n=1}^{M_t} \left\{ \prod_{i=0}^{N} [\beta G(t_n^-|h_{n-1})(1-M_{i1}(t_{n-1})]^{j_n(i,1)} [\gamma C_i(t_{n-1})]^{j_n(i,2)} \right\} \times$$
$$\exp\{-\int_0^t \Lambda(\tau)d\tau\}dt_1\cdots dt_{X_t}, \tag{20}$$

for $0 < t_1 < t_2 < \ldots < t_{X_t} \le t$, $j_n(i,\ell) = 0$ or 1 subject to $\sum_{i=0}^{n} \sum_{\ell=1}^{2} j_n(i,\ell) = 1$ as in (18). For $X_t = 0$, we have

$$P[X_t = 0] = \exp\{-\int_0^t \Lambda(\tau)d\tau\}. \tag{21}$$

When f is the function defined in Example 2, the joint density in (20) reduces to the density of $T_1,\ldots,T_{X_t}$, the J's and X_t of the general epidemic model (1). It is given by

$$\prod_{n=1}^{X_t} \beta^{\delta_n}\gamma^{1-\delta_n} \prod_{i=0}^{N} [G(t_n^-|h_{n-1})(1-M_{i1}(t_{n-1}))]^{j_n(i,1)} [C_i(t_{n-1})]^{j_n(i,2)} \times$$
$$\exp\{-\int_0^t [\beta(N-M_1(\tau))+\gamma]I_\tau d\tau\}dt_1\cdots dt_{X_t}, \quad \text{if } X_t > 0, \tag{22}$$

for $0 < t_1 < t_2 < \ldots < t_{X_t} < t$, $\delta_n = \sum_{i=0}^{N} j_n(i,1) = 1$ or 0, and I_τ as given in (1).

For $X_t = 0$, we have

$$P[X_t = 0] = \exp\{-(\beta N + \gamma)t\}. \tag{23}$$

From the joint density (20) and (21) we see that the likelihood for $\beta > 0$ and $\gamma > 0$ is proportional to L,

$$\begin{aligned} L &= \prod_{n=1}^{X_t} \beta^{\delta_n} \gamma^{1-\delta_n} \exp\{-\int_0^t [\Lambda_1(\tau) + \Lambda_2(\tau)]d\tau\}, \quad \text{if} \quad X_t > 0 \\ &= \exp\{-\int_0^t \Lambda(\tau)d\tau, \quad \text{if} \quad X_t = 0. \end{aligned} \tag{24}$$

The maximum likelihood estimates $\hat{\beta}$ and $\hat{\gamma}$ for β and γ are given by

$$\hat{\beta} = \frac{M_1(t)}{\frac{1}{\beta}\int_0^t \Lambda_1(\tau)d\tau} \quad \text{and} \quad \hat{\gamma} = \frac{M_2(t)}{\frac{1}{\gamma}\int_0^t \Lambda_2(\tau)d\tau}. \tag{25}$$

The denominators are independent of β and γ because of the definitions of Λ_1 and Λ_2 (see (19)). In view of (3) and (7), we have $\sum_{n=1}^{X_t} \delta_n = M_1(t)$ and $\sum_{n=1}^{X_t} (1-\delta_n) = M_2(t)$.

It would be very difficult to get the exact distribution of the estimates. We shall instead discuss an asymptotic approach to finding the distributions of $\hat{\beta}$ and $\hat{\gamma}$ as N tends to infinity.

Since we are dealing with a finite population, the number of infective cases (not including the initial case) and removals are bounded,

$$EM_k(t) \le N + 1, \quad \text{for} \quad k = 1,2 \qquad \text{(cf. (7))}.$$

The processes $M_1(t)$ and $M_2(t)$ are submartingales with respect to the σ-fields $\{A_t,\ t \ge 0\}$. We apply the statistical method developed in Aalen (1978) for counting processes; the method was used by Becker, 1981, to study the household infectiousness.

Let

$$Z_k(t) = M_k(t) - \int_0^t \Lambda_k(\tau)d\tau, \qquad k = 1,2. \tag{26}$$

Then the $Z_k(t)$'s are orthogonal martingales with respect to A_t. A proof of (26) for general multivariate counting processes may be found in Jacobsen (p. 70).

From (25), we write $\hat{\beta} - \beta$ and $\hat{\gamma} - \gamma$ in terms of $Z_k(t)$ as

$$\hat{\beta} - \beta = \frac{1}{B_{1t}}[M_1(t) - \int_0^t \Lambda_1(\tau)d\tau] = \frac{Z_1(t)}{B_{1t}}$$

and (27)

$$\hat{\gamma} - \gamma = \frac{1}{B_{2t}}[M_2(t) - \int_0^t \Lambda_2(\tau)d\tau] = \frac{Z_2(t)}{B_{2t}},$$

where $B_{1t} = \frac{1}{\beta}\int_0^t \Lambda_1(\tau)d\tau$ and $B_{2t} = \frac{1}{\gamma}\int_0^t \Lambda_2(\tau)d\tau$.

Theorem 1.3 (p. 163) in Jacobsen gives conditions for a multivariate martingale sequence $(M_N)_{N\geq 1}$ to converge to a normal process. One of the conditions is the existence of two nondecreasing, continuous functions ϕ_1 and ϕ_2 on $[0,\infty) \to [0,\infty)$ with $\phi_k(0) = 0$ such that $\forall\varepsilon > 0$, $t \geq 0$, $k = 1,2$,

$$P_N[|\int_0^t \Lambda_k(\tau)d\tau - \phi_k(t)| > \varepsilon] \to 0 \quad \text{as} \quad N \to \infty. \tag{28}$$

In terms of B_{1t} and B_{2t} the probability in (28) is

$$P_N[|\beta B_{kt} - \phi_k(t)| > \varepsilon] \to 0.$$

If the conditions of Theorem 1.3 are met, then as $N \to \infty$

$$\begin{pmatrix} B_{1t}^{1/2}(\hat{\beta}-\beta) \\ B_{2t}^{1/2}(\hat{\gamma}-\gamma) \end{pmatrix} \xrightarrow{L} N\left(\begin{pmatrix}0\\0\end{pmatrix}, \Gamma\right) \quad \text{where} \quad \Gamma = \begin{pmatrix}\beta & 0 \\ 0 & \gamma\end{pmatrix}. \tag{29}$$

In general it is not easy to verify the conditions of Theorem 1.3 for Λ_1 and Λ_2. Clearly the results hold for the simple epidemic model discussed in Example 1 with $\gamma = 0$. Another related result is given in Kurtz (1981) and Wang (1977). They obtained the asymptotic normality for (S_t, I_t) in model (1) with transition rates redefined as $\beta I_t S_t/N$ and γI_t. Based on their results, the asymptotic distribution in (29) can be established. For the general model, the problem is being investigated.

References

Aalen, O. (1977). Weak convergence of stochastic integrals related to counting process. Z. Wahrsheinlichkeitstheorie und verw. Gebiete 38, 261-277.

Aalen,). (1978). Nonparametric inference for a family of counting processes. Ann. of Statist. 6, 701-726.

Bailey, N. T. J. (1953). The total size of a general stochastic epidemic. Biometrika 40, 177-185.

Bailey, N. T. J. (1975). The Mathematical Theory of Infectious Disease. Griffin, London.

Becker, N. (1981). The infectiousness of a disease within households. Biometrika 68, 133-141.

Kermack, W. O. and McKendrick, A. G. (1927). A contribution to the mathematical theory of epidemics. Proc. Royal Soc. London Ser. A, 115, 700-721.

Kurtz, T. (1981). Approximation of population processes. CBMS-NSF Regional Conference Series in Applied Mathematics. 36, SIAM.

Jacobsen, M. (1982). Statistical Analysis of Counting Processes. Springer-Verlag, New York.

Nagaev, A. V. and Startsev, A. N. (1970). Asymptotic analysis of a stochastic epidemic model. Theor. Verojat. Primen. 15, 97-105.

Von Bahr, B. and Martin-Löf, A. (1980). Threshold limit theorems for some epidemic processes. Adv. Appl. Prob. 12, 319-349.

Wang, F. J. S. (1977). Gaussian approximation of some closed stochastic epidemic models. J. Appl. Prob. 14, 221-231.

Whittle, P. (1955). The outcome of a stochastic epidemic - a note on Bailey's paper. Biometrika 42, 116-122.

Yang, G. L. (1968). Contagion in stochastic models for epidemics. Ann. Math. Statist. 39, 1863-1889.

Yang, G. L. (1972). Empirical study of a non-Markovian epidemic model. Math. Biosci. 14, 65-84.

A COMPUTER SIMULATION PROGRAM FOR THE ASSESSMENT OF MULTISTATE EPIDEMIOLOGICAL MODELS OF INFECTIOUS DISEASES

J. Božikov, B. Cvjetanović and Gj. Deželić

Andrija Štampar School of Public Health, Medical School,
University of Zagreb, Zagreb, Yugoslavia

SUMMARY

The computer simulation of multistate epidemiological models is presented and the conditions for performing simulation experiments are discussed. A general purpose interactive software package for continuous simulation developed on this basis is implemented. Successful tests were made with several models of infectious diseases. The need for the introduction of sensitivity analysis into simulation experiments is emphasized.

1. INTRODUCTION

Computerized models in epidemiology are nowadays an often-used tool for studying various epidemiological phenomena. They served to perform simulation experiments in various fields, as seen in the everyday growing literature. Among them several epidemiological models of various bacterial and viral diseases (1) as well as parasitic diseases (2) have been developed. The design of many of the infectious diseases models started from the picture of the natural history of the disease, consisting of a flowchart showing compartments linked by streams of the flow of the population through the compartments. This kind of modeling has been adopted e.g. in the papers of one of us (3).

Much of the modeling work in epidemiology today is performed by developing multistate (or multicompartment) models with appropriate software for computer simulation. This software consists of specialized programs which fit a certain disease problem or type of disease at best. Such computer programs are rigid in structure and use, since changes in the model demand changes in the symbolic code, recompilation of program elements and production of new object code suitable for execution.

During the modeling procedure it is often necessary to introduce new compartments and the interactions among them. Since the epidemiologist tries to add piece-by-piece the necessary epidemiological, biological and clinical knowledge to the model, and to transit from more heuristic constructions to more realistic situations, one has to develop software suitable for a selfstanding work by the epidemiologist himself. The software needed for such model development should allow interactive processing and should avoid use of computer programing code. The introduction of changes should be easy and the response of simulation results should be fast.

In the present paper a general program is described which permits easy implementation of new epidemiological multistate models and performance of simulation experiments. The program has been tested with various existing models and the results are shown for the model of typhoid fever (4) and shigellosis (5).

2. SIMULATION METHOD

The theoretical basis of the simulation method has been described in detail elsewhere (6). The equations describing the rate of change of N state variables x_i can be written as

$$dx_i/dt = b_i + \sum_{j=1}^{N} F_{ij}(x_j) + \sum_{k=1}^{M} G_{ik}(y_k), \quad i=1,2,\dots,N \tag{1}$$

where b_i is the base rate of change of x_i (trend), F_{ij} is the interaction of x_j on dx_i/dt in time $(t-D_{ij})$, G_{ik} is the interaction of y_k on dx_i/dt in time $(t-D_{ik})$, D_{ij} and D_{ik} are the delays of the interactions of x_j on dx_i/dt and y_k on dx_i/dt, respectively. Here y_k, $k=1,2,\dots,M$ are auxiliary variables which stand for interactions between more than two variables.

The numerical computation of Eq. (1) is performed by using the simple Euler approximation:

$$x_i(t+\Delta t) = x_i(t) + \Delta t \times \{dx_i(t)/dt\}. \tag{2}$$

The computer simulation is performed as a continuous process, i.e. the state variables are interacting in small but finite equal time increments, and the computations are performed in each of these increments.

A general software package, the Quick Interactive Simulation System (SISSY) developed by J. Retti (7) and implemented in its SISSY3 version written in FORTRAN, was used in our work. A computer graphics feature has been added to the package in our laboratory. The simulations were performed on a UNIVAC 1100/42 computing system with Hewlett-Packard 2648A Graphics.

3. RESULTS

The Zagreb version of the SISSY software package was tested on several models and the results with the model of shigellosis are shown as examples.

The shigellosis model (5) was taken as a 8-compartment model (the vaccination compartments left out) with the classes: susceptible, incubating, clinical sick, subclinical sick, short-term carriers, long-term carriers, resistant, deaths. For the first seven classes the age groups have been defined as state variables. Each class contains seven age groups as subclasses. A total of 49 state variables reflecting the age-group structure, and the state variables "deaths" and "time" complete the model.

The interaction parameters, the so-called intercompartmental transfer rates, have been taken from epidemiological knowledge as described (5). The force of infection (3) was taken as a time dependent interaction parameter, and its values have been chosen arbitrarily in order to render suitable simulation results which match at best the observed data.

The shigellosis model was tested with the bacillary dysentery data for Roumania (8,9) which comprise annual incidence rates in dependence of standard age groups (from zero to 65+ years) as well as the seasonal distribution of bacillary dysentery cases in 1976. The simulation results are shown in Fig. 1 and 2. Fig. 1 is obtained by taking the force of infection as a function of time defined by an analytical expression of type (10): $F=F_0\exp\{A\times\sin(t+B)\}$, where F_0, A and B are constants and F_0 is different for each age group. Our software allows easy implementation of such function into the computerized model. Fig. 1 shows the incidence of cases in two most sensitive age groups (0 years, 1-4 years) and in the total population during two years. If the F is presented by a piecewise linear function of time, which is also possible with SISSY, the incidence of cases in the total population is obtained as shown in Fig. 2. Here the shape of the maximum is better adjusted to the natural situation.

The simulation of the typhoid fever model showed excellent agreement with the results obtained earlier (4) with a specially designed computer program. It was possible to obtain single and periodic mass vaccination curves on the same figure by simulating two separate models by SISSY at the same time. This shows the great potential of this software package in epidemiological simulation.

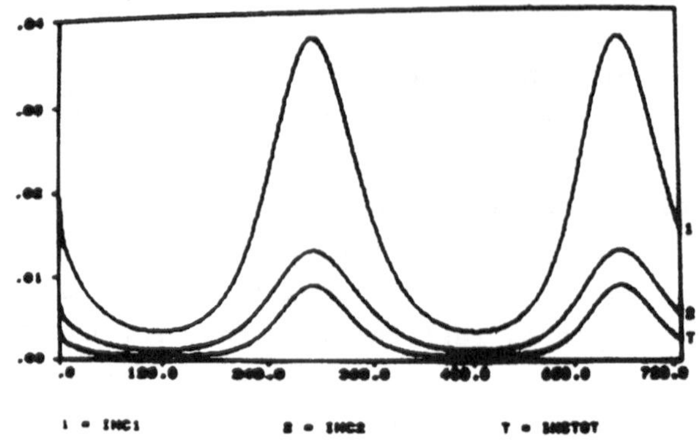

Figure 1. Simulation results obtained by taking the force of infection as an analytical function of time during two years (starting from october);
abscissa - time (in days), ordinate - daily incidence of bacillary dysentery cases per 1000 population;
curves 1 and 2 are incidence rates per age groups 1 (0 years) and 2 (1-4 years), respectively, T denotes incidence rate in the total population

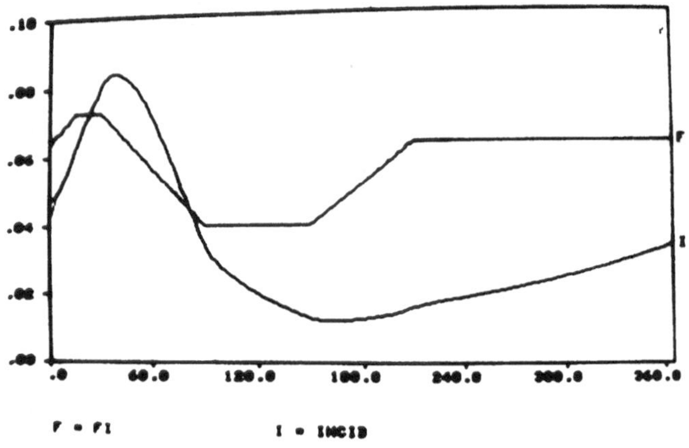

Figure 2. Simulation results obtained with the force of infection as a piecewise linear function of time during one year (starting from july);
abscissa - time (in days), ordinate - (i) daily incidence rates of bacillary dysentery per 10000 population (I), and (ii) force of infection (F)

4. DISCUSSION

In designing multistate epidemiological models one frequently has to start with predominanthly hypothetical constructions since the real system elements are often not well known. From the practical point of view, in epidemiological studies accurate prediction of events is generally less important than the rapid estimation of possible effects of planned actions. Computer simulation experiments with epidemiological models are, therefore, an important tool for the refinement of models and are needed to bring them nearer to the real situation.

The simulation software package SISSY, with its graphic output feature, proved to satisfy well the needs of epidemiological modeling, by allowing simple changes in the model during the interactive session. It is easy to implement a model by defining the state and auxiliary variables and functions in the model, and it is possible to change both the model parameters as well as its structure.

The simulation algorithms give satisfactory results when compared with results from specially developed programs. The computing time, both for model implementation and for its simulation was satisfactory, depending of the complexity of the model and the time increments.

The functions in SISSY can be defined in various manner: as algebraic expressions or in a tabular form. Introduction of stochastic variables is also allowed. It is thus possible to design models satisfying a rather broad range of structural types.

There is an important extension which has to be made in modeling with SISSY: one has to introduce algorithms for sensitivity analysis which appears to be so important in the study of multistate epidemiological models (11). This kind of analysis gives insight into the variation of model variables and should be regarded as an essential step in taking epidemiological modeling as a tool in real-life situations. Investigations in this direction are planned in the future.

Acknowledgment. The authors are indebted to Mr. J. Retti, Institut für medizinische Kybernetik der Universität Wien, for making the SISSY software package available for implementation.

REFERENCES:

1. Bailey, N.T.J., The Mathematical Theory of Infectious Diseases and its Applications, Ch. Griffin & Co. Ltd, London and High Wycombe, 1975.
2. Anderson, R.M. (Editor), The Population Dynamics of Infectious Diseases: Theory and Applications, Chapman and Hall, London-New York, 1982.

3. Cvjetanović, B., Grab, B., Uemura, K., Dynamics of Acute Bacterial Diseases, Supplement No. 1 to Vol. 56 of the Bull. WHO, Geneva, 1978.

4. *loc. cit.* 3, pp. 45-63.

5. Božikov, J., Deželić, Gj., Cvjetanović, B., Computerized Epidemiometric Model of Shigellosis and Its Use in Assessing Potential Usefulness of New Tools for Disease Control, Proceedings Medical Informatics Europe 82, Dublin March 21-25, 1982, Springer-Verlag, Berlin-Heidelberg-New York, 1982, pp. 611-617.

6. Deželić, Gj., Božikov, J., Development of Computerized Epidemiological Models of Infectious Diseases, Proceedings 4th World Congress on Medical Informatics MEDINFO 83, Amsterdam August 21-26, 1983, North-Holland, Amsterdam 1983, pp. 1218-1221.

7. Retti, J., SISSY, Ein schnelles interactives Simulationssystem, Berichte der österreichischen Studiengesellschaft für Kybernetik, Wien, 1979.

8. World Health Stat. Ann., Vol. 1 (1978), WHO, Geneva, 1978, pp. 394-395.

9. World Health Stat. Ann., Vol. 2 (1978), WHO, Geneva, 1978, p. 137, p. 195.

10. *loc. cit.* 3, p. 87.

11. Bailey, N.T.J., Duppenthaler, J., Sensitivity Analysis in the Modelling of Infectious Disease Dynamics, J. Math. Biology 1980; 10: 113-131.

MATHEMATHICAL MODEL OF HEPATITIS B

N. Delimar, M. Košiček, B. Cvjetanović and B. Špoljarić

Imunološki zavod, Zagreb, Yugoslavia

SUMMARY

A multistate epidemiological model is constructed on the ground of natural history of the disease. The transfers from one class to another are governed by the mathematical relationship between classesof population of the dynamics of hepatitis B is expressed in the system of equations. All rates of transfers are calculated on a daily basis. The model is used for simulating the natural course of the infection as well as the effects of various public health interventions, such as passive and active immunization and/or application of appropriate sanitary and hygienic measures. The model is evaluated through above simulations of actual and hypothetical situations.

INTRODUCTION

Multistate epidemiological model with (1) and without age structure (2) is constructed on the grounds of natural history of the hepatitis B virus (3,4).

The model was used for simulation of endemic and epidemic situations. It was applied for simulating the natural course of infection as well as the effects of various public health interventions, such as passive and active immunization and/or application of appropriate sanitary and hygienic measures.

The model is evaluated through the above simulations of actual and hypothetical situations.

MATHEMATICAL MODEL

We started from the assumption that the dynamics of the natural course of virus hepatitis B corresponds to Fig. 1 (5,6).

The following variables are introduced:

$X_i^k = X_i^k\ (t)$ i-th epidemiological class of k-th age group at time t

$X^k = \sum_{i=1}^{11} X_i^k$ k-th age group

$X_i = \sum_{k=1}^{N} X_i^k$ i-th epidemiological class

$X = \sum_i \sum_k X_i^k$ population

N = No. of age groups, $N \begin{cases} = 1, & \text{model without age structure} \\ > 1, & \text{model with age structure} \end{cases}$

B = birth rate, $0 < B < 1$

D^k = death rate of k-th age groups; $0 < D^k < 1$, $X_0 = B \cdot X$ newborn

L_i = duration of stay in i-th epidemiological class

$XI_i^k\ (t)$ = No. of people who enter the i-th epidemiological class of k-th age group at time t

$XO_i^k\ (t)$ = No. of people who leave i-th epidemiological class of k-th age group at time t

RI^k = force of infection, is the probability of the susceptible person to get the infection when in contact with the infected person.

The transfer from one epidemiological class to the other is determined by coefficients of transfer. Coefficient of transfer from i-th to j-th epidemiological class of k-th age group is defined as follows:

$$R_{ij}^k = P_{ij}^k / L_i; \quad i,j = 0,1,\dots,13; \quad k = 1,\dots,N$$

where P_{ij}^k is the probability of transfer from i-th to j-th epidemiological class of k-th age group.

Coefficient R^k_{23} characterizes the spreading power of the infection and is defined as follows:

$$R^k_{23} = RI^k \, (X_4 + X_5 + X_6 + X_7 + X_8) \, / \, X$$

Transfers from class X_0 are defined as follows:

$$R^k_{01} = X_9/X, \quad R^k_{02} = 1-(R^k_{01} + R^k_{03} + R^k_{0.11}); \quad k = 1$$

$$R^k_{01} = 0, \quad R^k_{02} = 0, \quad k = 2,3,\ldots,N$$

$$R^k_{03} = (X_4 + X_5 + X_6 + X_7 + X_8) \, / \, X, \quad k = 1$$

$$R^k_{03} = 0, \quad k = 2,3,\ldots,N$$

$R_{0.11}$ depends on the immunization programme and so do coefficients $R^k_{2.10}$, $R^k_{2,11}$, $R^k_{3.11}$.

The dynamics of the disease is completely Jefined with epidemiological classes, duration of stay in the epidemiological class and coefficients of transfer.

Mathematically this dynamics is described with the following system of differential equations:

$$
\begin{aligned}
\dot{X}_1 &= R_{01}X_0 - (R_{12} + R) \cdot X_1 \\
\dot{X}_2 &= R_{02}X_0 + R_{12}X_1 + R_{72}X_7 + R_{92}X_9 + R_{10.2}X_{10} RR_{11.2}X_{11} - (R_{23} + R_{2.10} + R_{2.11} + R) \cdot X_2 \\
\dot{X}_3 &= R_{03}X_0 + R_{23}X_2 - (R_{34} + R_{3.11} + R) \cdot X_3 \\
\dot{X}_4 &= R_{34}X_3 - (R_{45} + R_{46} + R) \cdot X_4 \\
\dot{X}_5 &= R_{45}X_4 - (R_{57} + R_{58} + R_{59} + R) \cdot X_5 \\
\dot{X}_6 &= R_{46}X_4 - (R_{67} + R_{68} + R_{69} + R_{6.12} + R) \cdot X_6 \qquad (1) \\
\dot{X}_7 &= R_{57}X_5 + R_{67}X_6 - (R_{72} + R_{78} + R_{79} + R_{7.12} + R) \cdot X_7 \\
\dot{X}_8 &= R_{58}X_5 + R_{68}X_6 + R_{78}X_7 - (R_{89} + R_{8.12} + R) \cdot X_8 \\
\dot{X}_9 &= R_{59}X_5 + R_{69}X_6 + R_{79}X_7 + R_{89}X_8 - (R_{92} + R) \cdot X_9 \\
\dot{X}_{10} &= R_{2.10}X_2 + R_{11.10}X_{11} - (R_{10.2} + R) \cdot X_{10} \\
\dot{X}_{11} &= R_{0.11}X_0 + R_{2.11}X_2 + R_{3.11}X_3 - (R_{11.2} + R_{11.10} + R) \cdot X_{11} \\
\dot{X}_{12} &= R_{6.12}X_6 + R_{7.12}X_7 + R_{8.12}X_8
\end{aligned}
$$

$X_{13} = D X - \dot{X}_{12}$,

for every $k = 1,2,\ldots,N.$, and $R = \dot{X}_{13}/X$

The above index k is omitted in order to simplify writing the equations and the equations are identical for every k.

Transfer to a higher age group is given approximately with the following iteration:

$$X^1 = X_0 \cdot (1-D^1)$$
$$X^k = X^{k-1}\ (1 - D^{k-1}), \quad k = 2,3,\ldots,N. \qquad (2)$$

To determine the No. of people who at time t enter and leave the i-th epidemiological class of k-th age group we used the following algorithm:

$$X_i^k\ (t) = X_i^k\ (t-1) + XI_i^k\ (t) - XO_i^k\ (t)$$
$$XI_i^k\ (t) = X_i^k + (\sum_j R_{ij}^k + R) \cdot X_i^k \qquad (3)$$
$$XO_i^k\ (t) = XI_i^{k-M_i}\ (t - L_i) \cdot \prod_{l=0}^{M_i} (1 - D^{k-1})$$

where M_i is the No. of age groups passed through in time t.

The proof of algorithm (3) follows directly from (1) and (2).

The existence of solving system (1) follows from theorems on the existence of solving the system of differential equations (Epidemiological classes X_i^k: $t \rightarrow X_i^k(t)$ are derivable functions of time)

For a sufficiently small period of time Δt we can make a discretization of system (1) which results in a system of finite differences.

$$\Delta X_i^k = \Delta t \cdot f_i (X_0, X_1^k, \ldots, X_{13}^k) \quad i = 1,2,\ldots,13; \quad k = 1,2,\ldots,N \qquad (4)$$

where $f_i(X_0, X_1^k, X_2^k, \ldots, X_{13}^k)$ are the right side of equation (1). System (4) is solved iteratively, if the initial value is known at time t (Euler's method for $\Delta t = 1$).

$$X_i^k(t+1) = X_i^k(t) + f_i\ (X_0, X_1^k, X_2^k, \ldots, X_{13}^k), \quad X_i^k(t_0) = \text{known} \qquad i = 1,2,\ldots,13$$

$k = 1,2,\ldots,N$

APPLICATION OF THE MODEL IN SIMULATIONS

Changing the value of epidemiological parameters enables the simulation of different situations of the natural history of the disease especially of the effects of different health interventions and preventive measures. Simulation has a practical use in public health and health economics (1,7-9).

1) Simulation of the natural history of the disease

a) Stable endemicity - corresponds to equilibrium solution of system (1); X_i^k = const. for every i, every k. Since then $\dot{X}_i^k = 0 \quad \forall i, \ \forall k$ we obtain the equilibrium solution by solving the system.

$$f_i(X_0, X_1^k, \ldots, X_{13}^k) = 0 \qquad i = 1,2,\ldots,13 \qquad k = 1,2,\ldots,N \qquad (5)$$

Equilibrium solution exists only in the case of constant size of population (X = const. $\Rightarrow$ natality=mortality).

b) State of epidemics - appearance and size of the epidemic is simulated by changing the value of force of infection (FI).

2) Simulation of health interventions

a) Active immunization is simulated by different values of coefficient $R_{2.10}^k$. $R_{2.10}^k$ = cov. eff , where cov is immunization coverage and eff is the efficiency of the vaccine.

b) Passive immunization depends on the value of coefficient, $R_{0.11}^1$, $R_{2.11}^k$, $R_{3.11}^k$, $R_{i.11}^k$ = cov. eff, $i = 0,2,3$.

c) Preventive measures are simulated by decreasing the force of infection (RI).

The model also simulates different combinations of these three measures and thus gives the possibility to find the most optimal strategies.

The structure of the model which effects the natural history of chronic infection is an important fraction of infectious population classes which persist in spite of interventions for log periods of time. This by itself limits the effect of control measures which otherwise are very effective in the control of acute infectious diseases without carrier state.

The computer model for the dynamics of virus hepatitis B is worked out by the same authors for the HP-9845 and APPLE II computers.

REFERENCES

1. Cvjetanović, B., Grab, B., Uemura, K.: Dynamics of Acute Bacterial Diseases; Epidemiological Models and Their Application in Public Health. Bull. Wld Hlth Org. 56; Supplement 1, 1978.

2. Bailey, N.T.J. and Duppenthaler, J.: Sensitivity Analysis in the Modelling of Infectious Disease Dynamics. J. Math. Biology 10, 113-131, 1980.

3. Sobeslavsky, O.: Prevalence of markers of hepatitis B virus infection in various countries: A WHO Collaborative study. Bull.Wld Hlth Org. 58, 621-628,1980.

4. McCollum, R.W., Zuckerman, A.J. (1981): Viral Hepatitis; Report of a WHO Informal Consultation. Journal of Med. Vir. 8:1029, 1981.

5. Mims, C.A.: Vertical transmission of viruses. Microbiological Review, p. 267, 281, 1981.

6. Szmuness, W., Stevens, C.E., Harley, E.J., Zang, E.A., Oleszko, W.R., William, D.C., Sadovsky, R., Morrison, J.M., Kellner, A.: Hepatitis B vaccine; demonstration of efficacy in a controlled clinical trial in a high-risk population in the United States, New England Journal of Medicine 303:833-841, 1980.

7. Parsonage, M.: Assessment of direct and indirect costs of viral hepatitis. ICP/ESD 003(I)12. World Health Organization, Regional Office for Europe. Mimiographed document, August 1981.

8. Mulley, A.G., Silverstein, M.D., Dienstag, J.L.: Indications for use of Hepatitis B Vaccine,based on cost-effectiveness analysis. New England Journal of Medicine 307:644-651, 1982.

9. Cvjetanović, B., Grab, B., Dixon, H.: Epidemiological Models of Poliomyelitis and Measles. Bull. Wld Hlth Org., 60:405-422, 1982.

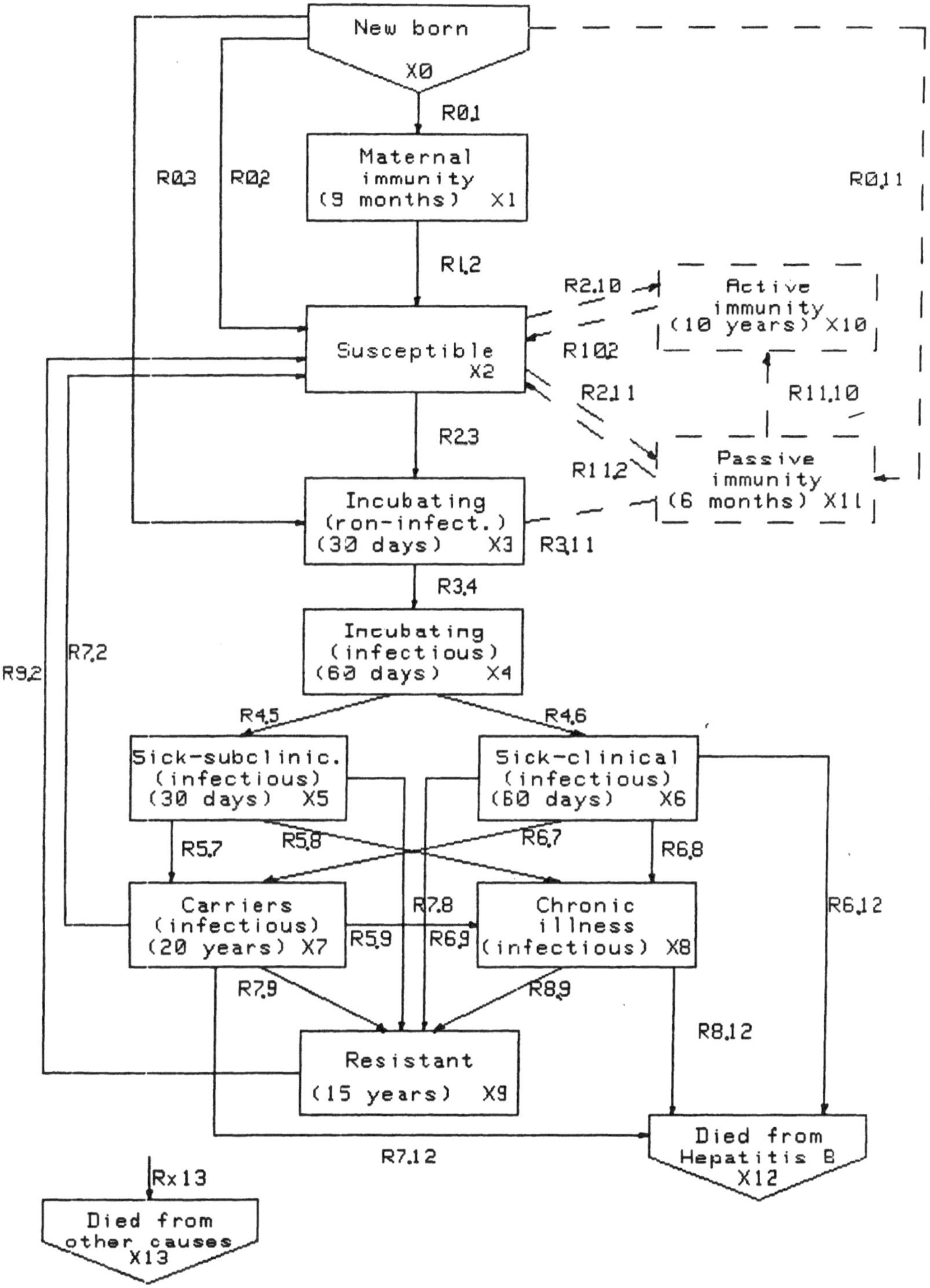

Fig. 1. Flow chart of the dynamics of Hepatitis B

MODELS OF THE INTERACTION OF HOST GENOTYPES AND INFECTIOUS DISEASE

Ira M. Longini, Jr.
Department of Biostatistics
University of Michigan
Ann Arbor, Michigan 48109/USA

Introduction

The proposition that infectious diseases act as important selective forces for maintaining polymorphisms in host populations is an intriguing part of theories of evolution. Moreover, a proposition of considerable epidemiological importance is that host genetic variability is partially responsible for the maintenance of infectious disease in populations.

In this paper, the evolution of the host, but not the pathogen, is considered. Non-synchronized generational models are developed where the generation time of the host (e.g., human) can be on the order of many years, while that of the pathogen (e.g., viruses) may be only a matter of a few days. Models of this form have been considered by Gillespie (1975), Lewis (1981) and Longini (1983). Analysis will be further restricted to those diseases that do not confer immunity following infection as analyzed by Longini (1983) for non-synchronized generations and by Kemper (1982) for synchronized generations. In addition, an example is worked out for data collected on the interaction of the malaria parasite P. Falciparum and the resistance of erythrocytes containing HbS (i.e., the sickle-cell trait) in human African populations (see Fleming, et al. [1979] and Molineaux and Gramiccia [1980]).

Genetic - Epidemic Model

An important assumption is that the generation time of the host is much longer than that of the pathogen. The host has discrete, non-overlapping generations such that the epidemic process runs its course within each host generation. Analysis will be limited to single-locus diallelic diploid populations. It is assumed that there is no migration or mutation and that the population size is large (conceptually infinite). Let the two alleles A and a occur with frequencies p and q = 1-p. Then using standard population genetic arguments (e.g., Crow and Kimura [1970]), the frequency of A in the next generation will be

$$p' = \frac{pw_1}{\bar{w}} , \tag{1}$$

where $\bar{w} = pw_1 + qw_2$, $w_1 = pw_{11} + qw_{12}$ and $w_2 = pw_{12} + qw_{22}$.

Then w_{11}, w_{12} and w_{22} represent the relative fitnesses of genotypes AA, Aa and aa, respectively.

The infectious process envisaged is the so-called S-I-S epidemic process, where infected individuals return to the susceptible state following their infection, i.e., there is no lasting infection acquired immunity. Examples of diseases that roughly follow this pattern are gonorrhea, meningitis, rhinovirus (common cold) and, to some extent, malaria.

Let S(t) and I(t) be the fractions of the genetically susceptible population in the susceptible and infected classes at time t, respectively. The differential equation for the S-I-S process is

$$\text{(2)}\qquad \frac{dI(t)}{dt} = S\bar{k}(p)\beta I - \gamma I\ ,\qquad S(0) = S_o > 0,\ I(0) = I_o > 0,\ \ S(t) = 1 - I(t),$$

where $\beta>0$ is the infectious contact rate and $\gamma>0$ is the rate of recovery from infection.

$$\text{(3)}\qquad \bar{k}(p) = k_{11}p^2 + 2k_{12}pq + k_{22}q^2$$

is defined as the average density of the susceptible population, where $0 \leq k_{11} \leq 1$, $0 \leq k_{12} \leq 1$ and $0 \leq k_{22} \leq 1$ are the coefficients of relative susceptibility of the three genotypes AA, Aa, aa, respectively. The asymptotic behavior of I(t), for a given frequency of p within a single host generation, is given by

$$\text{(4)}\qquad \lim_{t\to\infty} I(t) = \mathcal{L}(\bar{k}(p),\ \sigma) = \begin{cases} 0 & \text{if } \bar{k}(p)\sigma \leq 1\ , \\ 1 - \dfrac{1}{\bar{k}\sigma} & \text{if } \bar{k}(p)\sigma > 1\ , \end{cases}$$

where $\sigma = \beta/\gamma$ is the infectious contact number defined in Hethcote (1975).

Completely Recessive Susceptibility

We first consider the case where the allele A, which confers susceptibility, is completely recessive with respect to the resistant allele a. Therefore, individuals of genotype AA are completely susceptible, while individuals of genotypes Aa and aa are completely resistant, i.e., $k_{11} = 1$, $k_{12} = k_{22} = 0$. Furthermore, without loss of generality, we allow a susceptible (but not infected) individual of genotype AA to have unit fitness, but an infected individual has reduced fitness $1 - \tau$, $0 < \tau \leq 1$. Genetically resistant individuals will have relative fitness $1 - s$, $0 < s \leq 1$. It follows immediately that

$$\text{(5)}\qquad w_{11} = 1 - \mathcal{L}\tau,\ w_{12} = w_{22} = 1\text{-}s \text{ and } \bar{k}(p) = p^2\ .$$

We are concerned with finding the conditions under which a polymorphic equilibrium, $p^* > 0$ and $\mathcal{L}^* > 0$, exists and is stable for the system given by (1), (2), (4) and (5). The nature of such a polymorphic equilibrium is given by the following theorem:

Theorem 1. If $\sigma > 1$, $s < (1-1/\sigma)\tau$, and $0 < s < \tau \leq 1$, the polymorphic equilibrium, $p^* = [\tau/(\tau-s)\sigma]^{\frac{1}{2}}$, $\ell^* = s/\tau$, for the system (1), (2), (4) and (5), exists and is locally asymptotically stable (LAS).

Proof. At equilibrium $w_1^* = \bar{w}^* = w_2^*$. Solving the system (1), (2), (4) and (5) at equilibrium, we find the unique solution, with $0 < p^* < 1$ and $0 < \ell^* < 1$, to be $p^* = [\tau/(\tau-s)\sigma]^{\frac{1}{2}}$ and $\ell^* = s/\tau$ only when $0 < s < \tau \leq 1$, $\sigma > 1$ and $s < (1-1/\sigma)\tau$. The equilibrium, *, is LAS if and only if

$$(6) \qquad -1 < \left.\frac{\partial p'}{\partial p}\right|_* = 1 + \frac{p}{\bar{w}}\left[\frac{\partial w_1}{\partial p} - \frac{\partial \bar{w}}{\partial p}\right]\Bigg|_* < 1 .$$

Evaluation of (6) yields

$$(7) \qquad -1 < 1 - \left[\frac{p^2 q\tau}{1-s}\right]\ell'\Big|_* < 1 ,$$

where $\ell' = \dfrac{d\ell}{dp} = \dfrac{2}{p^3\sigma}$. The right-hand side of inequality (7) is trivally satisfied, while the left-hand side reduces to

$$(8) \qquad \sigma^2(1-s)^2 + \sigma\tau(2-\tau-s) > -\tau^2 ,$$

which is satisfied for $0 < s < \tau \leq 1$, $\sigma > 1$. This completes the proof.

Biocorollary 1. In the case of completely recessive susceptibility, a polymorphic equilibrium exists and is stable if an infected individual of genotype AA is less fit than a resistant individual of genotype Aa or aa who, in turn, is less fit than a susceptible individual of genotype AA.

Completely Dominant Susceptibility

We now consider the case where the allele A is completely dominant with respect to allele a. Now, individuals of genotype AA and Aa are completely susceptible, while individuals of genotype aa are completely resistant, i.e., $k_{11} = k_{12} = 1$ and $k_{22} = 0$. Individuals of genotype AA and Aa have unit fitness when susceptible, but a fitness of $1-\tau$ when infected; while individuals of genotype aa have a fitness of $1-s$. It follows immediately that

$$(9) \qquad w_{11} = w_{12} = 1 - \ell\tau,\ w_{22} = 1 - s \text{ and } \bar{k}(p) = p(1+q).$$

The nature of the polymorphic equilibrium is given by the following theorem:

Theorem 2. If $\sigma > 1$, $s < (1-1/\sigma)\tau$, and $0 < s < \tau \leq 1$, then the polymorphic equilibrium, $p^* = 1 - [1 - \tau/(\tau-s)\sigma]^{\frac{1}{2}}$, $\ell^* = s/\tau$, for the system (1), (2), (4) and (9), exists and is LAS.

Proof. Solving the system (1), (2), (4) and (9) at equilibrium, we find the unique solution, with $0 < p^* < 1$ and $0 < \ell^* < 1$, to be $p^* = 1 - \sqrt{1-\theta}$ and $\ell^* = s/\tau$, where $\theta = \tau/(\tau-s)\sigma$, only when $0 < s < \tau \leq 1$, $\sigma > 1$ and $s < (1-1/\sigma)\tau$. Evaluation of (6) yields

(10) $$-1 < 1 - \left[\frac{pq^2\tau}{1-s}\right] \mathcal{L}'|_* < 1 ,$$

where $\mathcal{L}' = \dfrac{2q}{p^2(1+q)^2\sigma}$. The right-hand side of inequality (10) is trivally satisfied, while the left-hand side reduces to

$$(1 - \theta)^{3/2} < \left(\frac{1-s}{\tau-s}\right) [1 + (1-\theta)^{1/2}].$$

Since $\left(\frac{1-s}{\tau-s}\right) \geq 1$ and $0 < \theta < 1$, it follows that inequality (10) is satisfied for $0 < s < \tau \leq 1$, using standard arguments. This completes the proof.

Biocorollary 2. In the case of completely dominant susceptibility, a polymorphic equilibrium exists and is stable if an infected individual of genotype AA or Aa is less fit than a resistant individual of genotype aa who, in turn, is less fit than a susceptible individual of genotype AA or Aa.

The trajectories of both the above systems have been analyzed in the $p - \mathcal{L}$ plane. Typical phase portraits have been given in Longini (1983) and for the analogous cases where genetic time is treated continuously in Kemper (1982). No limit cycles or chaotic behavior have been revealed in either investigation.

The Sickle-Cell Trait and Malaria -- Partially Dominant Susceptibility

A good area for application is the case of malaria p. falciparum and the sickle-cell trait. The correlation between the sickle-cell trait and the prevalence of p. falciparum has been well-studied by Birdsell (1972). In this case, we have partially dominant susceptibility, where a is the resistant sickle-cell allele. Then the three genotypes are AA, Aa and aa, whose individuals are totally susceptible, partially susceptible, and resistant, respectively, i.e., $k_{11} = 1$, $k_{12} = \alpha$ and $k_{22} = 0$, where $0 < \alpha \leq 1$. Furthermore, without loss of generality, genotype Aa (the heterozygote) has unit fitness regardless of infection status. A susceptible individual of genotype AA has unit fitness, but an infected individual of this genotype has fitness $1 - \tau$. Individuals of the completely resistant genotype aa develop severe anemia and rarely survive to child-bearing age. Thus, the fitness of this genotype is zero. It follows immediately that

(11) $$w_{11} = 1 - \mathcal{L}\tau,\ w_{12} = 1,\ w_{22} = 0 \text{ and } \bar{k}(p) = p^2 + 2\alpha pq.$$

The endemic level of infection in the host (human) population depends on parameters for that population as well as on parameters from the vector (mosquito) population. The asymptotic behavior of $I(t)$ for a given frequency of p, within a single host generation, is found from the so-called Ross-MacDonald model (see MacDonald [1952], Ross [1911], Aron and May [1981] and Dietz, et al. [1974]) given by

(12) $$\lim_{t\to\infty} I(t) = \mathscr{I} = \begin{cases} 0 & \text{if} \quad z_o \le 1 , \\ \dfrac{z_o - 1}{z_o + a/\mu} & \text{if} \quad z_o > 1 , \end{cases}$$

where

(13) $z_o = \bar{k}(p)ma^2b/\mu r$ is known as the basic reproductive rate and,

m = number of female mosquitoes per human host,

a = mosquito biting rate on man (# bites/time unit),

b = proportion of infected bites on man that produce an infection,

r = rate of recovery from infection for human host,

μ = mortality rate for mosquitoes.

Solving (1), (11) and (12) at equilibrium, we find the value of $\mathscr{I}^*$, where $0 < p^* < 1$, to be

(14) $$\mathscr{I}^* = \frac{q^*}{\tau p^*}$$

Evaluation of (6) and (11) yields the following condition for (14) with $0 < p^* < 1$ to be LAS:

(15) $$-1 < 1 - q[\frac{1}{p} + p\tau\mathscr{I}']|_* < 1 ,$$

where

(16) $$\mathscr{I}' = \frac{2ma^2b[a/\mu + 1][p + \alpha(1-2p)]}{\mu r[z_o + a/\mu]^2}$$

The right-hand side of inequality (15) is trivally satisfied for all $\mathscr{I}'|_* \ge 0$, while the left-hand side is satisfied if

(17) $$q\ [\frac{1}{p} + p\tau\mathscr{I}']|_* < 2 .$$

Fleming, et al. (1979) and Molineaux and Gramiccia (1980) have measured the frequency of the sickle-cell trait, the prevalence of malaria and appropriate entomological parameters in populations of the Sudan savanna of Nigeria. They have determined the frequency of the a allele to be $q = .146$. Assuming equilibrium, they have calculated the value of $\mathscr{I}^*\tau$ as

(18) $$\mathscr{I}^*\tau = q^*/p^* = \frac{.146}{1-.146} = .171$$

From their data, we estimate $\mathscr{I}^* = .915$ (see Longini [1983]) and substituting into (18) we obtain $\tau = .187$. Other parameter estimates from their work are $\alpha = .8$, $a = .44$, $b = .097$, $r = .0125$ and $\mu = .185$. Given these parameter estimates and $\mathscr{I}^* = .915$, $p^* = .854$, an estimate of $m = 4.953$ is found from (13). Substitution of the above equilibrium values and parameter estimates into (16) and (17) yields

$$q\left[\frac{1}{p} + p\tau\mathcal{L}'\right]\Big|_{*} = .171 < 2,$$

indicating that the sickle-cell allele is probably maintained as a stable polymorphism ($p^* = .854$) at the sampled endemic prevalence rate of malaria p. falciparum ($\mathcal{L}^* = .915$) in this particular population. Clearly, an understanding of the dynamic interaction of the sickle-cell trait and p. falciparum in indigenous populations should be an integral part of malaria control strategies. Further results concerning other epidemic-genetic processes can be found in Kemper (1982) and Longini (1983). The latter reference includes analysis of S-I-R epidemic processes, where infected individuals become immune or resistant to further infections upon recovery.

References

Aron, J.L. and R.M. May. (1981). The Population Dynamics of Malaria. Population Dynamics of Infectious Diseases, ed. R. Anderson, pp. 139-179. Chapman and Hall, Ltd.

Birdsell, J.B. (1972). Human Evolution. Chicago: Rand McNally.

Crow, J.F. and M. Kimura. (1970). An Introduction to Population Genetics Theory. New York: Harper and Row.

Dietz, K., Molineaux, L. and A. Thomas. (1974). A Malaria Model Tested in the African Savannah. Bull. Wld. Hlth. Org. 50, pp. 347-57.

Fleming, A.F., Storey, J., Molineaux, L., et al. (1979). Abnormal Haemoglobins in the Sudan Savanna of Nigeria. Ann. Trop. Med. Parasit. 73, pp. 161-173.

Gillespie, J.H. (1975). Natural Selection for Resistance to Epidemics. Ecology 56, pp. 493-495.

Hethcote, H.W. (1976). Qualitative Analyses of Communicable Disease Models. Math. Biosc. 28, pp. 335-356.

Kemper, J.T. (1982). The Evolutionary Effect of Endemic Infectious Disease: Continuous Models for an Invariant Pathogen. J. Math. Biol. 15, pp. 65-77.

Lewis, J.W. (1981). On the Evolution of Pathogen and Host: II. Selfing Hosts and Haploid Pathogens. J. Theoret. Biol. 93, pp. 953-985.

Longini, I.M. (1983). Models of Epidemics and Endemicity in Genetically Variable Host Populations. J. Math. Biol. 17, pp. 289-304.

MacDonald, G. (1952). The Analysis of Equilibrium in Malaria. Trop. Dis. Bull. 49, pp. 813-828.

Molineaux, L. and G. Gramiccia. (1980). The Garki Project. Geneva: World Health Organization.

Ross, R. (1911). The Prevention of Malaria (2nd. Ed.). London: Murray.

ON THE GENERAL EPIDEMIC MODEL IN DISCRETE TIME

M-P. MALICE and C. LEFEVRE
Université Libre de Bruxelles
Campus Plaine, C.P.210
1050 Bruxelles - Belgique

This paper is concerned with the analysis of the discrete time version of the general epidemic model. Both deterministic and stochastic models are discussed. Moreover, using an heuristic argument an approximating system is proposed for the case where the population contains initially many susceptibles and little infectives.

1. INTRODUCTION

Over the last thirty years there has been a great deal of effort devoted to developing mathematical models to describe the spread of epidemics. The book of Bailey [2] provides a good account of the literature up to 1975.

Most of the research works, however, deal with continuous time models, and until now, discrete time models have been relatively little studied and applied. In fact, their use is confined to Greenwood's and Reed-Frost's chain-binomial models (Abbey [1], Gani and Jerwood [5]), and various model extensions (Ludwig [11], Von Bahr and Martin-Löf [13]) and modifications (Dayananda [3], Longini [10], Lefèvre [8]). This can seem rather surprising. As pointed out by Gani [6], a discrete time approach may be more appropriate for different reasons, of course when it is closer to reality, but also when it is more convenient. For more information on this point, we refer the reader to the discussion which followed Gani's paper.

Among the continuous time epidemic models, the basic one is the so called general epidemic. This model is concerned with a closed population subdivided into three compartments containing susceptible, infectious and removed individuals respectively; it is assumed that only two types of transition can occur : infections of susceptibles by contact with infectives, and removals of infectives from circulation. This is perhaps the most famous epidemic process, and it has been extensively studied (see, e.g., Bailey [2]).

In this paper, we construct and analyse the discrete time version of the general epidemic. The stochastic model is formulated in Section 2. In Section 3, we present and study the associated deterministic model. In Section 4, we indicate how to compute the transient distribution of the state of the population, and we then propose heuristically an approximating system for the case where the population contains initially many susceptibles and little infectives. In Section 5, we present a method for computing the distribution of the total size of the epidemic, and we then determine that distribution for the approximating system proposed in Section 4. Approximate results and simulations of the true process are compared with some examples.

2. FORMULATION OF THE STOCHASTIC MODEL

Let us consider a closed population of size N consisting of susceptibles, infectives and removals, and assume a discrete time scale $t = 0,1,2,\dots$. Let $S(t)$, $I(t)$ and $R(t)$ denote the numbers of susceptibles, infectives and removals at time t respectively. As $S(t) + I(t) + R(t) = N$, the state of the population can be described by the bidimensional vector $[S(t), I(t)]$. The initial values are denoted by $[S(0), I(0)] = [n,m]$, with $n+m = N$. To construct the discrete time version of the general stochastic epidemic, we adopt the suggestions made in the discussion of Gani's paper [6], and we suppose that given the state $[S(t), I(t)]$, the evolution of the epidemic at time t+1, $t \geq 0$, is governed by the two following rules.

On the one hand, each of the S(t) susceptibles can become infected at t+1 after an infectious contact during unit time; the probability that a susceptible escapes such a contact with any specified infective is denoted $\exp(-\Lambda), 0 < \Lambda < \infty$. On the other hand, each of the I(t) infectives can be removed from circulation at t+1 with probability $1-\Theta$, $0 \leq \Theta < 1$. With the usual assumption of an homogeneously mixing population, S(t+1) and I(t+1) are thus distributed as follows

$$\begin{cases} S(t+1) \sim \text{Bin}[S(t);\, e^{-\Lambda I(t)}], \\ I(t+1) \sim \text{Bin}[S(t);\, 1 - e^{-\Lambda I(t)}] + \text{Bin}[I(t),\Theta], \end{cases} \tag{1}$$

where the notation Bin $[a;b]$ denotes a binomial variable of exponent a and parameter b.

We note that the classical Reed-Frost model corresponds to the particular case where $\Theta = 0$, that is, the infectives present at t are all removed at t+1.

3. ANALYSIS OF THE ASSOCIATED DETERMINISTIC MODEL

The previous model suggests an analogous deterministic model defined by the difference equations

$$\begin{cases} s(t+1) = s(t)\, e^{-\Lambda i(t)}, \\ i(t+1) = s(t)[1 - e^{-\Lambda i(t)}] + \Theta\, i(t) , \end{cases} \tag{2}$$

with the initial conditions $[s(0), i(0)] = [n,m]$, $n+m = N$.

A link between the stochastic and deterministic versions (1) and (2) is provided by an asymptotic theorem of Lehoczky [9]. Roughly, it can be shown that under certain conditions, the stochastic process converges to the deterministic process almost surely as the population size N becomes infinite. This question is not examined here but will be discussed in a future paper.

Now, it is easily seen that the analysis of the model (2) is very similar to that of the continuous time version (for the latter, see, e.g., Bailey [2]). So, on defining by $\rho \equiv (1-\Theta)/\Lambda$ the relative removal rate, we find here also that $s(t) \exp\{[N-s(t) - i(t)]/\rho\}$ remains constant over time, and that as $t \to \infty$, $i(t) \to 0$

and $s(t) \to s > 0$, where s is the root of the equation $x \exp(-x/\rho) = n \exp(-N/\rho)$, $0 < x < n$. This leads to the following threshold property.

Threshold property. An epidemic can develop only if $\rho < n$. When $\rho \geqslant n$, $s \to n$ as $m \to 0$; an approximate value of s if ρ is not near n is $s \cong n - nm/(\rho-n)$. When $\rho < n$, $s \to s^* < \rho$ as $m \to 0$, where s^* is the root of the equation $x \exp(-x/\rho) = n \exp(-n/\rho)$, $0 < x < n$; an approximate value of s if ρ is not near s^* is $s \cong s^* - s^*m/(\rho-s^*)$.

4. THE TRANSIENT BEHAVIOUR OF THE STOCHASTIC MODEL

4.1. The exact model

It is clear that $[S(t), I(t)]$ jointly form a finite bivariate Markov chain, and that the epidemic process will always terminate after a finite number of steps by the extinction of the infection. As a consequence, Markov chain methods similar to those used by Gani and Jerwood [5] for the Reed-Frost model can be applied here also to obtain probabilities for the duration time and the state of the population. For the sake of brevity, we don't report the results in this paper and we just mention that their derivation can be somewhat facilitated by exploiting the particular structure of the transition matrix.

4.2. An approximating system

Very often, the population contains initially little infectives but many susceptibles. The computation of the transient distribution of the state of the population requires then a great number of calculations (the Markov chain contains $(n+1)(m+1+n/2)$ states !), and it is necessary, in practice, to use an approximation method. We present below an approximating stochastic system which is the discrete time version of that developed by Kendall [7], and completed by Faddy [4], for the continuous time model.

Let us consider the situation where $m < \infty$ and $n \to \infty$. Moreover, we suppose that each individual meets on the average a limited number of others, so that the parameter Λ in the probability of no encounter is here written as $\Lambda \equiv \lambda/n$ with λ independent of n. Now, we remark that during its initial stages, the behaviour of the population of infectives may be approximated by a branching process with m ancestors and offspring generating function $f(z) = [1 - \Theta + \Theta z] \exp[\lambda(z-1)]$. The probability of extinction of this process is equal to 1 if $\rho \geqslant n$ and to σ^m if $\rho < n$, where σ is the root of the equation $x = f(x)$, $0 < x < 1$. Imitating Kendall [7], we then propose the following approximating system.

An approximating system.

Case (i). If $\rho \geqslant n$, when ultimate extinction is certain, the process $I(t)$ behaves as the branching process described above.

Case (ii). If $\rho < n$, when extinction occurs with probability σ^m, there are two modes of behaviour A and B, with $P(A) = 1 - P(B) = \sigma^m$. In mode A, the process $I(t)$ behaves like case (i) subject to the condition of ultimate extinction. Using a theorem of Waugh [14], this conditional process is found to be a branching process with m ancestors and offspring generating function $\tilde{f}(z) = (1/\sigma)\ f(\sigma z)$. In mode B, the epidemic behaves as the associated deterministic model, so that a major epidemic occurs.

Imitating Faddy [4], we propose in mode B to substitute for the deterministic approximation a linear compartmental approximation obtained by replacing the stochastic variable I(t) by its deterministic equivalent i(t) in the probability of no encounter. After some manipulations, the approximate distributions of S(t) and I(t) can then be rewritten as follows

$$\begin{cases} S(t+1) \sim \text{Bin}[n;\ s(t+1)/n], \\ I(t+1) \sim \text{Bin}[n;\ (i(t+1) - m\ \Theta^{t+1})/n] + \text{Bin}[m;\ \Theta^{t+1}]. \end{cases} \tag{3}$$

For a general analysis of linear compartmental models in discrete time, the reader is referred to a recent paper by Malice and Lefèvre [12].

In order to see how good the compartmental approximation is, we compare below, for various values of the parameters, the means and standard deviations of I(t) from this approximation with the corresponding means and standard deviations from 500 simulations of the true process.

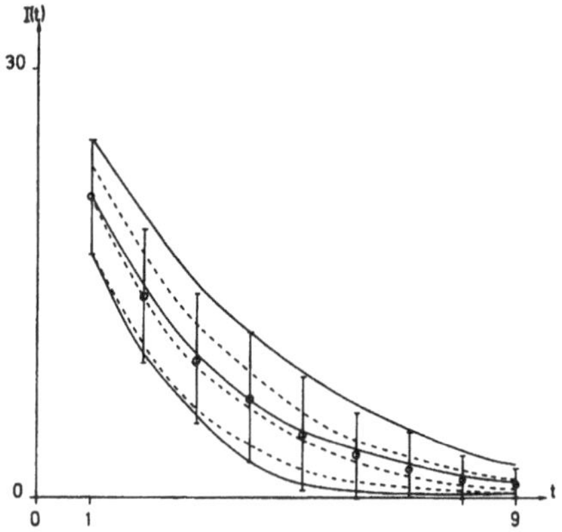

Figure 1 - Mean and mean ± 1 standard deviation from the compartmental approximation (---), the approximation by the branching process (——), and 500 simulations of the true process (Φ), when n = 500, m = 30, Λ = 0.001 and Θ = 0.2.

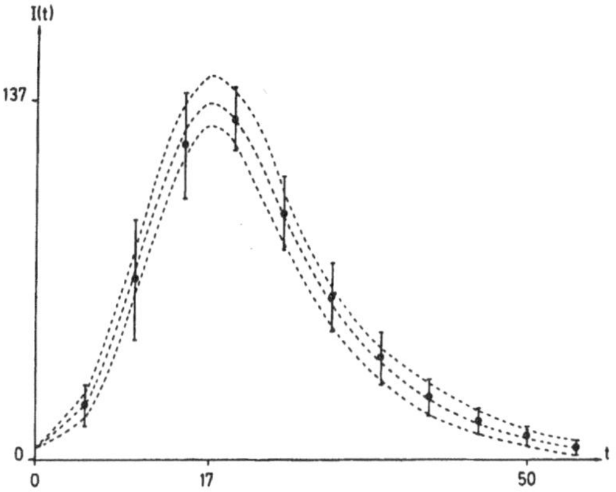

Figure 2 - Mean and mean ± 1 standard deviation from the compartmental approximation (---) and 500 simulations of the true process (Φ), when n = 300, m = 5, Λ = 0.0015 and Θ = 0.9.

In Figure 1, where $\rho = 800 > n = 300$, the compartmental approximation is satisfying but less precise than the approximation by the branching process. In Figure 2, where $\rho = 66.67 < n = 300$ and σ^m is very small ($= 0.19^5 = 24 \times 10^{-4}$), the agreement is very good.

equal to 1 if $\rho \geq n$ and to σ^m if $\rho < n$. The approximating system of 4.2 leads then to the following distribution of T.

An approximating result.

Case (i). If $\rho \geq n$, the generating function of T is given by $[h(z)/z]^m$, where $h(z)$ is the root of the equation $x = z\, g(x)$, $0 < x \leq 1$.

Case (ii). If $\rho < n$, the two modes A and B are again discerned. In mode A, the generating function of T is given by $[\tilde{h}(z)/z]^m$, where $\tilde{h}(z)$ is the root of the equation $x = (z/\sigma)\, g(\sigma x)$, $0 < x \leq 1$. In mode B, $T = n-s$ in the deterministic approximation and $T \sim \text{Bin}(n; 1 - s/n)$ in the linear compartmental approximation.

We mention that the limit distribution of T obtained heuristically for case (i) and case (ii), mode A can be established rigorously from a general theorem proved by Von Bahr and Martin-Löf [13] for the randomized Reed-Frost models. For case (ii), mode B, this theorem completes the limit result derived here by stating that T has an approximately Gaussian distribution around n-s.

Table 4 lists, for various values of the parameters, the expected total size of the epidemic obtained from 500 simulations of the true process and that computed from the approximating system.

Λ	Θ	$\rho > n$	$E_A(T)$	$E_S(T)$
0.0003	0.40	2000	1	1
0.0003	0.80	666.67	4	4
0.0003	0.88	400	15	13

Λ	Θ	$\rho < n$	σ	σ^5	n-s	$E_A(T)$	$E_S(T)$
0.0003	0.92	266.67	0.84	0.418	80	54	51
0.0003	0.95	166.67	0.52	0.038	223	214	211
0.0015	0.65	233.33	0.73	0.210	129	105	98
0.0015	0.80	133.33	0.39	0.009	258	255	255
0.0030	0.90	33.33	0.07	0.000	0	0	0
0.0030	0.50	166.66	0.43	0.015	223	220	217

Table 4 - Expected total size of the epidemic from the approximating system $[E_A(T)]$ and 500 simulations of the true process $[E_S(T)]$, when $n = 300$, $m = 5$ and for various values of Λ and Θ.

REFERENCES

[1] ABBEY, H. (1952) An examination of the Reed-Frost theory of epidemics. *Hum. Biol.* 24, 201-233.

[2] BAILEY, N.T.J. (1975) *The Mathematical Theory of Infectious Diseases and its Applications*. Griffin, London.

[3] DAYANANDA, P.W.A. (1974) An approximate chain-binomial model for simple epidemics. *Biometrics* 30, 705-708

[4] FADDY, M.J. (1977) Stochastic compartmental models as approximations to more general stochastic systems with the general stochastic epidemic as an example. *Adv. Appl. Prob.* 9, 448-461.

[5] GANI, J. AND JERWOOD, D. (1971) Markov chain methods in chain binomial epidemic models. *Biometrics* 27, 591-604.

[6] GANI, J. (1978) Some problems of epidemic theory. *J.R. Statist. Soc.* A, 140, 323-347.

t	$E_C[I(t)]$	$S.d_C[I(t)]$	$E_T[I(t)]$	$S.d_T[I(t)]$	$E_A[I(t)]$	$E_S[I(t)]$
1	2	1	2	1	2	2
3	7	3	8	6	6	7
5	25	5	26	16	21	23
7	68	7	65	26	57	56
9	99	8	93	22	84	78
11	80	8	75	18	67	63
13	42	6	44	16	35	37
15	19	4	22	14	16	18

Table 3 - Mean and standard deviation of I(t) from the compartmental approximation ($E_C[I(t)]$ and $S.d_C[I(t)]$), and the 422 simulated epidemics which did "take off" ($E_T[I(t)]$ and $S.d_T[I(t)]$), as well as the mean of I(t) from the approximating system ($E_A[I(t)]$) and the 500 simulated epidemics ($E_S[I(t)]$), when $n = 300$, $m = 1$, $\Lambda = 0.0045$ and $\Theta = 0.6$.

In Table 3, where $\rho = 88.89 < n = 300$ and σ^m is not near 0 (= 0.16), the comparison is made by considering only the 422 ($\approx 500\,[1-0.16]$) simulated epidemics which did "take off". The agreement is not too bad but less satisfying than in the two other cases. Note also that the expected number of infectives from the 500 simulations is rather well approximated by the corresponding expected value obtained from the complete approximating system.

5. THE STATIONARY BEHAVIOUR OF THE STOCHASTIC MODEL

5.1. The exact model

When $I(t) = 0$ for the first time, the epidemic terminates and S(t) remains at its final value denoted S, $0 \leq S \leq n$. The distribution of S can be computed rather easily by following an ingenious method developed by Ludwig [11] in particular for the continuous time model. This method is based upon the notion of "randomized" Reed-Frost models, that is, classical Reed-Frost models where the probabilities of no infectious contact between two specified individuals during unit time are no more constant but i.i.d. random variables. Now, it can be seen that for the model (1), the distribution of S is the same as for a randomized Reed-Frost model where for each i-infective, the probability of no encounter with any fixed susceptible is $\exp(-\Lambda Y_i)$, where the Y_i are i.i.d. with a truncated geometric distribution of parameter $1-\Theta$. Consequently, it suffices to compute the distribution of S for the latter, which can be achieved by a simple iterative procedure built up by Ludwig [11].

5.2. An approximating result

Let us reconsider the case where $m < \infty$, $\Lambda = \lambda/n$ and $n \to \infty$. Using the approach indicated in 5.1 , we now derive the limit distribution of the total size of the epidemic, $T = n-S$, for the approximating system proposed in 4.2.

We first remark that for n large and approximately constant, the distribution of T is the same as for a "randomized" branching process with m ancestors and offspring generating function $g(s) = E\{\exp[\lambda(s-1)Y_i]\}$, with Y_i defined as above. We note that as expected, the probability of extinction of this process is, here also,

[7] KENDALL, D. (1956) Deterministic and stochastic epidemics in closed populations. *Proc. 3rd. Berkeley Symp. Math. Statist. Prob.* 4, 149-165.

[8] LEFEVRE, C. (1982) Un modèle d'épidémie simple à temps discret. *Cahiers du C.E.R.O.* 24, 313-320.

[9] LEHOCZKY, J.P. (1980) Approximations for interactive Markov chains in discrete and continuous time. *J. Math. Sociology* 7, 139-157.

[10] LONGINI, I.R. (1980) A chain binomial model of endemnity. *Math. Biosci.*50, 83-93.

[11] LUDWIG, D. (1975) Final size distributions for epidemics. *Math. Biosci.*23, 47-73.

[12] MALICE, M-P. AND LEFEVRE, C. (1983) On linear stochastic compartmental models in discrete time. Submitted for publication.

[13] VON BAHR, B. AND MARTIN-LÖF, A. (1980). Threshold limit theorems for some epidemic processes. *Adv. Appl. Prob.* 12, 319-349.

[14] WAUGH, W.A. O'N. (1958) Conditioned Markov processes. *Biometrika* 45, 241-249.

CONTROL OF VIRUS TRANSMISSION IN AGE-STRUCTURED POPULATIONS

D. Schenzle
Department of Medical Biometry
Tübingen University
D7400 Tübingen, FRG

The purpose of this note is to demonstrate the need of using models with explicit age-structure when dealing with questions of control of virus transmission.

Consider an isolated population of constant size n with stable age-structure. With respect to a given virus infection, each individual is supposed to be either susceptible, infectious or permanently immune, the respective proportions in the population being denoted by u, v and w. Moreover, a random proportion p of "newborns" may be immunized by vaccination, whereas the remaining proportion $q = 1-p$ remains unvaccinated. Then the "standard" global mass action model (Martini, 1921; Dietz, 1975) reads:

$$\begin{aligned} \dot{u} &= q\mu-\beta vu-\mu u, \\ \dot{v} &= \beta vu-(\gamma+\mu)v, \\ \dot{w} &= p\mu+\gamma v-\mu w, \end{aligned} \tag{1}$$

where μ denotes a constant birth- and death rate, infected individuals are removed after an average time span $\Gamma = 1/\gamma$, and the basic parameter β denotes an effective contact rate, which is population- and infection-specific. Any endemic equilibrium with a proportion $\bar{v} > 0$ of infectives is stable and has the same proportion of susceptibles given by

$$u_0 = (\gamma+\mu)/\beta. \tag{2}$$

In turn, to any infection-free stationary state with a proportion q of susceptibles there corresponds an initial infection reproduction rate

$$\rho(q) = \beta q, \tag{3}$$

and the condition

$$\rho(q) < \gamma+\mu \tag{4}$$

apparently implies stability of this infection-free state. Hence within the framework of global mass action theory, where the special relationship (2) holds, one may formulate: the incidence rate of an infection approaches zero if $q \leq u_0$, whereas for $q > u_0$ the ultimate post-vaccination incidence rate of infection is given by

$$\bar{I}(q) = \bar{I}(0)(q-u_0)/(1-u_0), \tag{5}$$

where $\bar{I}(0)$ denotes the pre-vaccination endemic incidence rate.

The crucial quantity u_0 may be estimated from age-specific prevalence data $P(a)$ of susceptibles (non-immunes), as described by Dietz (1975), and in the case of measles various authors arrived at values in the range 0.05-0.08 (Anderson and May, 1982, 1983; Hethcote, 1983; Yorke et al., 1979). However one has to keep in mind that age-specific prevalence data of susceptibles do not decrease exponentially according to a constant endemic force of infection $\lambda = \beta\bar{v}$, as it is postulated in model (1). In fact the force of infection for measles (and other virus infections) shows a pronounced age-dependency, which is probably related to increased infection transmission among school children (Fine and Clarkson, 1982; Horwitz et al., 1974; Paule et al., 1979; Griffiths, 1974). Then, contrary to a recent claim by Anderson and May (1982, 1983), the endemic proportion $\bar{u}$ of susceptibles is no longer constant and independent of q, and therefore the stability criterion formulated above ceases to be applicable.

Proper consideration of control of infection transmission in the presence of age-dependencies requires generalization of the system of integro-differential equations proposed by Dietz (1975), allowing for differential contact rates $\beta(a,\alpha) > 0$ between infectives of age α and susceptibles of age a. Then, denoting with $u(a,t)$, $v(a,t)$ and $w(a,t)$ density functions of susceptibles, infectives and immunes of age a at time t, the model reads

$$\begin{aligned}\frac{\partial u(a,t)}{\partial a} + \frac{\partial u(a,t)}{\partial t} &= -[\lambda(a,t)+\mu(a)]u(a,t),\\ \frac{\partial v(a,t)}{\partial a} + \frac{\partial v(a,t)}{\partial t} &= \lambda(a,t)u(a,t)-[\gamma+\mu(a)]v(a,t),\\ \frac{\partial w(a,t)}{\partial a} + \frac{\partial w(a,t)}{\partial t} &= \gamma v(a,t)-\mu(a)w(a,t),\end{aligned} \tag{6}$$

where

$$\lambda(a,t) = \int_0^\infty \beta(a,\alpha)v(\alpha,t)d\alpha$$

denotes the age-specific force of infection and $\mu(a)$ the age-specific death rate. The initial and boundary conditions are specified as

$$u(a,0) = \tilde{u}(a), \quad v(a,0) = \tilde{v}(a), \quad w(a,0) = \tilde{w}(a)$$

and

$$u(0,t) = q\nu, \quad v(0,t) = 0, \quad w(0,t) = p\nu.$$

Here too only immunization "at birth" is considered, but of course more realistic vaccination schedules could be incorporated into the model equations (6).

Again there exist stationary infection-free states characterized by $v(a,t) \equiv 0$ and

$$u(a,t) \equiv u^*(a) = q\nu\exp[-\int_0^a \mu(\alpha)d\alpha],$$

but apart from some special cases it is difficult to specify sharp conditions on the stability of these states. One such case would be obtained if the death rate were constant, $\mu(a) \equiv \mu_0$, and if the age-specific force of infection could be written as

$$\lambda(a,t) = \bar{\beta}(a)v(t), \tag{7}$$

with $v(t)$ denoting the total proportion of infectives. Then the condition

$$\int_0^\infty \bar{\beta}(\alpha)u^*(\alpha)d\alpha > \gamma+\mu$$

is necessary and sufficient for an endemic state to exist. However the choice (7) suggested by Anderson and May (1982, 1983) and also by Cvjetanović et al. (1982) is epidemiologically unrealistic. It implies for example in the case of measles, on the basis of observed forces of infection, that for an adult infective the chance of infecting a given school child would be about ten-fold higher than the chance for a school child to infect a given adult.

The complications of dealing with partial differential equations of type (6) are avoided if aging is simulated by letting individuals pass through a number of "age-compartments". This yields a straightforward generalization of model (1) and allows application of results from the theory of ordinary differential equations. Using L compartments, where $L = 75$ corresponds to the number of annual birth cohorts in the population, and denoting with u_i and v_i ($i = 1,\ldots,75$) the proportions of individuals who are susceptible or infectious in compartment i, we write:

$$
\begin{aligned}
\dot{u}_1 &= q\mu \quad -(\lambda_1+\mu L)u_1, & \dot{v}_1 &= \lambda_1 u_1 \quad -(\gamma+\mu L)v_1, \\
\dot{u}_2 &= \mu L u_1 \quad -(\lambda_2+\mu L)u_2, & \dot{v}_2 &= \mu L v_1 \quad +\lambda_2 u_2-(\gamma+\mu L)v_2, \\
&\cdots\cdots\cdots & &\cdots\cdots\cdots \\
\dot{u}_L &= \mu L u_{L-1}-(\lambda_L+\mu L)u_L, & \dot{v}_L &= \mu L v_{L-1}+\lambda_L u_L-(\gamma+\mu L)v_L,
\end{aligned}
\tag{8}
$$

where
$$\lambda_i = \sum_{j=1}^{L} \beta_{ij} v_j$$

for $i = 1,\ldots,L$, with L^2 differential contact rates $\beta_{ij} > 0$. The proportion of individuals who are immune in compartment i is of course given as $w_i = 1/L-u_i-v_i$. This model yields a gamma survival curve

$$S(a) = \left[\sum_{m=0}^{L-1} \frac{(\mu L a)^m}{m!}\right]\exp(-\mu L a), \tag{9}$$

life expectancy being L, with standard deviation $\sqrt{L}$ in the case $\mu = 1/L$.

Of primary interest are the infection-free stationary states with state vector $u^* = (q/L,\ldots, q/L, 0,\ldots,0)$. To each of these states there corresponds a matrix

$$
\begin{bmatrix}
\rho_{11}-\gamma-\mu L & \rho_{12} & \cdots & \rho_{1L-1} & \rho_{1L} \\
\rho_{21}+\mu L & \rho_{22}-\gamma-\mu L & \cdots & \rho_{2L-1} & \rho_{2L} \\
\vdots & \vdots & \vdots & \vdots & \vdots \\
\rho_{L1} & \rho_{L2} & \cdots & \rho_{LL-1}+\mu L & \rho_{LL}-\gamma-\mu L
\end{bmatrix},
\tag{10}
$$

the quantities

$$\rho_{ij}(u^*) = \rho_{ij}(q) = \beta_{ij} q/L \tag{11}$$

denoting differential initial infection reproduction rates in the state u^*. The sum

$$\rho_i = \sum_{j=1}^{L} \rho_{ji} \tag{12}$$

yields the initial infection production rate of infection in "age-compartment" i. Then according to standard stability theory the stationary infection-free state u^* is stable if all eigenvalues of the matrix (10) have negative real parts. This criterion therefore generalizes criterion (4) of global mass action theory to models of type (8). If the maximum real part exceeds zero, then a stable endemic equilibrium state

$$(\bar{u}_1, \ldots, \bar{u}_L \,,\, \bar{v}_1, \ldots, \bar{v}_L)$$

exists, which however can hardly be expressed analytically in terms of the model parameters.

Unfortunately now a problem of model identification arises, since practically nothing is known empirically about the parameters β_{ij} or ρ_{ij}. The best information currently available is in the form of age-specific forces of infection and prevalences of susceptibles (immunes). This is not enough, since one really needs to know "who infects whom" in a population, where some infectives reproduce infection systematically more often than others. This problem has not been appreciated in the recent literature on control of infection. Here, for the purpose of providing an illustrative numerical example, the structure of the β_{ij}-matrix shall be drastically reduced, putting:

$$\beta_{ij} = \begin{cases} \beta_A & \text{if} \quad i > 10 \quad \text{or} \quad j > 10, \\ \beta_S & \text{if} \quad 6 \leq i \leq 10 \quad \text{and} \quad 6 \leq j \leq 10, \\ \beta_E & \text{if} \quad i,j \leq 10 \quad \text{and} \quad \min(i,j) \leq 5. \end{cases} \tag{13}$$

These three model parameters are to be estimated from three corresponding forces of infection $\lambda_S > \lambda_E > \lambda_A$ on individuals representing respectively "school children", "pre-schoolers" and "adolescents-adults", i.e.

$$\lambda_i = \begin{cases} \lambda_E & \text{if} \quad i \leq 5, \\ \lambda_S & \text{if} \quad 6 \leq i \_ 10, \\ \lambda_A & \text{if} \quad i > 10. \end{cases} \tag{14}$$

Hence school-children are singled out as a core group with intense intra-group transmission of infection, which also largely contributes to the force of infection on the rest of the population, e.g. by family contacts. For the so-called childhood infections this idea is in line with common epidemiologic wisdom. Now, for a pre-vaccination endemic equilibrium ($q = 1$) with specified values of λ_E, λ_S and λ_A (as well as of μ and γ) the proportions $\bar{u}_i$ and $\bar{v}_i$ for $i = 1,\ldots,L$ are completely determined from model (8), since

$$\begin{aligned} \bar{u}_1 &= \mu/(\lambda_1+\mu L), \\ \bar{v}_1 &= \lambda_1 \bar{u}_1/(\gamma+\mu L), \end{aligned} \tag{15}$$

and for $i = 2,\ldots,L$:

$$\begin{aligned} u_i &= \mu L\bar{u}_{i-1}/(\lambda_i+\mu L), \\ v_i &= (\mu L\bar{v}_{i-1}+\lambda_i\bar{u}_i)/(\gamma+\mu L). \end{aligned}$$

For the special choices made here one therefore obtains

$$\begin{aligned} \beta_A &= \lambda_A/\sum_{i=1}^{L} \bar{v}_i \,, \\ \beta_E &= (\lambda_E-\beta_A \sum_{i>10} \bar{v}_i)/\sum_{i\leq 10} \bar{v}_i \,, \end{aligned} \tag{16}$$

and

$$\beta_S = (\lambda_S-\beta_A \sum_{i>10} \bar{v}_i-\beta_S \sum_{i\leq 5} \bar{v}_i)/\sum_{i=6}^{10} \bar{v}_i.$$

Finally the prevalence $P(a)$ of susceptibles among individuals of age a is determined as

$$P(a) = \sum_{i=1}^{L} x_i(a), \tag{17}$$

where $x_i(a)$ is to be calculated from the following system of differential equations with initial condition $x_1(0) = 1$, $x_2(0) = \ldots x_L(0) = 0$:

$$
\begin{aligned}
dx_1/da &= -(\lambda_1+\mu L)x_1 \\
dx_2/da &= \mu L x_1-(\lambda_2+\mu L)x_2 \\
&\cdots \\
dx_L/da &= \mu L x_{L-1}-\lambda_L x_L .
\end{aligned}
\tag{18}
$$

The reported sample calculation uses forces of infection $\lambda_E = 0.08$, $\lambda_S = 0.5$ and $\lambda_A = 0.05$ per year, which are thought to reflect pre-vaccination conditions for measles. The resulting prevalences of immunes (non-susceptibles), $1-P(a)$, for ages 1,2,...,15 are calculated as: .077, .155, .248, .365, .492, .613, .713, .789, .842, .878, .900, .915, .924, .930, .935. These figures agree fairly well with available data (Fine and Clarkson, 1982; Horwitz et al., 1974; Griffiths, 1974; Hedrich, 1933). The calculated total proportion of susceptibles in the population is $\bar{u} = 0.09$, which value is slightly larger than the previously derived values quoted above. Here one should keep in mind that the value of $\bar{u}$ depends critically on the proportion of susceptible adults, for which no precise estimates exist. Now, the resulting values for the parameters β_E, β_S and β_A are 0.62, 5.88 and 0.37. Then all elements in the matrix (10) are specified and one may localize numerically the maximum value of q yielding negative eigenvalues. In the present example this value was found to be about $q = 0.21$, which is thus substantially larger than the critical q-values of 0.05-0.08 predicted from global mass action theory. Moreover this example clearly demonstrates that in general the value of the pre-vaccination endemic proportion $\bar{u}$ of susceptibles cannot be used to predict effects of mass immunization. Also with the present model (8) the proportion of susceptibles is predicted to rise, from a pre-vaccination endemic value of 0.09 to a maximum possible level of 0.21 in the case $q = 0.21$. It may be instructive finally to look at the basic initial infection reproduction rates per infectious period $1/(\gamma+\mu)$ for the three subpopulations under consideration. These are: $\rho_E = 3.9$, $\rho_S = 7.3$ and $\rho_A = 3.6$. The highest value applies to individuals in compartments 6-10, i.e. to young children in kindergarten and elementary school. But of the 7.3 secondary cases one child could produce in the average, 3.8 occur among "schoolmates", whereas only 3.5 cases occur among the remaining 70 annual birth cohorts. Apparently the single effective contact rate in global mass action models represents an average value over widely varying differential contact rates.

The work reported here represents only an illustrative exercise. The problem of evaluating mass immunization programs in actual populations is of course much more complicated. But on the basis of the sample results given one may question the reliability of some current theoretical predictions on the impact of vaccination against measles and rubella (Anderson and May, 1982, 1983; Hethcote, 1983). Certainly it is premature to state "that epidemic theory has clarified the target", as an Editorial (1982) in the *Lancet* recently asserted. More work needs to be done in studying how results from age-structured models depend on the various differential contact rates, which cannot be estimated unambigously from pre-vaccination age-specific data. Moreover, in an attempt to identify these models more precisely, one should collect and analyze as much information as possible about current vaccination policies and their effects in the population.

REFERENCES

ANDERSON, R.M. and MAY, R.M. (1982). Directly transmitted diseases: control by vaccination. *Science*, 215, 1053-60.

ANDERSON, R.M. and MAY, R.M. (1983). Vaccination against rubella and measles: quantitative investigations of different policies. *J. Hyg. Camb.*, 90, 259-325.

CVJETANOVIĆ, B., GRAB, B. and DIXON, H. (1982). Epidemiological models of poliomyelitis and measles and their application in the planning of immunization programmes. *Bull. WHO*, 60, 405-22.

DIETZ, K. (1975). Transmission and control of arbovirus diseases. In: *Proceedings of a SIMS Conference on Epidemiology*. (Editors: Ludwig, D. and Cooke, K.L., Philadelphia, Society for Industrial and Applied Mathematics), 104-21.

EDITORIAL (1982). Mathematics and measles. *Lancet*, II, 248-9.

FINE, P.E.M. and CLARKSON, J.A. (1982). Measles in England and Wales - II: The impact of the measles vaccination programme on the distribution of immunity in the population. *Int. J. Epidemiol.*, 11, 15-25.

GRIFFITHS, D.A. (1974). A catalytic model of infection for measles. *Appl. Statist.*, 23, 330-9.

HEDRICH, A.W. (1933). Monthly estimates of the child population susceptible to measles, 1900-1931, Baltimore, Md. *Am. J. Hyg.*, 17, 613-636.

HETHCOTE, H.W. (1983). Measles and rubella in the United States. *Am. J. Epidemiol.* , 117, 2-13.

HORWITZ, O., GRÜNFELD, K., LYSGAARD-HANSEN, B. and KJELDSEN, K. (1974). The epidemiology and natural history of measles in Denmark. *Am. J. Epidemiol.*, 100, 136-149.

MARTINI, E. (1921). *Berechnungen und Beobachtungen zur Epidemiologie und Bekämpfung der Malaria*. Hamburg: Gente.

PAULE, C.L., BEAN, J.A., BURMEISTER, L.F. and ISACSON. P. (1979). Post-vaccine era measles epidemiology. *JAMA*, 241, 1474-6.

YORKE, J.A., NATHANSON, N., PIANIGIANI, G. and MARTIN, J. (1979). Seasonality and the requirements for perpetuation and eradication of viruses in populations. *Am. J. Epidemiol.*, 109, 103-23.

PARAMETER ESTIMATION AND VALIDATION FOR AN EPIDEMIC MODEL

Ronald Shonkwiler
School of Mathematics
Georgia Institute of Technology
Atlanta, Georgia 30332 USA

and

Maynard Thompson
Department of Mathematics
Indiana University
Bloomington, Indiana 47405 USA

1. Introduction

In an earlier report [2] a general model for common source epidemics was developed and subsequently [3] specialized to an outbreak of toxoplasmosis. The model for infection developed in [4] can be used to refine the results in [3] and provide a stronger physiological basis for the study. In this paper we use epidemiological data collected in the toxoplasmosis outbreak, in particular the daily onset data, to estimate the experimental parameters of the model. We then validate the model with respect to its predictive ability.

The population as described in [3] can be partitioned into 20 subgroups with respect to attendance pattern and physical location. The model of [2] as specialized in [3] has two components. The first describes the introduction of the pathogens into the environment and their diffusion. The output of the first component is an exposure history for each of the 20 groups. As applied to the toxoplasmosis outbreak, the exposure component of the model involves one experimental parameter, a diffusion velocity. Other important parameters can be determined from studies reported in the literature.

The second component of the model takes an exposure history and an attendance/location group (assumed consisting entirely of susceptibles) as input and provides as output a probability that an individual in that group becomes ill. It is convenient to view this infection process as consisting of two phases. First the pathogens must penetrate an individual's physical and immuniological barriers to reach target tissue, and second, the pathogens must survive in a hostile environment. This infection component of the model involves two experimental parameters, the transmission rate for pathogen invasion, and the net rate of increase, the difference of the birth and death rates of the pathogen in vivo. By fixing the birth rate (using reasonable biological data), the difference can be replaced by a single experimental parameter, the death rate.

In Section 2 we summarize the basic equations of the model, in Section 3 we describe the parameter estimation process, and in Section 4 we discuss the validation question.

2. Fundamental Equations of the Model

Suppose that the time horizon of interest is of length T, and take the initial reference time to be 0. Partition $[0,T)$ into n cells $J_k = [t_{k-1},t_k)$, $k = 1,2,\ldots,n$, where $0 = t_0 < t_1 < \ldots < t_n = T$. Let $C(k)$ denote the probability that a randomly selected individual from the m-th attendance/location group becomes infected during $[0,t_k)$. Thus C is the cumulative distribution function for the time of infection for the m-th subgroup. The fundamental equation for C is

$$C(k) = \sum_{i=1}^{k} (1-C(i-1))\Pr[\text{infected in } J_i \mid \text{uninfected at } t_{i-1}].$$

Consequently, it is clearly sufficient to determine, for each subgroup, $\Pr[\text{infected in } J_i \mid \text{uninfected at } t_{i-1}]$. Referring back to the discussion of Section 1, we consider first the exposure component of this term. Pathogens introduced into the environment at a fixed location diffuse according to a two dimensional diffusion equation with radial symmetry. The concentration at distance r and time t of a "pulse" of pathogens is given by

$$u(r,t) = \frac{1}{4\pi ct} \exp(-r^2/4ct)$$

where c is the diffusion velocity, taken to be an experimental parameter. Using this and data on toxoplasma gondii (see [3] for details), we can determine the intensity of pathogens at each location and for each time.

We are now ready to turn to the infection component of the model. Let $N = N(t)$ denote the number of pathogens in an individual at time t. Following [4], we assume that the number of pathogenic agents which invade an individual in a time interval Δt is proportional to the intensity of pathogens in the immediate environment. The proportionality constant, a penetration rate which we refer to as transmissivity, is an experimental parameter. If we denote it by τ, then our basic relation is $\Delta N = \tau(\text{Intensity})\Delta t$.

Once pathogens enter a host, we assume that they evolve according to a birth-death process. Using reasonable biological data we fix the birth rate λ, and we retain the death rate μ as an experimental parameter. Let $P_t[X = x]$ be the probability that the random variable X for the number of pathogens has value x at time t given a single pathogen at time 0. Also, let $E[X](t)$ be the expected number of pathogens at time t. Then we have for $\lambda \neq \mu$

$$E[X](t) = e^{(\lambda-\mu)t}, \quad P_t[X=0] = \alpha, \quad P_t[X=x] = (1-\alpha)(1-\beta)\beta^{x-1}$$

where

$$\alpha = \frac{\mu(e^{(\lambda-\mu)t}-1)}{\lambda e^{(\lambda-\mu)t}-\mu}, \quad \beta = \frac{\lambda}{\mu}\alpha$$

and if $\lambda = \mu$, $E[X](t) = 1$,

$$P_t[X=0] = \alpha = \beta = \frac{1}{1+(1/\lambda t)}, \quad P_t[X=x] = (1-\alpha)^2\alpha^{x-1}.$$

3. Estimation of Experimental Parameters

We estimate the experimental parameters using the method of maximum likelihood. That is, for each individual we determine, as a function of the experimental parameters,

$$\Pr[\text{becomes ill}]\ \Pr[\text{onset day is } D \mid \text{becomes ill}],$$

and our likelihood function is the product of these expressions for all individuals. The onset day in this expression is the day for which the criteria for illness is first met. We take as the criteria for illness a pathogen population size exceeding a fixed, specified value. The cumulative distribution function C of Section 2 is the distribution of onset days for the mth subgroup. Due to the complexity of the model, it is not feasible to derive an analytic expression for the likelihood function. Instead we obtain an approximate density function for each parameter set by Monte Carlo simulation. In particular, 100 simulations were run for each parameter set and for each of the 15 groups actually represented in the population. Based on preliminary analysis the following parameter values were selected for full simulation:

values for c: 0.5, 1, 3, 6, 10;

values for τ: 1, 3, 5, 7, 9;

values for μ: 2.90, 2.94, 2.98, 3.0, 3.02, 3.04, 3.10.

Information recorded for each simulation run included attack rate (fraction of each group which became ill), and the first and second statistical moments of onset of illness for those who became ill. In addition, for selected runs the group of the probability density function was printed.

These graphs suggested that the onset day is a Gamma distributed random variable. A Kolmogorov-Smirnov test was performed sustaining this hypothesis at the 35% level (acceptance is usually taken at the 5% level).

Using the simulation results we determined the probability that the data reported in the actual epidemic would occur as an outcome of the simulation. The maximum likelihood choice of the parameters are those for which this probability is largest. They are: $c = 3.0$, $\tau = 3.0$, $\mu = 3.0$. Note that this set is well inside the lattice of parameter values considered.

4. Model Validation

We determine the extend to which the predictions based on the model are consistent with the observed toxoplasmosis data by using a chi-square test for goodness of fit. Thus, we partition the possible outcomes into a number of cells. The use of the test is justified (see [1], p. 426) by the following observations: First, as a result of the nature of the expressions summarized in Section 2 of this paper, the cell probabilities are twice continuously differentiable functions of the parameters (for $c \geq c_0 > 0$), and the independence condition (condition d) on p. 427 in [1]) follows from the analytic form of the functions arising in the birth-death process. It should be pointed out that since the maximum likelihood values of the experimental parameters were determined based upon individuals, the chi-square statistic may not be asymptotically chi-square (see [5]). Consequently, the type I error may be under-estimated.

The normal requirement that there be at least 5 expected outcomes in each cell is easily met. We have 67 susceptibles and we partition them into 10 cells.

Our adaptation of the chi-square test is somewhat unusual in that the model makes two very different kinds of predictions. First, it assesses the likelihood of a given individual becoming ill. Second, in the case of illness, it predicts onset day. In order that there be equal weight assigned to each prediction, we use an equal number of cells, 5, for each. We discuss each contribution separately.

Chi-Square for Illness. The individuals were labelled and then assigned randomly into 5 cells. Next, using the group-specific attack rates generated by the simulations (with parameters determined by maximum likelihood) the expected numbers of individuals avoiding illness in each cell were calculated. The contribution to the chi-square statistic relating to illness/no illness was computed as shown in Table 1.

Chi-Square for Onset Day. Thirty-three of the 67 individuals included in the study became ill. Corresponding to each group there

TABLE 1.

$c = 3.0,\ \tau = 3.0,\ \mu = 3.0.$

Cell Number	Number of Susceptibles	Expected Number not Sick E	Actual Number not Sick A	Chi-Square $(E-A)^2/E$
1	15	5.61	7	0.344
2	12	7.19	6	0.197
3	17	6.43	7	0.051
4	12	5.45	7	0.441
5	11	5.11	7	0.699
TOTALS	67	29.79	34	1.732

is a specific Gamma distribution for onset day of illness. Partitioning the probability mass of this distribution into 5 quintiles gives rise to 5 equally likely onset categories: very early, early, expected, late, and very late. Therefore, we have constructed 5 onset cells with the expected number of outcomes in each equal to 33/5 = 6.6.

On the other hand, from the data of the epidemic we know the actual onset days for those who became ill, and consequently, the actual number in each cell. The results are shown in Table 2.

TABLE 2.

$c = 3.0,\ \tau = 3.0,\ \mu = 3.0$

Onset Category	Expected Number E	Actual Number A	Chi-Square $(E-A)^2/E$
Very Early	6.6	4	1.024
Early	6.6	6	0.055
Expected	6.6	6	0.055
Late	6.6	7	0.024
Very Late	6.6	10	1.752
TOTAL	33	33	2.910

Summary of Validation Results. Combining the contributions which assess the predictions of attack rate and onset of illness we have a chi-square statistic of 1.73 + 2.91 = 4.64. Since there are 10 cells and we have estimated 3 parameters from the data, there are 6 degrees of freedom. Referring to a chi-square table, we find the level of significance to be 59%. However, in view of the earlier discussion, this figure cannot be entirely trusted.

5. Discussion and Conclusions

The toxoplasmosis epidemic provided a stringent test of the model. Indeed, several aspects of this application should be noted. First, the phenomena represented by the data are very complex, as indicated by the varying attack rates and distribution of onset days, yet the model requires only 3 experimental parameters. Next, while we were fortunate in having fairly detailed epidemiological data, certain matters remain unclear. For instance, the amount of time a specific individual spent in his/her primary location cannot be precisely determined. Likewise, the onset days were reported retrospectively and may be subject to recollection errors. It is interesting that though independent reasoning along evolutionary lines would lead to the expectation that pathogen birthrate and deathrate would be close, such a conclusion emerges from this study. Thus, in many respects the results of this effort are quite encouraging.

References

[1] H. Cramer, Mathematical Methods of Statistics, Princeton Universt University Press, Princeton, 1946.

[2] R. Shonkwiler and M. Thompson, Common Source Epidemics I: A Stochastic Model, Bull. Math. Biol. 44 (1982), 259-269.

[3] ———, Common Source Epidemics II: Toxoplasmosis in Atlanta, Bull. Math. Biol. 44 (1982), 377-398.

[4] ———, A Model for Infection, in Population Biology Proceedings, Edmonton, 1982, H. I. Freedman and C. Strobeck, eds., Lecture Notes in Biomathematics No. 52, Springer-Verlag, New York, 1983, 429-432.

[5] H. Chernoff and E. L. Lehman, The Use of Maximum Likelihood Estimates in χ^2 Tests for Goodness of Fit, 25 (1954), 579-586.

LOCAL STABILITY IN EPIDEMIC MODELS FOR HETEROGENEOUS POPULATIONS

Horst R. Thieme

Universität Heidelberg, SFB 123

D-6900 Heidelberg, BR Deutschland

The dynamics of many infectious diseases cannot be properly described by a mathematical model, if it does not take the structure of the host population into account. Typical examples are venereal diseases, where the population has to be divided into several subgroups according to sex, symptomatic and asymptomatic infection etc., and diseases also involving intermediate hosts.

In order to obtain some insight into how the prevalence of an endemic disease depends on the parameters of the model it is useful to study special solutions to the model equations, in particular stationary states (equilibrium points). So Hethcote, Yorke and Nold [2], in order to evaluate the effectivity of different control methods for gonorrhea, analysed their impact on the stationary endemic state (endemic equilibrium). This procedure, however, is only appropriate, if the special solutions describe a typical situation of the endemic. So Hethcote et al. knew [2] from the work of Lajmanovich and Yorke [3] that the stationary endemic state is globally stable, i.e. all solutions starting from a non-zero initial state converge to the endemic equilibrium. The minimum requirement for the stationary endemic state to be epidemiologically relevant is local asymptotic stability, i.e. if the endemic is driven away from the equilibrium by a small perturbation it remains in a close neighbourhood and finally returns to the equilibrium state.

A lot of local asymptotic and also some global stability results are known for epidemics in homogeneous populations (see the paper by Hethcote, Stech and van den Driessche [1] for a survey). Apart from the global stability result for S-I-S models in [3] and its generalisation to S-E-I-R-S models with short incubation and immunity periods by the author [6] few results seem to be known on local or global stability in models for heterogeneous populations. The main difficulty arises from the fact that, in general, the stationary endemic state can no longer be determined explicitly. This paper shows how this problem can be overcome by using the spectral properties of irreducible non-negative matrices. The arguments apply to various epidemic models, but in order to be specific we consider an S-E-I-R-S model consisting of o.d.es..

The model can be represented schematically in the following way:

$$S_j \longrightarrow E_j \longrightarrow I_j \longrightarrow R_j \longrightarrow S_j$$

$$S_k \longrightarrow E_k \longrightarrow I_k \longrightarrow R_k \longrightarrow S_k \quad .$$

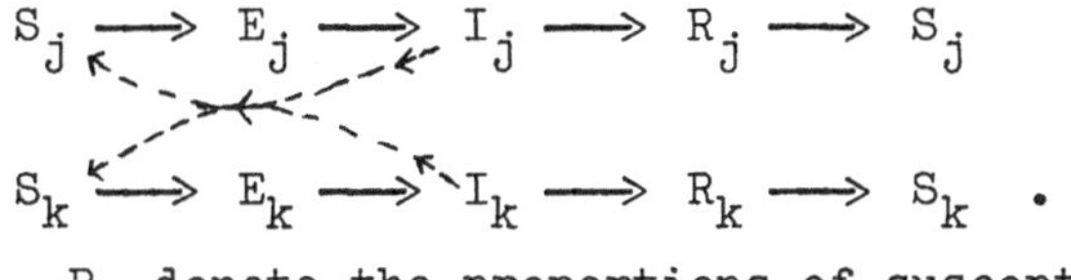

S_j, E_j, I_j, R_j denote the proportions of susceptible, exposed (incubating the disease), infected, removed (immune) individuals in the jth of m subpopulations which form the host population. The epidemic interaction between the subpopulations is due to the ability of infective individuals to infect susceptibles in others than their own subpopulation. We confine our consideration to non-lethal diseases, but we include some vital dynamics in so far as individuals may enter or leave the subpopulations (by birth, death or comparable processes); we assume, however, that the subpopulations remain constant in size. The size of every subpopulation has been normalized to unity. (See [6]).

The model equations for every subpopulation have the form:

$$(1)\qquad \begin{aligned} E_j' &= (1-E_j-I_j-R_j)\,Q_j - (\beta_j+\mu_j)\,E_j \\ I_j' &= \beta_j E_j - (\gamma_j+\mu_j)\,I_j \\ R_j' &= \gamma_j I_j - (\rho_j+\mu_j)\,R_j \end{aligned}$$

We remark that $S_j = 1-E_j-I_j-R_j$ gives the proportion of susceptible individuals in the jth subpopulation. The epidemic interaction between the different subpopulations is described by Q, namely

$$(2)\qquad Q_j = \sum_{k=1}^{m} \alpha_{jk}\, I_k$$

Further we impose non-negative initial conditions satisfying

$$(3)\qquad E_j(0) + I_j(0) + R_j(0) < 1$$

in every subpopulation j. Infection is described by the usual law of mass action with Q_j denoting the infective impact on subpopulation j. α_{jk} gives the infective impact on susceptibles in subpopulation j by infectives in subpopulation k. The durations of the periods of incubation, infectivity and immunity are assumed to be exponentially distributed with mean durations $1/\beta_j$, $1/\gamma_j$ and $1/\rho_j$. μ_j indicates the rate of entering and leaving subpopulation j with all individuals that enter being susceptible.

It follows by standard arguments (Banach's fixed point theorem) that there exists a unique non-negative solution to (1), (2), (3) globally for all $t \geq 0$.

Before we study the existence and local asymptotic stability of stationary endemic (i.e. non-negative, non-zero) states, we make the following

<u>ASSUMPTIONS</u> a) $\mu_j, \gamma_j, \rho_j \geq 0$; $\beta_j > 0$.

b) $\mu_j > 0$ or $\gamma_j, \rho_j > 0$.

c) The matrix $A = (\alpha_{jk})$ is non-negative and irreducible, i.e.
$\sum_1^p A^q$ is a strictly positive matrix for some natural number p.

Assumption b) implies that $\gamma_j + \mu_j > 0$ and $\rho_j + \mu_j > 0$, i.e. no infective individual stays infective forever and the reservoir of susceptibles is replenished by newborn individuals and/or removed individuals becoming susceptible again. Assumption c) means that the disease finally affects all subpopulations regardless of the subpopulation in which it first appears.

A sufficient and necessary condition for the existence and uniqueness of a stationary endemic (i.e. non-negative, non-zero) state (endemic equilibrium) follows essentially in the same way as in [3,4].

THEOREM 1. *There exists an endemic equilibrium of* (1) *iff the spectral radius of the following matrix* $B = (b_{jk})$ *exceeds one*:

$$(4) \qquad b_{jk} = (\beta_j + \mu_j)^{-1} \alpha_{jk} \beta_k (\gamma_k + \mu_k)^{-1}$$

The stationary endemic equilibrium is uniquely determined and given by a strictly positive vector.

It follows from (1.1) that

$$(5) \qquad \tilde{E} = C\tilde{E}$$

with the vector $\tilde{E} = (\tilde{E}_1, \dots, \tilde{E}_m)$ and the matrix

$$(6) \qquad C = ([1 - \tilde{E}_j - \tilde{I}_j - \tilde{R}_j]\, b_{jk})$$

and $\tilde{E}_1,\dots,\tilde{E}_m$, $\tilde{I}_1,\dots,\tilde{I}_m$, $\tilde{R}_1,\dots,\tilde{R}_m$ denoting the stationary endemic state. It follows from the proof of theorem 1 that C is a non-negative matrix.

As usually the local asymptotic stability of the stationary endemic state is studied by linearizing system (1) around that state. It is well-known that the stationary endemic state is asymptotically stable if there are no solutions to the linearized system having the form

$$e^{zt}\,(u_1,\dots,u_m,\ v_1,\dots,v_m,\ w_1,\dots,w_m)$$

with complex numbers z, u_j, v_j, w_j and the real part $\operatorname{Re} z$ of z being non-negative. In order to check this condition we eliminate v_j and w_j from the equations. Making the transformation $u_j = \tilde{u}_j(1 + z/(\gamma_j + \mu_j))$ we obtain

$$(7) \qquad (1 + \tilde{\varepsilon}_j(z))\,(1 + \varepsilon_j(z))\,\tilde{u}_j = (C\tilde{u})_j$$

with the matrix C defined in (6) and complex numbers $\tilde{\varepsilon}_j(z)$, $\varepsilon_j(z)$ satisfying

$$(8) \qquad \operatorname{Re} \tilde{\varepsilon}_j(z) \geq 0 \quad \text{and} \quad \operatorname{Re} \varepsilon_j(z) > 0 \qquad \text{if} \quad \operatorname{Re} z \geq 0 .$$

We suppose that $\operatorname{Re} z \geq 0$. With $\varepsilon(z)$ being the minimum of the real parts of $\varepsilon_j(z)$, $j = 1,\ldots,m$, $|\tilde{u}| = (|\tilde{u}_1|,\ldots,|\tilde{u}_m|)$, and '$\leq$' being the canonical ordering in $\mathbb{R}^m$ we obtain from (7) and (8) that

$$(9) \qquad (1 + \varepsilon(z))\;|\tilde{u}| \;\leq\; C\;|\tilde{u}| .$$

We define η to be the minimum number such that $|\tilde{u}| \leq \eta \tilde{E}$. Since $\tilde{E}$ is a strictly positive vector by theorem 1, $\eta < \infty$. Now $(1 + \varepsilon(z))\,|\tilde{u}| \leq C\,|\tilde{u}| \leq \eta C\tilde{E} \leq \eta \tilde{E}$ by (9) and (5). Since $\varepsilon(z) > 0$ by (8) this inequality contradicts the minimality of η; hence $\operatorname{Re} z < 0$. So we have proved

THEOREM 2. *The endemic equilibrium of* (1) (*if it exists*) *is locally asymptotically stable*.

This method also applies to epidemic models which handle incubation or immunity in a different way, e.g. by assuming periods of fixed length. Such a model has been considered in [6] where global stability is studied. In this example one cannot prove local asymptotic stability for all parameter values by the above method, but it is possible to give sufficient conditions which are considerably weaker than the conditions found for global stability. If the discrete subdivision of the population is replaced by a continuous one (spatial distribution, e.g.), this method still works. It is restricted, however, to epidemic models the equations of which contain the heterogeneous interaction in a monotone increasing way.

REFERENCES

[1] Hethcote, H.W.; Stech, H.W; van den Driessche, P.: Periodicity and stability in epidemic models: a survey. Differential Equations and Applications in Ecology, Epidemics and Population Dynamics (S.N. Busenberg, K.L. Cooke, eds.). New York - London - Toronto - Sydney - San Francisco: Academic Press 1981, 65-82

[2] Hethcote, H.W.; Yorke, J.A.; Nold, A.: Gonorrhea modelling: a comparison of control methods. Math. Biosci. 58 (1982), 93-109

[3] Lajmanovich, A; Yorke, J.A.: A deterministic model for gonorrhea in a nonhomogeneous population. Math. Biosci. 28 (1976), 221-236

[4] Nold, A.: Heterogeneity in disease transmission modeling. Math. Biosci. 52 (1980), 227-240

[5] Krasnosel'skii, M.A.: Positive Solutions of Operator Equations. Groningen: Noordhoff 1964

[6] Thieme, H.R.: Global asymptotic stability in epidemic models. Equadiff 1982

PART III

RESOURCE MANAGEMENT

BAYESIAN METHODS IN ECOLOGY AND RESOURCE MANAGEMENT

Colin W. Clark
Department of Mathematics
University of British Columbia
Vancouver V6T 1W5 CANADA

The behavior of predators, both natural and human, must be influenced to some extent by the need to obtain information pertaining to the location and abundance of their prey. If prey items are uniformly distributed at random throughout the entire environment, then no particular search strategy is of any avail. (If predation depletes the prey in localized areas, then the uniform distribution is upset, and strategy may become important.)

In the vast majority of cases, however, the distribution of prey is patchy, and this fact opens up a number of questions of strategic nature for the predator. (The possibility of aggregation of prey also opens up strategic considerations for the prey, but this is not our present concern.) For example, how much time should the predator spend feeding in a given patch? -- this question has been addressed in the literature on optimal foraging theory (Charnov 1976, Oaten 1977, Green 1980, Iwasa et al 1981, McNair 1982). If patch locations are known to the predator, but the quality of the patches is uncertain, in what order should the patches be sampled (Mangel and Clark 1983, Clark and Mangel 1983)? If patch location or quality is uncertain, would cooperative searching be more efficient than independent search (ibid)?

These questions are primarily of a short-term or intermediate-term nature. For human resource managers, more long-term questions may also arise, related to the estimation of biological rates and other important model parameters (Ludwig 1982).

A common characteristic of many of these situations is that the actual process of exploitation of the prey resource provides, besides the immediate capture of prey, information which the predator can use to improve his knowledge of resource location, abundance, and other properties. This knowledge allows the predator to improve his future prospects. It is then an interesting question whether the appropriate predator strategy is a "passive" one, in which information is accumulated merely as a by-product of the predation, or an "active" one, in which the predation pattern is itself influenced by the information requirement. The answer should clearly depend on how valuable the information is, and this depends in turn on the degree of uncertainty facing the predator at any given time. As information is accumulated and uncertainty reduced, one might expect to see predation strategies progress from an active to a passive mode.

Although problems of this kind appear interesting and important, it is only recently that they have attracted much attention. Some recent work will be described in the following pages.

Search Theory and Fisheries

The following model is discussed by Mangel and Clark (1983). We imagine two fishing grounds A_1, A_2, on each of which the abundance of fish varies from year to year. The fluctuations are independent of one another, and from one year to the next. (The model can be extended in many ways.)

A vessel fishing on ground A_i encounters schools of fish according to the Poisson distribution:

$$\text{prob (encounter one school in } t,\ t+dt) = \lambda_i dt \tag{1}$$

The encounter rate λ_i is proportional to the average density of fish on ground A_i. Thus each λ_i is a random variable, varying from year to year. At the start of a given fishing season, the fisherman does not know the values of λ_1, λ_2, but he does know (from experience) their probability distributions

$$f_i(\lambda_i)d\lambda_i = \text{prob } (\lambda_i \in (\lambda_i, \lambda_i + d\lambda_i)) \tag{2}$$

For reasons which will soon become clear we assume that f_i is a gamma distribution, $f_i(\lambda_i) = \gamma(\lambda_i;\nu_i,\alpha_i)$, where

$$\gamma(\lambda;\nu,\alpha) = \frac{\alpha^\nu}{\Gamma(\nu)} \lambda^{\nu-1} e^{-\alpha\lambda} \qquad (\lambda \ge 0) \tag{3}$$

The mean and variance of this distribution are given by

$$\begin{aligned} \bar{\lambda} &= \nu/\alpha \\ \sigma_\lambda^2 &= \nu/\alpha^2 \end{aligned} \tag{4}$$

The coefficient of variation is thus $\sigma_\lambda/\bar{\lambda} = 1/\sqrt{\nu}$.

If k vessels search independently on ground A (subscripts suppressed for simplicity) for time t, and if n denotes the total number of schools encountered, we have

$$\text{prob } (n) = \frac{(k\lambda t)^n e^{-k\lambda t}}{n!} \tag{5}$$

(Note that the effect of <u>removal</u> of the schools is ignored here: see Mangel and Clark 1983.) We are interested, however, in the inverse problem: <u>given</u> that n schools are encountered, by k vessels in time t, how does this fact affect the probability distribution for λ? This question can be answered by means of Bayes' theorem:

$$f(\lambda|n) = \frac{\text{prob }(n|\lambda)f(\lambda)}{\int \text{prob }(n|\mu)f(\mu)d\mu} \tag{6}$$

where prob $(n|\lambda)$ is given by (5).

A simple calculation shows that, when the prior distribution $f(\lambda)$ is a gamma distribution $\gamma(\lambda;\nu,\alpha)$, then the updated, or posterior distribution $f(\lambda|n)$ is also a gamma distribution:

$$f(\lambda|n) = \Gamma(\lambda,\gamma+n,\alpha+kt) \tag{7}$$

(The gamma and Poisson distributions are said to be <u>conjugate</u> distributions.) The mean and coefficient of variation of the updated distribution are therefore

$$\begin{aligned} \bar{\lambda}' &= \frac{\nu + n}{\alpha + kt} \\ CV' &= 1/\sqrt{\nu+n} \end{aligned} \tag{8}$$

Since the coefficient of variation measures the degree of uncertainty in the estimate of the abundance parameter λ, we see that the data generated by the fishery progressively reduces this uncertainty -- as one would expect. Also, the rate at which this uncertainty can be expected to decrease obviously increases with the number of participating vessels.

We can now pose our strategic problem. Assume that the fishing fleet, consisting of M vessels, can each make N trips of unit length $t = 1$, to either fishing ground A_i, in the year. How many vessels should be allocated to each ground for each trip, in order to maximize the year's expected total catch?

To address this question we use the method of dynamic programming. Suppose there are n trips left to be made, and let ν_i,α_i denote the currently estimated parameter values for ground A_i. Let $J_n(\nu_1,\alpha_1,\nu_2,\alpha_2)$ denote the maximum expected catch for the remaining n trips.

For n =1 we have simply

$$\begin{aligned} J_1(\nu_1,\alpha_1,\nu_2,\alpha_2) &= \max_{k_1+k_2=N} E\{k_1 n_1 + k_2 n_2\} \\ &= \max_{k_1+k_2=N} \left(k_1 \frac{\nu_1}{\alpha_1} + k_2 \frac{\nu_2}{\alpha_2}\right) \\ &= N \max_i \left(\frac{\nu_i}{\alpha_i}\right) \end{aligned} \tag{9}$$

and all N vessels should be sent to the apparently most productive ground. With only one trip remaining, nothing is to be gained from updating.

For general n the dynamic programming equation is

$$J_{n+1}(\nu_1,\alpha_1,\nu_2,\alpha_2) = \max_{k_1+k_2=N} \left[k_1\nu_1/\alpha_1 + k_2\nu_2/\alpha_2 + E\{J_n(\nu_1+n_1,\alpha_1+k_1,\nu_2+n_2,\alpha_2+k_2)\}\right] \quad (10)$$

where the expectation is over the random variables n_1,n_2:

$$E\{\ldots\} = \sum_0^\infty \sum_0^\infty J_n(\cdots)\ \text{prob}\ (n_1)\ \text{prob}\ (n_2) \quad (11)$$

where

$$\text{prob}\ (n) = \int_0^\infty \text{prob}\ (n|\lambda)\gamma(\lambda;\nu,\alpha)d\lambda$$

$$= \frac{k^n}{n!} \frac{\alpha^\nu}{(\alpha+k)^{n+\nu}} \frac{\Gamma(n+\nu)}{\Gamma(\nu)} \qquad (k \neq 0) \quad (12)$$

and

$$\text{prob}\ (n) = \begin{cases} 1 & n = 0 \\ 0 & n > 0 \end{cases} \qquad (k = 0) \quad (13)$$

In principle we can now proceed to determine the optimal allocation k_i for every n. The computations are distinctly nontrivial, however, for n > 2. For n = 2 the computation can be done on a microcomputer. A typical result is shown in Figure 1, for which the following parameter values were used (Mangel and Clark 1983):

$$N = 10,\ \alpha_1 = \alpha_2 = 0.1,\quad \gamma_1 = \gamma_2 = 1.0$$

The figure demonstrates the importance of allocating at least some search effort to each ground, although in this case it makes little difference how many vessels are actually used on each ground. Similar results are obtained with other choices of parameter values. Sampling is particularly important on grounds on which the prior coefficient of variation $1/\sqrt{\nu_i}$ is large.

The above model does not allow for progressive depletion of the fish stock as fishing proceeds. Clearly such depletion is possible, especially for small "patches," or fishing areas, but the formulas and computations become somewhat more complicated (Mangel and Clark 1983). Other developments of the model appear in Mangel and Beder (1983) and Mangel and Plant (1983).

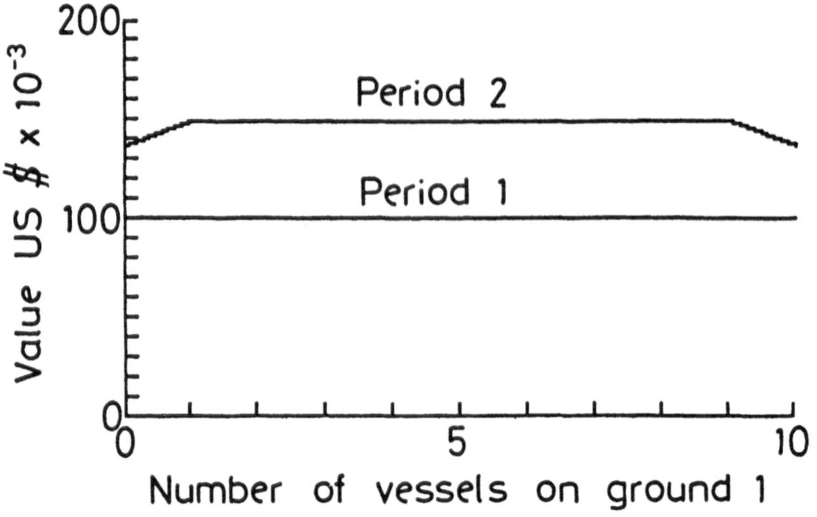

Figure 1. Expected first and second year returns J_1, J_2 for fishery search model (Mangel and Clark 1983).

Foraging and Flocking Strategies

The method of Bayesian updating has been applied to optimal foraging theory by Green (1980), and Iwasa et al (1981). The latter paper employs a Poisson model (with depletion) to model search within a patch, and considers three different prior distributions for the number of prey within a patch: negative binomial, binomial, and Poisson. For these distributions the updating formulas for the case of stock depletion are simple. Iwasa et al show that the Bayesian predator's estimate of the expected number of prey left in the patch goes up after each encounter if the prior distribution is negative binomial (a contagious distribution), and down if it is binomial (a regular distribution).

Clark and Mangel (1983) discuss the informational advantages of cooperative foraging. They pose a variety of models to analyze the tradeoff between the increased rate of accumulation of information and the increased rate of resource depletion, as the foraging group size increases. They also argue, on game-theoretic grounds, that foragers may form groups which are much larger than the optimal size.

This work leads one to suspect that the Bayesian view of uncertainty and information might be used to develop new insights into various other aspects of animal behavior.

Optimal Fishing Capacity

Investment decisions must often be made in the presence of major uncertainty. If the initial investment is too small, profitable opportunities may be foregone; in addition the rate of accumulation of information about these opportunities will be small. But financial losses will also occur if the initial investment is too large.

The problem of investment in fishing capacity for a developing fishery is discussed by Clark et al (1983). The simplest model, which is subject to analytic solution, can be described as follows. Recruitment R_k to the fish stock in year k is a log-normally distributed random variable, $\log R_k \sim N(\mu,\sigma^2)$. Thus

$$\bar{R} = e^{\mu+\sigma^2/2}, \quad \sigma_R^2 = \bar{R}^2(e^{\sigma^2}-1) \tag{14}$$

The variance σ^2 is assumed to be known, but the mean μ is uncertain, being estimated from the available historical record of recruitment: $\hat{\mu}_0 = \overline{\log R_k}$. It is appropriate to use a normal distribution for the prior distribution π_0 of μ:

$$\pi_0(\mu) \sim N(\hat{\mu}_0,\sigma^2/N_0) \tag{15}$$

where N_0 = number of historical records. For the case of a developing fishery N_0 will be a small number.

Generations do not overlap, and there is no statistically significant stock-recruitment relationship. (Tropical penaeid shrimps fit the model quite well.)

Let F denote annual fishing mortality, so that

$$C_k = (1-e^{-F})R_k \tag{16}$$

represents the total catch in year k. We assume that F is directly proportional to fishing capacity, e.g. the number of vessels.

The first problem considered is the one-stage decision of choosing capacity F_0 at year k = 0 so as to maximize expected discounted net economic revenue:

$$\underset{F_0}{\text{maximize}}\ J(F_0) = E\left\{\sum_{k=0}^{H} \alpha^k pC_k - \gamma F_0\right\} \tag{17}$$

where p denotes the price of landed fish, γ the cost of capacity (in terms of F), and H is the time horizon. The expectation in (17) is a double expectation, with respect to both the sequence $\{R_k\}$ of future recruitments (assumed i.i.d.) and the prior distribution π_0.

Since these expectations can be easily calculated explicitly, this first problem is trivial:

$$E\{\ldots\} = pA(H)(1 - e^{-F_0})E_{\pi_0}\{\bar{R}\} - \gamma F_0$$

where $A(H) = \sum_0^H \alpha^k$. This gives

$$F_0^* = \log \left[pA_H E_{\pi_0}\{\bar{R}\}/\gamma\right] \quad (\text{if} > 0) \tag{18}$$

The value of F_0^* has the expected sensitivity to cost, price, and productivity parameters. Since $E_{\pi_0}(\bar{R}) = \exp\,(\hat{\mu}_0 + \sigma^2/2N_0)$ we see also that F_0^* is an increasing function of the degree of prior uncertainty, as represented by the term $\sigma^2/2N_0$.

A less trivial problem arises if we allow capacity investment decisions to be delayed to future years, when increased experience will have reduced the uncertainty about average recruitment. Specifically, suppose that a single decision on increasing capacity will be taken after N_1 (fixed) years of operation of the fishery. We then obtain the two-stage optimization problem

$$\underset{F_0 \geq 0}{\text{maximize}} \left[pA(N_1)e^{\hat{\mu}_0 + \sigma^2/2 + \sigma^2/2N_0}(1 - e^{-F_0}) - \gamma F_0 + \alpha^{N_1} E\{J_1(F_0, \hat{\mu}')\}\right] \tag{19}$$

where

$$J_1(F_0, \hat{\mu}') = \max_{F_1 \geq 0} \left[pA(H - N_1)e^{\hat{\mu}' + \sigma^2/2 + \sigma'^2/2}(1 - e^{-(F_0 + F_1)}) - \gamma F_1\right] \tag{20}$$

with $\sigma'^2 = \sigma^2/(N_0 + N_1)$. The expectation in (19) is again a double expectation, with respect to the prior π_0 and with respect to the random variable $\hat{\mu}'$, the updated estimate of recruitment after N_1 additional samples. (We do not consider the possibility of errors in the samples.)

This double expectation can be reduced to a single integral, which itself can be expressed as a sum of error functions. Hence an efficient computer routine can be used to completely solve the above optimization problem -- no explicit dynamic programming is required. Details and numerical results are given in the reference (see Fig. 2).

The numerical results suggest that, unless the initial uncertainty is very large ($N_0 = 1$, CV $\geq$ 100%), little is gained by delaying the investment decision. Further work is urgently needed to test the robustness of this finding. Substantial public funds are currently being devoted to stock assessments in fisheries. Few if any attempts have been made to assess the economic value of this improved accuracy; according to Cyert et al (1975), managers usually tend to overestimate the amount of information needed in order to reach an investment decision.

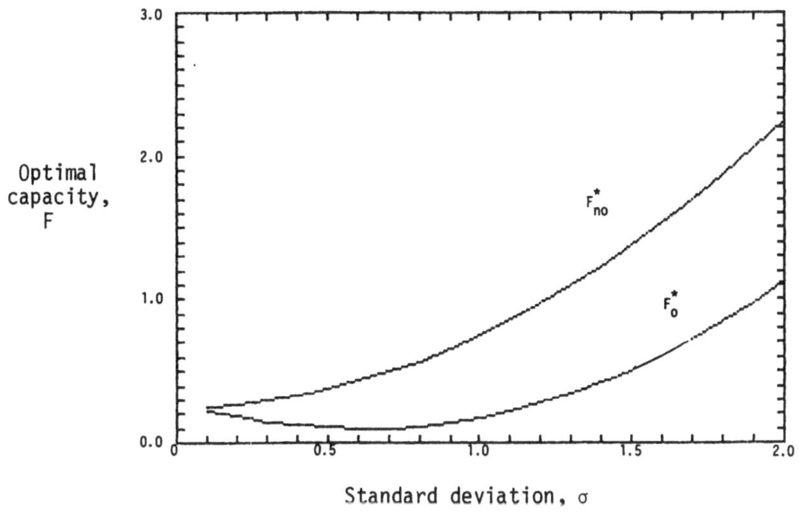

Figure 2. Dependence of optimal fishing capacity F on standard deviation σ (Clark et al 1983).

Other Applications

Significant uncertainty is the rule, rather than the exception, in the resource management field. Current practice, however, is universally to make decisions on the basis of deterministic models, using the "best available" estimates of variables and parameter values. Sometimes – but certainly not commonly – these estimates are accompanied by statistical error estimates, but these are almost never used to influence management decisons in any systematic way. (Indeed, the existence of uncertainty may be invoked by industry spokesmen to plead for higher quotas, which may cause the resource manager to suppress his error estimates.) In principle, at least, the methods of Bayesian decision analysis could be brought to bear in these areas.

Progress has remained slow, however, no doubt partly because of the technical difficulties involved in the dynamic setting characteristic of resource management (see Ludwig 1982), and also because of lack of appropriate training among resource managers. Perhaps with the growing public awareness of the economic importance of resource industries, a more professional approach to the handling of uncertainty will start to develop.

REFERENCES

Charnov, E.L. 1976. Optimal foraging: the marginal value theorem. Theor. Popul. Biol. 9:129-136.

Clark, C.W., A. Charles, J.R. Beddington, and M. Mangel. 1983. Optimal capacity decisions in a developing fishery (m/s).

Clark, C.W. and M. Mangel. 1983. Foraging and flocking strategies: information in an uncertain environment. Amer. Nat. (in press).

Cyert, R.M., M.H. deGroot, and C.A. Holt. 1978. Sequential investment decisions with Bayesian learning. Manag. Sci. 24:712-718.

Green, R.F. 1980. Bayesian birds: a simple example of Oaten's stochastic model of optimal foraging. Theor. Popul. Biol. 18:244-256.

Iwasa, Y., M. Higashi, and N. Yamamura. 1981. Prey distribution as a factor determining the choice of optimal foraging strategy. Amer. Nat. 117:710-723.

Ludwig, D.A. 1982. Optimal harvesting with imprecise parameter estimates. Ecol. Modelling 14:273-292.

Mangel, M., and J.H. Beder. 1983. Random search with depletion. Oper. Res. (in press)

Mangel, M., and C.W. Clark. 1983. Uncertainty, search, and information in fisheries. J. du Cons. Intern. Expl. Mer (in press).

Mangel, M., and R.E. Plant. 1983. Regulation and uncertainty in fisheries: quotas, profit maximizing, and fishermen behavior. J. Marine Econ. (in press)

McNair, J.N. 1982. Optimal giving-up times and the marginal value theorem. Amer. Nat. 119:511-529.

Oaten, A. 1977. Optimal foraging in patches: a case for stochasticity. Theor. Popul. Biol. 12:263-285.

STOCHASTIC DIFFERENTIAL EQUATION MODELS OF FISHERIES IN AN UNCERTAIN WORLD: EXTINCTION PROBABILITIES, OPTIMAL FISHING EFFORT, AND PARAMETER ESTIMATION

Carlos A. Braumann
Universidade de Évora
7000 Évora, Portugal

Abstract. Logistic and Gompertz models of population growth, with an extra term to allow for fishing under constant quotas, constant effort, and mixed policies, are considered. Stochastic fluctuations of environmental and fishing conditions are added as noise terms. Ultimate extinction occurs with probability one for constant quotas or mixed policies, but not for constant moderate effort policies. For constant effort policies, we show that the fishing efforts that maximize the expected yield are close to the ones obtained in the deterministic models if the noise fluctuations are small. Conditions on the effort in order to control the probability of the population size dropping below some critical threshold are studied. Maximum likelihood techniques of parameter estimation based on yield data are developed for constant effort models.

1 - INTRODUCTION

Let $N(t)$ be the fish population size at time t. The growth models are of the form $dN/dt = r\,N\,f(N)$, where $r > 0$ is a growth parameter. Among the several available forms of $f(N)$ we shall consider the logistic model

$$f(N) = (1 - N/K) \tag{1}$$

and the Gompertz model, also known as Fox (1970) model,

$$f(N) = \ln K - \ln N. \tag{2}$$

Here $K > 0$ is the carrying capacity of the environment.

If the population is under fishing pressure there is a fishing removal, which we assume to be of the form $\alpha + \beta N$ ($\alpha, \beta > 0$). The policy of constant quotas corresponds to $\beta = 0$, $\alpha > 0$. Constant effort policies correspond to $\alpha = 0$, $\beta > 0$ (β is the effort). If $\alpha > 0$ and $\beta > 0$, we have a mixed policy. In practice, a rigid constant quota can not be maintained if the population size is low. Actual policies are likely to be much more complex, somewhat like $\alpha N/(N + \varepsilon K)$ ($\varepsilon > 0$), as suggested by Beddington and May (1977), but the optimization and the estimation of parameters, as well as the extinction probabilities are, for these policies, quite difficult to analyse and shall not be considered in this paper.

Random environmental fluctuations can be introduced by an extra term $h(N)\sigma\,\varepsilon(t)$, where $\varepsilon(t)$ can be taken, with good approximation, to be standard white noise and $\sigma > 0$ is a noise intensity parameter. If the noise represents fluctuations on the growth parameter r, we have

$$h(N) = N\,f(N). \tag{3}$$

If the noise represents fluctuations of the per capita growth rate $(1/N)(dN/dt)$ or of the effort parameter β under a constant effort policy we have

$$h(N) = N \tag{4}$$

The model of fishing in a randomly fluctuating environment so obtained is therefore of the form

$$dN/dt = r\,N\,f(N) - (\alpha + \beta N) + h(N)\,\sigma\,\varepsilon(t), \tag{5}$$

where $f(N)$ can be of the form (1) or (2) and $h(N)$ can be of the form (3) or (4). We can interpret this stochastic differential equation (SDE) either in the Ito or in the Stratonovich sense. Braumann (1979, 1980, 1983 b) and citations therein deal with the (mostly apparent) difference between the Ito and the Stratonovich calculus.

We now proceed briefly to the study of the models referred to above with respect to stationary densities, boundary classification, and circumstances under which there is ultimate extinction with probability one.

Optimization of the expected yield is studied next. Many of these questions have been studied in variable depth. See Beddington and May (1977), Gleit (1978), May *et al* (1978), Braumann (1981), Dennis and Patil (1983), and citations therein.

We then study the question of controlling the probability of the population size dropping below some critical threshold. Finally, the problem of parameter estimation based on yield data, which is of great practical importance, is addressed.

2 - CONDITIONS FOR NON-EXTINCTION

The solution $N(t)$ of the SDE (5) is a diffusion process with drift coefficient

$$M(N) = r\,N\,f(N) - (\alpha + \beta N) \tag{6}$$

or

$$M(N) = r\,N\,f(N) - (\alpha + \beta N) + (\sigma^2/2)\,h(N)\,h'(N), \tag{7}$$

according to whether one uses Ito or Stratonovich calculus. The diffusion coefficient is in either case

$$V(N) = \sigma^2\,h^2(N). \tag{8}$$

If there is a stationary density, it is of the form

$$p(N) = \frac{C}{V(N)} \exp\left(\int^N \frac{2M(n)}{V(n)}\,dn\right) \tag{9}$$

for $0 < N < K$ if $h(N) = N\,f(N)$, or for $0 < N < +\infty$ if $h(N) = N$. This can be easily computed (see, for example, Braumann, 1981) and, assumming there is some fishing ($\alpha > 0$ or $\beta > 0$ or both), $p(N)$ is, in every case considered, always integrable at the upper boundary (K or $+\infty$, according to the form of $h(N)$).

If $f(N) = 1 - N/K$ (logistic), $p(N)$ is integrable at the lower boundary $N = 0$ iff $\alpha = 0$ and $r' - \beta > 0$, where $r' = r - \sigma^2/2$ (Ito) or $r' = r$ (Stratonovich). Braumann (1979, 1980, 1983 b) shows that r' means, under either calculus, the expected arithmetic mean of sample values of the intrinsic growth rate. If $f(N)=\ell n\, K - \ell n\, N$ (Gompertz), $p(N)$ is integrable at $N = 0$ only if $\alpha = 0$ and, in the case $h(N)=N\, f(N)$, if, in addition, we use Stratonovich calculus. If these conditions of integrability at $N = 0$ are not met, the boundary $N = 0$ is an atractive or an absorbing boundary and, since the upper boundary is natural, this means that extinction occurs with probability one.

Therefore, if a constant quota ($\alpha > 0$, $\beta = 0$) or a mixed fishing policy ($\alpha>0,\beta>0$) is used, extinction will occur with probability one, although it may take quite a while. On the contrary, under a constant effort policy ($\alpha = 0$, $\beta > 0$), extinction may occur, but (unless the effort is higher than the growth parameter) not with probability one. Beddington and May (1977) have already pointed out, for the logistic model with $h(N) = N\, f(N)$ and using Ito calculus, that a constant quota policy would lead to extinction; they have also studied the "stability" of the steady-state distributions for the logistic model and several fishing policies by means of the characteristic return times.

3 - OPTIMAL FISHING EFFORT

Consider now a constant effort policy ($\alpha = 0$, $\beta > 0$) and, when using the logistic model, assume that the effort β is smaller than r'.

Now $p(N)$ is integrable at both boundaries and there is a stationary probability given by (9), where C is such that $\int_0^{K \text{ or } +\infty} p(N)\, dN = 1$. We have:

A - Logistic model, $h(N) = N$. In this case $p(N)$ is gamma

$$p(N) = \Theta^{\phi}\, N^{\phi-1}\, e^{-\Theta N}/\Gamma(\phi) \quad (0 < N < +\infty) \tag{10}$$

with $\Theta = 2r/(K\sigma^2)$ and $\phi = 2(r'-\beta)/\sigma^2$. The yield from fishing is βN and the expected yield is

$$E\,[\beta N] = \beta\, \phi/\Theta = \beta(r'-\beta)\, K/r.$$

The optimal expected yield is $(K\, r/4).(r'/r)^2$ and the optimal effort is $\beta^* = r'/2$. ∎

B - Logistic model, $h(N) = N\, f(N)$. We have

$$p(N) = \phi\Theta\, [K\Gamma(\Psi)]^{-1}\, (\Theta N/(K-N))^{\Psi-1}\, (1-N/K)^{-\varepsilon}\, .\exp(-\Theta N/(K-N)) \quad (0 < N < K), \tag{11}$$

where $\phi = \beta^2/[r^2-\sigma^2(r+\beta)/2]$, $\varepsilon = 4$ for Ito calculus and $\phi = 1$, $\varepsilon = 2$ for Stratonovich calculus. Also, $\Theta = 2\beta/\sigma^2$ and $\Psi = 2(r'-\beta)/\sigma^2$. Then

$$E\,[\beta N] = K\, \phi\, r\, \Psi/(2\beta)$$

or

$$E\,[\beta N] = K\, \beta\, \Psi\, \Theta^{\Psi}\, U(\Psi-1,\, \Psi-1, \Theta),$$

according to whether one uses Ito or Stratonovich calculus. Here U is a confluent hypergeomeotric function (see Abramowitz and Stegun, 1965, p. 505). For Ito calculus, the optimal expected yield is $K(r-\sigma^2/4)/4$ and the optimal effort is

$\beta^* = r(2r'/\sigma^2)(1-\sqrt{r'/r}) \approx r/2 - 3\sigma^2/16$ (for σ^2 small compared to r). For Stratonovich calculus, numerical methods are required.■

C - Gompertz model, $h(N) = N$. In this case, $p(N)$ is lognormal

$$p(N) = (\Theta\sqrt{2\pi}\,N)^{-1} \exp(-(\ell n N-\mu)^2/(2\sigma^2)) \quad (0 < N < +\infty), \tag{12}$$

with $\Theta = \sigma/\sqrt{2r}$ and $\mu = \ell n\, K-\beta'/r$, where $\beta' = \beta + \sigma^2/2$ (Ito) or $\beta' = \beta$(Stratonovich). The optimal expected yield is $(Kr/e)\,\psi$ with $\psi = \exp\,(\sigma^2/(4r))$ (Ito) or $\psi = 1$ (Stratonovich) and the optimal effort is $\beta^* = r$.■

D - Gompertz model, $h(N) = N\,f(N)$, Stratonovich Calculus. We have

$$p(N) = \Theta^{\phi}\,(K\Gamma(\phi))^{-1}\; y^{-\phi-1} \exp(y - \Theta/y) \quad (0 < N < K), \tag{13}$$

where $y = \ell n(K/N)$, $\Theta = 2\beta/\sigma^2$ and $\phi = 2r/\sigma^2$. The computations of the optimal expected yield and the optimal effort require numerical methods.■

The general conclusion is that the optimal effort is close to the one we obtain for deterministic models if the noise intensity σ is small compared to the growth parameter r.

4 - PROBABILITY OF CROSSING A CRITICAL THRESHOLD

Extinction shall not be taken as meaning $N = 0$. Besides the fact that a population with $N < 1$ individual is extinct, these models do not consider Allee effects, demographic fluctuations, and other phenomena that can lead to extinction when $N < N_o$, where $N_o > 0$ is a critical threshold below which extinction occurs or fishing is not profitable. Let us call $P = \text{Prob}\,(N < N_o)$ the "quasi-extinction probability". Braumann (1979, 1983 a) considers, for models with no fishing, the probabilities Prob($N(t) < N_o$ for some t) and Prob ($N(t) < N_o$) at a fixed time t. Here, only the steady-state situation will be considered. We have

$$P = \int_0^{N_o} p(N)\, dN.$$

Under the conditions and with the notations of section 3, we have for the constant effort models:

A) $\quad P = \gamma(\phi,\Theta N_o)/\Gamma(\phi)$;

B Ito) $\quad P = \sigma^4\left[4(r^2-\sigma^2(r+\beta)/2)\Gamma(\Psi)\right]^{-1}\cdot\left[(\Psi+(\Psi-\Theta)^2)\gamma(\Psi,\Theta z_o)-(1+\Psi+2\Theta+\Theta z_o)\Theta^{\Psi} z_o^{\Psi} e^{-\Theta z_o}\right]$;

B Strat.) $\quad P = \gamma(\Psi,\,\Theta z_o)/\Gamma(\Psi)$;

C) $\quad P = 1 - \Phi\left[(-\beta' + r\,\ell n\,(K/N_o))/(r\Theta)\right]$;

D) $\quad P = 1 - \gamma\,(\phi,\,\Theta\,\ell n\,(K/N_o))/\Gamma(\phi)$.

Here $\gamma(\phi,x) = \int_0^x e^{-t}\,t^{\phi-1}\,dt$ is an incomplete gamma function, $z_o = N_o/(K-N_o)$, and $\Phi(x) = \int_{-\infty}^{x} \exp(-t^2/2)/\sqrt{2\pi}\,dt$ is the cumulative d.f. of the standard normal distribution.

One can either compute P for several β values or, using tables of critical values of probability distributions, compute β such that P becomes equal to some chosen value

P_0. For example, in model C one obtains $\beta' = r\,[\ell n(K/N_0) - \sigma\,\Phi^{-1}(1-P_0)/\sqrt{2r}]$. Either way, we can exert some control on the quasi-extinction probability by an adequate choice of the fishing effort.

5 - PARAMETER ESTIMATIONS

For models with no fishing and noise of type (3), Braumann (1979, 1980, 1983 a) presents moment and maximum likelihood (ML) methods of parameter estimation. This requires the transient p.d.f.'s of N(t) which are not available for the fishery models considered. Soong (1973) presents truncation techniques of approximate moment estimation. Another approach is to assume that the distribution of the population size is at the steady-state or very close to it and use the stationary p.d.f.'s p(N) given in section 3., which is reasonable if a policy of constant effort β is in effect for a long time. We will also assume that expected time-averages coincide with expected ensemble-averages due to the ergodicity of the process.

We shall only analyse constant effort models.

Let us consider the case where the available information are the yields $y_1,\ldots,y_n$, with $y_i = \beta N_i$, of n different years. In the particular case where h(N) = N, the effort β need not be exactly constant over those years, for any random fluctuations of β can be absorbed into the noise term. The log-likelihood function is, in any case,

$$L\,(y_1,\ldots,y_n) = \sum_{i=1}^{n} \ell n\,[p(y_i/\beta)/\beta]$$

and depends on the unknown parameters. Equating the partial derivatives of L with respect to unknown parameters to zero provides a system of equations. Solving that system w.r.t. the unknown parameters, if possible, gives us their ML estimators. The knowledge of the parameters is important to study possible improvements of fishing policies. We are even able to compare alternative fitted models using likelihood ratio tests.

As an example, consider the model A of section 3. We have

$$L(y_1,\ldots,y_n) = n\,[\phi\,\ell n(\varepsilon) - \ell n\,\Gamma(\phi) + (\phi-1)\,\overline{\ell n y} - \varepsilon\,\overline{y}],$$

where $\varepsilon = \Theta/\beta$, $\overline{\ell n\, y} = \Sigma\,(\ell n\, y_i)/n$, and $\overline{y} = \Sigma y_i/n$. The equations $\partial L/\partial\varepsilon = 0$ and $\partial L/\partial\phi = 0$ give the system

$$\varepsilon = \phi/\,\overline{y} \tag{14}$$

$$\psi(\phi) - \ell n\phi = \overline{\ell n y} - \ell n\overline{y}, \tag{15}$$

where $\psi(\phi) = d\,\ell n\,\Gamma(\phi)/d\phi$ is the digamma function. Solving (14) by numerical techniques gives an estimator of ϕ. Substituting into (15) gives an estimator of ε. The original parameters were K, σ, r, β, and so the estimators of ϕ and ε allow us to estimate two of the original parameters only if we know the other two. If, however, we have a long fishing period with a given (unknown) effort β_1 followed by another with an effort $\beta_2 = q\,\beta_1$ (and q can be easily known), we can obtain estimates of ϕ_1, ε_1 and ϕ_2, ε_2 from which we can estimate the four unknown parameters K, σ, r, β_1.

Moment estimators can also be obtained using sample averages, variances, etc., of yields $y_i = \beta N_i$ to estimate E $[\beta N]$, Var$[\beta N]$, etc. and then using the expressions of E$[\beta N]$ obtained in section 3. (and of Var$[\beta N]$, etc. that can be obtained similarly) to estimate the parameters.

6 - CONCLUSION

The parameter estimation techniques give a real possibility of applying stochastic models to actual fisheries where records on yields and efforts have been kept for some time. Nature is not, or at least does not look to us, deterministic. The scattered points of fishing data graphs can only reinforce this view. Stochastic modelling seems therefore a more appropriate tool of analysis. Although there is no pure constant quota policy, there are situations that come close to it. The stochastic models have shown us the danger of such policies and also of constant effort policies with excessively large efforts. Optimal efforts may also deviate, although usually not very much, from the ones obtained under deterministic models and that may improve the yields. The effort can also be adjusted to take into account the quasi-extinction probabilities, enabling us to exert some control over the risks of population extinction.

I believe stochastic fishery models can be quite useful.

REFERENCES

Abramowitz, M. and I. A. Stegun, eds. (1965). *A handbook of mathematical functions.* Dover, N.Y.

Beddington, J.R. and R.M. May (1977). Harvesting natural populations in a randomly fluctuating environment. *Science 197 (4302)*: 463-465.

Braumann, C.A. (1979). *Population growth in random environments.* Ph. D. thesis, S.U. N.Y., Stony Brook, N.Y.

Braumann, C.A. (1980). Time-average methods for estimation of parameters and prediction for some stochastic differential equation models. Invited talk at the *Int. Summer School on Statistical Distributions in Scientific Work*, Ms. 93, Trieste.

Braumann, C.A. (1981). Pescar num mundo aleatório: um modelo usando equações diferenciais estocásticas. *Proceedings of the XII Congresso Luso-Espanhol de Matemática*, pp. 301-308, Coimbra

Braumann, C.A. (1983 a). Population extinction probabilities and methods of estimation for population stochastic differential equation models. In *Nonlinear Stochastic Problems*, R.S. Bucy and J.M.F. Moura (eds.), pp. 553-559, D. Reidel Publ. Co.

Braumann, C.A. (1983 b). Population growth in random environments. *Bull. of Math. Biology 45(4)*: 635-641.

Dennis, B. and G.P. Patil (1983). The gamma distribution and weighted multimodal gamma distributions as models of population abundance. (submited)

Gleit, A. (1978). Optimal harvesting in continuous time with stochastic growth. *Math. Biosc. 41.*

May, R.M., J.R. Beddington, J.W. Horwood, and J.G. Shepherd (1978). Exploiting natural populations in an uncertain world. *Math. Biosc. 42*: 219-252.

Soong, T.T. (1973). *Random differential equations in Science and Engineering.* Academic Press, N.Y.

PART IV

PHYSIOLOGY AND MEDICINE

EXPERIMENT DESIGN: IDENTIFICATION AND VALIDATION OF SIMPLE AND COMPLEX MODELS OF ENDOCRINE-METABOLIC SYSTEMS

Claudio Cobelli and Karl Thomaseth
Istituto di Elettrotecnica e di Elettronica, Università di Padova
and LADSEB-CNR, Padova, Italy

1. INTRODUCTION

Mathematical models in conjunction with dynamic input-output data are increasingly used in quantitative studies of endocrine-metabolic systems both in physiology and clinical medicine /1/The purpose of modeling includes understanding, estimation of internal (non-measurable) parameters, diagnosis and control. Depending on the purpose of the study, models of different complexity are developed, and an essential role is played by identification and validation, respectively the determination of structure and parameter values and the assessment of whether or not the postulated model is adequate for its intended purpose.

We discuss here some basic aspects of model identification and validation, particularly in relation to model complexity. Reflecting this, procedures are discussed for two well-defined classes of models. Emphasis in what follows is placed on dynamic processes, particularly at the whole-organism level, described by models which reflect the underlying physiological structure, usually described by differential equations. Some examples will be used to add substance to the discussion of methodological issues.

2. CLASS OF MODELS

We consider models described by general nonlinear state variable equations of the form:

$$\dot{x}(t,p) = f\left[x(t,p);\ u(t),t;\ p\right], \quad x_0=x(t_0,p) \tag{1}$$

$$y(t,p) = g\left[x(t,p);\ p\right] \tag{2}$$

$$h\left[x(t,p),\ u(t),\ p\right] \geq 0 \tag{3}$$

$$z(t_k,p) = y(t_k,p) + e(t_k) \tag{4}$$

where $x=[x_1\ x_2\ \dots\ x_n]^T$ is the state vector (e.g. masses or concentrations); $u=[u_1\ u_2\ \dots\ u_n]^T$ is the input vector (e.g. endogenous or exogenous test inputs); $y=[y_1\ y_2\ \dots\ y_m]^T$ is the output vector prior to measurement (e.g. concentrations); f is a non-linear function which defines the postulated structure of the model, parametrised by a parameter vector $p=[p_1\ p_2\ \dots\ p_P]^T$; g is a non-linear vector function which describes the known measurement process; h represents all additional and independent algebraic equality or inequality constraints known a priori relating x, u and p; $z=[z_1\ z_2\ \dots\ z_m]^T$ is the discrete time measurement vector ($k=1,2\dots N_1$;

$l=1,2,...m$), and $e=[e_1\ e_2\ \dots\ e_m]^T$ is the measurement error assumed to be additive.

The model (1)-(4) is a fairly general description of an endocrine-metabolic system and input-output experiment designed for its identification, when for purpose of study a structure needs to be postulated based on all a priori information on the system.

In many circumstances the observation equation is linear, in this case (2) reduces to

$$y(t,p) = C(p)x(t,p) \tag{5}$$

where C is an m x n matrix.

Often, a linear structure may be adequate, in this case (1) becomes

$$\dot{x}(t,p) = A(p)x(t,p) + B(p)u(t) \tag{6}$$

where A and B are respectively of dimension n x n and n x r.

A particular case of (6) is the strictly compartmental model /1/, where structural constraints are imposed on A; i.e. (3) includes:

$$\begin{aligned} &\cdot\ a_{ij} \geq 0 \qquad (i \neq j) \\ &\cdot\ a_{ii} \leq 0 \text{ and } a_{ii} \geq \sum_{\substack{j=1 \\ j\neq i}}^{n} a_{ji} \end{aligned} \tag{7}$$

Examples of models described by (1),(3),(4),(5) and (3),(4),(5),(6),(7) are shown in Fig.1 and 2 respectively. The first is a nonlinear control model of glucose-insulin dynamic interaction /2/ and the second is a linear compartmental model of ketone body metabolism /3/.

3. IDENTIFICATION AND VALIDATION OF SIMPLE MODELS

The purpose of study , e.g. model-based measurement in a clinical application, often requires all unknown parameters of the model to be estimated from input-output data. Models all of whose unknown parameters can be estimated using formal identification techniques (e.g. least squares and maximum likelihood estimation /1/),i.e. theoretically identifiable models ,will for convenience be regarded as simple models in contrast to the complex forms discussed later. Whilst this class of model will usually have a relatively simple structure, it should be realised that it includes all models where there is no mismatch between the structure postulated and the measurements which are available for model identification. The validation strategy for this class of model involves assessment in terms of both numerical and statistical criteria arising from the identification procedures and also examination of the plausibility of the model in relation to current physiological thinking.

3.1. Theoretical (a priori) identifiability

Suppose we have postulated one or more mathematical models (Eq.1), and know what sites are available, u, for the application of test signals and what measurable variables, y (Eq.2) are available. Before actually proceeding with the experiment (Eq.4)

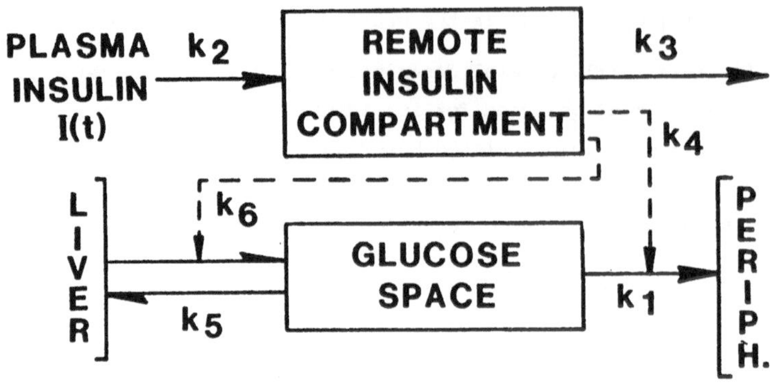

Fig.1. A model of the control of insulin on glucose kinetics /2,5/ (continuous line: material flux; dashed line: control signal). Remote insulin controls (k_4,k_6, min^{-1}) glucose fluxes. The other k_i are transfer rate parameters (min^{-1}) for glucose and insulin. The model will be discussed in Sect.3.5.

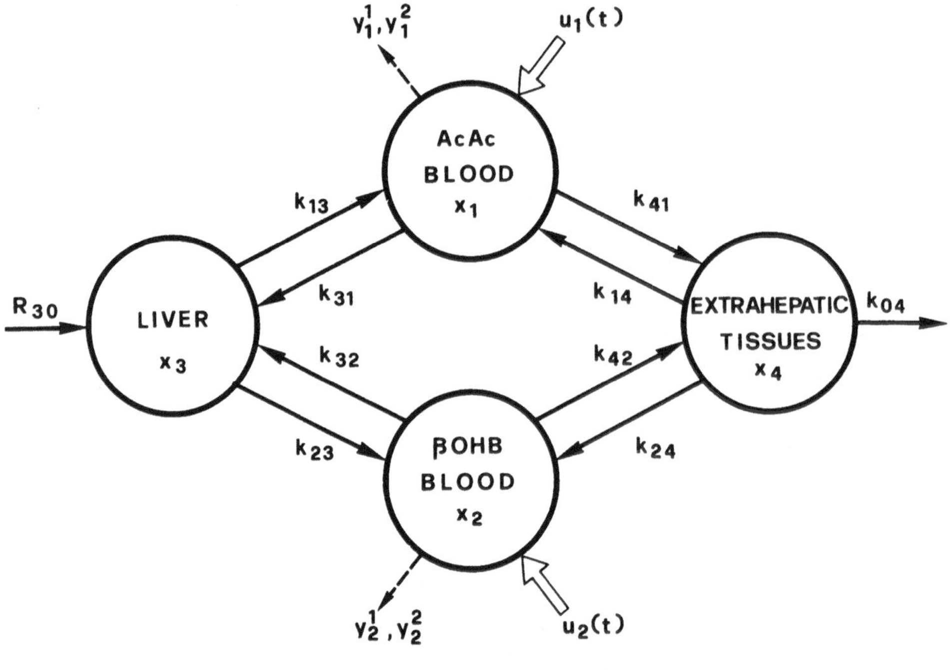

Fig.2. A four compartment model of ketone body metabolism /3/. AcAc is acetoacetate; BOHB is 3hydroxybutyrate; R_{30} is endogenous rate of production of ketone bodies; k_{ij} are transfer rate parameters, and u-y denotes the two input -four output experiment.

it is necessary to examine whether, from the data which the experiment would generate (assuming the data to be ideal, i.e. Eq.2), it is theoretically possible to make unique estimates of all the unknown model parameters. This question is often referred to as that of theoretical,a priori,identifiability. See /1/ for formal definitions and available techniques to test structural identifiability of linear and non-linear models described in Section 2.

Theoretical identifiability analysis allows three cases to be distinguished: unique identifiability, in which it is theoretically possible to make unique estimates of all the model parameters; non-unique identifiability, in which one or more of the parameters has two or more (but a finite number of) possible values; and unidentifiability, in which one or more of the parameters could have one of an infinite number of values. Normally a model in which the parameters are to be estimated using identification techniques, if it is to be deemed valid, must satisfy the criterion of unique identifiability. In the case of a model that is only non-uniquely identifiable, one set of parameter values may be found to be valid on the basis of examination of their plausibility as outlined below. A model that is unidentifiable must be rejected unless it is possible and appropriate to reparameterise it, that is, by considering only identifiable combinations of parameters.

3.2. Quantitative criteria based on the results of identification

Having tested for theoretical identifiability, the actual experiment (Eq.4) can be carried out and parameter estimation techniques, e.g. least squares and/or maximum likelihood,can be used to obtain numerical estimates of unknown parameters. The identification-based quantitative criteria for judging model validity are practical identifiability, goodness of fit and statistics of residual errors.

Practical a posteriori identifiability analysis involves examining the achievable accuracy of the parameter estimates obtained from the given, noisy experimental data (Eq.4). A lower bound of the covariance matrix of parameter estimates can be obtained by computing the inverse of the Fisher information matrix, according to the Cramer-Rao lower bound theorem. For instance (see /1/ for details), for the single input -single output case, assuming the measurement error to be white and Gaussian,the generic element j_{ij} of the (PxP) Fisher information matrix J is given by

$$j_{ij} = \sum_{k=1}^{N} \frac{1}{\sigma^2(t_k)} \frac{\partial y(t_k)}{\partial p_j} \frac{\partial y(t_k)}{\partial p_i} \qquad (8)$$

where $\sigma^2(t_k)$ is the variance of measurement error at time t_k, and derivatives are evaluated at parameter estimates.

Where the standard deviations of the estimated parameter values are unreasonably large, the model will normally be considered invalid. This situation can arise either as a result of shortcomings in the test data (e.g. paucity of measurements, measurement errors) or as a result of having postulated a model the structure of which is false or too complex for the available test data. However, in some circumstances a model with relatively poor parameter accuracy may be retained. This might occur, for example, if the structure of such a model has a sounder basis than a competing one that, although yielding more accurate parameter estimates, is physiologically less plausible.

The next criterion to be satisfied if the model is to be deemed valid is the goodness of fit, which can be evaluated through the residual sum of squares. Where a single model has been postulated, for it to be valid its response must fit the experimental test data with sufficient accuracy, given the errors inherent in the measurement process. If there is a set of two or more candidate models, comparison between them should be made of their ability to fit the available data as one measure of model validity. In this way the best model, in terms of goodness of fit, may be selected. However, the improvement in predictability of the data by one model over a competing model should not be obtained purely as a result of an increase in the number of parameters, but must truly reflect a more accurate representation of data corrected for the decrease in the degrees of freedom in fitting (principle of parsimony). For linear dynamic models the Akaike Information Criterion implements this principle.

Finally examination of the residual errors arising from the fitting process enables the assumptions made regarding the measurement process to be tested and any systematic deviation between data and model prediction to be detected (classical statistical tests can be used e.g. to test normality and uncorrelation). In this case it is most likely that model structure is invalid. This criterion should be adopted both for the case where just a single model has been postulated and also as a means of comparing alternative models.

3.3. Model plausibility

After examining the fit of the model, its plausibility should be examined. In essence this involves examining the correspondence of the model with all other relevant theoretical and empirical information. The two criteria are plausibility of the parameters estimated by formal identification procedures and the plausibility of other features of structure, parameters and behaviour.

3.4. Optimal experiment design

In many instances parameters resulting from the estimation process may be deemed adequate for the intended purpose. In other cases, further work may be required. This may be due to the fact that the previously designed experiment has resulted in parameters whose accuracy is judged inadequate for the intended purpose (it should be noted, however, that this lack of accuracy might have arisen as a result of having postulated an inappropriate model structure; this particular problem is essentially of a different type). Alternatively, whilst the parameter estimates might of themselves be adequate, the experimental designs providing data for their evaluation might not be appropriate, for instance in clinical studies where it is desirable to minimize the experimental effort such as by minimizing the number of blood samples collected. In both situations, there are a number of experimental design variables which need to be considered: (i) the form of the test signal, u; (ii) the duration of the time over which output variables are measured, T; (iii) the number of samples, N, obtained during T; (iv) the sampling schedule, SS, that is where the N samples are located in T (i.e. the choice of $t_1, t_2, \ldots, t_k, \ldots, t_N$), and (v) the variance of the measurement errors, $\sigma^2 = [\sigma^2(t_1) \ldots \sigma^2(t_N)]^T$.

The approach to be adopted in optimizing the variables listed above in a given situation again makes use of the Fisher information matrix, J. This information matrix can be viewed as a function of the experimental design variables listed above, that is

$$J = J(u, T, N, SS, \sigma^2) \tag{9}$$

and thus the aim is, by suitable choice of these five variables, to maximize the expected accuracy of the parameter estimates, by optimizing an appropriate function of J, for instance maximization of the determinant of J.

In practice it is extremely difficult to consider simultaneously more than one or two of the five design variables. Where it is required to maximize accuracy of parameter estimates one important aspect is the determination of the optimal input u; another is that of seeking to optimize SS (u and N fixed). Where one is attempting to obtain cost-effective estimates of parameters in the clinical situation, it is often necessary to attempt to reduce the number of samples, N, to a minimum, where u and T are fixed by choosing an appropriate sampling schedule.

3.5. A minimal glucose-insulin model for non-invasive measurement: model selection and optimal experiment design

The purpose of model here is to extract from a clinical test, the intravenous glucose tolerance test (IVGTT; injection of glucose and subsequent discrete-time measurement of glucose insulin concentrations) information on the sensitivity of glucose utilization to insulin. In contrast, other approaches to evaluating this clinical parameter require more sophisticated and invasive experimental procedures which cannot be performed routinely in a clinical environment. Thus there is the need to select the richest model structure which can be quantified from the IVGTT and at the same time contains an adequate physiological basis.

Due to ethical and practical severe constraints which are imposed at the experiment design level, clinical models inherently tend to be simple, but care must be taken in oversimplifying the physiological reality thus arriving at simplistic model structures. Paradigmatic in this sense is the linear glucose-insulin model which is used by a number of clinical groups: if appropriate identification and validation strategies are applied to this model it can soon be shown to be inadequate /4/.

In order to avoid these problems the <u>minimal</u> (<u>parsimonious</u> or <u>optimal complexity</u>) <u>modelling approach</u> has been proposed /2,5/ whereby a model has to be selected from a series of rival models by applying the set of criteria discussed previously. A single member of the proposed series of models can be chosen as the best or minimal (for the particular data set) and the parameters of that model will probably be capable of quantitating specific unit processes of the system. By using such minimal representation, the frequent problem of non-unique physiological models which leads to multiple possible interpretations of plasma dynamics can be avoided.

Seven dynamic models of increasing structural (physiological) complexity have been proposed /2/. They have been numerically identified and successively the set of models has been compared on the basis of the described validity criteria /5/. The results are summarized in Table 1. From our systematic model comparison, one model (Fig.1) emerged as superior and a useful parameterisation of it is given by

$$\dot{G} = (p_1 - X)\,G + p_0 + u\ ,\quad G(0) = G_{SS} \tag{10}$$

$$\dot{X} = p_2(X - S_I\,I)\ ,\qquad X(0) = 0 \tag{11}$$

where G is plasma glucose (G_{SS} is the steady state value); I is plasma insulin (deviation from steady state); X is proportional to insulin in a remote compartment I' ($X=(k_4+k_6)I'$); u is the impulsive test input; p_0, p_1 and p_2 are uniquely identifiable constant parameters ($p_0=-p_1\,G_{SS}$; $p_1=-(k_1+k_5)$; $p_2=-k_3$); and S_I is the insulin sensitivity index defined as the quantitative influence of insulin in enhancing glucose disappearance ($S_I=k_2(k_4+k_6)/k_3$).

Table 1 Selection of the Minimal Model

Model	Theoretical Identifi-ability	Practical Identifi-ability	Statistics of Residual Errors	Goodness of Fit and Number of Parameters	Acceptance on the Basis of Identifi-cation Results	Overall Physiological Plausibility
I	Uniquely identif-iable	Acceptable	Acceptable	Unaccept-able	No	Not considered
II	Uniquely identif-iable	Unaccept-able	Not consid-ered	Not considered	No	Not considered
III	Uniquely identif-iable	Unaccept-able	Not consid-ered	Not considered	No	Not considered
IV	Uniquely identif-iable	Unaccept-able	Not consid-ered	Not considered	No	Not considered
V	Uniquely identif-iable	Acceptable	Acceptable	Acceptable	Yes	Marginally acceptable
VI	Uniquely identif-iable	Acceptable	Acceptable	Acceptable	Yes	Acceptable
VII	Identif-iable (two solutions)	Unaccept-able	Not con-sidered	Not considered	No	Not considered

The model allows in vivo estimation of glucose effectiveness (p_1) and particularly of the insulin sensitivity index (S_I) from the IVGTT test. It should be noted that the estimation of insulin sensitivity is a difficult task in experimental medicine and requires invasive and sophisticated labour-intensive techniques (see review in /6/). A typical mean value of S_I obtained in normal dogs is 7.0×10^{-4} ($min^{-1}/(\mu U/ml)$). This carefully validated model has been also applied in glucose intolerant humans /7/, where the model can also be used as a basis for optimal treatment.

At present some aspects related to optimal experiment design of the model (10)-(11) are under investigation, more precisely the determination of the optimal input u(t) for estimating p_1 and S_I (in (10) u was assumed to be of the impulsive type), and the discrete-time optimal sampling schedule for such an optimal input. The optimal input problem is studied by maximizing a performance index which contains a measure of the information matrix J, subject to various input constraints via the Pontryagin maximum principle. The problem is complicated here by having de facto opened the glucose-insulin feedback loop (insulin is assumed known in (10); see /2,5/ for details), so that one has to resort to a rather robust model of the glucose regulation system to generate the necessary insulin pattern to different glucose inputs. The approach we are using is an iterative one. The performance index at iteration k+1 is:

$$\Phi_{k+1} = \frac{1}{2} \int_0^T \left[\operatorname{tr}\left(J^{-1}(u_k)\ J(u)\right) - q\, u^2(t) \right] dt \tag{12}$$

(q is the Lagrange multiplier),
thus giving the optimal input as:

$$u_{k+1}(t) = \arg \max_{u \in \Omega} \Phi_{k+1} \tag{13}$$

where J is the Fisher information matrix and Ω is the class of admissible inputs. An exploratory study on the optimal sampling schedule for the impulsive input has been performed for the normal and obesity state model /7/. A D-optimality criterion has been used, i.e. the optimal sampling schedule is chosen from:

$$\max_{(t_k : k=1,2\ldots N)} \det J \tag{14}$$

A new version of an algorithm already applied in other optimal sampling studies /8/ has been developed. Mean parameter values of the normal and obesity state were chosen as reference, and measurement error of glucose has been assumed white, normal of zero mean and $\sigma^2(t_k)=(0.02\ G(t_k))^2$. The optimal sampling schedule has been reached in the set:

$$I = \left\{ t_k : t_k = t_o + k\ d\ ;\ k=1,\ldots 360 \right\} \tag{15}$$

where d=1 min and t_o=1 min, and a total number of seven samples have been considered. The results are shown in Table 2 where also the achievable accuracy (from J^{-1}) of parameter estimates is reported. An aggregation of samples towards a situation where the number of optimal samples equals the number of estimated parameters, a situation already noted in /8/, can be observed. Differences in the sampling schedule for the normal as compared to the obesity state may be noted.

Table 2 Optimal sampling schedule (OSS) and parameter estimates achievable accuracy (percent coefficient of variation, CV%) for the minimal model

OSS (min)						
Normal state	(23,24)		(63,64,65)		(138,140)	
Obesity state		(36,37)		(99,100)		(226,227,228)

CV%	p_1	p_2	S_I
Normal state	10	34	10
Obesity state	15	23	4

4. IDENTIFICATION AND VALIDATION OF COMPLEX MODELS

Often the purpose of study requires a more isomorphic model structure, e.g. in a hypothesis testing physiological investigation. In these cases usually complex models,still described by (1)÷(4), are developed by which it is intended that they are theoretically (a priori) unidentifiable (not all the unknown parameters can be estimated using formal identification procedures). Such complex models are essentially incomplete as there must be a very high degree of uncertainty with respect to both structure and parameters.

4.1. Increasing model testability

The approach adopted involves first seeking to enhance model testability through model simplification, improved experimental design and model decomposition. The resulting model may be theoretically identifiable in which case validation can proceed as outlined above. If the model is still unidentifiable, the enhancement of testability will have reduced its overall uncertainty. In this case, or even if approaches to increasing testability prove not to be feasible, the validation procedure should continue by way of adaptive fitting.

4.2. Adaptive fitting

This involves, first, seeking a set of parameter values in the model such that its response for one input/output experiment, say for a normal physiological condition, adequately matches the corresponding experimental test data. If this parameter set is not plausible another must be sought which is within the physiologically-feasible range. If this can be achieved the model is trained on this input/output experiment. The model incorporating these parameter values is then tested against all other input/output experiments corresponding to normal physiology and other relevant data. This testing by computer simulation should include the examination of model predictions for a wide range of test signals corresponding to both physiological and abnormal conditions. In all these tests the model must match the experimental data if it is to be deemed empirically valid in terms of the following criteria : qualitative feature comparison; quantitative feature comparison, and time course prediction.

4.3. Model plausibility

Having carried out the adaptive fitting procedures outlined earlier and having obtained a model fit that is deemed adequate, it is now necessary to examine the plausibility of the model. This involves examining both structure and parameters in relation to a number of factors such as model complexity, sensitivity of model outputs to uncertainty in model parameters and the plausibility of the parameter values for any particular model structure.

Concerning the plausibility of parameter values for a given structure of model, there is the need to examine whether the values the parameters must take to produce a good model fit are reasonably constrained about the values adopted for the model test program described above. This is particularly important since in large models, considerable uncertainty may be associated with many of the parameter values.

A number of approaches are available for making a quantitative assessment of the plausibility of complex physiological models. The first of these is sensitivity analysis (differential, incremental or Monte Carlo type). In order to examine the effects of parameter uncertainty on model plausibility, it is necessary to test the sensitivity of the predicted model responses to the numerical values of the parameters. In this way, possible deficiencies in the model can be revealed if, for example, small changes in a parameter from its nominal value result in large, improbable changes in patterns of model prediction. Equally, sensitivity analysis can indicate ranges of parameters that are compatible with the known ranges of behaviour likely to be observed in the population under study.

Furthermore, sensitivity analysis can provide insight as to which are the particular parameters that need to be specified accurately if the model is to reproduce reliably the patterns of response observed in a particular subject; in other words, it can specify the conditions necessary to tune the model or some part of it to that particular individual. In a related manner sensitivity analysis may provide some guidelines for the reduction of complex models by indicating those variables and parameters that determine the essential behaviour of the system and hence must be retained in any simpler model.

In some cases with large-scale models it may be possible on the basis of results obtained from sensitivity analysis, to fix many of the parameters at their nominal values and to use formal identification techniques to estimate a subset of the parameters. One situation in which this arises is the case where from sensitivity analysis a particular parameter subset has been shown to require accurate specification. This enables the model to be tuned to particular experimental test data corresponding to the individual subject. Another situation in which this arises is when only a particular subset of the parameters is of interest to the investigator.

4.4. A comprehensive model of glucose regulation to understand diabetes control and to assess domain of validity of clinical models

An identification and validation program of the type outlined above has been used to develop a rather complex model of the short-term glucose regulation system. The model is shown in Fig.3 and consists of 10 non-linear differential equations with approximately 50 parameters. For details on the identification and validation strategies the reader is referred to /9,10,12/.

Briefly on three uses of the model. First, alternative algorithms and routes of infusing insulin for achieving optimal control of metabolism in insulin-dependent diabetes (no endogenous insulin available for secretion) via closed-loop mechanical external devices have been compared /11/. This kind of quantitative evaluation is not feasible in vivo on a diabetic subject for obvious ethical and practical reasons, and thus a model-based approach is essential. The study has shown the heuristic potential and pragmatic utility of the model, e.g. it predicts relative unimportance of type of algorithm and importance of the route of infusion of insulin.

Secondly, the model of Fig.3 has been used as a test-bed for examining the empirical validity of a number of simpler models /12/ intended for clinical application, the purpose of which is quantifying turnover of endogenous substances in the non-steady state using data from tracer experiments. The study has revealed that these simple models have a restricted range of applicability. In fact the over-simplified physiological representation incorporated gives rise to the uncertainty as to its validity in the wider domain, so that new richer model structures are needed in order to quantify with reliability the complex non-steady state situation.

Thirdly the model provides the robust glucose-insulin mathematical description necessary to undertake the optimal input study on the minimal model, as discussed in Section 3.5.

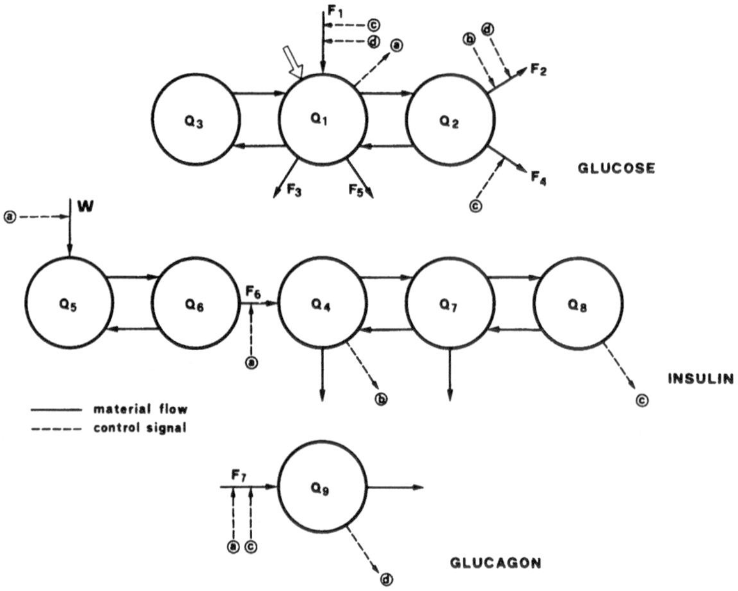

Fig.3. Block diagram of the comprehensive blood glucose regulatory system model. Glucose, insulin and glucagon subsystems are shown with their interactions. Glucose subsystem: Q_1 is blood glucose mass; Q_2 and Q_3 are a rapidly and a slowly equilibrating compartment; F_1 is endogenous production of glucose; F_3 is renal excretion of glucose; F_5 is insulin-independent glucose utilization; F_2 and F_4 are insulin-dependent glucose utilizations in liver and muscle and adipose tissues. F_1, F_2, F_4 are nonlinearly controlled by insulin and glucagon. Insulin subsystem: stored (Q_5) and promptly releasable (Q_6) insulin in the pancreas; liver and portal insulin (Q_4), plasma insulin (Q_7) and insulin in the interstitial fluids (Q_8). Insulin synthesis (W) and secretion (F_6) are nonlinearly controlled by plasma glucose. Glucagon subsystem: Q_9 is extracellular glucagon; its secretion (F_7) is controlled by glucose and insulin levels.

5. CONCLUSIONS

Differential equation models are increasingly used as an investigative tool in dynamic studies of endocrine-metabolic systems both in physiology and clinical medicine. A general class of models has been presented and the essential role of experiment design, model identification and validation in developing useful models has been discussed. Particular emphasis has been given to the notions of theoretical (a priori) and practical (a posteriori) identifiability and to approaches to optimally design an identification experiment (optimal input, optimal sampling schedule). A formal framework for model identification and validation has been proposed and validation strategies to be adopted have been outlined for simple and complex models, these two classes being operationally defined on the basis of a priori identifiability. Various models of the glucose regulation system have been used to illustrate different model purposes and to provide a concrete counterpart of the discussed theoretical aspects. The examples show that simple and complex models each have a role both in physiology and clinical application, if an integrated methodology for system identification and validation is used.

ACKNOWLEDGMENT

This research was supported in part by a grant of the Ministero della Pubblica Istruzione.

REFERENCES

1. Carson E.R., Cobelli C. and Finkelstein L. The Mathematical Modeling of Metabolic and Endocrine Systems. Model Formulation, Identification and Validation. J.Wiley, New York, 1983
2. Bergman R.N., Ider Y.Z., Bowden C.R. and Cobelli C. Quantitative estimation of insulin sensitivity. Am. J. Physiol. 236, E667-E677, 1979.
3. Cobelli C., Toffolo G. and Nosadini R. Mathematical models of ketone bodies kinetics in the human. Experiment design and their identification:validation. In Proc. 6th IFAC Symp. on Identif. & Syst. Param. Estimation. Ed.G.A. Bekey and G.N.Saridis, Oxford: Pergamon 1983, vol.1, 222-227.
4. Cobelli C., Mari A., Thomaseth,K. and Bergman,R.N. Simple vs. comprehensive models of glucose/insulin dynamic interactions in the whole organism. Ibid.,vol.2,886-901.
5. Bergman,R.N., Bortolan G., Cobelli C. and Toffolo G. Identification of a minimal model of glucose disappearance for estimating insulin sensitivity. Proc. 5th IFAC Symp. Ident. System Param. Estim.,Ed.R.Isermann,Oxford:Pergamon 1980,vol.2,883-890.
6. Bergman,R.N., Bowden C.R. and Cobelli C. The minimal model approach to quantification of factors controlling glucose disposal in man. In Carbohydrate Metabolism.Eds C. Cobelli and R.N. Bergman, Chichester: Wiley 1981, 269-296.
7. Bergman R.N., Philipps L.S. and Cobelli C. Physiologic evaluation of factors controlling glucose tolerance in man:measurement of insulin sensitivity and β-cell glucose sensitivity from the response to intravenous glucose.J.Clin.Inv.68,1456-67,1981
8. Cobelli C., DiStefanoIII J.J. and Ruggeri A. Minimal sampling schedules for identification of dynamic models of metabolic systems of clinical interest: case studies for 2 liver function tests. Math.Biosci.63, 173-186, 1983
9. Cobelli C., Federspil G., Pacini G., Salvan A. and Scandellari C. An integrated mathematical model of the dynamics of blood glucose and its hormonal control. Math. Biosci. 58, 27-60, 1982
10. Cobelli C. and Mari.A. Validation of mathematical models of complex endocrine-metabolic systems. A case study on a model of glucose regulation. Med. Biol. Eng.Comp. 21, 390-399, 1983
11. Cobelli C. and Ruggeri A. Evaluation of portal/peripheral route and of algorithms for insulin delivery in the closed-loop control of glucose in diabetes. A modeling study. IEEE Trans. Biomed. Eng. 30, 93-103, 1983
12. Cobelli C., Ruggeri A., Toffolo G., Avogaro A. and Nosadini R. Is the "pool-fraction" paradigm a valid model for assessment of in vivo turnover in non-steady state? Am. J. Physiol., November 1983.

A MATHEMATICAL MODEL FOR CARDIAC ELECTRIC SOURCES AND RELATED POTENTIAL FIELDS

P. Colli Franzone
Istituto di Matematica, Informatica e Sistemistica
dell'Università di Udine, UDINE and
Istituto di Analisi Numerica del CNR, PAVIA (Italy)

1 - Mathematical formulation of a model of the cardiac electric sources for the forward problem.

The electrocardiographic forward problem consists in the prediction of the potential distribution on the body surface given the geometry and conductivity of the cardiac tissue and the thorax and the knowledge of the cardiac electric events, i.e., electric changes occurring in the cardiac cells during the depolarization and repolarization processes and the activation sequence of the cardiac tissue. We shall investigate a model concerning the potential generated only in the depolarization phase and, as in the majority of attempts made to solve the forward problem, we confine ourselves to the ventricular depolarization which is the origin of the QRS complex in the electrocardiogram. We denote by H the ventricular heart muscle and in the following we shall refer only to this part of the heart. During this phase of the heart beat a "thin" layer of cardiac cells undergoes the depolarization process, i.e., a change of the intracellular electric potential w of biochemical origin occurs in a very "short" time interval from a resting value u_r to a plateau value u_a with $u_r<u_a$; moreover since during a heart beat any cell depolarizes only once and in the QRS phase the repolarization effects are negligible, the following assumptions are usually made:

(1.1) The depolarization of a cell is an instantaneous process and it is the same for all ventricular cells (i.e. u_a and u_r are constant in H).

(1.2) At any time instant t of the QRS heart beat complex, setting

$H^a = \{x \quad H: w(x) = u_a\}$, the activated heart tissue,

$H^r = \{x \quad H: w(x) = u_r\}$, the resting heart tissue,

we have $\bar{H} = \bar{H}^a \quad \bar{H}^r$ and the so called excitation wavefront surface $S = \partial H^a \quad \partial H^r$ is a "regular" surface.

Almost all the attempts to explain the QRS complex have used, for describing the potential generated by the excitation wavefront S, the following model that we call "classical model"; the cardiac sources are represented by means of a uniform "dipole layer" on S, normal to S and oriented towards the resting heart tissue with dipole moment $u_a - u_r$. Although good geometrical data and data about depolarization are available, the several attempts to explain the QRS complex have not been successful; moreover recent experiments on animals have shown a marked discrepancy between data and prediction of the classical model revealing a strong influence of the cardiac fiber direction on the potential distribution (see [7], [1], [11], [3], [10], [5]). In [3], [4], [6] we developed a new intracellular current model of the cardiac sources based on hypotheses (1.1)(1.2) which we now shall formulate.

Let Ω be the human body volume and H be the ventricular heart muscle. We assume Ω and H to be open connected sets of R^3 of class C^2 and $\bar{H} \subset \Omega$; the unit vectors normal to $\partial\Omega$ and ∂H are oriented outwards relative to Ω and H respectively. The excitable cardiac muscle H is conceived as a tissue composed by the superposition of two continuous anisotropic conducting media: the extracellular (e) and the intracellular (i) media with electric conductivity tensors D_e and D_i respectively. The anisotropic electric conduction properties characterized by these tensors are related to the fiber structure of the heart muscle. The structure of the cardiac fiber system is very complicated (see e.g. [2]), however, interpreting H as only the lower part of the heart which contains the two ventricles, the following physiologically reasonable assumptions on H result

I) The boundary ∂H of H consists of two parts $\partial_1 H$ and $\partial_2 H$ such that:

$\partial_1 H$ is closed in ∂H, $\partial_2 H$ is open in ∂H, $\partial_1 H \cap \partial_2 H = \phi$ moreover, the system of the muscle fibers is assumed to be the support of a family F of C^1 simple and open curves L such that:

a) the extreme points of L are on $\partial_1 H$
b) any point $x \in \bar{H} - \partial_1 H$ belongs to one and only one curve L
c) if L has a point on $\partial_2 H$, L is tangent at this point to $\partial_2 H$
d) the orientation of the curves L is "coherent" (i.e. the unit vector tangent to the curve is continuous in $\bar{H} - \partial_1 H$)

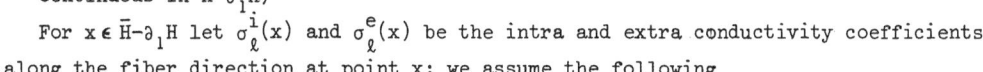

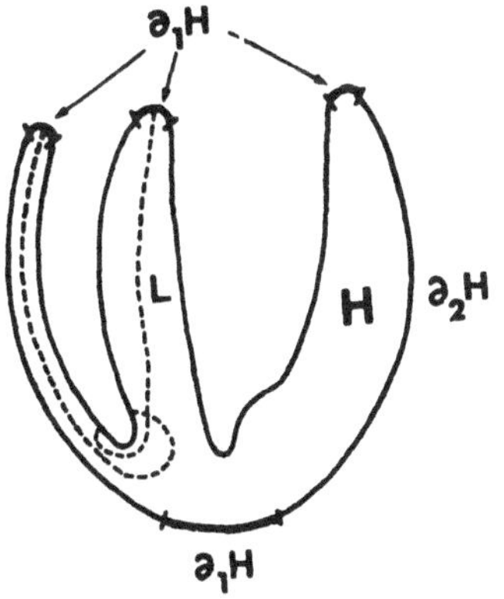

For $x \in \bar{H} - \partial_1 H$ let $\sigma_\ell^i(x)$ and $\sigma_\ell^e(x)$ be the intra and extra conductivity coefficients along the fiber direction at point x; we assume the following

II) Hypothesis of axial symmetry around the fiber direction for the electric conductivity.

Let $\sigma_t^i(x)$ and $\sigma_t^e(x)$ be the intra and extra conductivity coefficients in any direction perpendicular to the fiber direction at point x. Let us consider the matrix $A = (\underline{a}_1, \underline{a}_2, \underline{a}_3)$ where $\{\underline{a}_1, \underline{a}_2, \underline{a}_3\}$ is a "local basis" at point $x \in \bar{H} - \partial_1 H$, with $\underline{a}_3$ parallel to the direction of the fiber passing through x and $\underline{a}_1, \underline{a}_2$ defined up to a rotation around the unit vector $\underline{a}_3$; then the conductivity tensors D_i and D_e are defined by:

(1.3) $D_i = A \, \mathrm{diag}(\sigma_t^i, \sigma_t^i, \sigma_\ell^i) A^T = \sigma_t^i (\underline{a}_1 \underline{a}_1^T + \underline{a}_2 \underline{a}_2^T) + \sigma_\ell^i \underline{a}_3 \underline{a}_3^T, \quad D_e = A \, \mathrm{diag}(\sigma_t^e, \sigma_t^e, \sigma_\ell^e) A^T$

We shall now specify the regularity hypothesis on the wavefront S:

III) S is a regular C^2 oriented connected surface, "open" (i.e. with boundary ∂S) or "closed" (i.e. without boundary), contained in H if "open", with boundary ∂S contained in $\partial_2 H$, or contained in $\bar{H}-\partial_1 H$ if "closed", moreover, S separates H in two disjoint connected open subsets H_S^a (activated heart tissue) and H_S^r (resting heart tissue), and the orientation of the unit normal $\underline{n}$ at each point $x \in S$ is taken towards the resting part H_S^r (if S is closed H_S^a is the "interior" of S).

We denote by u(x) the potential in Ω (i.e. in the extracellular heart medium H and in the extracardiac medium $\Omega-\bar{H}$) and by w(x) the potential in the intracellular heart medium that due to (1.1),(1.2) is given by:

$$w(x)=\begin{cases} u_a & \text{in} \quad H_S^a \\ u_r & \text{in} \quad H_S^r \end{cases}$$

with u_r and u_a constant and $u_r<u_a$. We now formally introduce the current densities $\underline{I}_e$ and $\underline{I}_i$, the first is in the extracardiac medium and in the extracellular medium while the second is in the intracellular medium:

$$\underline{I}_e=-D \text{ grad } u \ , \ \underline{I}_i=-D_i \text{ grad } w \ , \quad \text{where} \quad D=\begin{cases} D_e & \text{in } H \\ \sigma_o I & \text{in } \Omega-\bar{H} \end{cases}$$

Applying the current conservation law between the intra and extra cellular media in H and taking into account that there are no sources outside H we obtain:

$$(1.4)\quad \begin{cases} \operatorname{div}(\underline{I}_e+\underline{I}_i) = 0 \ , & \text{in } H \\ \operatorname{div} \underline{I}_e=0 \ , & \text{in } \Omega-\bar{H} \\ \underline{n}.\underline{I}_e=0 \ , & \text{on } \partial\Omega \end{cases}$$

the boundary condition is due to the fact that the body surface $\partial\Omega$ is surrounded by air which is an insulating medium.

In the following we make the simplifying assumption:

IV) the extracellular cardiac medium and the body volume are isotropic and homogeneous with the same constant conductivity coefficients σ_o, i.e., $D_e=\sigma_o I$ with σ_o constant. Since grad $w=-(u_a-u_r)\delta_S \underline{n}_S$, where δ_S is the Dirac measure on S and $\underline{n}_S$ is the normal to S directed toward H_S^r; then (1.4) implies the following relationship in terms of the potential u:

$$(1.5)\quad \begin{cases} \Delta u = \begin{cases} \operatorname{div} M(\delta_S \underline{n}_S) & , \quad \text{in } H \\ 0 & , \quad \text{in } \Omega-\bar{H} \end{cases} \\ \dfrac{\partial u}{\partial n}=0 & , \quad \text{on } \partial\Omega. \end{cases}$$

where $M = \dfrac{u_a-u_r}{\sigma_o} D_i$, with eigenvalues $m_t(x)= \dfrac{u_a-u_r}{\sigma_o} \sigma_t^i$, $m_\ell(x) = \dfrac{u_a-u_r}{\sigma_o} \sigma_\ell^i$.

Applying results of the elliptic boundary values problems (see e.g. [9]) we obtain in [6] the following:

Theorem 1. Under hypotheses I),II),III) and the following assumptions:

V) S (if "open") is a part of the boundary $\partial\Omega^*$ of an open set Ω^* of class C^2 such that $\bar{\Omega}^*\subset\Omega$, $\partial\Omega^*\cap\partial_1 H=\phi$.

VI) the matrix M can be extended in $\Omega-\partial_1 H$ so that its coefficients belong to $C^1(\Omega-\partial_1 H)$ and the extended matrix results symmetric positive definite having at most two distinct eigenvalues, $m_t(x)$, $m_\ell(x)$.

There exists a solution u of problem (1.5), unique up to an additive constant, which belongs to $H^{\frac{1}{2}-\varepsilon}(\Omega)$, $\forall\,\varepsilon>0$ and satisfies the following jump relationships:

$$(1.6)\quad \begin{aligned} &[u]_S=\alpha, \text{ in the sense of } H^{-\varepsilon}(S), \text{ with } \alpha=\underline{n}^T M\underline{n} \\ &\left[\frac{\partial u}{\partial n}\right]_S = \operatorname{div}^S\underline{\beta}, \text{ in the sense of } H^{-1-\varepsilon}(S), \text{ with } \underline{\beta}=M\underline{n}-\alpha\underline{n} \end{aligned}$$

where $[v]_S$ means the "jump" v^r-v^a between the trace of v on S from H^r_S and from H^a_S and div^S is the divergence operator on S.

Moreover it is possible to prove (see [6]) the following representation theorem for our solution which is connected with another possible approach to the forward problem (see [4]).

Theorem 2. Under the same hypotheses of Theorem 1 the solution of problem (1.5) admits the following boundary integral representation

$$(1.7)\qquad u(x)=-\int_{\partial\Omega} v(\xi)\underline{n}_\xi\cdot\operatorname{grad}_\xi s(x,\xi)d\sigma_\xi+\int_S \underline{n}_\xi^T M(\xi)\operatorname{grad}_\xi s(x,\xi)d\sigma_\xi,\ \forall\, x\in\Omega-\bar{S}$$

where v is the solution in $C^0(\partial\Omega)$, unique up to a constant, of the following integral equation on $\partial\Omega$:

$$(1.8)\quad \tfrac{1}{2}v(x)+\int_{\partial\Omega} v(\xi)\underline{n}_\xi\cdot\operatorname{grad}_\xi s(x,\xi)d\sigma_\xi=\int_S \underline{n}_\xi^T M(\xi)\operatorname{grad}_\xi s(x,\xi)d\sigma_\xi \qquad \text{on } \partial\Omega$$

and $s(x,\xi) = \frac{1}{4\pi|x-\xi|}$ is the fundamental solution of the Laplace operator in $\mathbb{R}^3$; moreover u=v on $\partial\Omega$.

We shall develop some consequences of the previous boundary integral representation (1.7). Setting

$$(1.10)\qquad p(x)=\int_S \underline{n}_\xi^T M(\xi)\operatorname{grad}_\xi s(x,\xi)d\sigma_\xi$$

from (1.7) we see that this term represents the source generating the potential field. In other words model (1.5) gives rise to a representation of the cardiac electric sources by means of an *oblique double layer on S* oriented as the vector field $\underline{\lambda}=M\underline{n}$ and having moment density $\phi=\underline{n}^T M\underline{n}$ on S, since $p(x)=\int_S \phi(\xi)\,\frac{1}{\underline{\nu}\cdot\underline{n}}\,\frac{\partial}{\partial\underline{\nu}_\xi}s\,(x,\xi)d\sigma_\xi$ with

$\underline{\nu}=\lambda/||M\underline{n}||$.

Using the decomposition $M\underline{n}=\alpha\underline{n}+\underline{\beta}$ and Green's formula we have

$$p(x)=\int_S \alpha(\xi)\underline{n}_\xi\cdot \mathrm{grad}^S_\xi s(x,\xi)d\sigma_\xi-\int_S(\mathrm{div}^S_\xi\underline{\beta}(\xi))s(x,\xi)d\sigma_\xi+ \tag{1.11}$$
$$+\int_{\partial S} s(x,\xi)\underline{\beta}(\xi)\cdot\underline{n}^b_\xi d\gamma_\xi$$

where grad ^{S} is the gradient operator on S, $\underline{n}^b$ is the unit vector tangent to S, perpendicular to the boundary ∂S of S (when S is an "open" surface) and directed outward with respect to S and the curvilinear integral does not appear if S is "closed".

Since $u(x)=\int_{\partial\Omega} v(\xi)\underline{n}_\xi\cdot\mathrm{grad}_\xi s(x,\xi)d\sigma_\xi+p(x)$ using (1.11) we obtain a representation of u as a sum of a normal double layer on $\partial\Omega$, a normal double layer on S, a single layer on S and a curvilinear single layer on ∂S.

Other useful decompositions of the source term p(x) are discussed in [4] which tie the model to the classical one and to the "axial" model proposed in [7]. The presence in (1.11) of the single layer on ∂S introduces a logarithmic singularity in the potential hence u can be unbounded in Ω; but on physical grounds a realistic model of the cardiac potential must give rise to a *bounded potential*. The examination of the terms of the representation (1.11) shows that u is bounded if S is closed or, when S is open, if and only if

$$\underline{\beta}\cdot\underline{n}^b = 0 \quad \text{for each point } x \text{ of } \partial S. \tag{1.12}$$

In the case of an open wavefront surface S the boundedness condition on u is equivalent to impose some constraints on the wavefront boundary ∂S with respect to the fiber orientation on $\partial_2 H$ as shown by the following theorem (see [4]).

Theorem 3. Under hypotheses I),...,IV) if S is an open surface with boundary ∂S and we assume that $m_t(x)\neq m_\ell(x)$, $\forall$ $x\in\partial S$, then $\underline{\beta}\cdot\underline{n}^b=0$ at $x\in\partial S$ if and only if one of the following conditions is satisfied at x:

(1.13) $\underline{a}_\ell \,||\, \underline{\tau}^b$ (i.e. $\underline{a}_\ell$ tangent to ∂S)

(1.14) $\underline{n}\,||\,\underline{n}_H$ (i.e. S is tangent to $\partial_2 H$)

(1.15) $\underline{n}\perp\underline{n}_H$ (i.e. S is perpendicular to $\partial_2 H$).

where $\underline{a}_\ell$ is a unit vector parallel to the fiber direction, $\underline{n}_H$ is the unit vector normal to ∂H and $\underline{\tau}^b$ is the unit vector tangent to ∂S and oriented in such a way that the triplet $(\underline{n},\underline{n}^b,\underline{\tau}^b)$ is orthogonal and right-handed and $||$ (or $\perp$) means that the two considered vectors are parallel (or orthogonal).

The previous proposition implies that the boundedness of the potential is equivalent to require that the wavefront surface S and the heart surface ∂H can intersect at an arbitrary angle in the point of ∂S only if $\underline{a}_\ell$ is tangent to ∂S, otherwise the two sur faces must be tangent or perpendicular.

Now, in order to assure that the solution u of problem (1.5) is bounded in Ω, we introduce, by virtue of Theorem 3, the following hypotheses

VII) $m_t(x) \neq m_\ell(x)$, $\forall x \in \bar{H}-\partial_1 H$

VIII) $\forall x \in \partial S$ one of the following conditions is satisfied $\underline{a}_\ell \| \underline{\tau}^b$, $\underline{n}_H \| \underline{n}$, $\underline{n}_H \perp \underline{n}$

The representation of the depolarization wavefront S by means of the previous oblique dipole layer model was validated in [3],[5] using this cardiac sources model to simulate the potential field elicited by isolated paced dog heart imbedded in a cylindrical homogeneous volume conductor. The simulations were obtained for suitable values of m_t, m_ℓ and proved that, even neglecting the anisotropy of the extracellular medium (hypothesis IV , see [12]) the oblique dipole layer model yielded results which reproduced the measured pattern, in three different types of ventricular stimulation, with a good quantitative agreement (see [3],[5]). Moreover the fiber direction and curvature were essential in obtaining the above results.

2 - Formulation of the inverse problem and uniqueness results

Under the hypotheses of the preceding sections, we showed that to any surface S corresponds a solution of problem (1.5), defined up to an additive constant; in the following we shall indicate this solution by u_S. The *inverse problem*, which we shall now deal with, can then be roughly formulated as follows: is it possible to determine S so that u_S takes on prescribed values on $\partial\Omega$? Obviously, S is allowed to vary in a suitable family of "admissible" surfaces and the problem can be more precisely stated as follows: given Ω, H,M and the family $\mathcal{S}$ of surfaces S satisfying certain suitable hypotheses (at least all those introduced in the previous §), is it possible, assigned g on $\partial\Omega$, to find $S \in \mathcal{S}$ such that $u_S = g$ on $\partial\Omega$?
We shall confine ourselves to the uniqueness problem i.e.:

(2.1) $\qquad u_{S_1} = u_{S_2}$ on $\partial\Omega$ implies $S_1 = S_2$

From the uniqueness of the Cauchy problem and the analytic continuation of harmonic functions, we obtain immediately that proposition (2.1) holds if

(2.2) $\qquad S_i$, $i=1,2$ are both open with ∂S_1 and ∂S_2 not identically coincident

or if

(2.3) $\qquad S_1$ is open and S_2 is closed with $\partial S_1 \cap S_2 \neq \phi$ or S_1 outside S_2

In order to reach a general uniqueness result, it is necessary to add other hypotheses on $\{\Omega, H, M, \mathcal{S}\}$ which on the one hand must assure that for any $S \in \mathcal{S}$ the potential u_S *is effectively an oblique* dipole layer potential and on the other hand must be in agreement with a more detailed information concerning the behaviour of the cardiac fiber system.

Let us consider now the following case

(2.4) S_1 and S_2 are both "open", with the same boundary Γ (i.e. $\partial S_1=\partial S_2=\Gamma$) and no interior common points

Then the surface $\Sigma=S_1\cup S_2\cup\Gamma$ can be considered as the boundary of an open and bounded set E and the following situations are possible

1) $E=H^a_{S_1}\cap H^r_{S_2}$, 2) $E=H^r_{S_1}\cap H^a_{S_2}$, 3) $E=H^r_{S_1}\cap H^r_{S_2}$, 4) $E=H^a_{S_1}\cap H^a_{S_2}$

The normal $\underline{n}_\Sigma$, where it exists ($\underline{n}_\Sigma$ is not defined at the points $x\in\Gamma$ in which $\overline{S}_1$ and $\overline{S_2}$ may have different tangent planes) will be directed out of E, which means referring to the case 1)

(2.5) $\underline{n}_\Sigma=\underline{n}_{S_1}$ on S_1 and $\underline{n}_\Sigma=-\underline{n}_{S_2}$ on S_2

Using representation formula (1.7) for both u_{S_1} and u_{S_2} and taking into account that $u_{S_1}=u_{S_2}$ on $\partial\Omega$, we may write ($\underline{n}=\underline{n}_\Sigma$ on Σ)

(2.6) $u(x)=u_{S_2}(x)-u_{S_1}(x)=-\int_{\Sigma-\Gamma}\underline{n}^T(\xi)M(\xi)\mathrm{grad}_\xi s(x,\xi)d\sigma_\xi,\quad \forall x\in\bar{\Omega}-\Sigma$

and since $u=\frac{\partial u}{\partial n}=0$ on $\partial\Omega$, we have

(2.7) $u(x)=0,\quad \forall x\in\bar{\Omega}-E$

For $\forall x\in\Sigma-\Gamma$ the vector $\underline{\lambda}=M\underline{n}$ is directed out of E and is never tangent to Σ because M is positive definite. Therefore using the "limit formulae" for the oblique double layer potential (2.6), we deduce, that u is, in a formal sense, an "eigenfunction" of the following "oblique derivative" boundary problem:

(2.8) $\Delta u=0$ in E , $\frac{1}{\underline{\ell}\cdot\underline{n}}\ \underline{\ell}\cdot\mathrm{grad}u-\beta^\ell u=0,\quad \forall x\in\Sigma-\Gamma$

where u and gradu denote the limit on $\Sigma-\Gamma$ from the interior of E , $\underline{\ell}$ is the vector symmetric of $\underline{\lambda}$ with respect to $\underline{n}$ and finally $\beta^\ell=\mathrm{div}^\Sigma(\underline{n}^TM\underline{n})^{-1}\underline{\beta}$, with $\underline{\beta}=M\underline{n}-(\underline{n}^TM\underline{n})\underline{n}$. Using again the limit formulae of the potential theory, from (2.6) and (2.7) it follows also that:

(2.9) $\Delta u=0$ in E , $u=\alpha=\underline{n}^TM\underline{n}$ on $\Sigma-\Gamma$, $\frac{1}{\underline{\ell}\cdot\underline{n}}\ \underline{\ell}\cdot\mathrm{grad}u=\beta^\ell\alpha$ on $\Sigma-\Gamma$

hence condition (2.7) can be considered as a "compatibility condition" on the Cauchy data (2.9) for the harmonic function u(x) given by (2.6).
We now pose the question: how can we exploit relations (2.8) and (2.9) in order to prove (2.1)? This question is connected with the "uniqueness properties" of the oblique derivative problem (2.8) or with the compatibility of the Cauchy problem (2.9). We remark that the domain E can be "irregular"; more precisely its boundary Σ, which is of class C^2 at every $x\in\Sigma-\Gamma$, can be irregular at $x\in\Gamma$, where $\bar{S}_1$ and $\bar{S}_2$ can have different tangent planes and Σ may also include cuspidal arcs (see hypothesis VIII). Moreover we remark that if the oblique double layer on Σ reduces to a normal one with constant density, statement (2.1) does not obviously hold. Applying results of potential theory (see [8]) and using Hopf's maximum or minimum principle it is possible

to prove (see [6]) the following:

Theorem 4. Under hypotheses I),,..VIII) and the following assumptions

(2.10) the eigenvalues m_t and m_ℓ of M are constant on $\bar{S}_1 \cup \bar{S}_2$ and $m_t < m_\ell$,

(2.11) at least on one of the surfaces $S_i (i=1,2)$, $\underline{a}_\ell$ is not identically // (or $\perp$) to $\underline{n}$, but $\exists\, x_i \in S_i$ such that $\underline{a}_\ell$ // (or $\perp$) $\underline{n}$ at x_i,

if S_1 and S_2 satisfy (2.4) then the uniqueness proposition (2.1) holds.

Assumption (2.10) is equivalent to considering a homogeneous intracellular anisotropy, i.e., $\sigma_t^i, \sigma_\ell^i$ positive different constants, while the first part of (2.11) aims at excluding normal double layer of constant density on $\bar{S}_1 \cup \bar{S}_2$. We remark that if Σ is regular (e.g. of class $C^{1,1}$) then the existence of at least one point $x_i \in \Sigma$ at which $\underline{a}_\ell$ is // (or $\perp$) to $\underline{n}$ derives from a well known topological theorem concerning vector fields on a regular closed surface; hence in this case (2.11) reduces only to the first part, i.e., to assume that on at least one of the surfaces S_i we have effectively an "oblique" double layer potential.

The second part of assumption (2.11), from a physiological point of view, means that there exists a fiber which is perpendicular or tangent to the wavefront surface.

A different technique can be used to obtain uniqueness results in the case of oblique double layers contained in a set $H_1 \subset H$ in which the variations of the fiber direction are small. This case is interesting from the physiological point of view when one considers ectopic stimulations, originating from a single point, and the excitation fronts are limited to a small region; a similar situation occurs for front perturbations in a region with a weak variation in the fiber direction. We consider

(2.12) a bundle of fibers between two parallel planes and we assume that the region occupied by the bundle is a simply connected region H_1 and that the fiber direction in H_1 differs "little" from that of the unit normal $\underline{k}$ to the planes, i.e., if $\underline{a}_\ell = \underline{a}_\ell(\xi)$, $\xi \in H_1$ is the unit vector tangent to the fiber of H_1 passing through ξ, oriented so that $\alpha = \underline{a}_\ell \cdot \underline{k} > 0$, then our assumption is that α is "near" to 1 and α' is "near" to zero where α' denotes the derivative along the fiber direction.

Surfaces which intersect once the same set or bundle of fibers are called sections of the bundle; an orthogonal section intersects the fibers of the bundle at a right angle. The positive direction of the unit normal $\underline{n}$ to a section of a bundle in H_1 is that for which: $\underline{a}_\ell \cdot \underline{n} \geq 0$ with $\underline{a}_\ell \cdot \underline{k} > 0$. If S_1, S_2 are sections of a bundle we say that S_2 follows S_1 if, moving along each fiber of the bundle in the positive direction, we find successively the intersection of the fiber with S_1 and S_2.

(2.13) We shall assume that there exists in H_1 a surface S_0 orthogonal to all the fibers of H_1

We consider now the"axial" potential

$$v_S(x) = \frac{1}{4\pi}\int_S (\underline{a}_\ell\cdot\underline{n})\underline{a}_\ell\cdot \text{grad } r^{-1} d\sigma$$

where $r = |x-\xi|$ and ξ describes S.

Then it is possible to prove (see [6]) the following:

Theorem 5. Under the assumptions (2.12), (2.13), if S_1, S_2 are sections of a bundle in H_1 and S_2 follows S_1 then there exist points x, not on the fiber segments intercepted by S_1 and S_2, such that

$$|v_{S_1}(x)| < |v_{S_2}(x)|.$$

Applying this result it is easy to establish the following:

Theorem 6. Under the same hypotheses of Th. 5, if

S_1, S_2 are sections of the same bundle with a common boundary, no other points in common

or if

S_1, S_2 are closed surfaces, S_1 inside S_2

then the uniqueness proposition (2.1)holds.

On the basis of the uniqueness results related to several important cases which have been investigated above and of others described in [6] we are led to conjecture that the mathematical model we formulated for describing the cardiac electric potential field, based on the representation of the cardiac sources as a particular oblique dipole layer on the excitation wavefront, incorporates those "essential" physiological constraints which should yield a unique solution of the inverse problem in "general".

Establishing the validity of the stated conjecture is notable conceptually since it means that the body surface potential distribution actually contains all the information concerning the excitation of the heart.

In practice the possibility of uniquely determining the excitation wavefront from the measured surface data will meet with mathematical and numerical difficulties due to the ill-posedness of the problem. In conclusion we remark that the effective numerical solution of the inverse problem is a relevant task since it should allow one to establish on objective grounds the correlation between features of the potential surface fields, like the number, location and time course of potential maxima and minima on the chest surface, and the location and shape of the depolarization wavefront surface in the heart.

REFERENCES

[1] S. BARUFFI, S. SPAGGIARI, D. STILLI, E. MUSSO, B. TACCARDI: The importance of fiber orientation in determining the features of cardiac electric field. In: Modern Electrocardiology, edited by Z. Antaloczy, Akademiai Kiado, Budapest (1978), 89-92.

[2] R.M. BERN (ed). Handbook of Physiology vol 1, "The Heart", Baltimore: Williams and Wilkens, (1980).

[3] P. COLLI FRANZONE, L. GUERRI, C. VIGANOTTI, E. MACCHI, S. BARUFFI, S. SPAGGIARI, B. TACCARDI: Potential fields generated by oblique dipole layers modeling excitation wavefronts in the anisotropic myocardium. Comparison with potential fields elicited by paced dog hearts in a volume conductor, Circulation Research, 51, 3(1982), 330-346.

[4] P. COLLI FRANZONE, L. GUERRI, C. VIGANOTTI: Oblique dipole layer potentials applied to electrocardiology, J. Math. Biology 17, (1983), 93-124.

[5] P. COLLI FRANZONE, L. GUERRI, B. TACCARDI, S. TENTONI, C. VIGANOTTI: Cardiac fibers orientation and potential fields: model studies. Jpn. Heart J., 23,I(1982), 279-281.

[6] P. COLLI FRANZONE, L. GUERRI, E. MAGENES: Oblique double layer potentials for the direct and inverse problems of electrocardiology, submitted to J. Math. Anal.Appl.

[7] L.V. CORBIN II, A.M. SCHER: The canine heart as an electrocardiographic generator: dependence on cardiac cell orientation. Circ. Res. 41(1977), 58-67.

[8] L.L. HELMS: Introduction to Potential Theory. Wiley, New York (1969).

[9] J.L. LIONS, E. MAGENES: Non Homogeneous Boundary Value Problems and Applications. I. Grundlehren 181, Springer-Verlag (1972).

[10] D.E. ROBERTS, A.M. SCHER: Effects of tissue anisotropy on extracellular potential fields in canine myocardium in situ. Circ. Res. 50,(1982), 342-351.

[11] M.S. SPACH, W.T. MILLER, E. MILLER-JONES, R.B. WARREN, R.C. BARR: Extracellular potentials related to intracellular action potentials during impulse conduction in anisotropic canine cardiac muscle. Circ. Res. 45, (1979), 188-204.

[12] D.B. GESELOWITZ, R.C. BARR, M.S. SPACH, W.T. MILLER III: The impact of adjacent isotropic fluids on electrocardiograms from anisotropic cardiac muscle, Circulation Research, 51, 5(1982), 602-613.

MATHEMATICAL ANALYSIS OF IMMOBILIZED ENZYME SYSTEMS

E.J. Doedel, Computer Science Department, Concordia University, 1455 de Maisonneuve Blvd. W. Montreal, Quebec H3G 1M8

J.P. Kernevez, Université de Technologie de Compiègne, BP 233, 60206, Compiègne, France

Abstract : Simple enzyme systems, with interacting diffusion and reaction, are modeled by O.D.E.s involving several parameters and showing, according to the parameter values, multiple steady states and/or periodic solutions. The 2-parameter continuation of singular points gives some insight into the behaviour of these systems. In particular it yields isola's.

Introduction : This paper deals with the application of continuation techniques to biochemical systems represented by autonomous O.D.E.s. We consider autonomous dynamical systems

$$u' = f(u, \lambda, \mu) \tag{1}$$

where λ, μ are scalar parameters, $u,f \in \mathbb{R}^n$ and $u' = du/dt$. We are interested in their steady state and time-periodic solutions.

Continuation and branch switching techniques have already been discussed elsewhere : treatments on continuation of steady state or periodic solutions [1,2,3,4], steady state and periodic solution limit points [4,5,6,7,10], Hopf bifurcations [3,5,11], cusps and Hopf bifurcation limit points [5] can be found in the literature. We first recall the basic ideas.

Enzyme systems have been described in 8 . We restrict our considerationshere to very simple models, with the aim to show the usefulness of continuation techniques in obtaining isola's of steady state and periodic solutions.

For related work on coupled cells we refer, among others, to [9,13].

Continuation of steady states : We consider the pairs (u,λ) such that

$$f(u,\lambda) = o, \quad u,f \in \mathbb{R}^n, \ \lambda \in \mathbb{R}.$$

Parameterizing a curve of solutions (u,λ) in $\mathbb{R}^{n+1}$ by arclength s, we have

$$f(u(s), \lambda(s)) = o, \quad |\dot{u}|^2 + \dot{\lambda}^2 = 1$$

where $\dot{u} = du/ds$ and $\dot{\lambda} = d\lambda/ds$. Starting with the solution (u_o, λ_o), the curve of continued solutions as λ is varied can be obtained as the solution of the O.D.E.s.

$$\dot{u} = \xi, \quad \dot{\lambda} = \eta, \quad u(o) = u_o, \quad \lambda(o) = \lambda_o \tag{2}$$

where (ξ, η) is given by

$$f_u \, \xi + f_\lambda \eta = o, \quad |\xi|^2 + \eta^2 = 1. \tag{3}$$

Here f_u and f_λ are, respectively, the Jacobian matrix of f and its derivative with respect to λ. For pseudo-arclength approximation of (2), (3) together with branch switching techniques, we refer to [1].

Continuation of limit points of steady state solutions : We consider now a second parameter μ and the equation

$$f(u, \lambda, \mu) = o, \quad u, f \in \mathbb{R}^n, \quad \lambda, \mu \in \mathbb{R}.$$

At a limit point we have

$$f(u, \lambda, \mu) = o, \quad f_u \, v = o, \quad v \in \mathbb{R}^n, \quad |v| = 1. \tag{4}$$

If w denotes the 2n+1 - vector (u, v, μ), equations (4) can be rewritten

$$\mathcal{F}(w, \lambda) = o$$

with $\mathcal{F} \in \mathbb{R}^{2n+1}$, which define an algebraic continuation (or bifurcation) problem of 2n+1 equations in 2n+2 variables, viz. u, v, λ and μ. Except for degenerate cases, one will have curves of solutions in 2n+2 - dimensional space [6,7,10].

Continuation of periodic solutions : After scaling the time variable t, periodic solutions of $u' = f(u, \lambda)$ can be defined by

$$\begin{cases} u' = Tf(u, \lambda) \\ u(o) = u(1) \\ \displaystyle\int_o^1 u(t)^* \, v'(t) \, dt = o \end{cases} \tag{5}$$

where $u, f \in \mathbb{R}^n$, $\lambda \in \mathbb{R}$, $u' = du/dt$, T is the period and the last equation is an "anchor" equation [2] defining the origin of time $t = o$ on the orbit by synchronization with a known solution $v(t)$. A more concise way of writing (5) is

$$\mathcal{F}(w,\lambda) = o, \quad w = (u,T)$$

and Keller's pseudo-arclength continuation and branch switching techniques can be applied to the problem of following (w,λ) as λ varies, subject to $\mathcal{F}(w,\lambda) = o$ [2].

Continuation of periodic solution limit points with a second parameter μ is, from a general point of view, identical to that of steady state limit points given above.

Continuation of Hopf bifurcation points : For a curve of steady states

$$f(u,\lambda,\mu) = o, \qquad u,f \in \mathbb{R}^n, \qquad \lambda,\mu \in \mathbb{R}$$

a Hopf bifurcation point is such that [5]

$$f(u,\lambda,\mu) = o, \qquad f_u(\xi + i\eta) = i\omega(\xi + i\eta), \qquad (\omega = 2\pi/T)$$

$$\xi^\star\xi + \eta^\star\eta = 1, \qquad \eta_0^\star\xi - \xi_0^\star\eta = o, \qquad \xi,\eta \in \mathbb{R}^n.$$

These equations define a curve in $\mathbb{R}^{3n+3}$

$$\mathcal{F}(w,\lambda) = o, \qquad \mathcal{F} \in \mathbb{R}^{3n+2}$$

if w denotes $w = (u,T,\xi,\eta,\mu)$ $(\in \mathbb{R}^{3n+2})$.

Isola's of steady state solutions : We consider the system

$$\begin{cases} s_1' = (s_o - s_1) + (s_2 - s_1) - \rho\, F(s_1) \\ s_2' = (s_o + \mu - s_2) + (s_1 - s_2) - \rho\, F(s_2) \end{cases} \tag{6}$$

where $$F(s) = s \,/\, (1 + s + k_2\, s^2). \tag{7}$$

Here s_1 and s_2 are substrate concentrations in 2 compartments communicating by diffusion with each other and with reservoirs where the concentrations are respectively s_o and $s_o+\mu$. An enzyme reaction with so-called substrate inhibited kinetics consumes the substrate at a rate proportional to $F(s)$.

Parameters k_2 and ρ have fixed values, $k_2 = 1.$ and $\rho = 100.$
We fix μ at $\mu = o$ and consider the steady states :

$$\begin{cases} s_o = 2s_1 - s_2 + \rho\, F(s_1) \\ s_o = 2s_2 - s_1 + \rho\, F(s_2). \end{cases} \tag{8}$$

A continuation analysis of these steady states as s_o varies yields the diagram of Figure 1, a), about which we note the following :

(i) the quantity represented on the ordinate is the Euclidean norm of the solution :

$$\text{norm} = (s_1^2 + s_2^2)^{1/2}$$

(ii) the full (resp. dashed) lines represent stable (resp. unstable) steady states.

(iii) the branch emanating from the origin corresponds to "symmetric" solutions :

$$\begin{cases} s_1 = s_2 = s \\ s_o = s + \rho\, F(s). \end{cases}$$

(iv) the bifurcated branch is a double branch since, due to the symmetry of the compartments, if $s_1 = u$, $s_2 = v$ is a solution to (8), so is $s_1 = v$, $s_2 = u$ with the same norm.

(v) there is an arc of stable nonsymmetric solutions on the bifurcated branch.

Now consider the steady states of (6) when $\mu \neq o$:

$$\begin{cases} s_o = 2s_1 - s_2 + \rho\, F(s_1) \\ s_o + \mu = 2s_2 - s_1 + \rho\, F(s_2). \end{cases}$$

As a starting point for the continuation process as s_o changes for a given value $\tilde{\mu}$ of μ, we no longer know a trivial solution (like $s_1 = s_2 = s_o = o$ in the case $\mu = o$). Instead we can follow a limit point as μ varies from o to $\tilde{\mu}$, thus obtaining $(\tilde{s}_1, \tilde{s}_2, \tilde{\lambda}, \tilde{\mu})$, then continue this point as s_o varies, μ being held fixed at $\tilde{\mu}$.

Continuing the limit points A, B, C and D of Figure 1, a) as μ varies ($\mu \geq o$) we

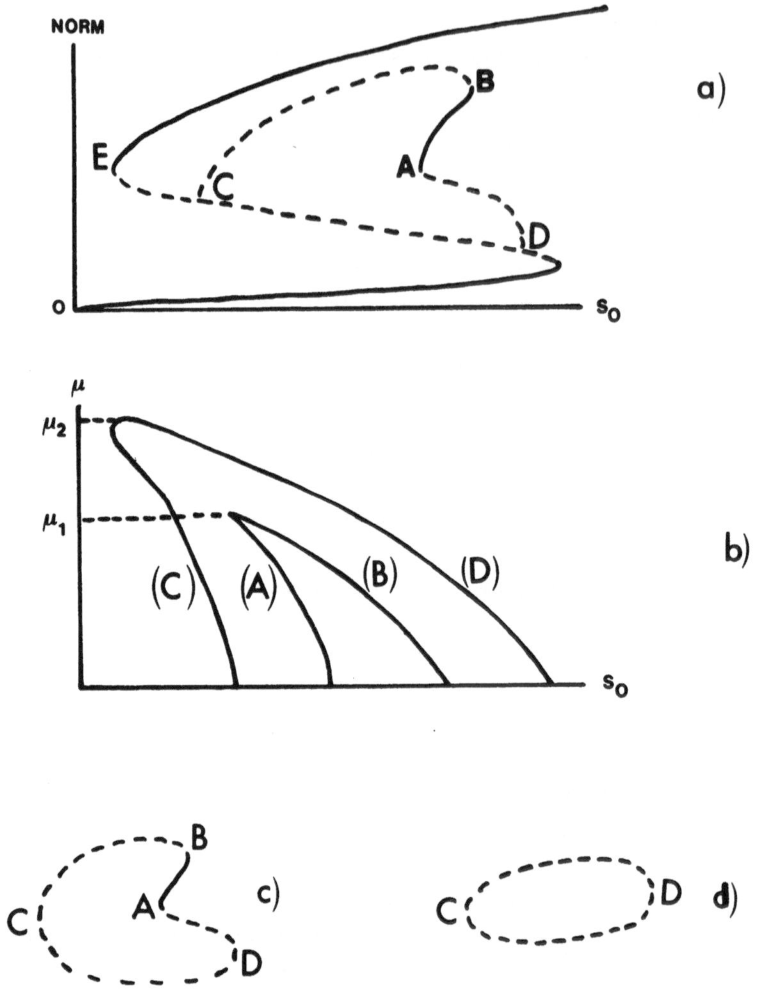

FIGURE 1

Isola's of steady state solutions

find curves whose projections (A), (B), (C) and (D) onto the (s_0,μ) - plane are as Figure 1, b). (More accurately, for (C) and (D), we continue the limit points that result from perturbing these pitch-fork bifurcations).

If $0 < \mu < \mu_1$ we find an isola (Figure 1, c)) when continuing B for example as s_0 changes. This isola comes from the closed curve ABCDA of Figure 1, a) and contains an arc AB of stable steady states. This arc vanishes as μ crosses μ_1 and, for $\mu_1 < \mu < \mu_2$, the isola consists of 2 arcs of unstable steady states with respectively 1 and 2 modes of instability, separated by the limit points C and D (Figure 1, d)). Finally the isola disappears past μ_2. Of course there is another branch of solutions, corresponding to the curve ODABCE... of Figure 1, a), and which can be obtained, for a given value $\tilde{\mu}$ of μ, by following the turning point E for example as μ varies from 0 to $\tilde{\mu}$, then by following the obtained point as s_0 varies, μ being held fixed at $\tilde{\mu}$.

Isola's of periodic solutions : We consider the system

$$\begin{cases} s' = s_0 - s - \rho\, a\, F(s) \\ a' = \alpha(a_0 - a) - \rho\, a\, F(s) \end{cases}$$

where s and a are the concentrations of 2 substrates S and A in a compartment separated by a diffusion membrane from a reservoir where they are at concentrations s_0 and a_0. The parameter α denotes the ratio of A and S diffusion coefficients in the membrane : $\alpha = 0.2$. In the compartment an enzyme reaction consumes the substrates S and A with a reaction rate proportional to $aF(s)$ where F is given by (7) with $k_2 = 0.1$. In the following the parameter a_0 is fixed at $a_0 = 500$. If we fix the parameter s_0 at $s_0 = 100.$ and leave the parameter ρ free, we find the bifurcation diagram of Figure 2, a), where steady states are represented with the same conventions as above, and the stable (resp. unstable) periodic solutions are represented by ● ● ● (resp. ○ ○ ○). For periodic solutions, their norm

$$\text{norm} = T^{-1} \left\{ \int_0^T [s^2(t) + a^2(t)]\, dt \right\}^{1/2},$$

is represented on the ordinate, while for steady state solutions the Euclidean norm of the steady state solution vector (s,a) is measured along the ordinate.

With these definitions of steady state and periodic solution norms, a Hopf bifurcation point occupies the same point in the bifurcation diagram, whether considered as a point on a steady state branch or as a point on a periodic branch.

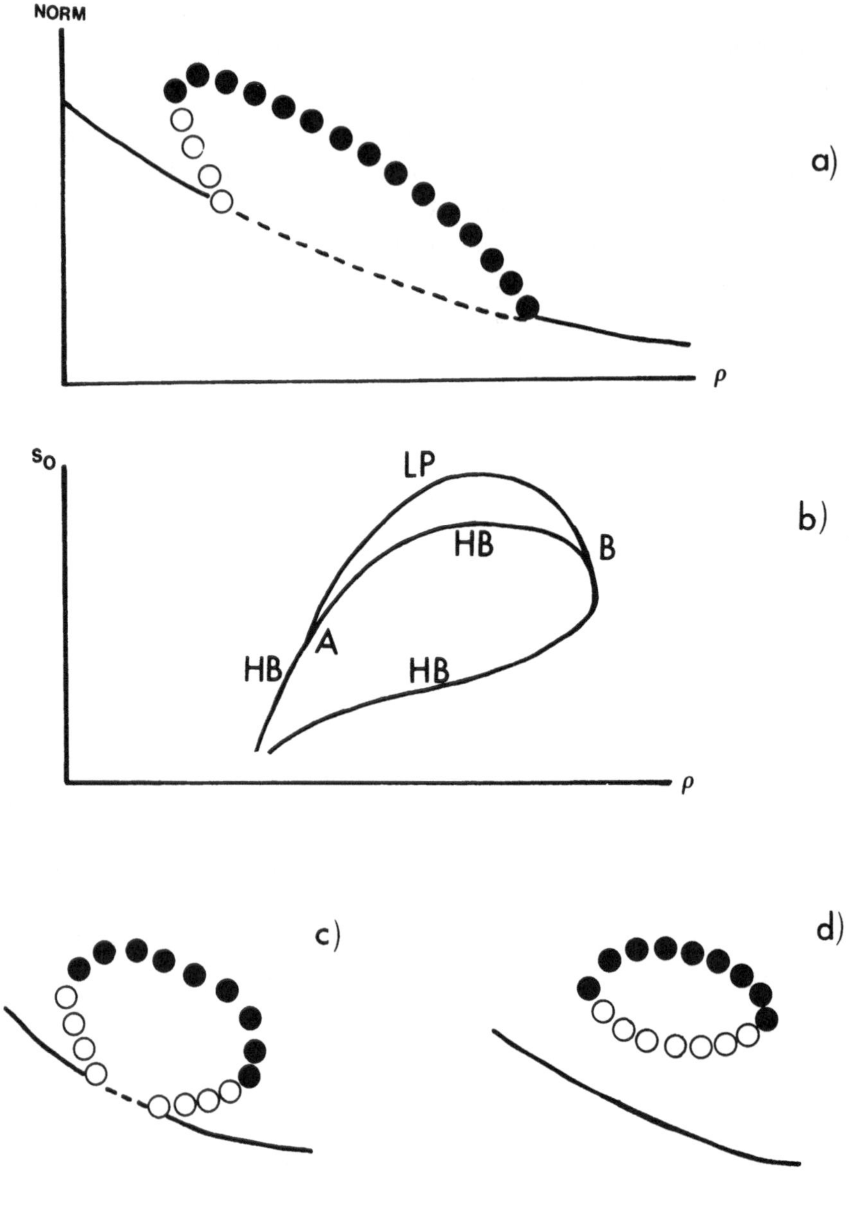

FIGURE 2

Isola's of periodic solutions

Now leave one more parameter free, namely s_0, and trace out the curves HB of Hopf bifurcation points and LP of periodic solution limit points (Figure 2, b)). It is seen that :

(i) if s_0 lies between the lower limit of the diagram and the ordinate of point A, there are 2 Hopf bifurcation points and no periodic solution limit point.

(ii) if s_0 is between the ordinates of A and B, there are 2 Hopf bifurcation points and 1 limit point (Figure 2, a)).

(iii) if s_0 is between the ordinate of B and the maximum ordinate on HB, there are 2 Hopf bifurcation points and 2 limit points (Figure 2, c)).

(iv) if s_0 lies between the maximum ordinate on HB and the maximum ordinate on LP, namely $110.5 < s_0 < 112.5$, there are no more Hopf bifurcations and we have an isola of periodic solutions (Figure 2, d)), with 2 limit points separating arcs of stable and unstable solutions. For a general treatment of such phenomena see [14].

Conclusion : Continuation of singular points gives more insight into the behaviour of systems modeled by autonomous O.D.E.s depending upon several parameters. These methods are general and can be applied to more complex models than those presented here, arising either in biology or in other fields.

References :

[1] Keller,H.B., Numerical solution of bifurcation and nonlinear eigenvalue problems, in : Applications of Bifurcation Theory, Rabinowitz, P.H., ed., Academic Press, 1977, 359 - 384.

[2] Doedel, E.J., AUTO : a program for the automatic bifurcation analysis of autonomous systems, Cong. Num. 30, 1981, 265 - 284.

[3] Jepson, A.D., Numerical Hopf Bifurcation,Thesis, California Institute of Technology, Pasadena, Ca., 1981.

[4] Rinzel, J., and Miller, R.N., Numerical calculation of stable and unstable periodic solutions to the Hodgkin-Huxley equations, Mathematical Biosciences 49, 1980, 27 - 59.

[5] Doedel, E.J., Continuation techniques in the study of chemical reactors, preprint.

[6] Jepson, A. and Spence, A., Folds in solutions of two parameter systems and their calculation : Part I, Numerical Analysis Project, Manuscript NA - 82 - 02, Computer Science Department, Stanford University, March 1982.

[7] Rheinboldt, W.C., Computation of critical boundaries on equilibrium manifolds, SIAM J. Numer. Anal. 19, n° 3, 1982, 653 - 669.

[8] Kernevez, J.P., Enzyme Mathematics, North-Holland, Amsterdam, 1980.

[9] Othmer, H.G., Synchronized and Differentiated Modes of Cellular Dynamics, in Systems, H. Haken ed., Springer, 1980.

[10] Griewank, A. and Reddien, G.W., Characterization and Computation of Generalized Turning Points, to appear in SIAM J. Numer. Anal.

[11] Griewank, A. and Reddien, G.W., The Calculation of Hopf points by a direct method, to appear.

[12] Jepson, A.D. and Spence, A., Paths of Singular Points and their Computation, to appear.

[13] Doedel, E.J., Kernevez, J.P. and Thomas, D., Bifurcation Behaviour of One- and Two-Compartment Enzyme Models, to appear.

[14] Golubitsky, M. and Langford, W.F., Classification and Unfoldings of Degenerate Hopf Bifurcations, J. Diff. Eqs. 41, n° 3, 1981, 375 - 415.

MODELING CONTROLS AND VARIABILITY OF THE CELL CYCLE

Lilia Alberghina*, Luigi Mariani° and Enzo Martegani*
*Cattedra di Biochimica Comparata, Facoltà di Scienze, Università di Milano, Milano.
°Istituto di Elettrotecnica e Elettronica dell'Università and LADSEB-CNR,Padova,Italy

1. Introduction

Cell cycle events (initiation of DNA replication, length of the G_2 phase, occurrence of mitosis and of cell division) follow temporal patterns which are characteristic of each cell type and growth condition, thus suggesting the existence of complex control mechanisms /1,2/. While considerable progress as been made in the description of macromolecular syntheses which characterize different growth conditions, our knowledge on the regulatory mechanisms that coordinate growth with nuclear and cell division is still poor. Clearly the reductionistic approach, which has been so useful for the description of the molecular components of a cell and of the more simple regulatory systems (regulation of enzymatic pathways, control of gene expression in bacteria), is not adequate for the study of highly integrated systems, such those controlling growth and cell division, for which it seems necessary to develop an integrated approach. Mathematical models may well be useful for this undertaking. In fact, they allow quantitative description of both the interrelations among the relevant variables of the phenomenon and the dynamics of the events under consideration. Many mathematical models have been developed in order to describe the dynamics of the cycle events, and their analytical solutions and the simulations, which are often required for a more accurate analysis of their predictions, may help to better understand the regulatory mechanisms of cell proliferation. Comparison of the predictions of the model with experimental results allows verification of the validity of the assumed functional or causal links. In any event, it is abvious that the isolation of the phenomenon from its context and the choice of the variables and their interactions are to a certain extent arbitrary. Different models can describe the same phenomenon, and each of them be more useful than others for a given specific purpose; thus a model is able only to give a partial and approximate representation of a real phenomenon and therefore it makes no sense to attempt to develop a true model of the cell cycle. Instead, when compared with other models, a given cell cycle model may be more complete, if it accounts for a larger number of aspects of cell growth and division, more accurate, if its predictions give a more close fitting of experimental results, and more useful, if it is able to suggest a new experimental design and even to rule out, although not to prove, a proposed control mechanism.

2. Controls of the cell cycle

A wealth of experimental findings (discussed in detail in ref. 1 to 16) suggest that the progression of the cycle of actively growing cells depends on growth dynamics (i.e. protein and RNA accumulations) since in unperturbed cell popolations the discontinuous events of DNA replication and of nuclear and cell division are activated when critical level (P_s and P_m respectively) of protein are sequentialy attained in the cell. Whereas the attainement of Ps is sufficient to initiate DNA replication, the occurence of nuclear division further requires the elapse of a G_2 period, which can not be shorter than a minimal incompressible period, characteristic of the organism and of the growth condition.

3. The model, its first subsystem and the growth condition

The model is divided in two subsystems: the first, which is common to all the three cases considered, describes the controls and the rates of cellular growth, and the second, whose structure is somewhere different, accounts for DNA replication and cell division. The upper part of fig.1 represents as a block diagram the first subsystem, i.e. the relations proposed /15,16/ for the rates of proteins (P) and ribosomes (R) synthesis, and connects them to the events of DNA replication and cell division, since when $P=2^kP_s$(P_s=cost. k=0,1,2,.. with k=1 in diploid cells, unless otherwise indicated) a round of DNA replication, with 2^k origins, starts and the second subsystem is activated, then it proceeds until cell division following specific rules (see next Sections). If a growth condition ($K_2\varrho-1/T_2>0$) is satisfied /15,16/, the protein and ribosome contents increase exponentially with rate a constant $\alpha=K_2\varrho-1/T_2$ and a duplication time $T=\ln 2/\alpha$.

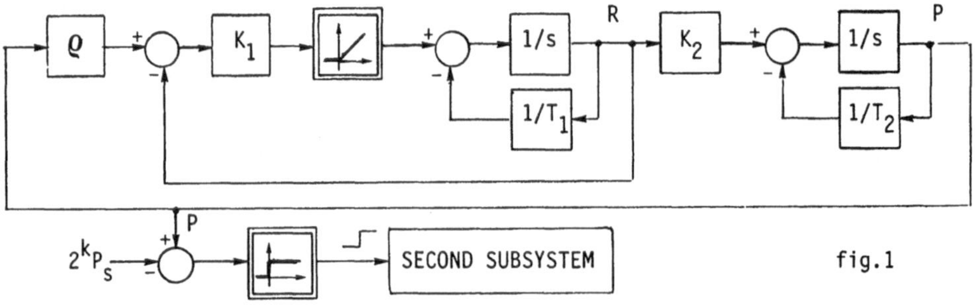

fig.1

4. The second subsystem and the cell cycle patterns in eukaryotes with equal division

In eukaryotic cells DNA is replicated during a discrete phase of its cycle, the S phase (of constant duration τ_R), and, as it has been illustrated in Section 2, one of the more important factors in determining the entrance in S is the attainment of a critical cells mass, (P_s). The occurrence of mitosis and cell division in some cell cycles also depends on reaching a given cell size (P_m) and on the elapse of a minimal time period ($\tau_{G\ min}$) from the end of DNA replication. An exact divi-

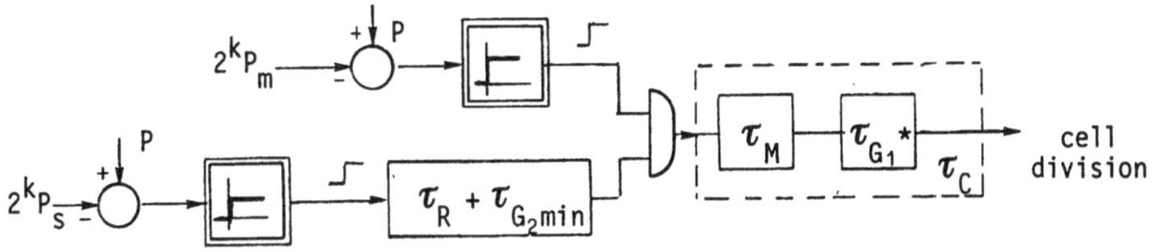

fig.2

sion in two equal daughter cells is assumed to occur after a time τ_C from the beginning of nuclear division. Depending on the values assumed in a given growth condition by the parameters P_s, P_m, τ_{G_2min}, and by T, whose duration is set, as indicated before, by the dynamics of the first subsystem, and under the further assumption that $T \geq \tau_R + \tau_{G_2min} + \tau_C = \tau_m$, three (A,B,C) different steady state patterns of growth may be obtained /14/. The first two patterns are characterized by the fact that during the cell cycle both threshold levels (P_s and P_m) are crossed, and cycles with all four classical phases (G_1, S, G_2 and M) are originated. The third pattern, which is less common, is characterized by the fact that the initial P_0 level is larger than the P_s level, and DNA replication therefore starts at the beginning of the cycle; a cell cycle lacking the G_1 phase is thus generated . All three patterns have experimental evidence. Analytical conditions have been found which can be condensed in a nomogram (fig.A1 of /14/) useful for estimating the type of steady state that takes place with a given set of values for the above-mentioned parameters.

As stated in the previous section, protein and ribosome contents of an individual cell increase exponentially during a cycle starting from initial values P_0 and $R_0 = \rho P_0$. The initial value P_0 is equal to 0.5 P(T), since it is assummed that cell division originates two identical daughter cells. It follows that P_0 is univocally determined for each steady state pattern of growth from P(T) which, in turns, for each regime can be derived from P_m or P_s. Furthermore if an ideal population in balanced exponential growth is considered, with its cells distributed along the cycle according to the age density function $n(a) = 2 \ln 2 \; 2^{-a}$ $(0 \leq a \leq 1)$, where a=t/T is the physiological age, the macromolecolar levels (i.e. the average protein, ribosome, and DNA contents per cell in the population) can be calculated as a function of the temporal parameters of the cell cycle. Conversely, the determination of these parameters can be done knowing the macromolecular levels and other experimental findings, such as the labelling index /14/. The agreement between experimental results and the prediction of the model has been found very satisfactory.

5. The second subsystems and the cell cycle patterns in unequally dividing cells

A similar structure of the second subsystem can be used to model the cycle when asymmetry in division occurs, so to produce two different populations, i.e. parent cells with $P_0 \cong P_s$ and daughter cells with $P_0 < P_s$. In balanced exponential growth, with mass duplication time $T=\ln 2/\alpha$ the population is therefore composed by two subpopulations with different cycle time, i.e. $T_P \leq T$ for parents and $T_D \geq T$ for daughters, that must satisfy the following relation /18/, which follows from the fact that birth rate is the same rate in the two subpopulations:

$$(1) \quad e^{-\alpha T_P} + e^{-\alpha T_D} = 2^{-T_P/T} + 2^{-T_D/T} = 1.$$

Except for the difference in the G_1 phase, parents and daughters follow the same cycle, which is imposed, as seen for equally dividing cells, by the relative value of the parameters P_s, P_m, τ_R, $\tau_{G_2 min}$, τ_C and T. The same three patterns (A,B,C) arise and the same boundary conditions hold, so that also the nomogram reported in fig.A1 of /14/ is still valid. The third (C) regime corresponds to the limit case, which occurs when the final protein content of the cell is larger than $2P_s$. In this case equal division occurs, so that parent and daughter cells follow the same cycle starting from an initial protein value greater than P_s, both lacking the G_1 phase and the population becomes homogeneous. For every regime the following values of T_P, T_D and P_o (the initial protein content of daughter cells) can be calculated:

$$(2) \quad T_P = \tau_m \; ; \quad T_D = \tau_m - T\,[\lg_2(2^{\tau_m/T} - 1)] \; ; \quad P_o = P_s(2^{\tau_m/T} - 1) \qquad \text{for regime A}$$

$$(3) \quad T_P = T\cdot \lg_2(P_m/P_s) + \tau_C \; ; \quad T_D = T\,\lg_2[P_m/(P_m 2^{\tau_C/T} - P_s)] + \tau_C \; ; \quad P_o = P_m 2^{\tau_C/T} - P_s \qquad \text{for regime B}$$

$$(4) \quad T_P = T_D = T \quad ; \quad P_o = P_m/(2^{1-\tau_C/T} - 1) \qquad \text{for regime C.}$$

In an ideal population in balanced exponential growth, cells are distributed along the cycle according to an age distribution function which coincides with $n(a) = 2 \ln 2\, 2^{-a}$ only for the regime C, whereas when T_P and T_D are different is given by:

$$(5) \quad n(a) = 2 \ln 2\, 2^{-a} \quad 0 \leq a < T_P/T \quad ; \quad n(a) = \ln 2\, 2^{-a} \quad T_P/T \leq a < T_D/T$$

where the variable $a=t/T$ in this case spans from 0 to $T_D/T>1$. Taking into account the appropriate age distribution functions, the following average protein, ribosomes and DNA contents per cell in the population can be derived:

$$(6) \quad \hat{P} = P_s[T_P - T_D + T_D\, e^{\alpha T_D}] = P_s \ln 2\, [T_P/T - T_D/T - (T_D/T)2^{T_P/T}] \qquad \text{for regime A or B}$$

$$(7) \quad \hat{P} = 2 \ln 2\, P_o = 2 \ln 2\, P_m/(2^{1-\tau_C/T} - 1) \qquad \text{for regime C}$$

$$(8) \quad \hat{R} = \rho \hat{P} \quad ; \quad (9) \quad \hat{D} = 1.44\, D_o\, 2^{T_P/T} \cdot (1 - 2^{-\tau_R/T})/(\tau_R/T)$$

where T_P and T_D are given by (2),(3). Relations (6),(7), as the companions presented in /14/, allow the determination of the macromolecular levels in the population, starting from the knowledge of cycle times, or, conversely, the determination of the length of the phases from the average macromolecular contents.Furthermore they give explanation of the experimentally observed change in size in populations of the same cells growing in different media, i.e. with different duplication time, change that are larger than those provided by a homogeneus population model. In fact,in these experiments, different regimes and/or modifications in the thereshold levels often occur.

6. The second subsystem and the DNA patterns in bacteria

The behaviour of bacterial cells as Escherichia coli can be modeled by a second subsystem like that of fig.3, i.e. with only the size control over the onset of DNA replication and constant times τ_R and τ_C to complete replication and cell division, which is assumed to occur exactly in two equal parts.

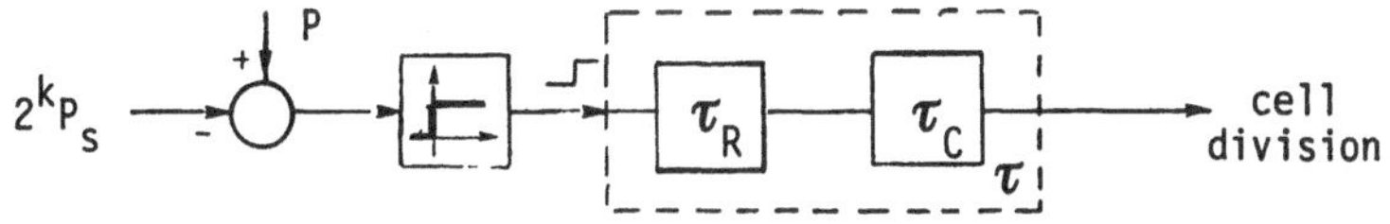

fig.3

A second, but very important difference with the types of cell previously considered, difference which is not evident from the figure, is that multiple DNA replications can start at each initiation and can proceed also in subsequent cycles, giving different patterns of DNA synthesis at the various growth rates and therefore T can be lower than $\tau_R+\tau_C=\tau$. The cell division cycle is completely characterized by the timing of the events of initiations of DNA replication and of terminations of previously started rounds of replication and by the fact that a fixed period (τ) has to elapse between the termination of DNA replication and the ensuing cell division /16/. Considering that in a steady state with a rate of growth high enough to have: $1<\tau_C/T<\tau/T$, a round of replication initiated at time t_1 stops at time $t_1+\tau_R$, and a cell division occurs at time $t_1+\tau$, before this division a number of complete cell cycles given by the integer number k that satisfies: $k\leq\tau/T<k+1$, has taken place, so that: $t_1=(k+1)T-\tau$. In a similar way, in the same cell cycle a termination of DNA replication must occur at the time t_2 given by: $t_2=(h+1)T-\tau_C$, where h is the integer number that satisfies: $h\leq\tau_C/T<h+1$ and should involve 2^h rounds of DNA. Thus the activity of DNA synthesis in a general cell cycle is characterized by an initiation of DNA replication with 2^k chromosomal origins at time t_1, a termination of 2^h rounds of DNA replication at time t_2, and a continuous activity of the relative value $2^{k-1}+2^{k-2}+\ldots+2^h=2^k-2^h$, so that at most two discontinuities of the rate of DNA synthesis can occur in any cell cycle, thus confirming a previous suggestion /19/.

If a population of bacterial cells is in balanced exponential growth, the average protein, ribosome and DNA contents per cell in the population can be calculated, as it has been shown for eukaryotes, considering the cells distributed according to the ideal age distribution function $n(a)=2\ \ln 2\ 2^{-a}$, and formulas result /16/ that confirm preceeding findings /19/ which were in very good agreement with experimental data.

A similar approach has been used by Nishimura and Bailey, which developed in an independent way a model of bacterial populations /20/.

7. Variability of cycle times

The length of cycle time of individual cells in a population is very variable
even when cell division is symmetrical, and also when the population as a whole is
growing exponentially in steady state /2,21,22/. Different interpretations have been
given for this very general observation and of course no cell cycle model may be
adequate, if it does not account for this variability, as well as for the other
features of cell proliferation we have discussed so far.

There are at least two sources of variability that are relevant for cell size
model of the cell cycle: (a) cells of the same age have not the same size; instead
their size is distributed following a momentary size distribution, which is generally
lower at the beginning of the S phase and which can be approximated by a gaussian distribution with a CV of 10-20% /23,24,25/; (b) in general the CV of newborn cells is
higher than that of mitotic cells, suggesting that cell separation at division is not
exactly in half /7,25/. The effects of these two sources of size variability (i.e.
variability of the threshold protein level P_s and the unequal division of cell mass)
on the variability of cycle times have been analyzed in the frame of our model /18/
and the predicted statistical properties of the populations with different sources of
variability are the following. With a variability of the threshold P_s alone or predo-
minant, the CV of cell size at mitosis is about the same as the CV of P_s and the CV
of intermitotic time distribution is about twice as the CV of P_s, as a number of
experimental observations have in fact shown /2,11,26/. The sister to sister correlation of intermitotic times is positive, as generally observed in experimental
system /21,27/. The mother to daughter correlation of intermitotic times is negative
as found in some experimental system /26,28/. The terminal parts of α and β curves
are parallel. There is a small initial downward curvature of the β curve, as observed
in lower eukaryotes /11/ and occasionally also in mammalian cells /29/. The homeo-
stasis of cell size and an approximately constant cell size distribution are maintained. Instead, the transition probability model is not able to account for the maintenance of a mean cell size /17/, if the rate of cell growth follows, as it does
/3,4,6/, an exponential kinetics, as acknowledged recently by Brooks /30/.

8. The acquisition of the competence as a source of variability in mammalian cells

Cultured normal mammalian cells require the presence of serum factors in the
medium for continuous proliferation. Serum factors have been fractioned into
competence and progression factors and cells have first to acquire competence, by
interacting with specific competence factors, to be able to proliferate /31/. It is
also known that at each new generation cells have first to acquire competence to
continue to proliferate /32/. Thus one of the rate limiting steps of proliferation
may be the acquisition of competence.

Kinetic analysis of competence induction in mouse fibroblasts shows that
competence is acquired with first order kinetics whose rate depends upon the
concentration of competence factor /33/. The values of the rate constant obtained for

different concentrations of competence factor are such to suggest that the acquisition of competence may easily introduce an exponential variability in the distribution of intermitotic times of the population, which compounds with the intrinsic variability in the progression of the cycle. If the available concentration of competence factors is low (as it is likely to be at low serum concentration), a newborn cell has a probability lower than one to interact soon with the competence factor and to progress directly into the cycle. The proliferation of mammalian cells may thus be modeled as reported in fig.4.

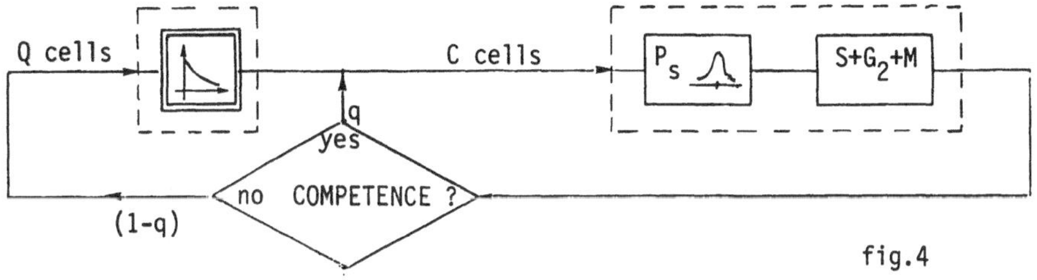

fig.4

A direct evidence that the percentage of incompetent cells increases during exponential growth at low serum concentration is reported in /34/, where the statistical properties of populations with increasing rate constant for the acquisition of the competence according to the model of fig.4 are also reported. It is apparent therefore that a reduction of the rate of acquisition of the competence (and hence an increase of the steady state level of incompetent cells in the population) is accompanied by an increase of the average intermitotic times and by an increase of CV of the age at division, both comparable with those observed experimentally.

The α and β curves of population, whose rate of acquisition of competence is such that the probability for a newborn cell to acquire competence after cell division is less than one, are reported in /34/ and they shown that a reduced probability to acquire competence brings upon a reduction of the downward bend at the top of the β curve. With a fairly low rate of acquisition of competence which has only a small effect on the lengthening of the average duplication time of the population (about 20 hr as compared to 18 hr for a population of totally competent cells as shown in /34/) an almost perfectly exponential β curve is obtained.

9. Concluding remarks

In this paper various aspects (including variability) of cell proliferation have been discussed. It is shown that a cell size model, in which the progression of the nuclear division cycle requires the sequential attainment of one or two protein thresholds, accounts for a number of features of cell growth and division.

Only moving from descriptive cell biology to molecular biology of the cell cycle regulation may lead to a real understanding of the problem. Also in these developments model building may be useful to clarify the regulatory properties of the proposed molecular mechanisms. For instance, several molecular mechanisms have been proposed for cell size controls /35/. One of the more interesting is a bimolecular model suggested by a number of observations. It is based on two molecules, one acts as an inhibitor of the entrance in the S phase and it is synthesized just after cell separation in a fixed amount per nucleus, the other is an activator of the S phase and it is synthesized at rate proportional to the overall protein accumulation. The activator reacts stoichiometrically with the inhibitor and, when it reaches a threshold value, it triggers the onset of DNA replication. This model has been tested /36/ and the noteworthy aspect of our analysis is that it demonstrates that such a model is able to predict with great accuracy several features observed experimentally in cell populations and, in the case of unequally dividing cells, it offers a plausible interpretation for many unusual properties of the cell cycle of budding yeast, including the increase of size of parent cells with genealogical age.

REFERENCES

1. Mitchinson,J.M.(1971) "The Biology of the Cell Cycle" Cambridge University Press.
2. Prescott,D.M.(1976) "Reproduction of Eukaryotic Cells", Academic Press, New York.
3. Elliot,S.G. and McLaughlin, C.S.(1978) Proc.Natl.Acad.Sci.U.S.A. 75, 4384-4388.
4. Ronning, O.W., Pettersen, E.O. and Seglen, P.O. (1979) Exp. Cell Res. 123, 63-72.
5. Baxter G.C. and Stanner, C.P. (1978) J. Cell Physiol. 96, 139-145.
6. Lindmo, T. (1981) Exp. Cell Res. 133, 237-245.
7. Killander, D. and Zetterberg, A. (1965) Exp. Cell Res. 40, 12-20.
8. Johnston,G.C.,Ehrhardt,C.W.,Lorincz,A.and Carter,B.L.A.(1979) J.Bacteriol.137,1-8.
9. Nurse, P. (1975) Nature 256, 547-551.
10. Fantes, P.A. and Nurse, P. (1977) Exp.Cell Res. 107, 377-387.
11. Fantes, P.A. (1977) J.Cell Sci. 24, 51.
12. Zippel R.,Martegani E.,Vanoni M.,Mazzini G.,Alberghina L.(1982)Cytometry 2,426-30.
13. Alberghina L.,Mariani L.,Martegani E. and Vanoni M.(1983)Biot.Bioeng.25,1295-1310.
14. Alberghina L., Mariani L. and Martegani E. (1980) J.Theor. Biol. 87, 171-188.
15. Alberghina L. and Mariani L. (1978) Biomathematics and Cell Kinetics (Valleron and Macdonald eds.) Elsevier/North Holland Biomedical Press, 89-102.
16. Alberghina L. and Mariani L., (1980) J. of Math. Biology, 9, 389-398.
17. Alberghina L., Mariani L., and Martegani E. (1981) Biomathematics and Cell Kinetics (Rotenberg Ed.) Elsevier/North Holland Biomedical Press, 295-309.
18. Hartwell L.H. and Unger M.W. (1977) J.Cell Biol. 75, 422-435.
19. Cooper S. and Helmstetter, C.E. (1968) J. Mol. Biol. 31, 519-540.
20. Nishimura Y. and Bailey J.E. (1980) Math. Biosc., 51, 305-328.
21. Shields R. (1980) Control Mechanisms in Animal Cells (De Asua L.J., Levi-Montalcini R., Shields R. and Jacobelli S.Eds.) Raven Press, New York, 157-164.
22. Dawson K.B., Madoc-Jones H. and Field E.O. (1964) Exp.Cell Res. 94, 267-277.
23. Steen H.B. and Lindmo T. (1978) Cell Tissue Kinet. 11, 69-81.
24. James T.W., Hemond P., Czer G. and Bohman R. (1975) Exp.Cell Res. 94, 267-277.
25. Darzynkiewicz Z.,Crissman H.,Traganos F.,Steinkamp J.(1982)J.Cell Phys.113,465-74.
26. Kock A.L. (1980) Nature 286, 80-82.
27. Brooks R.F., Bennett D.C. and Smith J.A. (1980) Cell 19, 493-504.
28. Collyn-d'Hooghe M.,Valleron A.J. and Maloise E.P.(1977) Exp.Cell Res.106,405-407.
29. Castor L.N. (1980) Nature 287, 857-859.
30. Brooks R.F. (1981) The Cell Cycle (P.C.L.John Ed.) Cambridge Univ. Press, 35-61.
31. Pledger A.J.,Stiles C.D.,Anroniades H.N.,Scher C.D.(1977)P.N.A.S. USA 74,4481-85.
32. Scher C.D., Stone M.E. and Stiles C.D. (1979) Nature 281, 390-392.
33. Zippel R.,Martegani E. and Alberghina L.(1981) Cell Biol.Intern.Reports 5,963-67.
34. Alberghina L. (1983) Cell Cycle Workshop. Praga, April 1983.
35. Fantes P.A.,Grant W.O.,Pritchard R.H.,Sudbery P.E. and Wheals A.E.(1975) J.Theor. Biol. 50, 213-244.
36. Alberghina L.,Martegani E.,Mariani L. and Bortolan E.(1983) Paper in preparation.

THE PHASE RESETTING CHARACTERISTICS OF ENDOGENOUSLY ACTIVE NEURONES

M. Barbi and A.V. Holden
Istituto di Biofisica, C.N.R., Via S. Lorenzo 26, 56100 Pisa
and
Dept. of Physiology, Univ. of Leeds, Leeds LS2 9JT, U.K.

Abstract. The first transient and steady-state phase transition curves of a model of a molluscan neurone are obtained numerically, and discussed in relation to experimental results.

Introduction. Voltage clamp analyses of the mechanisms of excitation of molluscan neurone cell bodies give complicated, high order, non-linear differential excitation systems. However, the behaviour of neurones may be described by simpler, phenomenological models. Hindmarsh and Rose (1982) account for repetitive activity in Lymnaea neurones induced by maintained depolarizing currents by two coupled differential equations:

$$\dot{x} = -a\,(f(x) - y - z)$$
$$\dot{y} = b\,(f(x) - q\,e^{rx} + s - y),$$

where x is the membrane potential, z the applied current, and $f(x) = cx^3 + dx^2 + ex + h$, $a = 5{,}400$ mV s^{-1}, $b = 30$ s^{-1}, $c = 1.7 \times 10^{-5}$, $d = -10^{-3}$, $e = -10^{-2}$, $h = -1$, $q = 0.024$, $r = 0.088$, $s = 0.046$. This Fitzhugh-like system simulates both the fast trajectory during an action potential and the slow trajectory during the interspike interval (but see Game, 1982), and does not simulate slow processes such as adaptation. We obtain numerically phase transition curves for this system, using a point-slope formula and constant step length $\Delta t = 10^{-4}$ s.

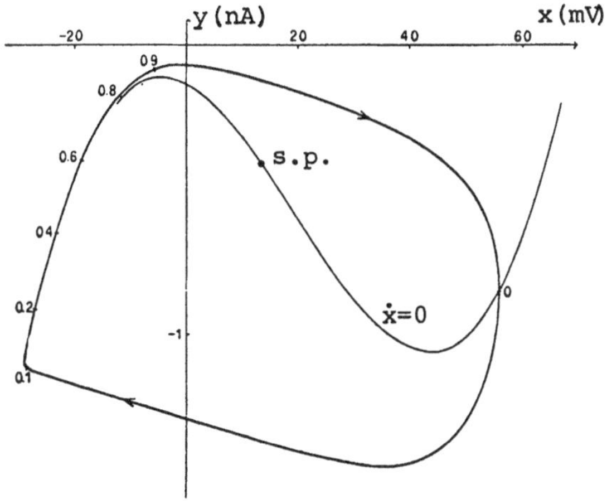

Figure 1: limit cycle of (1).

Steady-state PTCs. For $z > -.026$ nA, the system has an elementary stable limit cycle, shown in Fig. 1 for $z = 0.033$ nA. The limit cycle contains the singular point (an unstable node) that lies on the $\dot{x} = 0$ isocline. Some values of the phase $\phi = t/T$ are shown, where t is the time elapsed

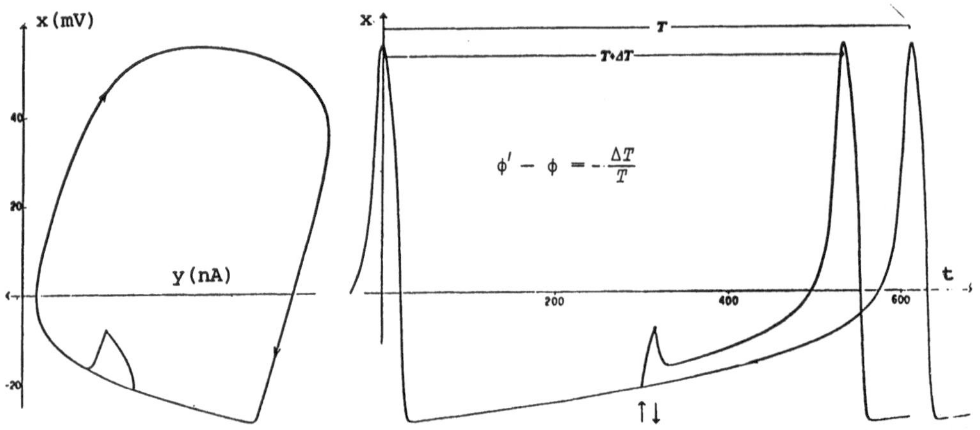

Figure 2: Effects of perturbation on limit cycle.

since the last maximum of potential (ϕ = 0) and T is the period (=0.61 s). The response of this periodic activity to current perturbation pulses of duration 0.015 s and magnitudes ± 0.4, 0.8 nA was characterized by computing the steady-state and first transient phase transition curves (PTC), i.e. plots of new phase ϕ' against old phase ϕ: see Fig. 2. Steady-state PTCs are shown in Figs. 3b - 6b, as the full lines. PTCs for μ = ±0.4 nA appear to be type 1, i.e. with an average slope of one, while for μ = ±0.8 nA the PTCs are type 0, with an average slope of zero. This could have been inferred on the basis of general topological properties (Kawato, 1981), from the positions of the closed locus C_μ of the representative state points at the end of the perturbation, relative to the limit cycle or the singular point (Figs. 3a - 6a). In fact, for μ = ±0.4 nA, C_μ contains the singular point, whereas for μ = ±0.8 nA, C_μ is displaced beyond it.

Some doubt could arise about the right continuation that we assumed for the PTCs for μ = ±0.4, 0.8 nA across the phase ϕ^* corresponding to the intersection Ω of C_μ with the $\dot{x} = 0$ isocline (Figs. 3a - 5a), where $\phi'(\phi)$ changes rapidly. Figs. 3a - 5a also show (by the dashed line) the trajectories during the perturbation just reaching Ω. Figs. 3c - 5c show, on a scale 10 times larger than in (a), some phase-plane trajectories starting (at the end of the perturbation) from points of C_μ close to Ω: these trajectories show that $\phi'(\phi)$ is a continuous function and so we have taken the correct value for $\phi'(\phi)$ for $\phi > \phi^*$. An erroneous addition/subtraction of 1 to the value of ϕ' would have altered the PTC classification.

In the two-dimensional system (1) the null space reduces to a single point, and so there is only a single value of μ at which the type 1 → type 0 transition occurs. At this critical magnitude, annihil-

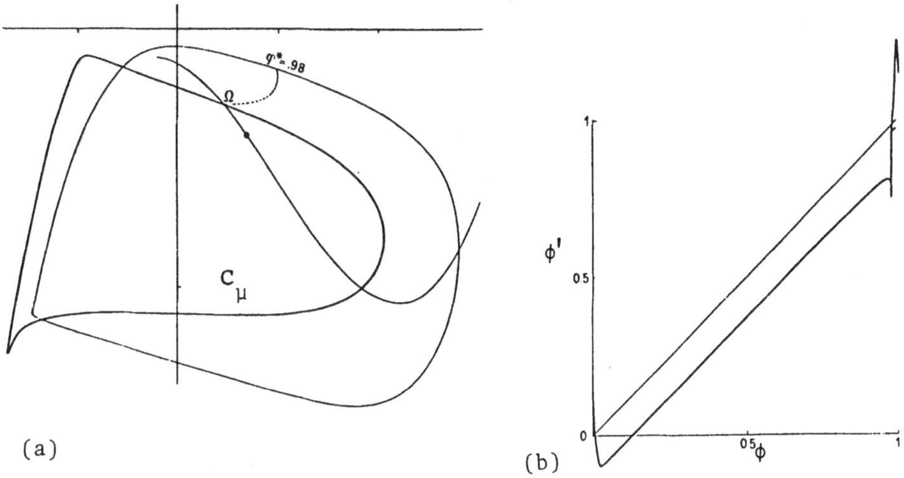

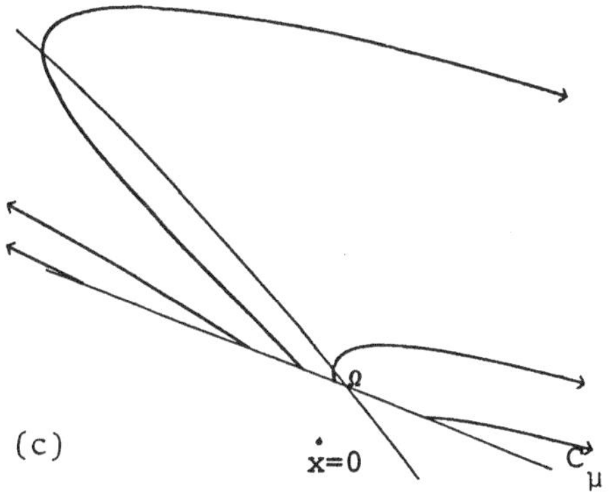

Figure 3: μ *= -0.4 nA. (a) limit cycle,* C_μ *(set of state points at end of perturbation) and isocline. (b) phase transition curve. (c) phase-plane trajectories starting at end of perturbation from points on* C_μ *close to* Ω*.*

ation of periodic activity should only occur at one precise value of ϕ. This differs from the 4-dimensional Hodgkin-Huxley excitation system that was considered by Best (1979), where annihilation may be produced by a range of values of the perturbation parameters.

Transient PTCs. In real experiments, transient and not steady-state PTCs are usually measured. This avoids slow processes like adaptation. We have obtained the first transient PTC by defining a suitable event line. A natural definition of the event line is the curve $\dot{x} = 0$, $\ddot{x} < 0$, i.e. the branch of the nullcline starting from the singular point and going left: this is the locus of the local maximum of the potential that occurs after the perturbation. We also accepted maxima occurring during the perturbation.

The first transient PTCs are shown in Figs.3b - 6b by dashed lines where they differ from the steady-state PTCs. The difference between

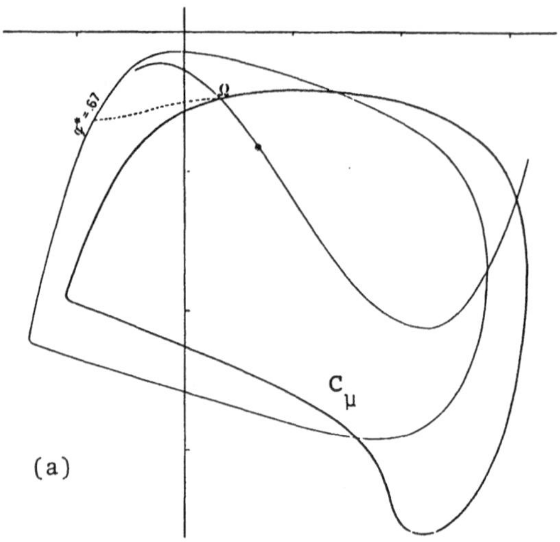

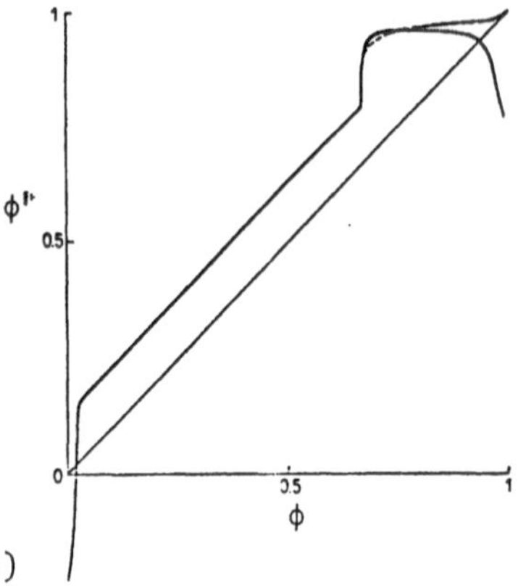

Figure 4: μ + 0.4 nA. (a) limit cycle, C_μ, isocline. (b) phase transition curve. (c) phase plane trajectories starting at end of perturbation from points on C_μ close to Ω.

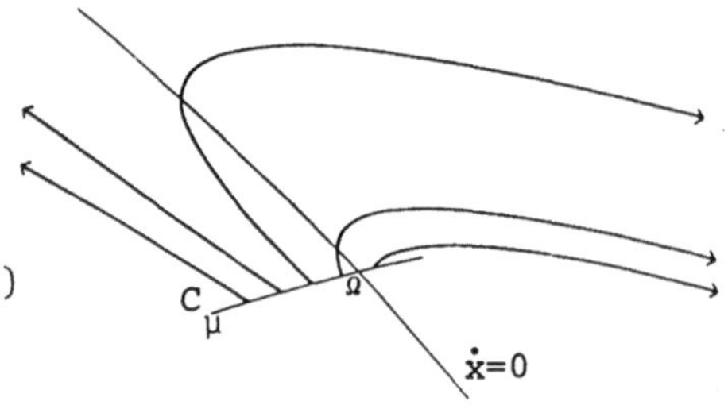

the two curves becomes more evident as the perturbation increases from -0.4 to +0.8 nA. Where the two curves differ, the shape and amplitude of the first action potential after the perturbation generally differs from the normal 56 mV action potential. Fig. 5d plots the amplitude of the first action potential after a 0.8 nA perturbation as a function of ϕ.

Discontinuities in the first transient PTCs are to be expected at ϕ values corresponding to the intersections of C_μ with the event line (Kawato, 1981): the most evident example is that near $\phi = 0.41$ for $\mu = 0.8$ nA. Thus transient PTCs cannot in general be classified as either type 1 or type 0.

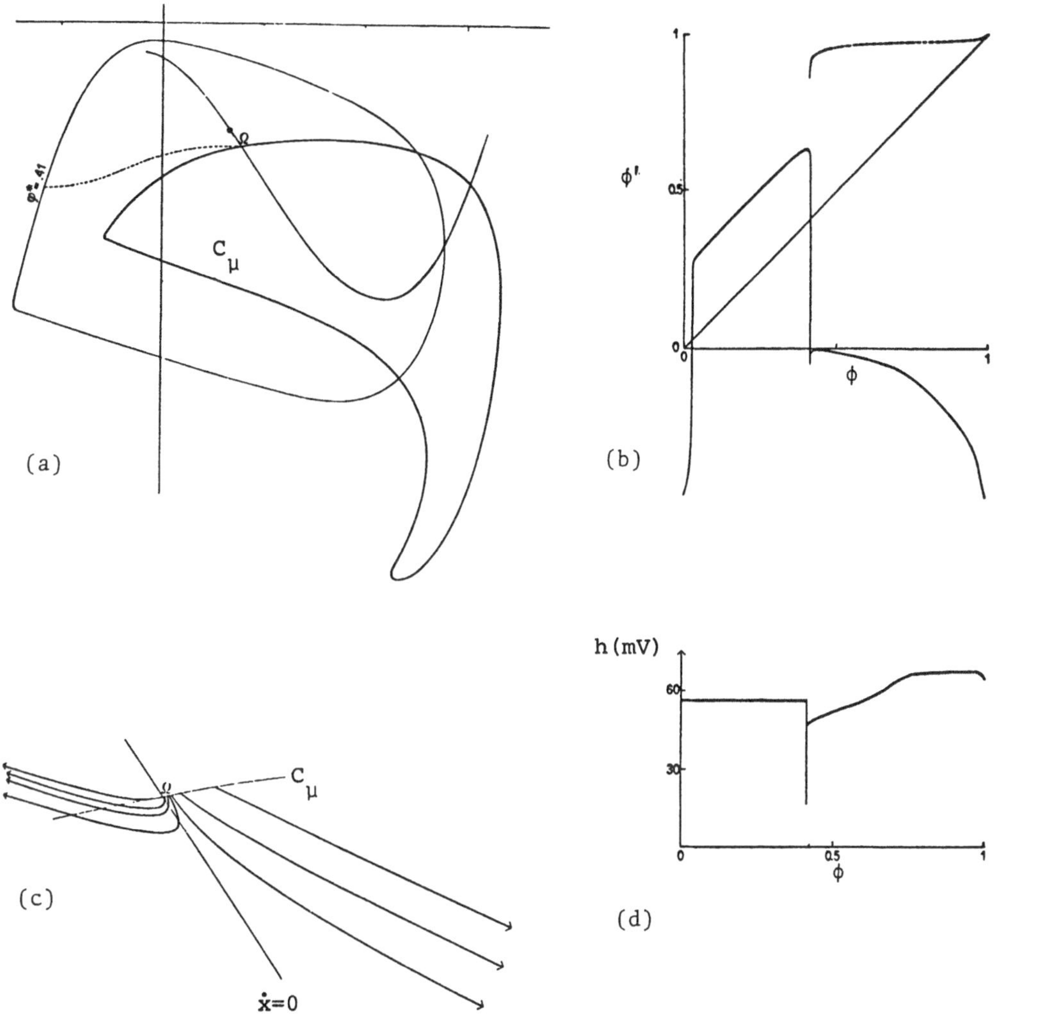

Figure 5: $\mu = +0.8$ nA. (a) limit cycle, C_μ and isocline. (b) PTCs. (c) trajectories at end of perturbation. (d) amplitude h of 1st action potential after perturbation as a function of phase of perturbation.

<u>Discussion</u>. It would be difficult to identify the steady state PTC for $\mu = \pm 0.8$ nA as type 0 in a real system, as this would require that the trend of the trajectories after the perturbation (as shown in Fig. 5c) were determined in a very narrow range and the noise always present in real systems would make the required precision unobtainable.

Further, for $\mu = -0.8$ nA, the experimental noise broadens the interspike interval distribution, with a coefficient of variation of up to 10%. Since the rapid change of $\phi'(\phi)$ occurs in a narrow range ($0.98 < \phi < 1$) near the end of the interval, the phase shift induced

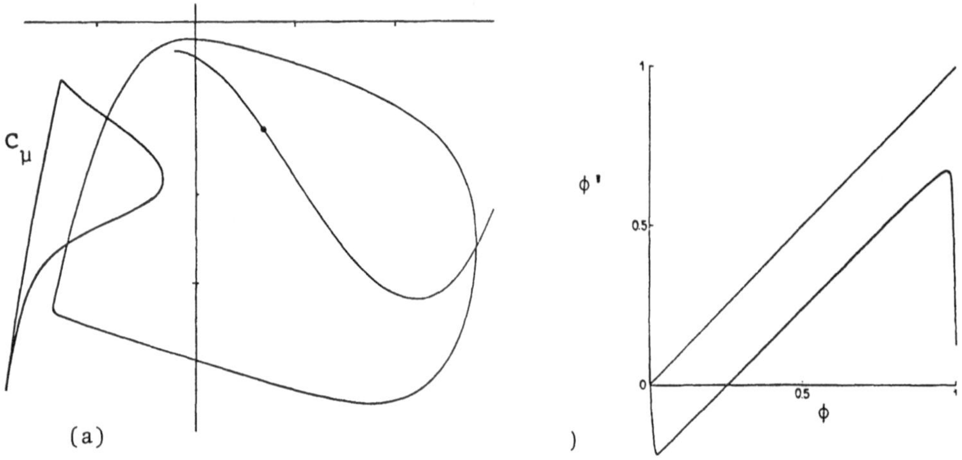

Figure 6: $\mu = -0.8$ *nA. (a) limit cycle,* C_μ *and singular point. (b) PTC.*

by a perturbation given near the end of the cycle should be measured with respect to a changed (lengthened) average interspike interval. Thus there are practical problems in the characterization of PTCs.

Acknowlegements. This work was partially supported by the CNR, grant N. 90701.

References

Best, E.N.: Null space in the Hodgkin-Huxley equations. Biophys. J. 27, 87-104 (1979).

Game, C.J.A.: BVP models of nerve membranes. Nature 299, 375 (1982).

Hindmarsh, J.L., Rose, R.M.: A model of the nerve impulse using two first order differential equations. Nature 296, 162-4 (1982).

Kawato, M.: Transient and steady-state phase response curves of limit cycle oscillators. J. Math. Biol. 12, 13-30 (1981).

THEORY OF DNA SUPERHELICITY

Craig J. Benham
Department of Mathematics
University of Kentucky
Lexington, KY, USA 40506

When a duplex DNA molecule is covalently closed into a circle, conservation of the antiparallel chemical (5'-3') orientations of the two strands dictates that the resulting ring molecule must have the topological structure of two closed, interlinked circles. This fixes the value of the molecular linking number Lk. This Lk is an integer topological invariant which measures the number of times either strand would have to be passed through the other before they could be physically disentangled (provided the molecule is not knotted). Its value remains fixed, independent of geometric deformation of the molecule, as long as both strands are covalently closed. Lk can only be varied by the introduction of transient strand breaks. A class of enzymes called DNA topoisomerases has been found whose function is to regulate Lk in this manner, suggesting a physiological function for this activity.[1,2]

The linking number may be decomposed as the sum of two geometric parameters, the total molecular twist Tw and the writhe Wr:[3,4,5]

$$Lk = Tw + Wr \ . \tag{1}$$

The twist Tw measures the number of times one strand rotates around the molecular central axis. If the molecule is entirely in the B-form conformation, Tw is the number of Watson-Crick helical turns present. The writhing number Wr is a poorly understood geometric parameter which is related to the large-scale shape of the central axis curve.[5,6] Equation (1) expresses Lk as the sum of contributions from twisting around the central axis, Tw, and from bending of the central axis, Wr. Both Tw and Wr may vary as the molecule is deformed, although their sum, Lk, must remain fixed. One may view the covalent ring closure of the duplex molecule as providing a topological coupling between twisting and bending.

A ring DNA molecule is said to be relaxed if it is closed into a geometric circle (having Wr=0) while its helix is twisted at a rate which is neither overwound nor underwound for the given environmental conditions. The relaxed twist for a molecule of N base pairs under conditions where the B-form is energetically favored is given by $Tw_o = N/10.4$ and its relaxed linking number is $Lk_o = Tw_o$. A molecule whose linking number deviates from Lk_o is said to be supercoiled or superhelical. *In vivo* DNA is commonly maintained in a negatively supercoiled state, $Lk < Lk_o$. Deviations of Lk from its relaxed value must induce twisting and/or bending deformations:

$$\Delta Lk = Lk - Lk_o = \Delta Tw + Wr \ . \tag{2}$$

This paper analyzes the various manifestations of these deformations which can occur at equilibrium.

Duplex DNA is a polymorphic substance. It can form several distinct types of helices which can be either right-handed (B-form, A-form) or, for appropriate sequences, left-handed (Z-form).[7] In addition, the base pairing may be disrupted in a region, causing the two strands to separate forming a bubble of local denaturation. Further, if two copies of a base sequence occur with opposite orientations, the resulting intra-strand base complementarity permits base pairing within each strand. This alternative secondary structure is called a cruciform or a hairpin. Each possible structure has an unstressed conformation as well as mechanical stiffnesses which resist deformations away from this unstressed conformation. For example unstressed B-form DNA is straight with a twist rate of about 10.4 bp/turn.[8] Its bending stiffness A and torsional stiffness C have been measured experimentally to be $A \simeq 2.15 \times 10^{-19}$ erg-cm, $C \simeq 1.43 \times 10^{-19}$ erg-cm.[9] Information regarding the unstressed conformations, sequence specificities and mechanical stiffnesses of various DNA conformations are summarized in Table I below.

A relaxed, closed circular DNA molecule under physiologically reasonable *in vitro* conditions will generally be in the B-form conformation.[8] When negatively supercoiled, several responses are possible consistent with equation (2) above. Some of the supercoiling deformation may be manifested as writhe resulting from bending deformations. Variation in the total twist, ΔTw, can occur in either of two ways. First, local conformational transitions may occur to structures having smaller unstressed twist rates. These can include transitions to Z-form, local denaturations and/or cruciform extrusions, depending upon details of base sequence. Although these transitions would be energetically disfavored in a relaxed or linear molecule under identical conditions, they may occur at equilibrium in a supercoiled molecule. This can happen because they localize undertwist at the site of transition, causing a corresponding decrease in strain in the rest of the molecule. Transition becomes favored when the strain energy which it relieves elsewhere in the molecule exceeds the (free) energy cost of performing it. Finally, ΔTw may also be manifested as a smooth torsional deformation of the various conformations along the molecule away from their respective unstressed conformations.

In the rest of this paper we analyze each of these possible behaviors in turn.

STRESSED ELASTIC EQUILIBRIA

We consider a closed circular B-form duplex DNA molecule of N base pairs having torsional stiffness C and (symmetric) bending stiffness A, which are presently assumed to be constant along the molecule. One may express the molecular shape in space by comparing the orientation of local x(s), y(s), z(s) axes at each distance s along the central axis to a reference set of space axes X, Y, Z.[10] This comparison is facilitated by the use of Euler angles $\theta(s)$, $\phi(s)$, $\psi(s)$. Expressed in this manner the curvature vector is the instantaneous rotation rate of the local axes relative to the fixed axes: $\vec{\kappa}(s) = \dot{\Phi}(s)$. This quantity may be decomposed into its components with respect to the local axes: $\vec{\kappa}(s) = (\kappa_x(s), \kappa_y(s), \tau(s))$. If one chooses the local system of axes for the relaxed state appropriately, τ is the torsional deformation rate at which the structure is locally twisted away from the unstressed B-form. The differential geometric curvature is $\kappa^2 = \kappa_x^2 + \kappa_y^2$. The curvature components may be expressed in terms of Euler angles as:

$$\begin{aligned} \kappa_x &= \dot{\phi} \sin\theta \sin\psi + \dot{\theta} \cos\psi \\ \kappa_y &= \dot{\phi} \sin\theta \cos\psi - \dot{\theta} \sin\psi \\ \tau &= \dot{\phi} \cos\theta + \dot{\psi} \end{aligned} \tag{3}$$

Kirchhoff has shown that the stressed equilibria of an elastic rod are those conformations which give extremals of the analogous action integral.[11] Expressed in terms of the Euler angles this action is

$$S = \int_0^L \frac{A}{2} (\dot{\phi}^2 \sin^2\theta + \dot{\theta}^2) + \frac{C}{2} (\dot{\phi} \cos\theta + \dot{\psi})^2 - N \cos\theta \, ds \, . \tag{4}$$

Here L is the length of the rod and N is the magnitude of the force component of the internal stress distribution. One may also express the superhelicity in terms of the Euler angles:[12]

$$\Delta Lk = \frac{1}{2\pi} \int_0^L \dot{\psi} + \frac{(\dot{\phi}-k)\cos\theta - \dot{\theta}\sin(ks-\phi)}{1 - \sin\theta\cos(ks-\phi)} \, ds \, , \tag{5}$$

where $k = 2\pi/L$. It follows that the elastic equilibrium conformations accessible to a superhelical DNA molecule are given by extremals of the integral in equation (4) subject to the constancy of expression (5) and the condition of smooth closure. The solution of the resulting isoperimetric problem is effected by standard techniques.[12] The constrained extremals are given by solutions of the resulting three second order Euler differential equations, one for each of the generalized coordinates θ, ϕ, ψ. After one integration and the substitution $u = \cos\theta$ these are

$$\dot{u}^2 = (2\alpha_3 - \eta u)(1-u^2) - (\alpha_2 - \gamma\alpha_1 u)^2 \, , \tag{6a}$$

$$\dot{\phi} = (\alpha_2 - \gamma\alpha_1 u) / (1-u^2) \, , \tag{6b}$$

$$\dot{\psi} = \alpha_1 - (\alpha_2 u - \gamma\alpha_1 u^2) / (1-u^2) \, , \tag{6c}$$

where α_i's are constants of integration, $\eta = 2N/A$, $\gamma = C/A$. The solution to equation (6a) involves the Jacobi elliptic function sn:[13]

$$u = u_1 + (u_2 - u_1)\,\mathrm{sn}^2(\beta s) \, . \tag{7}$$

From this expression the solutions for ϕ and ψ may be found by quadratures.

The qualitative conclusions of this analysis are as follows. First, the elastic equilibrium conformations for a closed circular, superhelical molecule are found to approximate toroidal helices. Second, for two molecules of different lengths which are supercoiled the same amount ΔLk, the equilibrium conformations will be geometrically similar. Finally, there will generally be multiple equilibria in any given instance, so exchanges of stability can occur.

These conclusions are strongly supported by the results of small angle x-ray scattering experiments from solutions of superhelical DNA molecules.[14] By fitting the experimental scattering curves one can determine the geometric parameters of the toroidal helix involved.[15] From this one can compute Wr. For natively supercoiled COP608 DNA having $\Delta Lk/Lk_0 \simeq -.055$, one finds that approximately two-thirds of ΔLk is manifested as twist, one-third as writhe. The portion which appears as twist may involve either smooth torsional deformations (such as the τ in the above analysis) or as local transition to alternative conformations, which we analyze next.

STRESS-INDUCED TRANSITIONS

Negative superhelicity can induce local transitions to alternative conformations having smaller unstressed twist rates (in the right-handed sense).[16,17] This behavior localizes undertwist at the site of transition, thereby permitting the rest of the molecule to partially relax. The possibilities include local denaturation,[18] cruciform extrusion at an inverted repeat sequence[19] or transition to Z-form[20] at a permissive sequence. The statistical mechanical analysis of paradigm cases of this phenomenon is presented below.

Consider a closed circular DNA duplex of N base pairs whose relaxed conformation in the given environment defines a reference state. In physiologically reasonable conditions this will usually be entirely B-form,[8] although the present theory is not limited to such cases. Each base pair can have one or several alternative conformations which it can also assume, although transition requires the input of (free) energy. For example, any base pair can denature. Specific sequences can assume the Z-form, while inverted repeats can extrude cruciforms. The free energy needed for such a transition may be decomposed into two contributions - the cooperativity free energy a required to initiate transition at a site, and the free energy b needed to transform one more base pair once initiation has occurred. These sequence specificities and free energies are given in Table I below for the paradigm transitions listed.

The equilibrium behavior of a population of identical molecules may be predicted from the governing partition function Z. In a system all of whose states are discrete the partition function is

$$Z = \sum_r \exp(-\beta E_r) \, , \qquad (8)$$

where r enumerates the states of the system and $\beta = 1/kT$. If the index i enumerates all states where a given event occurs then the equilibrium probability (i.e. frequency) of this event is

$$p = \sum_i \exp(-\beta E_i) \, / \, Z \, . \qquad (9)$$

If a parameter n has value n_r in state r, then its average value at equilibrium is

TABLE I.

Form	Sequence Specificity	Relaxed Conformation A (bp/turn)	Absorbed Undertwist (deg/bp)	Torsional Stiffness C (erg-cm)	Transition Energies* (kcal/mole) a	b
B	none	10.4	---	1.43×10^{-19}	---	---
Z	alternating purine-pyrimidine	-12.0	64.0	? (stiff)	5.0	0.5
A	prefers G+C-rich	11.5	3.0	? (stiff)	3.6	1.0
denatured	prefers A+T-rich	untwisted	>34.0	6.10×10^{-21}	8.0	1. for AT; 2. for GC
cruciform	inverted repeat	untwisted	34.0	undeformed	?	0.0

* These values assume physiologically reasonable conditions of temperature and ionic strength, $T = 300^\circ K$, $[Na^+] = .2M$.

$$\bar{n} = \sum_r n_r \exp(-\beta E_r) / Z \,. \tag{10}$$

Thin analysis permits the computation of equilibrium quantities of practical importance from the partition function Z. These may include the probability of transition of a given base pair, the expected number of transformed base pairs and the expected number of runs of transformation.

Suppose the molecule is superhelically twisted away from its relaxed structure an amount $\Delta Tw = q$. This deformation is partitioned between local transitions to alternative conformations and smooth twisting of all regions away from their unstressed structures. A given state of this system is a specific sequence of transformed and untransformed base pairs. The energy involved includes contributions arising from the transformation and also from the residual smooth deformation. For instance, if a given state has n specific transformed bases in r runs, all of which occur in the same alternative conformation, and if the two secondary structures present are residually twisted at rates τ_1 and τ_2 away from their unstressed conformations, then the free energy of this state is

$$E = ar + bn + \frac{C_1 (N-n) \tau_1^2}{2} + \frac{C_2 n \tau_2^2}{2} \,. \tag{11}$$

Because the total torsional deformation $\Delta Tw = q$ is fixed, one has a relationship among τ_1, τ_2, n and q:

$$\tau_1 = 2\pi \left[q + n \left(\frac{1}{A_1} - \frac{1}{A_2} - \frac{\tau_2}{2\pi} \right) \right] / (N-n) \,, \tag{12}$$

where A_i is the number of base pairs per turn in the unstressed conformation i. For a fixed n and q one may use this equation to determine the strains τ_1 and τ_2 which minimize the quadratic terms in equation (11) above.[21] (Alternatively, one may include fluctuations in τ_i by a slight extension of the theory, as is done in reference 18.) By performing this calculation for every possible sequence of transformed and untransformed base pairs one deduces a partition function for a given value of $\Delta Tw = q$, from which equilibrium properties may be determined as described in equations (8)-(10) above.

We indicate briefly the important results arising from the analysis of various transitions. In all cases we assumed a physiologically reasonable local environment (i.e. T=300°K, $[Na^+]$=.2M, etc.) in which the relaxed molecule is entirely B-form.

First, consider local denaturation. Calculations show that a sufficiently under-twisted sequence can locally denature at the A+T-richest sites, although the high co-operativity free energy involved confines transition to a small number of the most susceptible sites.[16,18] If there is a local disruption of base pairing (such as occurs at a pyrimidine dimer) then all further denaturation occurs at the site of this lesion because there the initiation has been performed already, whereas elsewhere it would cost the cooperativity free energy.[17] Such a lesion makes local denaturation there energetically preferred to virtually any alternative transition. If superhelically induced local transitions (to Z-form or other conformations) serve regulatory functions, then a lesion of this sort will disrupt this regulation throughout the supercoiled domain where it occurs.

Calculations show that extrusion of a cruciform and transition to Z-form can also occur at physiologically reasonable twist densities $\Delta Tw/Tw_o$.[19,20] In general the twist density at which transition occurs is less extreme for longer sequences, other factors remaining fixed. If the region susceptible to transition is relatively long states of partial transition can occur. If the susceptible site is short this transition has an all-or-nothing character, with intermediate states not occupied.[22]

In cases where multiple transitions are possible, the competition between them can be quite intricate.[19,22] Specific transitions may occur at equilibrium only in narrow regions of superhelicity, their reversion being coupled to other transitions occurring elsewhere. If Z-susceptible sites are present, their transition will occur before others because they absorb more undertwist per base pair, thereby affording more relaxation elsewhere.[22]

The results of these analyses are in good qualitative accord with experimental observations.[23-27] Stress-induced conformational transitions provide one mechanism by which superhelicity may exert its known physiological functions, most notably influencing the initiation of transcription.[28]

REFERENCES

1. Gellert, M.: Ann. Rev. Biochem. 50, 879-910 (1981).
2. Cozzarelli, N.: Science 207, 953-960 (1980).
3. White, J.H.: Am. J. Math. 91, 693-727 (1969).
4. Vinograd, J., Lebowitz, J. and Watson, R.: J. Mol. Biol. 33, 173-197 (1968).
5. Fuller, F.B.: Proc. Natl. Acad. Sci. USA 68, 815-819 (1971).
6. Fuller, F.B.: Proc. Natl. Acad. Sci. USA 75, 3557-3561 (1978).
7. Zimmerman, S.B.: Ann. Rev. Biochem. 51, 395-427 (1982).
8. Wang, J.C.: Proc. Natl. Acad. Sci. USA 76, 200-204 (1979).
9. Millar, D.P., Robbins, R.J. and Zewail, A.H.: J. Chem. Phys. 76, 2080-2094 (1982).
10. Benham, C.J.: Proc. Natl. Acad. Sci. USA 74, 2397-2401 (1977).
11. Kirchhoff, G.: J.f. Math (Crelle) 50, 285-313 (1859).

12. Benham, C.J.: "Geometry and Mechanics of DNA Superhelicity," to appear in Biopolymers.
13. Benham, C.J.: Biopolymers 18, 609-623 (1979).
14. Brady, G.W., Fein, D.B., Lambertson, H., Grassian, V., Foos, D. and Benham, C.J.: Proc. Natl. Acad. Sci. USA 80, 741-744 (1983).
15. Benham, C.J., Brady, G.W. and Fein, D.B.: Biophysical J. 29, 351-366 (1980).
16. Benham, C.J.: Proc. Natl. Acad. Sci. USA 76, 3870-3874 (1979).
17. Benham, C.J.: J. Mol. Biol. 150, 43-68 (1981).
18. Benham, C.J.: J. Chem. Phys. 72, 3633-3639 (1980).
19. Benham, C.J.: Biopolymers 21, 679-696 (1982).
20. Benham, C.J.: Nature 286, 637-638 (1980).
21. Benham, C.J.: Biopolymers 19, 2143-2164 (1980).
22. Benham, C.J.: "Statistical Mechanical Analysis of Competing Conformational Transitions in Superhelical DNA," to appear in Proceedings of the 47th Cold Spring Harbor Symposium on Quantitative Biology.
23. Dean, W.W. and Lebowitz, J.: Nature New Biology 231, 5-8 (1971).
24. Beerman, T.A. and Lebowitz, J.: J. Mol. Biol. 79, 451-470 (1973).
25. Lilley, D.M.: Proc. Natl. Acad. Sci. USA 77, 6468-6472 (1980).
26. Panayotatos, N. and Wells, R.D.: Nature 289, 466-470 (1981).
27. Singleton, C., Klysik, J., Stirdivant, S.M. and Wells, R.D.: Nature 299, 312-316 (1982).
28. Richardson, J.P.: J. Mol. Biol. 91, 477-487 (1975).

This work was supported by grant #PCM 80-02814 from the National Science Foundation and by a Research Fellowship from the Alfred P. Sloan Foundation.

DETERMINATION OF DNA SYNTHESIS RATE IN CELL POPULATIONS WITH LOCALLY EXPONENTIAL S-PHASE INFLUX

A. BERTUZZI, R. CLERICO, A. GANDOLFI, A. GERMANI, R. VITELLI
Istituto di Analisi dei Sistemi ed Informatica del CNR,
Viale Manzoni 30, 00185 Roma, Italy.

1. INTRODUCTION

Flow microfluorometry is a valuable tool for cell kinetics investigations [1]. Several mathematical methods are available to evaluate the fraction of cells in cycle phases from a flow-cytometric (FCM) histogram [2]. A few methods allow to reconstruct the DNA synthesis rate from a histogram measured in balanced exponential growth [3,4] or from histograms measured during desynchronization [5]. In a recent paper [6], an expression was given to determine the rate of DNA synthesis from a single histogram in balanced growth or from a couple of histograms when the S-phase influx has locally an exponential behaviour. With the aim of evaluating the reliability of the method, in the present paper the rate of DNA synthesis is reconstructed from couples of histograms in saturated stages of growth, obtained by computer simulation.

2. MATHEMATICAL MODEL

A mathematical model and a computational procedure for determining the DNA synthesis rate were discussed in detail in [6,7]. The model equations are the following:

$$\frac{d}{dt} N_1(t) = -f(t) + 2g(t) \tag{1}$$

$$\frac{\partial}{\partial t} n(x,t) + \frac{\partial}{\partial x}(v(x)n(x,t)) = 0, \quad f(t) = n(1,t)v(1) \tag{2}$$

$$\frac{d}{dt} N_2(t) = f(t-T_s) - g(t) \tag{3}$$

where $N_1(t)$ and $N_2(t)$ denote the number of cells in G_o/G_1 and G_2M respectively at time t, $n(x,t)dx$ the number of cells in S with DNA content between x and x+dx with $x \in (1,2)$, $f(t)$ the S-phase influx,

$g(t)$ the mitotic rate, $v(x)$ the DNA synthesis rate, and $T_S = \int_1^2 dx/v(x)$ the transit-time in S. This model, together with the fundamental equation of cytofluorometry [1], gives an expression for sequences of FCM histograms in terms of $f(t)$ and $v(x)$ that can be used for simulation as well as for identification purposes [7].

In particular, the following equation can be obtained for the distribution $\nu(\xi,t)$ of the fluorescence ξ measured by the flow-cytometer at time t:

$$\nu(\xi,t)=2K(\xi,1)-K(\xi,2)+(2-\theta_1(t))\int_1^2 \frac{\partial}{\partial x} K(\xi,x)\, e^{-\int_1^x dz/w(z;t)}\, dx \tag{4}$$

where the gaussian kernel $K(\xi,x)$ represents the dispersion of the fluorescence ξ for a cell with DNA content x, and $\theta_1(t)$ is the G_o/G_1 fraction. A piecewise constant approximation of $w(x;t)$ can be evaluated, together with $\theta_1(t)$ and the parameters of $K(\xi,x)$ [8], from a FCM histogram measured at t through a suitable identification algorithm [6]. This approximation allows to find the following quantities:

1) the rate of DNA synthesis, given by $v(x) = \alpha w(x;t)$, if the histogram is measured from a population in balanced exponential growth with growth rate α;
2) the rate $v(x)$ from a couple of histograms in t_1 and t_2, if the S-influx $f(t)$ can be approximated as

$$f(t) = f(t_1-T_s)e^{\beta(t-t_1+T_s)}, \quad t \in [t_1-T_s,t_2], \tag{5}$$

by the equation:

$$v(x) = \frac{1}{t_2-t_1}\ln \frac{\theta_s(t_2)N(t_2)}{\theta_s(t_1)N(t_1)}\left[1 + \frac{\dfrac{\theta_s(t_2)}{\theta_s(t_1)} - \dfrac{2-\theta_1(t_2)}{2-\theta_1(t_1)}}{\dfrac{N(t_1)}{N(t_2)} - \dfrac{\theta_s(t_2)}{\theta_s(t_1)}}\, e^{\int_1^x \frac{dz}{w(z;t_1)}}\right]w(x;t_1) \tag{6}$$

where $\theta_s(t)$ is the S-phase fraction and $N(t)$ the population size. The rate $v(x)$ is required to be time-independent in $[t_1-T_s,t_2]$;
3) the DNA distribution in S, $\tilde{n}(x,t) = n(x,t)/N(t)$, from an arbitrary histogram (even if model assumptions are not valid) by the equation:

$$\tilde{n}(x,t) = \frac{2-\theta_1(t)}{w(x;t)}\, e^{-\int_1^x dz/w(z;t)} . \tag{7}$$

It is important to note that the condition (5) of locally exponential S-influx holds if and only if $\tilde{n}(x,t_1)$ is proportional to $\tilde{n}(x,t_2)$,

$x \in (1,2)$, $t_2 - t_1 \leq T_s$ [6].

3. RESULTS

In order to test the reliability of the method for reconstructing the rate $v(x)$ in the case of locally exponential S-phase influx,couples of histograms were obtained by simulation of the model, and the phase fractions and the function $w(x;t)$ were estimated by (4) for each histo gram. Equation (6) gives then the rate $v(x)$.

Two growth conditions were considered. In the first condition the model (1),(2),(3) was simulated with $f(t) = \lambda_1 N_1(t)$, $g(t) = \lambda_2 N_2(t)$ and $v(x) = v$, starting from a population synchronized in G_0/G_1. The popula tion was allowed to arrive at a balanced exponential state with $\lambda_1 = 0.15h^{-1}$, $\lambda_2 = 0.4h^{-1}$ and $v = 0.15h^{-1}$ ($T_s = 6.67h$), then at a time $\bar{t} = 48h$ λ_1 was set to $0.05h^{-1}$ and λ_2 to zero obtaining an immediate saturation of $N(t)$ and an exponential decrease of $f(t)$. A first couple of fluorescence distributions was obtained in the balanced growth ($t_1 = 41h$, $t_2 = 48h$), and a second couple at $t_3 = 55h$, $t_4 = 62h$ ($t_3,t_4 > \bar{t} + T_s$). From each fluorescence distribution a histogram was generated, simulating by a Monte-Carlo approach the statistical variation due to the finite number of cells ($5 \cdot 10^4$ cells per sample). Each histogram was analyzed by the identification algorithm, obtaining estimates of $\theta_1(t)$, $\tilde{n}(x,t)$ and $w(x;t)$. Figure 1 shows the estimated DNA distribution in S at t_1 (A) and the rate reconstructed by (6) (B) for the couple in balanced growth; the same quantities are shown for the couple in saturation (C,D). The distributions in C satisfy the proportionality condition with reasonable error. The reconstructed rate (estimated T_s are 6.56h in B, 6.74h in D) is also reasonably accurate, but the deviations increase for histograms generated with 10^4 cells because of the higher sampling errors (estimated $T_s = 6.83h$).

In the second growth condition a milder saturation effect was simulated with $f(t) = \lambda_1 N_1(t)$, $g(t) = \lambda_2 N_2(t)/[1+\varepsilon(N(t)/N_0)^2]$, $v(x)=v$, (N_0 is the initial population size) with $\lambda_1 = 0.15h^{-1}$, $\lambda_2 = 0.4h^{-1}$, $v = 0.15h^{-1}$ and $\varepsilon = 0.01$. Figure 2A shows the computed $f(t)$ and $N(t)$ as function of time. Three couples of histograms were obtained at the times indicated in the Figure with $5 \cdot 10^4$ cells per sample; one histogram for each couple is reported in Fig. 2B to show how the fluorescence distribution changes in the various growth stages. The condition of exponential behaviour of $f(t)$ is only approximately satisfied in this case, expecially for the second and the third couple. However, the rate $v(x)$ is still well reconstructed from these couples (relative

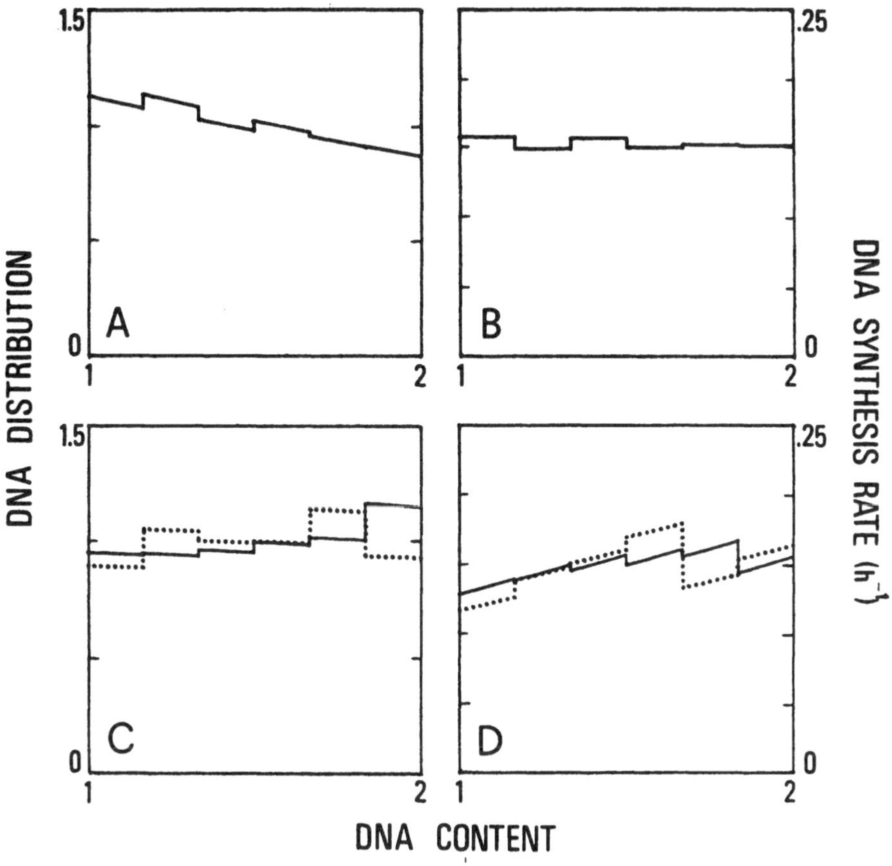

Figure 1. A) Estimated DNA distribution in S (six compartments) during balanced growth (t = 41h); the true DNA distribution is 1.164 exp(-0.312(x-1)). B) DNA synthesis rate reconstructed from histograms at 41 and 48h (true rate $0.15h^{-1}$). C) Estimated DNA distribution in S during saturation (t = 55h continuous line, t = 62h dotted line); the true distribution during saturation is 0.842 exp(0.333(x-1)). D) DNA synthesis rate reconstructed from histograms at 55 and 62h (continuous line, true rate $0.15h^{-1}$); reconstruction of the rate from histograms with 10^4 cells (dotted line).

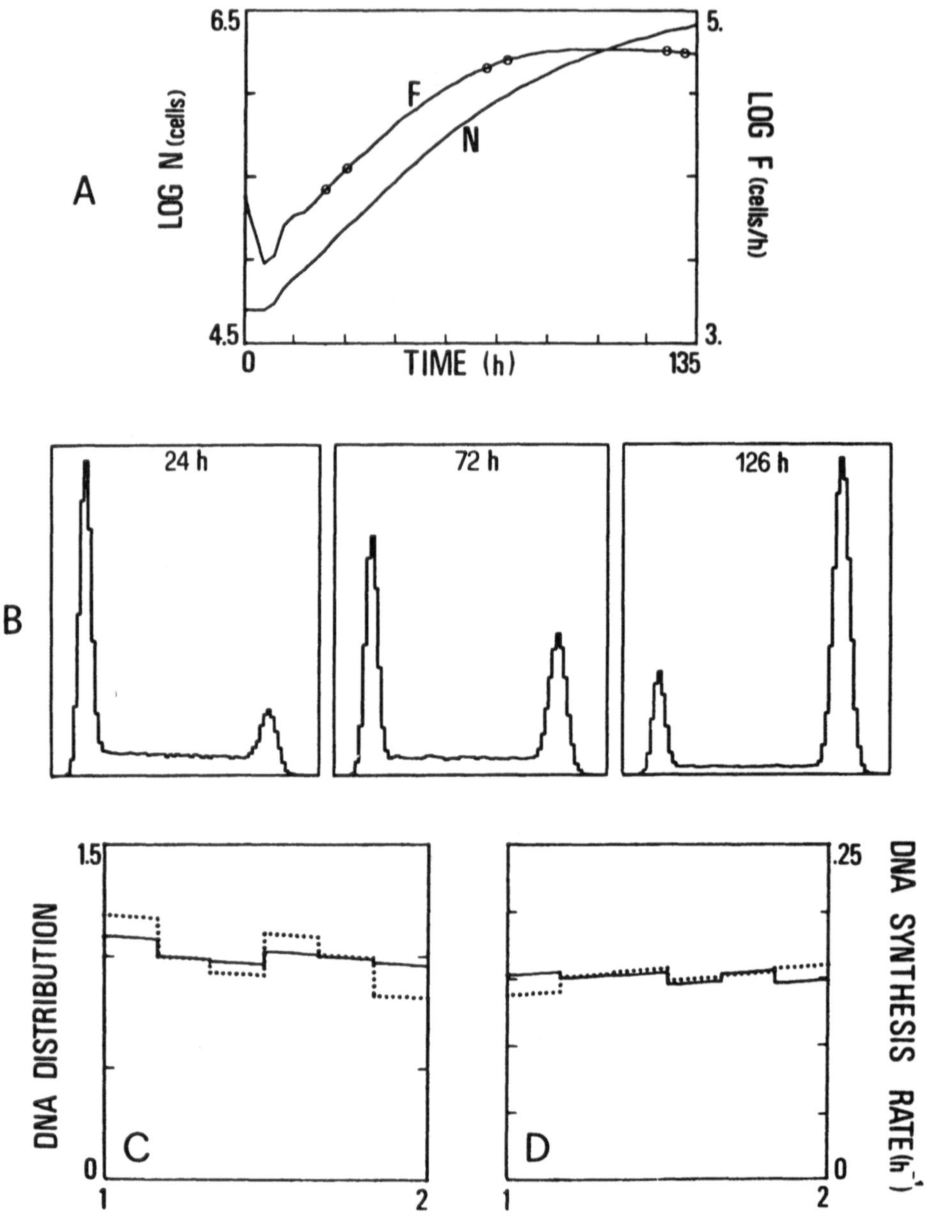

Figure 2. A) N(t) and f(t) as a function of time. The f values corresponding to the three couples of histograms (24,30h; 72,78h; 126, 132h) are indicated on the f curve. B) simulated fluorescence histograms at the time indicated. C) Estimated DNA distributions in S at 126h (continuous) and 132h (dotted). D) DNA synthesis rate reconstructed from the second couple (continuous) and the third couple (dotted) ; true rate $0.15h^{-1}$.

deviation < 10%, as shown in panel D (estimated T_s is 6.61h for both couples). Panel C depicts the S-phase DNA distributions for the third couple.

4. CONCLUSION

The present method for determining the rate of DNA synthesis from FCM histograms measured at various stages of a population growth appears to be accurate enough, at least when the rate of DNA synthesis does not change rapidly during growth. If the rate and therefore T_s should sensibly change in the time interval considered, a sort of time-averaged synthesis rate could possibly be obtained. This result can be of interest in the study of perturbed cultures by the analysis of sequences of FCM histograms.

ACKNOWLEDGEMENT

This research is supported by the "Progetto Finalizzato Controllo della Crescita Neoplastica" of CNR (National Research Council of Italy) under contract No. 8203150.96.

REFERENCES

[1] ZIETZ,S. and NICOLINI,C. (1978), in *Biomathematics and Cell Kinetics* (Valleron,A.J., and Macdonald,P.D.M., Eds.), Elsevier/ North-Holland Biomedical Press, Amsterdam, pp. 357-394.

[2] BAISCH,H., BECK,H.P., CHRISTENSEN,I.J., HARTMANN,N.R., FRIED,J., DEAN,P.N., GRAY,J.W., JETT,J.H., JOHNSTON,D.A., WHITE,R.A., NICOLINI,C., ZIETZ,S., and WATSON,J.V. (1982), *Cell Tissue Kinet.* 15, 235.

[3] DEAN,P.N., and ANDERSON,E.C. (1975), in *Pulse Cytophotometry, Part I* (Haaven,C.A.M., Hiller,H.F.P., and Wessels,J.M.C., Eds.), European Press Medikon, Ghent (Belgium), pp. 77-86.

[4] WHITE,R.A. (1980), *J. theor. Biol.* 85, 53.

[5] BECK,H.P. (1978), *Cell Tissue Kinet.* 11, 139.

[6] BERTUZZI,A., GANDOLFI,A., GERMANI,A., and VITELLI,R. (1982), Report of the Istituto di Analisi dei Sistemi ed Informatica del CNR R.53, Rome.

[7] BERTUZZI,A., GANDOLFI,A., GERMANI,A., and VITELLI,R. (1983), *J. theor. Biol.* 102, 55.

[8] BRUNI,C., KOCH,G., and ROSSI,C. (1983), *Cell Biophys.* (in press).

A NUMERICAL STUDY FOR A CELL SYSTEM

E. Bohl

University of Konstanz
Konstanz, West Germany

Introduction

In this paper we study an assemblage of N cells in which a chemical reaction takes place whose generation term is described by a function $g(\lambda,c_j)$ where λ stands for a control parameter and c_j denotes the concentration of a substrate in the j-th cell. The cells communicate via membranes M (compare Fig.1) admitting diffusive transport with diffusion coefficients D_j. The j-th cell may be connected to an outside reservoir which feeds constant concentration α into the system through a membrane with the diffusion constant E_j. If a cell is not connected to the reservoir we put $E_j=0$. Next we introduce the new variable $x_j=\alpha-c_j$ $(j=1,\ldots,N)$ and define the function $f(\lambda,x)=g(\lambda,\alpha-x)$. With this notation the steady states of our assemblage of cells are the solutions of the system

(1a) $\quad (D_2+E_1)x_1-D_2x_2 = f(\lambda,x_1)$

(1b) $\quad -D_jx_{j-1}+(D_j+D_{j+1}+E_j)x_j-D_{j+1}x_{j+1} = f(\lambda,x_j) \quad (j=2,\ldots,N-1)$

(1c) $\quad -D_Nx_{N-1}+(D_N+E_N)x_N = f(\lambda,x_N).$

Bunow and Colton [4] have studied (1) under the assumptions

(2a) $\quad D_j = D \quad (j=2,\ldots,N), \quad E_j = 0 \quad (j=2,\ldots,N-1),$

(2b) $\quad g(\lambda,c) = \lambda_1 c(1+c+\lambda_2 c^2)^{-1}, \quad \lambda = (\lambda_1,\lambda_2).$

In this paper we consider (1) in its full generality with a generation term f qualitatively given by Fig.2. In particular this includes the special case (2).

The organization of this paper is as follows: In section 1 we discuss the case where only the first cell is connected to the outside reser-

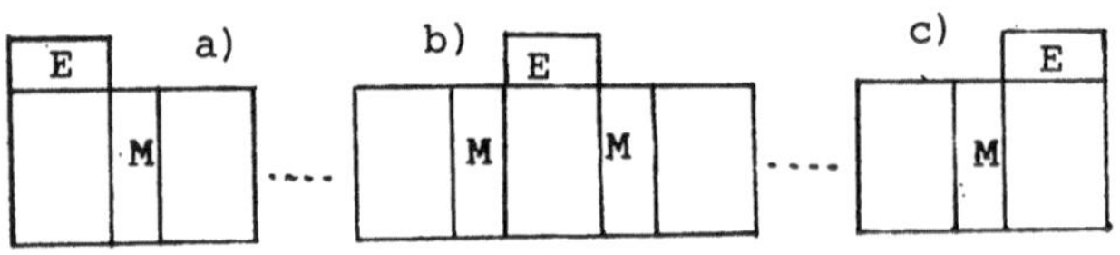

Fig.1

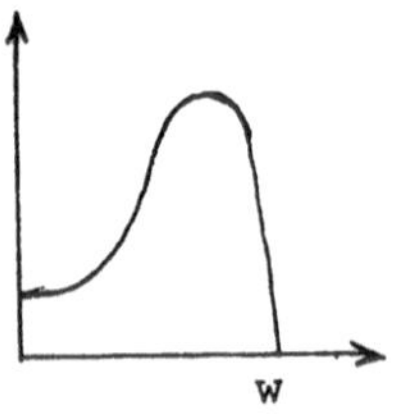

Fig.2

voir. Joining systems of that type results in the case where the two end cells are connected to the outside reservoir. This is the subject of section 2. Finally in section 3 we comment on the general case. In particular, we take a closer look at a special system with 3 cells connected to the outside reservoir. The resulting solution set of (1) can be very complex. We try to understand its complexity via singularity theory. Our study is an extension of the papers [1,2,3].

1. The cell system with one end open.

In this section we consider (1) with $E_j=0$ $(j=2,\dots,N)$ and f as qualitatively given by Fig.2. In the case of just one cell we have the equation

(3) $\quad E_1 x = f(\lambda,x).$

We assume that we can solve (3) for λ yielding the funciton $\lambda(x,E_1)$. It describes the intersections of the straight lines through O with the graph of the function f (see Fig.2 and (3)). Let $\lambda(x,E_1)$ qualitatively be given in a λ-x-plane as shown in Fig.3a.

We now turn to the case $N\geq 1$, $E_j=0$ $(j=2,\dots,N)$. For $D_j=0$ $(j=2,\dots,N)$ all cells are separated and we have the solution branch

(4) $\quad (\lambda(x,E_1),x,w,\dots,w)\in \mathbb{R}^{N+1}, \quad x\in[0,w].$

Note that w is the only zero of $f(\lambda,x)$ as is easily seen from Fig.2. This solution branch consists of only regular points. It is qualitatively given in Fig.3a). To persue its fate if the D_j grow we consider the most simple algebraic equation with a solution set of this type

(5) $\quad h(x,\lambda)+\alpha x = 0, \quad h(x,\lambda) = x^3-\lambda,$

where α is an arbitrary real constant. We may view (5) as a perturbation of $h(x,\lambda)=0$. If the parameter α varies in the interval $[-1,1]$, then the series of pictures given in Fig.3 appear. If α grows, the two turning points converge to each other and the hysteresis loop disappears for the ciritcal parameter $\alpha_k=0$. If α grows beyond that point, no turning point is left.

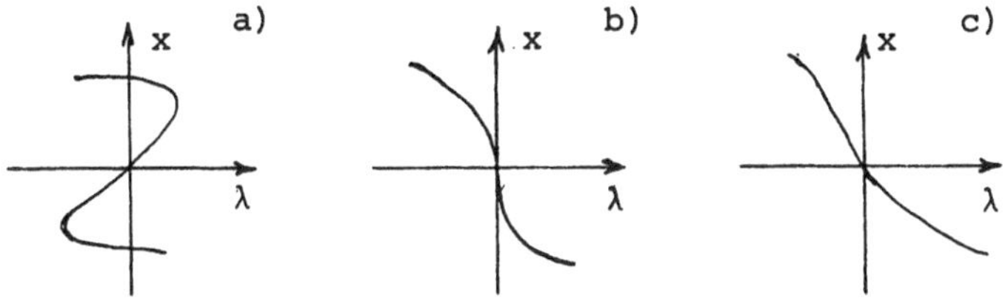

Fig.3

D	1.59	1.594	1.5943	6.0	8.0	10.0
x_1	3.4989	3.4892	3.4935	3.4818	3.5884	3.6498
x_2	3.9689	3.9682	3.9685	3.8328	3.8441	3.8500
λ	12.2402	12.2518	12.2527	29.5809	31.2178	32.5249
y_1	3.4739	3.4842	3.4880	1.9873	2.0144	2.0298
y_2	3.9673	3.9679	3.9681	2.1598	2.1443	2.1338
μ	12.2398	12.2518	12.2527	12.3551	12.2049	12.1175

Tab.1

Let us test if our branch (4) behaves the same way as we increase the diffusion parameter $D=D_j$ $(j=2,\dots,N)$. To this end we present some numerical results. Our figures given in the first 3 columns of Table 1 are based on the generation function (2b) with $16\lambda_1=10^{1.4}$, $\lambda_2=30$, $E_1=1$, $\alpha=4$, $N=2$. The (x,λ)-vector and the (y,μ)-vector denote the two turning points of the hysteresis loop. The figures show that both points converge to each other if the diffusion parameter D reaches a critical value D_k. This is in striking agreement with the behaviour of the solution branch of our model equation (5). We henceforth view (5) as a model to understand the behaviour of the assemblage of cells in the region of the critical parameter D_k. Further numerical studies also reveal that the branch (5) develops not only one but N hysteresis loops where N is the number of cells. If the diffusion D grows, the first N-1 hysteresis loops vanish according to the above pattern. This behaviour seems not to be true for the last hysteresis loop as is shown by the numerical study given in the last three columns of Table 1. Hence finally a branch with only one hysteresis loop emerges. Fig.4 shows the global deformation of the branch (4) for N=3 as D grows. Locally we observe the behaviour as given in the series in Fig.3.

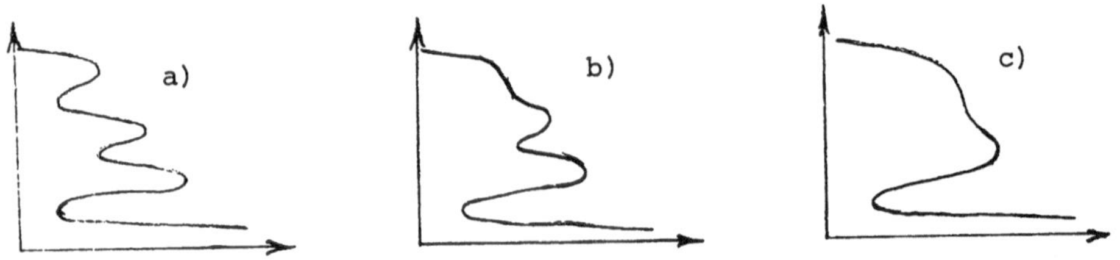

Fig.4

2. The cell system with two ends open.

In this section we are going to combine units of $M \geq 2$ cells of the form given in Fig.1a and 1b for $M=2$. We join them and obtain a system of $N=2M$ cells whose steady states are solutions of (1) where $E_1>0$, $E_N>0$, $E_j=0$ $(j=2,\dots,N-1)$. We first assume $D_j=D$ $(j=2,\dots,N=2M)$, $E_1=E_N$. Then our system consists of two identical units of length M. Choose its k-th hysteresis loop $(1 \leq k \leq M-1)$. Then there exists a critical value D_k such that this hysteresis loop disappears at D_k as described in section 1. For D nearby but less than D_k we find three different solutions $(\lambda_D, x(\sigma_j,D))$ $(j=1,2,3)$ for the control parameter λ_D between the control parameter corresponding to the turning points of the hysteresis loop (see Fig.3a). Since D is close to D_k the vectors $x(\sigma_j,D)=(x_1(\sigma_j,D),\dots,x_M(\sigma_j,D))$ are close together. Hence any vector

(6) $\quad (\lambda_D, x_1(\sigma_j,D),\dots,x_M(\sigma_j,D), x_M(\sigma_i,D),\dots,x_1(\sigma_i,D)),\quad i,j=1,2,3$

almost satisfies the system (1) with $E_N=E_1$ and $N=2M$. Indeed, if Δ_{ij} denotes the defect of (1) at the vector (6), then it easily follows that

(7) $\quad \|\Delta_{ij}\|=D_{M+1}|x_M(\sigma_j,D)-x_M(\sigma_i,D)|,\quad i,j=1,2,3,$ (e.g. $D_{M+1}=D$),

holds. Here we have used the maximum norm. Since the right hand side of (7) is small, the defect has a small norm. Hence Newton's Method applied to (1) with the initial approximation (6) will converge to a solution of (1) [6]. In this way we can construct 9 solutions for the combined assemblage of cells out of three solutions for the separated units. We note in passing that in principle we can join any two solutions of the two units if the diffusion coefficient $\tilde{D}$ joining both units is small enough. Indeed the defect is then given by (7) with $D_{M+1}=\tilde{D}$. It is small if $\tilde{D}$ is small enough. The procedure just described obviously results in a symmetric solution if i=j in (6). However, the solution will be unsymmetric if $i \neq j$. In their paper [4] Bunow and Colton point out the possibility of unsymmetric solutions for our system.

In an attempt to understand the solution set of (1) we construct a model along the lines of section 1. To this end we combine two equations of the form (5) for each unit yielding the set of equations

(8) $\quad h(x,\lambda)+\alpha x = 0,\quad h(y,\lambda)+\alpha y = 0$.

By just combining two copies of the corresponding hysteresis loop (see Fig.3a) we obtain the solution set of (8) in any x-y-plane. The result is Fig.5. Note that we have two branches: a main branch consisting of symmetric solutions (z,z) which forms the hysteresis loop and a branch

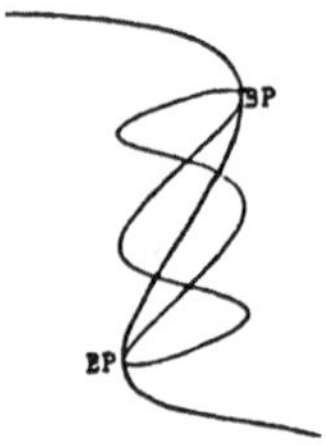

Fig.5

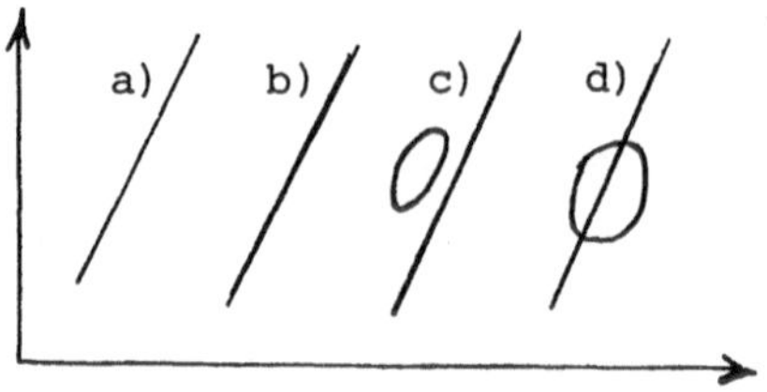

Fig.6

of unsymmetric solutions (x,y) ($x \neq y$) which forms the double-figure-eight. Both branches are joined via two bifurcation points BP which are created by the turning points of Fig.3. We can regard (8) as an unfolding of the singularity defined by $h(x,\lambda)=0=h(y,\lambda)$ which after elimination of λ yields

(9) $x^3-y^3 = 0.$

The complete picture of a universal unfolding of the singularity (9) is given in W.-J. Beyn [1]. The result is that essentially only the four possibilities given in Fig.6 can occur. Here circles indicate closed branches and straight lines symbolize branches which are not closed. A dot refers to an isolated solution. For details we refer to [1]. The possibility d) of Fig.6 corresponds to Fig.5.

Returning to our cell system we take a symmetric and an unsymmetric solution

(10) $(\lambda_D, s(\sigma))$, $(\lambda_D, u(\sigma))$

constructed as outlined above. We start a continuation process at each of these solutions and find Fig.5. The double-figure-eight consists of only unsymmetric solutions and is the result of the continuation process started at the second solution given in (10). The remaining branch consists of symmetric solutions and results from the continuation process initiated at the first solution of (10). This completes the discussion of (1) if $D_j=D$ $(j=2,\ldots,N)$, $E_1=E_N$, $E_j=0$ $(j=2,\ldots,N-1)$. We now perturb D into D_j (possibly different for different membranes) and E_1, E_N into $\bar{E}_1$, $\bar{E}_N$ (possibly $\bar{E}_1 \neq \bar{E}_N$). We, however, keep the assumption $E_j=0$ $(j=2, \ldots,N-1)$. The numerical study results in the possibilities symbolized by Fig.6. Hence, the forcast of the singularity (9) is fully observed. Note that the double-figure-eight structure and its unfolding pictures appear at different places of the bifurcation diagram of (1). In particular this is true if we view λ and D_j as control parameters. Locally we may regard (9) the singularity which governs our cell system. We remark that the discussion in [4] also considers the case $D_j=D$ $(j=2,\ldots,N)$, $E_j=0$ $(j=2,\ldots,N-1)$ but $E_1\neq E_N$. Then the unsymmetric branch moves away from the symmetric branch thereby breaking the bifurcation points. This just means that we go from d) to c) in Fig.6.

3. The cell system with more than two cells open.

Consider a cell system which is constructed out of end units of the form as given in Fig.1a), 1c) and middle units as in Fig.1b). For any unit the number of cells which are not connected to the outside reservoir is arbitrary but at least one for end units and at least two for middle units. Hence, the smallest assemblage constructed in this way consists of 7 cells and is shown in Fig.1. We assume (and numerical evidences support this) that the middle units work the same way as described in section 1 for the end units. Hence if again the generation function f is qualitatively given by Fig.2, we understand the performance of the separated units and we only have to join them the way we have done it in section 2 with two end units.

As an example let us take the assemblage of 7 cells given in Fig.1. The corresponding system (1) yields a bifurcation diagram which shows locally a structure resulting from the combination of 3 hysteresis phenomena given in Fig.3. In the first place this can produce all the pictures of Fig.6 (e.g. $E_2=0$). But in addition we get more complicated situations: Fig.7 supplements Fig.6 and shows symbolically possibilities of that type. Here, all intersection points stand for bifurcation points. The pictures a), b), c) of Fig.7 are perturbations of d). The complete structure may be understood as the unfolding of an underlying singularity which is described by the equations $x^3-y^3=0$, $y^3-z^3=0$. The discussion of a universal unfolding of this singularity needs 28 parameters and is very complex. Of course all unfolding pictures of the singularity (9) arise in the unfolding of the more general singularity given above. It will be partly analysed in a forthcoming paper of W.-J. Beyn and E.Bohl. Also the substructures given in Fig.7 occur in its unfolding.

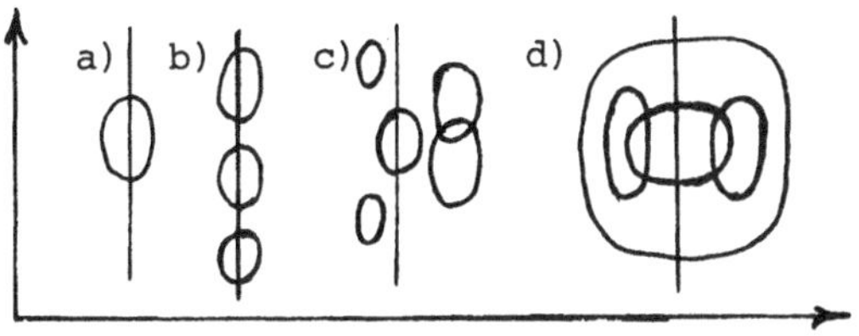

Fig.7

We have tested the predictions of the singularity on the system (1) numerically and have found qualitative agreement. This leads us to conjecture that locally this singularity governs a system (1) which results as a model of the assemblage of cells given in Fig.1. Any of the constructed solutions of our system is the result of the combination of solutions of the three subunits. The combination of stable solutions of the subunits yield a stable solution of the full assemblage. Hence any of the resulting branches has parts consisting of stable solutions. Note that the special case treated in section 2 may be regarded as a finite difference analogue to a boundary value problem which describes a continuous spacially distributed chemical system. It is impossible that the solution set corresponding to the boundary value problem consists of branches with bifurcation points. Hence in the passage from the discrete model to the continuous model the bifurcation points found in section 2 must disappear. At this point a fundamental difference between continuous and discrete modeling emerges.

References

[1] Beyn, W.-J., Numerical analysis of singularities in a diffusion reaction model. To appear in the Conference Proceedings of the EQUADIFF 82.

[2a] Bigge, J., Bohl, E., Deformations of the bifurcation diagram due to discretization. Submitted for publication.

[2b] Bigge, J., Bohl, E., On the steady states of finitely many chemical cells. Submitted for publication.

[3a] Bohl, E., A numerical procedure to compute many solutions of diffusion- reaction systems. ISNM 62, 25-36, 1983.

[3b] Bohl, E., Verzweigungsbilder diskreter Transportmodelle. To appear in the Proceedings of the Conference on Numerical Methods for Differential Equations, Halle 1983.

[4] Bunow, B., Colton, C.K., Substrate inhibition kinetics in assemblages of cells. BioSystems 7, 160-171, 1975.

[5] Kernevez, J.-P., Enzyme mathematics. Studies in mathematics and its applications Vol.10. Amsterdam, New York, Oxford, 1980.

[6] Ortega, J.M., Rheinboldt, W.C., Iterative solution of nonlinear equations in several variables. New York, San Francisco, London, 1970.

EXPERIMENTAL VALIDATION OF A MATHEMATICAL PROCEDURE FOR THE ANALYSIS OF FLOW CYTOMETRIC DATA(•)

C. Bruni[1], A. Costa[2], G. Del Bino[2], G. Koch[3], G. Mazzini[4], and R. Silvestrini[2]

[1]Istituto di Automatica, Università di Roma;
[2]Oncologia Sperimentale C, Istituto Nazionale per lo Studio e la Cura dei Tumori, Milano; [3]Istituto Matematico G. Castelnuovo, Università di Roma; [4]Centro di Studio per l'Istochimica del CNR, Pavia

1. INTRODUCTION

As well known, most types of cells undergo a growth and proliferation cycle, along which the following three main phases can be identified:

- the $G_{0/1}$ phase, during which the cell either rests or synthesizes RNA and proteins
- the S phase, in which DNA content increases from the resting value up to twice that
- the G_2+M phase, in which DNA content is twice the resting value, cell completes its growth and finally divides in two daughter cells (mythosis).

To the actual percentages of cells in the three above mentioned phases is attached information of obvious biological and medical interest, since they describe how the proliferative process evolves in time. Flow cytometry was recently developed as an experimental technique for investigating DNA distribution of a cell population (for a general reference, see for instance |25|).

Different methods have been proposed for quantifying cell cycle parameters from cytofluorometric histograms. Some authors have suggested empirical graphic procedures based on the decomposition of the experimental distributions in few simple geometrically shaped terms |2|, |4|, |5|, |18|, and others |12|, |16|, |23| |24| have modelled the $G_{0/1}$ and G_2+M phases by gaussian distributions and interpolated the S-phase by a low degree polynomial or by an exponential function, then exploiting least square and deconvolution techniques. For a comprehensive survey see for instance |3|, |25|.

The aim of present study is to verify the reliability of an automatic procedure recently proposed by Bruni et al. |10| to determine the distribution of cells in the different cell cycle phases, by exploiting cytofluorometric histograms. This procedure, which follows a detailed mathematical analysis of the problem |9|, was already checked against simulated data |11| and is tested in the present paper by means of experimental data. More specifically, this paper addresses three problems. The first one is how the proposed method compares with that of Fried |20|, |21| which appears to be one of the most widely adopted and sophisticated procedures. The second problem is how the S-phase percentages (S%), as given by the considered methods, relate to the labeling index (L.I.) values for a set of non-Hodgkin lymphoma (NHL). The L.I. is a more traditional measurement of percentage of cells which are synthesizing DNA, obtained by well established autoradiographic techniques. The final problem is to define a rapid and effective test on S% alternative to the autoradiographic index of proved clinical relevance in evaluating the aggressiveness of non-Hodgkin lymphoma |13|, |28|.

•Work supported by Progetto Finalizzato "Controllo della Crescita Neoplastica", CNR under grants No 81.01387.96 and 81.01388.96.

2. DATA PROCESSING AND STATISTICAL ANALYSIS

Pathological lymph nodes of 69 adult patients were analyzed by autoradiographic techinques in order to determine the |³H| thymidine Labeling Index. The same samples were analyzed by flow cytometric technique. All technical details on both experimental procedures are described in |15|.

To recover the distribution of cells over the three cycle phases. flow cytometric data were processed by two different mathematical methods, both of which based on a least square technique approach. The first one |20| is designed to yield the unknown samples of DNA distribution in the various cycle phases, assuming for the instrumental spreading density function, due to the measurement technique, a gaussian form with a constant coefficient of variation. Since the samples of the unknown distribution are dealt with, a suitable regularization procedure may be called for and positivity is guaranteed by an ad hoc mechanism. In the second method |8|, |10| the DNA distribution is parametrized by means of few quantities with clear physical meaning (percentages of cells in the different phases, location of G_0/G_1 and G_2+M peaks). For the above mentioned spreading density, a gaussian form in again assumed, but with the coefficient of variation decreasing with the square root of the channel number, as suggested by analysis of the binding mechanism of the fluorescent stain |10|. Finally, suitable weights are introduced to account for the statistics of the error term due to the finite size of cell sample and constraints for physical significance are included in an easier and more natural way.

Due to unavoidable experimental and numerical errors, comparison between the outcomes of the two analysis procedures for flow cytometric data, as well as between S% and L.I., must be carried out on statistical bases, that is by computing appropriate statistical tests.

To compare phase percentages obtained by the two methods, the Spearman rank test r_s |22| was used. Thus we test the hypothesis that percentages for the same phase obtained by the two methods, besides possible independent computational errors, do not contain any common term. This test appears to be more significant than a linear correlation test; moreover, it does not require any knowledge of the error statistics.

A further test was performed on the same data, assuming independence and identical zero mean gaussian distribution for the errors. The test index was:

$$s = \frac{n \sum_{1}^{n} {}_i (x_i - y_i)^2}{\sum_1^n {}_i (x_i - \bar{x})^2 + \sum_1^n {}_i (y_i - \bar{y})^2}, \qquad \bar{x} = \frac{1}{n} \sum_1^n {}_i\, x_i \qquad \bar{y} = \frac{1}{n} \sum_1^n {}_i\, x_i$$

where, for each phase, x_i, y_i are the percentages obtained by the two methods from the i-th sample, i=1,2,...n. By means of s we test the hypothesis that x_i. y_i have the same mean value; the significance of the test may be evaluated from the (approximately) χ^2 distribution of s.

To investigate the relationship between S% and L.I. we again test whether or not they contain a common term (which, besides measurement errors, differs from the S% value by a possible non cycling population) by use of the Spearman index.

Finally, it has been shown |13| that the L.I. threshold value of 4% has a clinical relevance.

The statistical dependence of this value on a similar cut-off value on S% (12%) was tested by the Cohen's K test |19| performed on the corresponding contingency tables.

For all statistical analysis, two-tailed p values were used; differences at a p level of 0.05 were considered significant.

3. RESULTS AND DISCUSSION

Comparison between the outcome of the two mathematical procedures

The outcomes and the test values of the percentage of cells in the different cycle phases evaluated by the two models on 42 cases are reported in Table I. An agreement was observed for the values of $G_{0/1}$ and S phases. On the contrary, the agreement was not significant for the G_2+M values. This is not surprising if we consider the greatest experimental variability which usually affects the G_2+M phase, as recently also pointed out in |16|.

Comparison between L.I. and S phase fraction

The values of L.I. and percentage of S phase cells determined by the two procedures on the same series of 42 patients are shown in Fig. 1. L.I. values appeared significantly related to S phase value sets determined by both procedures. The findings obtained by use of the procedure of Bruni et al. were consistently observed when the analysis was extended to a larger series including 69 cases (Fig.2).

This study confirms on a large series of NHL, and significantly for the proposed procedure |8|,|10|, the correlation already observed |6|, |7|, |14| between L.I. and S-cell fraction.

From an overall analysis, it appears that the S-cell fraction systematically overestimates L.I. Almost all data fall above the bisecting line of the L.I.-S% plane. This finding, which has been consistently observed in all studies of this type |7|, |26|, cannot be due to the effect of a zero mean experimental or computational error, but likely to the degree of accuracy of fluorescent staining in resolving DNA histograms |6| or to an actual biological difference between the two kinetic parameters. The latter hypothesis could be reasonably explained by the presence of a fraction of cells that presents a DNA content characteristic of the S phase but which is not in proliferative activity |1|, |26|, |27|.

Definition of an S-phase cut-off index

We looked for an S% cut-off value by using marginal distributions for L.I. and S% (obtained by interpolation of experimental histograms), assuming the latter one as a convolution of the former one with the distribution of the previously mentioned, non-proliferating population. A simple application of decision theory to those distributions yielded the value of 12% for the aforementioned S% cut-off.

The value of 12% of S-phase cells proved to be the value that maximizes the number of cases that fall above and below both cut-off values of L.I. and S.

Using this cut-off S phase index, it is possible to discriminate NHL at low and high biological aggressiveness, stressing the importance of a reproducible indicator of cell proliferation whose determination is not time consuming.

However, the relevance of the S-phase cut-off value as discriminant of tumors with a low or high L.I. is different for the sets of data obtained by the procedures of Fried and Bruni et al. (Table II). With the latter elaboration, a low S-cell fraction was predictive of a low L.I. in 97% of the cases, and a high S-cell fraction was predictive of high L.I. in 77% of the cases. From an overall analysis an agreement of the two kinetic parameters was observed in 91% of the cases. These findings relative to the series of 42 patients were confirmed on the larger series of 69 tumors. From the Fried elaboration, the low fraction of S cell was similarly highly indicative of the low L.I. (91% of the cases), whereas a high fraction of S cells was not discriminant of the L.I.

These findings make the procedure of Bruni et al. a valid mean, as an alternative to L.I. determination, for automatically and quickly identifying slow and fast proliferating tumors.

Table I. Distribution of cells in the different cycle phases evaluated by the mathematical procedures of Fried and Bruni et al. (on 42 cases).

	$G_{0/1}$ Mean SD (range)	S Mean SD (range)	G_2+M Mean SD (range)
Fried	77.6 $\pm$ 11.8 (51 - 94.5)	11.5 $\pm$ 9.5 (0.3 - 38.6)	10.9 $\pm$ 5.7 (2.2 - 25.7)
Bruni et al.	80.7 $\pm$ 11.7 (51.7-94.5)	12.5 $\pm$ 13.3 (0 - 45.7)	6.7 $\pm$ 4.8 (0 - 17.5)
r_s (p)	0.869 ($<$0.001)	0.669 ($<$0.001)	0.207 (not significant)
s	5.47 (significant)	14.0 (significant)	46.3 (not significant)

Table II. Accuracy of the S phase value, according to Fried and to Bruni et al., in predicting the labeling index on a series of 42 NHL(*)

	Bruni et al.		Fried	
	Low LI	High LI	Low LI	High LI
Low S	28	1	21	2
High S	3	10	10	9
K (P)	0.768	(0.00006)	0.40 (0.005)	

* L.I. cut-off: 4%; S index cut-off: 12%.

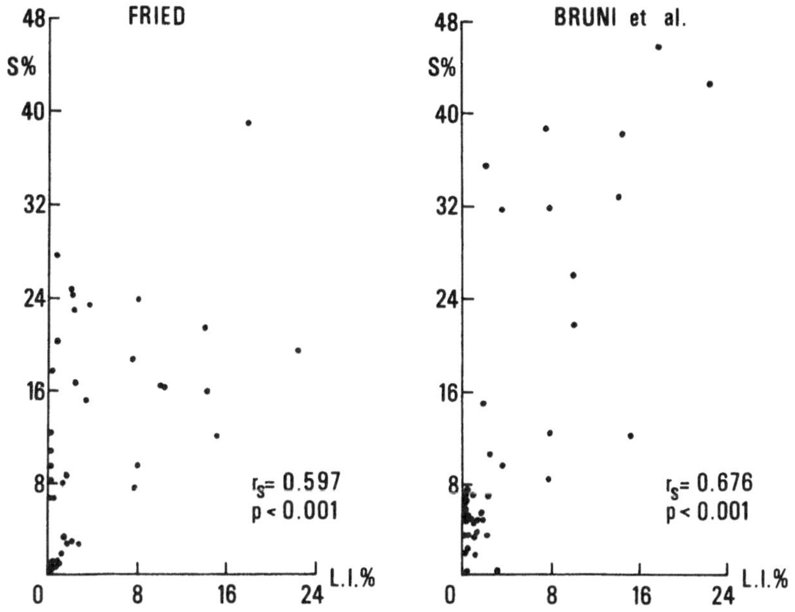

Fig.1. Comparative analysis of the correlation between LI and percentage of S phase cells determined by the procedures of Fried and of Bruni et al. on the same series of patients (42 cases).

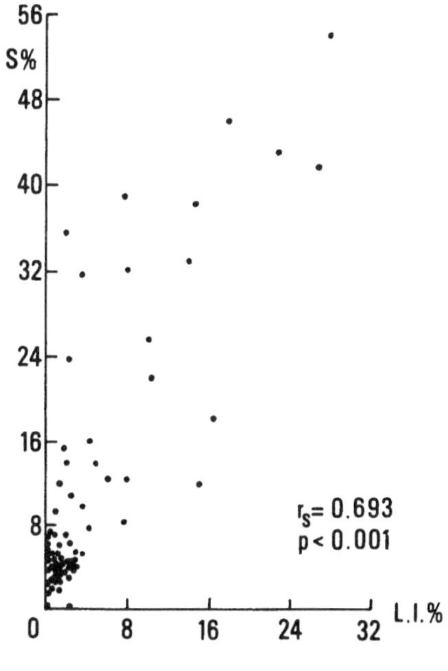

Fig.2. Relationship between LI and percentage of S phase cells obtained on a large series of patients (69 cases) by the proposed method.

|1| Alabaster O., Bunnag B.; Cancer Res. 36: 2744-2749, 1976.

|2| Baish H.: In: Mathematical Models in Cell Kinetics, Valleron A.J. (ed.): European Press Medikon, Ghent, 1975, p. 65-67.

|3| Baish H. et al.: Cell Tissue Kinet. 15:235-249, 1982.

|4| Barlogie B. et al.: Cancer Res. 36:1176-1181, 1976.

|5| Barlogie B. et al.: Cancer Res. 38:3333-3339, 1978.

|6| Barlogie B. et al.: Blood 48:245-258, 1976.

|7| Braylan R.C., Diamond L.V., Powell M.L., Harty-Golder B.: Cytometry 1:171-174, 1980.

|8| Bruni C., Curzi L., Koch G., Rossi C.: Biological aspects and mathematical analysis of the inverse problem in flow cytometry. Rept. IASI R - 16, July 1981.

|9| Bruni C., Koch G.: Identifiability of continuous mixtures of unknown gaussian distributions. Rept. Ist. Automatica R. 82-13, Dec. 1982 and Ist. Matematico, University of Rome.

|10| Bruni C., Koch G., Rossi C.: Cell Biophys 5, 1983.

|11| Bruni C., Koch G., Lucidi S.: Analysis of DNA distributions from flow cytometry: validation of an automatic procedure against simulated data. Rep. Ist. Automatica R. 83-06, Feb. 1983, University of Rome.

|12| Christensen I. et al.: In: Third International Symposium Cytophotometry, Lutz D. (Ed.): European Press Medikon, Ghent, 1978, p. 71-78.

|13| Costa A., Bonadonna G., Villa E., Valagussa S., Silvestrini R.: J. Natl. Cancer Inst. 66:1-5, 1981.

|14| Costa A., Mazzini G., Del Bino G., Silvestrini R.: Cytometry 2:185-188, 1981.

|15| Del Bino G., Bruni C., Koch G., Mazzini G., Costa A., Silvestrini R.: A mathematical procedure for the computer analysis of flow cytometric DNA data in human tumors. Rept. Istituto di Automatica R. 83-05, febbr. 1983.

|16| Dean N.: In: Third International Symposium on Pulse Cytophotometry, Lutz D. (Ed.): European Press Medikon, Ghent, 1978, p. 63-68.

|17| Dean P.N., Gray J.W. Dolbeare F.A.: Cytometry, 3:188-195, 1982.

|18| Drewinko B. et al.: Cell Tissue Kinet. 11:177-191, 1978.

|19| Fleiss J.L.: Statistical Method for Rates and Proportions. John Wiley, 1973.

|20| Fried J.: Comput. Biol. Res. 9:263-276, 1976.

|21| Fried J.: J. Histochem. Cytochem. 25:942-951, 1977.

|22| Hàjek J., Sidàk Z.: Theory of Rank Tests. Academic Press, 1967.

|23| Jett J.H.: In: Third International Symposium on Pulse Cytophotometry, Lutz D. (Ed.): European Press Medikon, Ghent, 1978, p. 93-102.

|24| Kim M., Perry S.: J. Theor. Biol. 68:27-42, 1977.

|25| Nicolini C., Zietz S.: In: Biomathematics and Cell Kinetics, Valleron A.J., Macdonald P.D.M. (Eds.), Elsevier/North Holland Biomedical Press, 1978, p. 357-394.

|26| Pallavicini M.G., Lalande M.E., Miller R.G., Hill R.P.: Cancer Res. 39:1891-1897, 1979.

|27| Preisler H.D.: Basic Applied Histochem. 24:383, 1980.

|28| Scarffe J.H., Crowther D.: Eur. J. Cancer 17:99-108, 1981.

A STOCHASTIC MODEL FOR DESCRIBING THE SPACE-TIME EVOLUTION OF THE EXTREME VALUES OF CARDIAC POTENTIALS ON THE CHEST SURFACE

C. Calvi Parisetti*, G. Di Cola*, E. Musso**, D. Stilli**, B. Taccardi**

Introduction

Body surface mapping is a new electrocardiographic method which consists in displaying the distribution of equipotential lines on the entire surface of the torso at a number of time instants during the cardiac cycle. It is widely recognised that cardiac electromaps provide more information on normal and abnormal intracardiac electrical events, than can be obtained from traditional 12-leads electrocardiograms.

In particular, body surface maps exhibit typical abnormalities in a number of heart conditions, even when the 12-leads electrocardiogram is within normal limits.

The physiological and clinical interpretation of body surface maps is based on visual inspection and on the measurement of significant parameters. Among the features that can be detected by visual inspection, the instantaneous location of potential maxima and minima and their trajectory on the body surface during cardiac excitation and recovery convey a considerable amount of useful information.

At present, the results of visual inspection are not expressed numerically and cannot be statistically analyzed. We considered that a statistical method for estimating the sequence of displacements of potential maxima and minima would be useful for the physiological and clinical interpretation of body surface maps. We therefore developed a mathematical model which expresses the successive displacements of the potential extrema on the body surface in probabilistic terms.

The model has been applied to a sample of 20 normal subjects and to a small group of patients with inferior myocardial infarction. The analysis was limited to the displacements of the potential minimum. The results showed a common trend of the displacements in normal subjects and a significantly different behaviour in the group of patients.

Mathematical Model

Given a cardiac electromap E, we consider a partition of E in a fixed number of rectangular subregions. We associate to each region a pair of integer numbers (n, m).

Let X(t), Y(t) represent the position of the minimum potential $V_m(t)$, at time t.

We assume that $\{ Z(t); t \geqslant 0 \}$, with $Z(t) = \{ X(t), Y(t) \}$, is a Markovian process, with state space

$$S = \{ r : r = (n, m), -N_x \leqslant n \leqslant N_x, -N_y \leqslant m \leqslant N_y \}$$

where N_x, N_y are assumed large enough so that the process can be considered unrestricted.

In order to determine the joint probability distribution $P_{n,m}(t) = P \{ X(t) = n, Y(t) = m; (n, m) \in S \}$, we suppose that the state of the system can be described by a bidimensional non-homogeneous Poisson process, that is in the time interval (t, t +h), the probabilities of transition are the following:

$$P(X \to X+1, Y \to Y) = \lambda(t)\, h + o(h)$$

$$P(X \to X-1, Y \to Y) = \mu(t)\, h + o(h)$$

$$P(X \to X, Y \to Y+1) = \nu(t)\, h + o(h)$$

$$P(X \to X, Y \to Y-1) = \rho(t)\, h + o(h)$$

$$P(X \to X, Y \to Y) = 1 - \{ \lambda(t) + \mu(t) + \nu(t) + \rho(t) \} h + o(h).$$

As the transition probability functions are only time dependent, the process evolution can be expressed by means of the two differential equations:

$$\frac{dP_n}{dt} = \lambda(t)\,P_{n-1}(t) - (\lambda(t) + \mu(t))\,P_n(t) + \mu(t)\,P_{n+1}(t)$$
$$\frac{dP_m}{dt} = \nu(t)\,P_{m-1}(t) - (\nu(t) + \rho(t))\,P_m(t) + \rho(t)\,P_{m+1}(t) \qquad (1)$$

where $P_n(t)$ and $P_m(t)$ are marginal probability densities.
The purpose of this work is to estimate the transition probability functions related to the processes X(t) and Y(t), starting from N realizations $x_{(n)}^{(j)} = (x_o^{(j)}, \ldots, x_n^{(j)})$ and $y_{(n)}^{(j)} = (y_o^{(j)}, y_1^{(j)}, \ldots, y_n^{(j)})$ $j = 1, \ldots, N$, where $(x_k^{(j)}, y_k^{(j)})$ is the minimum potential position at time t_k, as observed on the maps of a homogeneous group of «normal» subjects.
The probem can be solved by means of the classical Maximum Likelihood method, by maximizing the L functional, taking into account the probability differential model (1).
If N subjects are observed at the times $t_o, t_1, \ldots, t_n$ the Likelihood functional L of the realizations $x_{(n)}^{(1)}, \ldots, x_{(n)}^{(N)}$ can be written:

$$L = L(x_{(n)}^{(1)}, \ldots, x_{(n)}^{(N)}; \lambda(t), \mu(t)) = \prod_{j=1}^{N} \prod_{i=o}^{n-1} P\,[\,X(t_{i+1}) = x_{i+1}^{(j)} \mid X(t_i) = x_i^{(j)}; \lambda(t), \mu(t)]$$

being $P\,[\,X(t_o) = o\,] = 1$.
The same procedure can be applied for estimating the parameters $\nu(t)$ and $\rho(t)$.
The solution of the maximum constrained problem, can be obtained by standard variational methods; then, parameter estimates are computed by means of iterative techniques which need the solution of the state and adjoint differential equations system and the gradient, with hard though straightforward computations.

Parameter Estimation

To overcome the numerical difficulties described above, we approximate the intensity functions $\lambda(t)$ and $\mu(t)$, related to the process X(t), with step functions

$$\tilde{\lambda}(t) = \sum_{k=o}^{n-1} \lambda_k\, \chi(t_k, t_{k+1}), \qquad \tilde{\mu}(t) = \sum_{k=o}^{n-1} \mu_k\, \chi(t_k, t_{k+1})$$

respectively, where $\chi(a, b) = 1$ if $t \in (a, b)$, $\chi(a, b) = 0$ otherwise.

The L function of the realizations $x_{(n)}^{(1)}, \ldots, x_{(n)}^{(N)}$ of the process X(t), if $t_o, t_1, \ldots t_n$ denote the epochs when the events occurred, is given by:

$$L(x_{(n)}^{(1)} \ldots, x_{(n)}^{(N)}; \tilde{\lambda}(t), \tilde{\mu}(t)) = \prod_{k=o}^{n-1} \prod_{j=1}^{N} \left(\frac{\lambda_k}{\mu_k}\right)^{\frac{x_{k+1}^{(j)} - x_k^{(j)}}{2}} I_{x_{k+1}^{(j)} - x_k^{(j)}}\left(2\,\Delta t_k\,(\lambda_k \mu_k)^{1/2}\right) \cdot$$
$$\cdot\, e^{-(\lambda_k + \mu_k)\,\Delta t_k}$$

where: $\Delta t_k = (t_{k+1} - t_k)$
$x_{k+1} - x_k$ is an integer
$I_n(x) = I_n(x)$ is the modified Bessel-function of order n.

We can observe that, for fixed Δt_k, when $x^{(j)}_{k+1} - x^{(j)}_k > 0$ (< 0), we have $\mu_k = 0$ ($\lambda_k = 0$).

Let $J_1^{(k)}$ bethe number of subjects such that, in the interval (t_k, t_{k+1}), holds $x^{(j)}_{k+1} - x^{(j)}_k > 0$ and $J_2^{(k)}$ is the number of those for which holds $x^{(j)}_{k+1} - x^{(j)}_k < 0$, the L function can be written more simply as:

$$L = \prod_{k=0}^{n-1} \left[\prod_{j=1}^{J_1^{(k)}} e^{-\lambda_k \Delta t_k} \frac{(\lambda_k \Delta t_k)^{(x^{(j)}_{k+1} - x^{(j)}_k)}}{(x^{(j)}_{k+1} - x^{(j)}_k)!} \cdot \prod_{j=J_1^{(k)}+1}^{J_2^{(k)}} e^{-\mu_k \Delta t_k} \frac{(\mu_k \Delta t_k)^{(x^{(j)}_k - x^{(j)}_{k+1})}}{(x^{(j)}_k - x^{(j)}_{k+1})!} \cdot \prod_{j=J_1^{(k)}+J_2^{(k)}+1}^{N} e^{-(\lambda_k + \mu_k)\Delta t_k} \right]$$

The support functional $S = -\log L$, is given by:

$$S = -\sum_{k=0}^{n-1} \left\{ \sum_{j=1}^{J_1^{(k)}} [-\lambda_k \Delta t_k + (x^{(j)}_{k+1} - x^{(j)}_k) \log(\lambda_k \Delta t_k)] + \sum_{j=J_1^{(k)}+1}^{J_2^{(k)}} [-\mu_k \Delta t_k + (x^{(j)}_k - x^{(j)}_{k+1}) \log(\mu_k \Delta t_k)] + \right.$$
$$\left. + \sum_{j=J_1^{(k)}+J_2^{(k)}+1}^{N} [-(\lambda_k + \mu_k)\Delta t_k] \right\}$$

From the system:

$$\frac{\partial S}{\partial \lambda_i} = 0 \qquad\qquad \frac{\partial S}{\partial \mu_i} = 0$$

we get the M. L. estimators:

$$\hat{\lambda}_i = \frac{\sum_{j=1}^{J_1^{(i)}} (x^{(j)}_{i+1} - x^{(j)}_i)}{(N - J_2^{(i)})\,\Delta t} \qquad\qquad \hat{\mu}_i = \frac{\sum_{j=1}^{J_2^{(i)}} (x^{(j)}_i - x^{(j)}_{i+1})}{(N - J_1^{(i)})\,\Delta t}$$

being $\Delta t_i = \Delta t$.

The same procedure is applied for estimating the parameters $\nu(t)$, $\rho(t)$ related to the process Y (t).

When time t_k is varying, $\sum_{i=1}^{k} (\hat{\lambda}_i - \hat{\mu}_i)\,\Delta t_i$, give an estimate of successive displacements of the minimum potential position.

Subjects and Methods

The analysis of the trajectories of the potential minimum was performed on a sample of 20 normal subjects. We limited our study to the time interval related to the ventricular excitation (QRS interval). Its duration ranged between 68 and 104 msec. and was standardized to a common value T, by means of a suitable compression.

Body surface potentials were measured every 2 msec. in 219 points on the chest surface. The values expressed in μV, were mapped and the position of the minimum potential was identified.

If we divide the map in 2 (N_x x N_y) regions numbered $n = -N_x, \ldots, N_x$ from left to right and $m = -N_y, \ldots, N_y$ from top to bottom, the position of the potential minimum at time t is defined by the pair (n, m), characterizing the region to which the minimum belongs.

A state variable $Z(t) = \{X(t), Y(t)\}$ is associated to the position of the potential minimum.

Starting from the realizations of the process, the Maximum Likelihood method allowed to determine the estimated sequence of displacements of the potential minimum.

Moreover, the estimation of the process parameters give rise to the identification of the full covariance process.

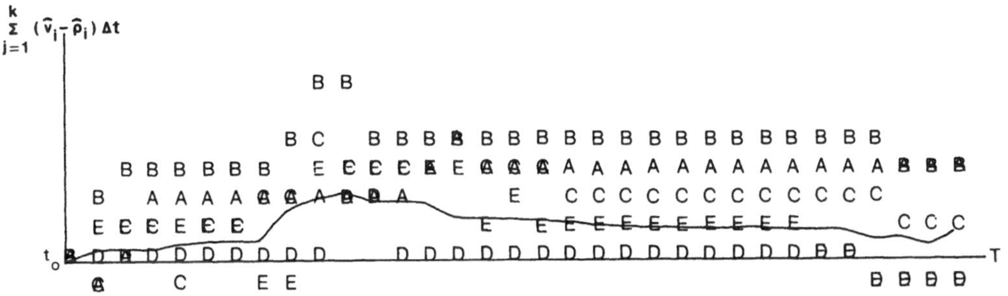

Fig. 1 ——— : Estimates of the sequence of vertical displacements for normal subjects.
Symbols: Realizations of Y(t) for some individual normal subjects.

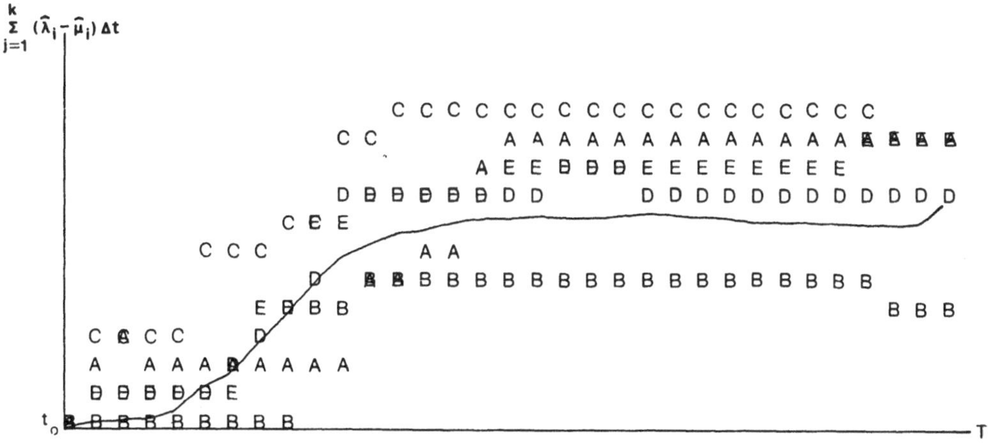

Fig. 2 ——— : Estimates of the sequence of horizontal displacements for normal subjects.
Symbols: Realizations of X(t) for some individual normal subjects.

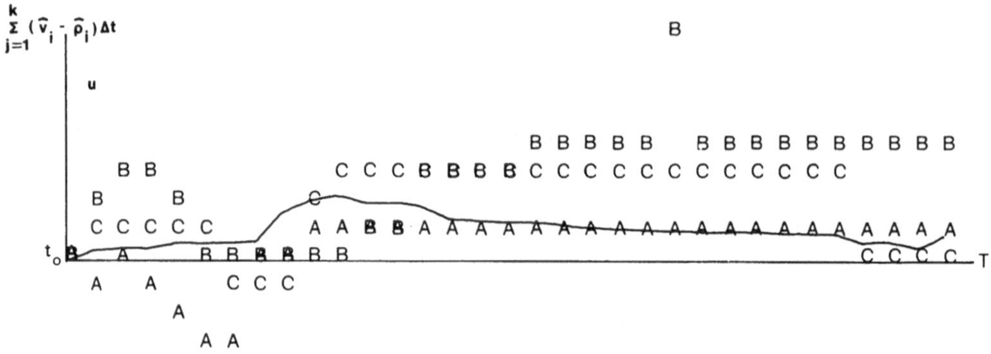

Fi g. 3 ——— : Estimates of the sequence of vertical displacements for normal subjects.
Symbols: Realizations of Y(t) for some individual patients (I.I.).

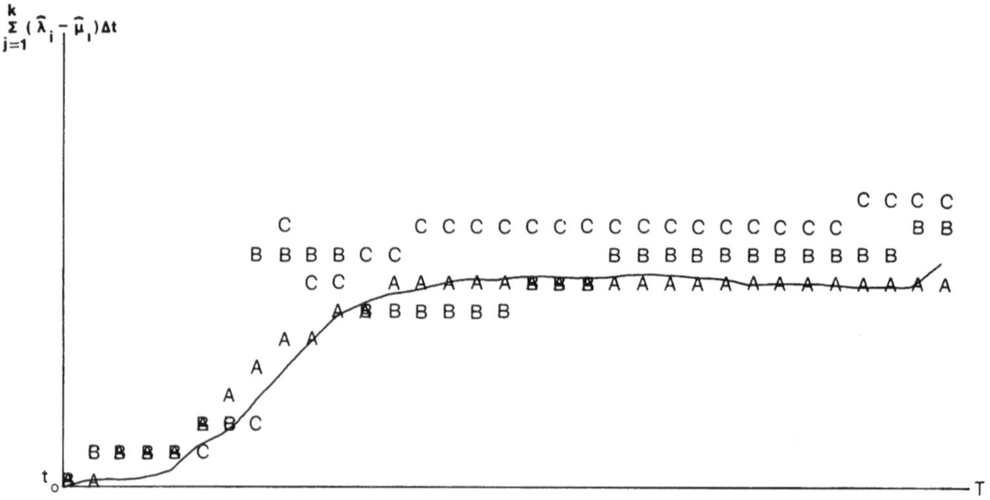

Fig. 4 ——— : Estimates of the sequence of horizontal displacements for normal subjects.
Symbols : Realizations of X(t) for some individual patients (I.I.).

Results

Fig. 1 - 2 show the estimated sequence of displacements of the potential minimum for the processes X(t), Y(t)respectively, with a few realizations from the normal sample.
The realizations exhibit a clear, common trend in normal subjects, whereas realizations from a sample of subjects affected by inferiori infarction (fig. 3 - 4) show an oscillatory trend, as compared with the estimated path.
The trajectories present a definite character only in the last part of QRS.

Conclusions

The mathematical model we developed provides a statistical tool for describing the trajectories of the potential minimum during the QRS interval in a group of normal subjects. The method can be applied to different groups of patients or different time intervals, with a view to quantifying the behaviour of the relevant variables in different heart conditions.
This procedure provides some insight into the physiological processes underlying the spatial distribution of heart potentials: moreover it may enable us to elicit a number of parameters that can be useful for multivariate discriminant analysis and automated diagnosis.

References

Basawa I. V. and Prakasa Rao B. L. S.: *Statistical inference for stochastic processes*, Academic Press, New York 1980.
Besozzi M., De Ambrogi L., Devizzi S., Taccardi B., Viganotti C.: *Analisi quantitativa di elettromappe cardiache normali*, Boll. Società Italiana di Cardiologia Vol. XVIII (1973).
Bharucha - Reid A. T.: *Elements of the theory of Markov processes and their applications*, Mc. Graw-Hill, New York 1960.
Calvi-Parisetti C.: *Analisi multivariata di un campione di elettromappe cardiache*, Istituto Applicazioni del Calcolo Quaderni IAC, Roma Serie III, n. 131 (1981).
Nanthi K.: *Statistical estimation for stochastic processes*, Queen's papers in Pure and Applied Mathematics No.62 Queen's University, Kingston, Ontario, Canada 1983.
Taccardi B., De Ambroggi L. and Viganotti C.: *Body-surface mapping of heart potential*, in: *The teoretical basis of electrocardiology*, ed. by C. V. Nelson and D. Geselowitz, Clarendon Press, Oxford, 1976.

* Istituto di Matematica, Università di Parma
** Istituto di Fisiologia Generale, Università di Parma

This research has been partially supported by M.P.I..

STATISTICAL ANALYSIS OF THE CODING CAPACITY OF COMPLEMENTARY DNA STRANDS

Cascino,A., Scarlato,V., Barni,N., Cipollaro,M.[**], Franzè,A., Macchiato,M.F.[*], Pierno,G. and Tramontano,A.

International Institute of Genetics and Biophysics, C.N.R.; [*]Institute of Physics and [**]III Servizio Analisi, 2nd Medical School, University of Naples, Naples, Italy.

ABSTRACT

A method that allows a statistical evaluation of the protein coding capacity of any sequenced DNA fragment is presented.

The method, applied to a limited number of protein coding gene sequences, allows the identification of some hypothetical protein genes which are coded by the the DNA strand complementary to the coding DNA strand. The finding that some proteins are indeed encoded on opposite DNA strands and that the two genes have complementary codon sequences, led us to investigate a much larger sample of DNA sequences, included in our DNA data bank.

The results of this exhaustive investigation demonstrate the existence of a new class of protein coding gene sequences that are coded by the complementary DNA strand.

INTRODUCTION

One of the more important consequences of the use of sequencing techniques is the discovery of the extreme complexity of genomes architecture. Overlapped genes, split genes and the finding that different products are made out of the same DNA sequence by using alternative splicing mechanisms are the most representative features arising from DNA sequencing data.

The availability of a large number of DNA sequences and their organization in computer data bank has also produced, for the first time in the field of molecular genetics, the possibility to analyse genotypic features under a variety of approaches. Taking advantage of these facilities we investigate the possibility that the DNA strand which is complementary to the DNA strand that codes for a known functional polypeptide is, as well, coding for a different product.

In previous papers (1,2,3) we performed a computer analysis first on 31 DNA sequences and then on 117 nuclear protein coding sequences. We were able to demonstrate that the two DNA strands can not be distinguished. In particular, we found that 50 protein coding sequences (longer than 100 sense codons) exist on the DNA strand complementary to the DNA sequence proved to code for 48 proteins. We called this new hypothetical class of genes "complementary inverted protein" or c.i.p. genes and we made the hypothesis that some of these c.i.p. genes might be translated into a functional product.

The finding that some proteins (4,5) are encoded on opposite DNA strands and that the two genes have complementary codon sequences, confirms our hypothesis.

In this paper we present a statistical evaluation of the more significant features of known and c.i.p. genes subdivided into three homogeneous pools: prokaryotic, eukaryotic and animal viruses genes.

METHODS

Statistical evaluation methods and criteria used to analyse DNA sequences have already been published (1). Fortran and PL/1 computer programs, implemented on an Univac 1100/80 system of the University of Naples, are available on tapes.

DNA sequences come mostly from CELIA, University of Bari DNA data bank. Some sequences were added by us directly.

All the 117 protein coding sequences under investigation have been checked to contain the 136 known translated sequences. Correct position of initiation and termination codons, exon positions of interrupted genes and other features of this gene pool are reported elsewhere (2,3). All immunoglobulin gene sequences were excluded from our pool since recombination at the level of DNA is involved in their expression.

The eukaryotic c.i.p. gene sequences found on the complementary DNA strands of known interrupted gene sequences, have been already investigated to locate their own introns (2).

RESULTS

a) reading frames length distribution

In order to analyse the coding capacity of both DNA strands, a Fortran computer algorithm has been implemented (1). The computer program analyses each

1

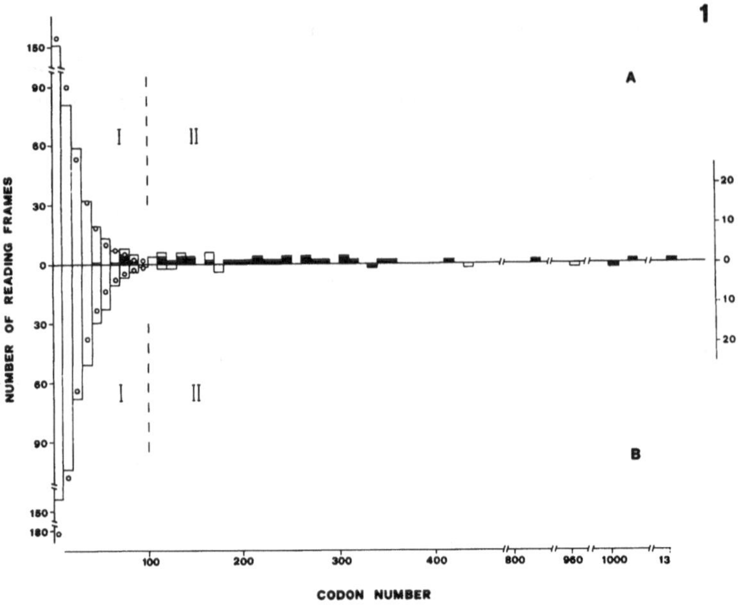
NUMBER OF READING FRAMES
CODON NUMBER
A
B
I
II
150
90
60
30
0
180
100
200
300
400
800
960
1000
13
20
10

2

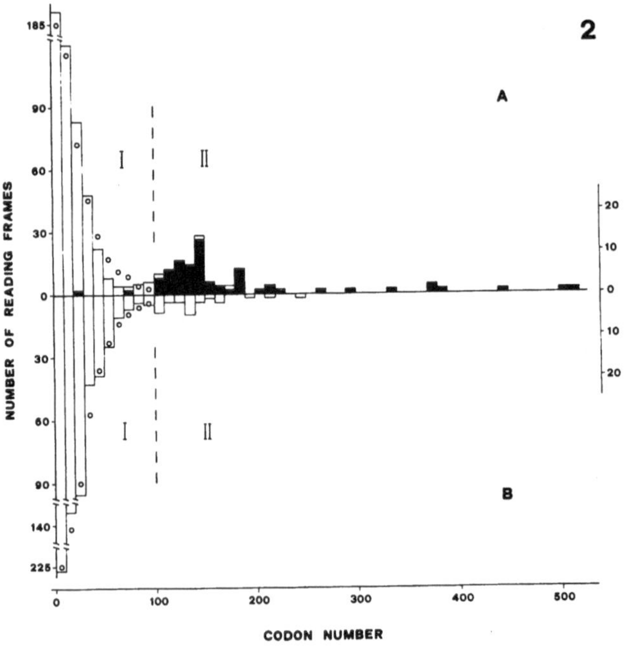
NUMBER OF READING FRAMES
CODON NUMBER
A
B
I
II
185
90
60
30
0
140
225
100
200
300
400
500
20
10

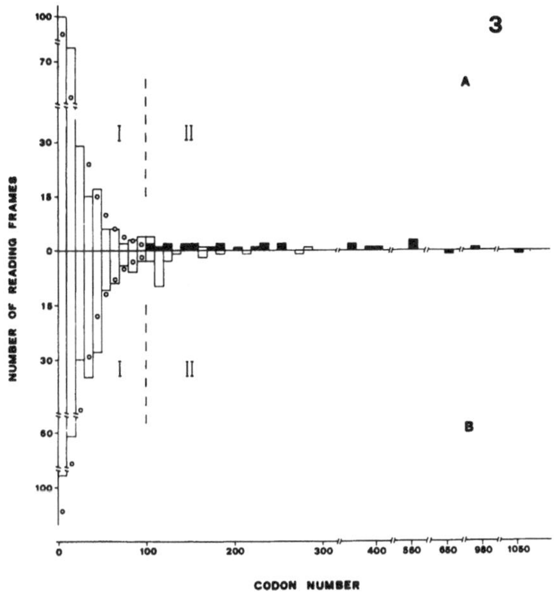

Figs. 1,2,3: reading frames distribution in the coding (panel A) and complementary (panel B) DNA strands. Prokaryotic gene pool (Fig. 1): 41 genes, 27 sequences, nucleotide number 46,630 (A = 12,520; T = 12,082; G = 11,291; C = 10,663; unidentified = 74); eukaryotic gene pool (Fig. 2): 67 genes, 68 sequences, nucleotide number 57,804 (A = 15,283; T = 13,939; G = 13,905; C = 14,048: unidentified = 629); animal virus gene pool (Fig. 3): 28 genes, 22 sequences, nucleotide number 34,146 (A = 8,574; T = 7,926; G = 8,938; C = 8,707; unidentified = 1). The 136 known genes are listed in Table I of Ref. (3). The 50 c.i.p. genes are listed in Table I of Ref. (2). Symbols: (▬) known genes; (▭) reading frames; (o) expected number of reading frames in each decade calculate according to equation 1).Region I refers to that part of distributions that contain reading frames shorter than 100 sense codons. Region II contains almost all known genes and the 50 c.i.p. genes. Left and right ordinates in Figs. 1 and 2 are for region I and II respectively.

sequence in all three possible reading frames and recognizes initiation and termination codons. When a reading frame is found, its length is computed starting from the first sense codon which follows AUG and terminating either to the first termination codon found in the same reading phase or to the end of the available sequence if no termination codon is found. The number of reading frames is then plotted as a function of length, for the two DNA strands, separately.

The result of this investigation are reported in Figures 1, 2 and 3 which show the reading frame length distributions of the three gene pools: Fig. 1 prokaryotes

(27 sequences); Fig. 2 eukaryotes (68 sequences); Fig. 3 animal viruses (22 sequences).

The expected distribution of reading frames is calculated as follows: once an initiation codon (AUG) is found in a sequence, if the appearance of one of the three termination codons (UGA, UAG, UAA) is a random event, the relative frequency q should be

$$q = q(U)\ q(A)\ \{2q(G) + q(A)\}$$

with q(N) being the frequency of the corresponding N base. Then the probability P(K) of finding a reading frame of K codons length is

$$P(K) = q\ (1 - q)^K \qquad 1)$$

The correspondent probability values have been plotted on the histograms shown in Figs. 1, 2 and 3 as open circles.

The hystograms show the existence of some reading frames longer than 100 sense codons on the coding DNA strand (overlapped genes) and of many reading frames longer than 100 sense codons on the complementary DNA strand (c.i.p. genes). Table I summarizes the comparison between the expected and the found figures of all reading frames present in region I and II for each gene pool.

		number of reading frames					
gene pool		prokaryotes		eukaryotes		animal viruses	
		E	F	E	F	E	F
region I	panel A	370	366	485	487	257	256
	panel B	446	443	605	591	297	284
region II	panel A	2	6	4	3	3	4
	panel B	3	6	6	21	4	17

Table I: Comparison between the expected (E) and the found (F) numbers of reading frames. Other symbols are as in Figs. 1, 2 and 3 legend.

The expected number of reading frames is calculated by summing the numbers obtained from equation 1) with the K value ranging between 1 and 99 or between 100 and infinite for region I and II respectively. The known genes, including the 5 translated c.i.p. genes (4,5), are obviously excluded from this evaluation.

The data show a very good agreement between expected and found numbers of reading frames in regions I on both DNA strands. Data referring to region II indicates that the finding of reading frames longer than 100 sense codons, overlapped to a known gene sequence (panel A), is consistent with a random event for eukaryotes and animal viruses gene pools. The finding of c.i.p. genes (panel B) is instead drastically in excess over the number predicted by chance alone, for all three gene pools.

b) codon usage pattern

In this section we analyse the preferential usage of codons of the c.i.p. genes. c.i.p. genes were pooled in the three classes, prokaryotes, eukaryotes and animal viruses, and compared to the pools of their counterposed known genes. The 5 translated complementary sequences (4,5) were excluded since we can not decide to which DNA strand these sequences belong.

This analysis, shown in Table II, was done by a computer program that sums and normalizes the codon numbers of each pool in four groups of codons, NNC, NNG, NNU and NNA.

These groups are not related to the amino acid-codon association. In this way we investigate a sequence coding feature almost independently from its correspondent average amino acid content.

gene pool	prokaryotes		eukaryotes		animal viruses	
codon group	known	c.i.p.	known	c.i.p.	known	c.i.p.
NNC	29.1	36.8	33.5	33.2	30.7	29.2
NNG	27.3	24.6	28.6	24.3	29.0	28.9
NNU	25.9	25.3	23.1	25.6	20.7	23.4
NNA	17.6	13.2	14.8	16.9	19.5	18.5
gene number	7	8	20	22	13	15
codon number	3,949	2,185	4,300	3,106	4,929	2,113

Table II: codon usage pattern in the known and c.i.p. gene pools.The discrepancies between the c.i.p. gene number presented in Table I with the number shown in this Table depends on the fact that many sequences contain known genes on both DNA strands.

The data show that there is a preferential usage of codons in all three pools of known genes, such that NNC is always the most represented group, NNA is the less aboundant group and the NNG group always predominates over the NNU group. Data referring to the c.i.p. gene pools indicate an analogous behaviour of the codon usage pattern for the NNC and NNA groups. NNG and NNU groups can not be distinguished in the prokariotic and eukariotic pools, while NNG is higher than NNU in the animal viruses pool.

No difference is observed when the average amino acid composition of the known genes translated sequence pools is compared to those of the c.i.p. genes (2).

DISCUSSION

On the basis of the statistical evaluation presented we deduce that some of the c.i.p. genes are candidated to be translated in functional products and that this new class of protein coding genes does not have significative differences when compared with the known counterposed gene class. We conclude that in the case of eukaryotic and animal viruses sequences the complementary translated sequence genetic contest is much more favoured than the overlapped one. In fact the number of overlapped genes found is close to that expected by chance alone, while the number of c.i.p. genes found is drastically in excess.

Finally the lack of formal genetic data on the existence of this new class of genes is explainable by assuming that the phenotype of a given c.i.p. gene is related to the phenotype of the correspondent known gene: i.e. the first regulates the expression of the latter.

AKNOWLEDGEMENTS We are deeply grateful to Pr. C. Saccone for the usage of the CELIA DNA data bank. We thank Drs C. Moscatelli, M. Candurro and L. Juliano for the computer programs implemented for our DNA data bank and for helpful discussion. Thanks are also due to Mrs. R. Vito and T. Vespa. This work was supported by the C.N.R. P.F. grant "Ingegneria Genetica e Basi Molecolari delle Malattie Ereditarie".

REEFERENCES

1. Cascino,A., Cipollaro,M., Guerrini,A.M., Spena,A. and Scarlato,V. (1981) Nucl. Acids Res. 9 (6), 1499-1518
2. Scarlato,V., Barni,N., Cipollaro,M., Franzè,A., Macchiato,M.F., Pierno,G., Tramontano,A. and Cascino,A. (1983) Nucl. Acids Res. (submitted)
3. Tramontano,A., Scarlato,V., Barni,N., Cipollaro,M., Franzè,A., Macchiato,M.F., and Cascino,A. (1983) Nucl. Acids Res. (submitted)
4. Rak,B., Lusky,M. and Hable,M. (1982) Nature 297, 124-159
5. Gingeras,T.R., Sciaky,D., Gelinas,R.E.,Bing-Dong,J., Yen,C.E., Kelly,P.A., Bullock,P.A., Parsons,B.L., O'Neill,K.E. and Roberts,R.J. (1982) J. Mol. Biol. 257 (22), 13475-13491.

MASS TRANSPORT ACROSS MULTIMEMBRANE SYSTEMS

A GENERAL ANALYSIS

F.C. Celentano and G. Monticelli

Dip. di Biologia and Dip. di Fisiologia e Biochimica Generali
Università degli Studi di Milano, Milano, Italy

The non-linear behaviour of a series array of two membranes has been first investigated by Kedem and Katchalsky (1) and by Patlak, Goldstein and Hoffman (2). This study has been subsequently resumed by several different Authors (3 - 11) and an extension to a series array of n membranes has been attempted by Ludwikõw (12), strictly following the approach by Kedem and Katchalsky.

Our generalization follows the suggestion by Patlak, Golstein and Hoffman (2), using the Kedem - Katchalsky linear "practical" equations for volumetric (J_v) and solute (J_s) flows in conditions where their intrinsic limitations vanish (10,11). These equations may be written in local form, for the i-th membrane in the array of fig. 1, as:

$$J_v = - L_p^{i\prime}\left(\frac{dp_i}{dx} - \sigma^i RT\phi \frac{dC_i}{dx}\right)$$

$$J_s = - \omega^{i\prime}\left[RT\phi \frac{dC_i}{dx} + (1 - \sigma^i)\, C_i(x)\, J_v\right] \qquad (1)$$

where J_v and J_s, as well as the driving pressure and concentration gradients, are assumed as positive in the x direction from membrane 1 towards membrane n, $L_p^{i\prime}$, σ^i and $\omega^{i\prime}$ are the filtration, reflection and solute permeability coefficients characterizing the local properties of each membrane and assumed as constant in the permeation direction, while $p_i(x)$ and $C_i(x)$ are the local hydrostatic pressure and solute concentration. R and T are, as usual, the gas constant and the absolute temperature and ϕ is the solute osmotic coefficient, also assumed as constant, needed to calculate the effective osmotic pressure of the permeating solute $\Pi = RT\phi C$.

The above equations are integrated along the x direction across each

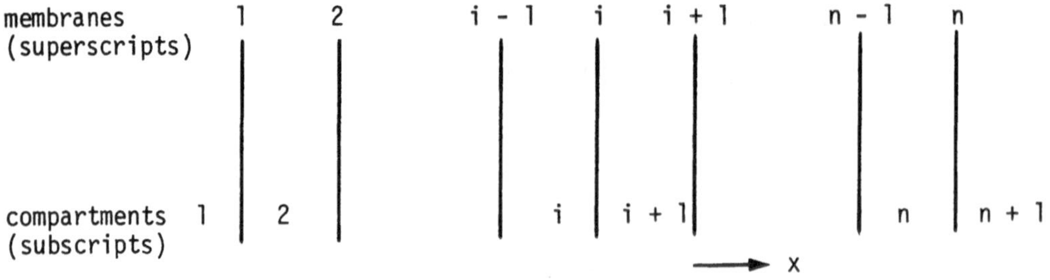

Fig. 1. Series array of n membranes. Superscripts indicate the membranes to which the system parameters are referred. Subscripts indicate the compartment number and the pertaining quantities.

membrane of the array with the boundary conditions $C_i(0) = C_i$, $C_i(d^i) = C_{i+1}$, $p_i(0) = p_i$, $p_i(d^i) = p_{i+1}$, where d^i is the thickness of the i-th membrane. One obtains, for membrane i:

$$C_{i+1} = C_i\, h^i + \frac{1 - h^i}{s^i}\,\frac{J_s}{J_v}$$

$$p_{i+1} = p_i - \frac{J_v}{L_p^i} + \sigma^i\, RT\phi\,(C_{i+1} - C_i) \qquad (2)$$

with $h^i = \exp(a^i J_v)$ and the positions $a^i = s^i/P^i$, $s^i = 1 - \sigma^i$, $L_p^i = L_p^{i\prime}/d^i$, $P^i = \omega^{i\prime} RT/d^i$.

In stationary state J_v and J_s are constant and time independent across the array and thus recursive substitutions of C_{i-1} into C_i and of p_{i-1} into p_i are possible, using the equations (2). One obtains:

$$C_{n+1} = C_1 \prod_{1}^{n}{}_i\, h^i + \frac{J_s}{J_v}\, A^n$$

$$p_{n+1} = p_1 - J_v\,\frac{1}{\Lambda} + RT\phi \sum_{1}^{n}{}_i\, C_i\,(s^i - s^{i-1}) \qquad (3)$$

where A^n is defined as

$$A^i = \prod_1^i{}_k h^k \sum_1^i{}_j \frac{1 - h^j}{s^j \prod_1^j{}_k h^k} \tag{4}$$

for i = n, and

$$\frac{1}{\Lambda} = \sum_1^n{}_i \frac{1}{L_p^i} \tag{5}$$

a relationship which generalizes the one previously obtained for two membranes (11) and would give the overall hydraulic permeability of the array according to the classical law by Darcy.

The two expressions (3) may be solved respectively for J_s and J_v, substituting C_i where needed and defining $\Delta C = C_{n+1} - C_1$ and $\Delta p = p_{n+1} - p_1$:

$$J_s = \frac{J_v}{A^n} \; [\Delta C + (1 - \prod_1^n{}_i h^i)\, C_1] \tag{6}$$

$$J_v = - \Lambda \left\{ \Delta p - \frac{RT\phi\Delta C}{A^n} \sum_1^n{}_i (s^{i+1} - s^i)\, A^i \right.$$
$$\left. - \frac{RT\phi C_1}{A^n} \sum_1^n{}_i [\, A^n \prod_1^i{}_j h^j + A^i \,(1 - \prod_1^n{}_j h^j)]\,(s^{i+1} - s^i) \right\} \tag{7}$$

remembering that $\sigma^{n+1} = 0$, as there is no selective membrane in the compartment n + 1, and thus $s^{n+1} = 1$.

The equations (6) and (7) are formally identical to the ones already published (10,11) for a system of two membranes. In particular, they show a non-linear dependence on C_1, that is the solute concentration above which the concentration difference ΔC is built up. The term in C_1 is added to the ones containing the driving forces ΔC and Δp and vanishes for vanishing C_1. It appears to be a correction term for the deviation of the behaviour of a solution from that of pure solvent. From the equations (6) and (7) the parameters describing the overall transport properties of the array can be obtained, by means of a proper redefinition of the corresponding parameters in the linear theory.

The solute permeability is defined in the linear theory as $\omega=(J_s/\Delta\Pi)_{J_v=0}$ A generalization to the n membranes array is possible, using the equation (6), as

$$\omega = \lim_{J_v \to 0} \frac{J_s}{\Delta\Pi} = \frac{1}{RT\phi \sum_1^n {}_i \, 1/P^i} \qquad (8)$$

which may be rearranged introducing the definition of P^i,yielding

$$\frac{1}{\omega} = \phi \sum_1^n {}_i \frac{d^i}{\omega^{i\prime}} \qquad (8')$$

This result for the overall system permeability ω generalizes the one previously obtained for two membranes (10) and is formally and conceptually analogous to the equation (5). In effect also $L_p^i = L_p^{i\prime}/d^i$, and in both cases we are involved with, in the electrical network jargon, resistances in series, that is conductances in parallel.

The volume flow equation (7) allows the determination of some other important transport parameters. The filtration (hydraulic conductivity) coefficient L_p may be defined as $(\partial\Delta p/\partial J_v)_{\Delta\Pi=0}$ and,being the (7) implicit in respect to J_v, its inverse may be computed

$$\frac{1}{L_p} = -\frac{1}{\Lambda} + RT\phi C_1 \sum_1^n {}_i \left[\frac{\partial}{\partial J_v} \prod_1^n {}_j h^j + \frac{\partial}{\partial J_v} \frac{A^i}{A^n}\left(1 - \prod_1^i {}_i h^j\right)\right] \cdot$$

$$\cdot (s^{i+1} - s^i) \qquad (9)$$

The non-linear term in C_1 confirms that the law by Darcy is not followed by a solution, but it is a limiting law for pure solvent transport, $C_1=0$. About this point it is interesting to note that the difference $J_v - \Lambda\Delta p$ at $\Delta\Pi = 0$ becomes a constant for $J_v \to \pm\infty$. This means that the slope of the volume flow curve, in presence of a solute, approaches the one predicted by Darcy, Λ, only for high flows, a behaviour which has been experimentally observed (14).

The definition of the osmotic flow coefficient L_{pd} strictly follows the procedure used for defining L_p:

$$\frac{1}{L_{pd}} = \left(\frac{\partial \Delta\Pi}{\partial J_v}\right)_{\Delta p=0} = \frac{1}{\Lambda}\frac{\partial}{\partial J_v}\frac{A^n}{\sum_1^n {}_i A^i (s^{i+1} - s^i)} - \tag{10}$$

$$- RT\phi C_1 \frac{\partial}{\partial J_v} \frac{\sum_1^n {}_i [A^n \prod_1^i {}_j h^j + A^i(1 - \prod_1^n {}_j h^j)] (s^{i+1}-s^i)}{\sum_1^n {}_i A^i (s^{i+1} - s^i)}$$

and again one obtains a non-linear term in C_1. The limiting values of the (9) and (10) for vanishing volume flow are

$$\frac{1}{L_p} = - \frac{1}{\Lambda} + RT\phi C_1 \sum_1^n {}_i \left[\sum_1^i {}_j a^j - \frac{\sum_1^i {}_j 1/P^i \sum_1^n {}_i a^i}{\sum_1^n {}_i 1/P^i}\right](s^{i+1} - s^i) \tag{10}$$

$$\frac{1}{L_{pd}} = \frac{\sum_1^n {}_i 1/P^i}{\sum_1^n {}_i \sigma^i/P^i}\left\{\frac{1}{\Lambda} - RT\phi C_1 \sum_1^n {}_i \left[\sum_1^i {}_j a^j - \frac{\sum_1^i {}_j 1/P^j \sum_1^n {}_i a^i}{\sum_1^n {}_j 1/P^j}\right]\cdot\right.$$

$$\cdot(s^{i+1} - s^i) \tag{11}$$

and their ratio, which should be the overall reflection coefficient of the array, turns out to be

$$- \frac{L_{pd}}{L_p} = \frac{\sum_1^n {}_i \sigma^i/P^i}{\sum_1^n {}_i 1/P^i} \tag{12}$$

In effect, the same ratio is obtained as limit for vanishing volume flow of the $\Delta p/\Delta\Pi$ ratio, which confirms that the (12) represents the system overall reflection coefficient, a weighed mean of the reflection coefficients characterizing the single membranes in the array.

It may be concluded that the equations describing the non-linear behaviour.

of a series array of n membranes are straightforward generalizations of the ones already obtained for two membranes (10,11) and that the general behaviour of the n membranes array remains qualitatively similar.

In particular it is confirmed that the law by Darcy requires a correction in the presence of a solute. This is a consequence of the solute accumulation in the inner compartments, and only at high volume flows the solution is carried across the array as a whole, with an increase of the slope of the J_v curve up to the limiting value Λ (11).

Increasing the number of the membranes, the number of the system parameters increases proportionally, while the number of the observable quantities and of the relationships between the latter and the system parameters remains constant. In effect, expecially in the presence of solute active transport, a remarkable redundancy of relationships has been found for the two membranes array (11), but this is certainly not the case in the present situation. It may be noticed, however, that when unstirred layers of known thickness are present, and also the solute diffusion coefficient D is known, $P^i = D/RTd^i$ (10), which allows to decrease the number of the unknowns when the array, as it happens in biological systems, can not be dismantled. At present it seems difficult to determine the number of membranes needed to simulate the flow-force behaviour of complex barriers, like the epithelia. The optimal degree of complexity of the system is likely to depend on the particular problem being studied and has to be determined by simulation and fitting of the esperimental data.

REFERENCES

1. Kedem o, Katchalsky A: Permeability of composite membranes. Part 3. Series arrays of elements. Trans Faraday Soc 42: 1491-1953, 1963

2. Patlak CS, Goldstein DA, Hoffman JF: The flow of solute and solvent across a two membrane system. J Theor Biol 5: 426-442, 1963

3. Tomicki B, Ludwikŏw F: Further properties of the two membrane system. Pressure in the middle compartment. Studia Biophys 26: 21-30, 1971

4. Celentano FC, Monticelli G: Non linear force-flow relationships in biological membranes. In Gomulkiewicz J and Tomicki B Eds. Biophy-

sics of membrane transport, Part 2, Wroclaw, Agricultural Academy, 1975, pp 1-23

5. Przestalsky S, Kargol M: Composite membranes. In Kuczera J and Przestalsky S Eds. Biophysics of Membrane Transport, Wroclaw, Agricultural Academy 1976, pp 53-69

6. Mikulecki DC: A simple network thermodynamic method for series - parallel coupled flows II. The non-linear theory with applications to coupled solute and volume flows through epithelial membranes. J Theor Biol 69: 511-541, 1977

7. Mikulecki DC, Thomas SR: A simple network thermodynamic method for series - parallel coupled flows III. Application to coupled solute and volume flows through epithelial membranes. J Theor Biol 73: 697-710, 1978

8. Mikulecki DC, Przestalski S, Kargol M: Nonlinear network analysis of graviosmotic flow in series membrane systems. In Kuczera J, Grigorczyk C, Przestalski S Eds. Biophysics of membrane transport, Wroclaw, Agricultural Academy, 1981, pp 147-167

9. Monticelli G, Celentano FC: Transport coefficients in a non linear series membranes array. In Borsellino A, Omodeo P, Strom R, Vecli A, Wanke E Eds. Developments in biophysical research. New York, Plenum, 1981, pp 157-164

10. Monticelli G, Celentano FC: Considerations on different thermodynamic models for mass transport across membranes. J Membr Sci, in press, 1983

11. Monticelli G, Celentano F: Further properties of the two-membrane model. Bull Math Biol, in press, 1983

12. Ludwików F: Transport substancji w ukladach wielu membran. Doctor Thesis, Wroclaw, Agricultural Academy, 1973

13. Kedem O, Katchalsky A: Thermodynamic analysis of the permeability of biological membranes to non-electrolytes. Biochim Biophys Acta 27, 229-246, 1958

14. Brodsky WA, Schilb TP: Osmotic properties of the isolated turtle bladder. Am J Physiol 208, 46-57, 1965

TIME HIERARCHY IN OSCILLATING METABOLIC SYSTEMS

I. Dvořák, L. Kubínová
Center of Biomathematics, Institute of Physiology

J. Šiška
General Computing Centre
Czechoslovak Academy of Sciences, Prague

TIME HIERARCHY IN METABOLIC SYSTEMS

Metabolic systems are composed of many biochemical reactions and transport processes with very different rates. Their dynamic interaction produces a temporal organization whose salient feature is the existence of a distinct time hierarchy. Only a subset of the dynamic variables of the system moves with a velocity comparable to the characteristic time scale of the whole metabolic system, while others move very quickly and produce a dynamic "rapid substructure". The mathematical description of this situation leads to differential equations with small parameters as multipliers of the derivatives of fast components (cf. Reich and Sel'kov 1981).

In the simplest case, when the time hierarchy has two levels only, the system of equations can be expressed in the following form

$$\frac{dx}{dt} = f(x,y), \qquad \varepsilon \frac{dy}{dt} = g(x,y). \tag{1}$$

Here $x \in R^m$, $y \in R^n$, f and g are smooth real vector functions and $\varepsilon > 0$ is a small parameter. In order to solve this system we specify initial conditions $x(0) = x_0$, $y(0) = y_0$.

Since f and g are usually nonlinear functions, the solution of (1) is a very complex task, especially when m, n > 1. When ε is sufficiently small it is reasonable to approximate this solution by the solution of the so called degenerate system

$$\frac{d\bar{x}}{dt} = f(\bar{x},\bar{y}), \qquad 0 = g(\bar{x},\bar{y}), \tag{2}$$

which is obtained from system (1) when $\varepsilon \to 0$. In the biochemical literature this is often called "quasi-steady approximation". The mathematical justification of this approximation is a theorem by Tikhonov (Tikhonov 1952).

A more detailed explanation can be found in our related paper (Šiška et al. in press). Closeness of the approximation of the solution $x_\varepsilon(t)$, $y_\varepsilon(t)$, for given ε, by the solution $\bar{x}(t)$, $\bar{y}(t)$ depends on the rate of convergence. Practical experience (cf. Dvořák et al. 1983) shows that for the good approximation ε, ($0 < \varepsilon << 1$), should be of the order of magnitude 10^{-2} or less.

OSCILLATING METABOLIC SYSTEMS

The steady states of metabolic systems may be unstable (cf. Reich and Sel'kov 1981). As a result all variables may undergo continuous oscillations of more or less irregular, nonlinear character. As an example let us consider the system depicted in Fig. 1, which may be encountered in metabolic pathways with product activation of a key reaction. Different variants of this system, namely kinetic induction and product activation, were studied in detail by Sel'kov and his co-workers (see Sel'kov et al. 1966, Sel'kov and Dynnik 1978, Sel'kov and Nazarenko 1980, 1981).

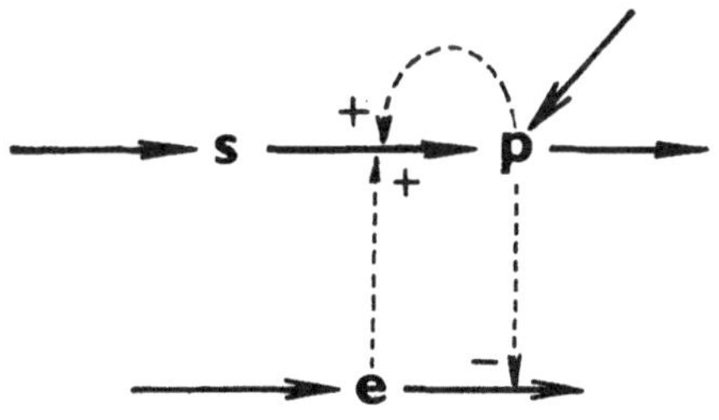

FIGURE 1. Scheme of product-activated metabolic system with slow enzyme turnover

The enzymic conversion of substrate s to product p is the central reaction of the system. The substrate as well as the product are supplied to the system at constant rate. The product can freely flow out of the system (the rate of outflow of the product is proportional to its concentration p), while the substrate cannot. Moreover, the product activates the rate of its own formation. The enzyme is supplied into the system with constant rate (e.g. by cellular biosynthesis of proteins). Its outflow from the system is inhibited by product p. We may imagine this outflow as a degradation of free enzyme molecules while those bound to the product are protected against degradation. The dynamics of such a system can be described by the following equations

$$\frac{de}{dt} = 1 - e\left(1 - \frac{p^2}{p^2 + 1}\right) ,$$

$$\varepsilon \frac{ds}{dt} = v_s - e\frac{sp^2}{p^2 + 1} , \qquad (3)$$

$$\varepsilon \frac{dp}{dt} = e \frac{sp^2}{p^2 + 1} - p + p_o .$$

Here v_s and p_o are normalized input flows of substrate and product, respectively. The input flow of the enzyme is normalized to unity. The coefficient ε is given by the ratio of characteristic times of substrate and enzyme turnover. As this is usually very small, system (3) may be decomposed into two subsystems: a slow subsystem, represented by the first equation, describing the changes of enzyme concentration, and a fast subsystem, described by remaining two equations, representing the motion of the normalized concentrations of substrate and product. Linear stability analysis of the fast subsystem on the assumption that e is constant shows that there exists a large area of values e and v_s, where the steady state of the fast subsystem is unstable.

As the diverging trajectories cannot leave a certain bounded region, a limit cycle appears in the dynamics of the fast subsystem.

System exhibiting a limit cycle in its dynamics does not fulfil the assumptions of the Tikhonov's theorem. A special study is therefore necessary for analysing the time hierarchy of such systems.

TIME HIERARCHY IN SYSTEMS WITH GENERAL ATTRACTORS

Let us turn our attention to those general systems (1) that are examplified by (3), i.e. to the systems whose trajectories of their fast subsystems are attracted by limit cycles or by more complicated attractors for any fixed value of slow variables. We shall suppose that for any fixed value of x the dynamic system generated by the fast equation

$$\varepsilon \frac{dy}{dt} = g(x,y)$$

obeys the A-axiom and that its trajectories approach a hyperbolic C^2-attractor A_x. This includes nearly all of the systems met with in the mathematical modelling of metabolism. The example studied above also belongs to them. A theorem formulated and proven by us recently makes it possible to find an analogon of the quasi-steady approximation for these systems.

The essence of the theorem is that under appropriate conditions the solution of system (1), as specified above, may be for $\varepsilon \to 0$ arbitrarily closely approximated by the solution $\overline{x}(t)$ of the equation

$$\frac{d\bar{x}}{dt} = \int_{A_x} f(\bar{x},y)\,d\mu_x(y), \tag{4}$$

with an initial condition $\bar{x}(0) = x_o$. The measures $\mu_x(y)$ in formula (4) are the Bowen-Ruelle measures on the attractors A_x. Although the construction of the Bowen-Ruelle measure for a general attractor is complicated, it is rather simple in the case of a limit cycle C_x. It is given by the formula

$$\mu(A) = \frac{1}{P_x} \int_0^{P_x} \chi_A(\phi_t(y))\ dt, \tag{5}$$

where P_x is the period (depending on the value of x), $\phi_t(y)$ is a flow on the limit cycle C_x, $y \in C_x$, and χ_A is a characteristic function of the set A.

The theorem follows from two deep theorems. From the theorem of Bogolyubov and Mitropol'ski (e.g. Hale 1969) it follows that for $\varepsilon << 1$ the system may be approximated by an averaged system where averaging is made over the trajectories of the fast subsystem. The theorem of Bowen-Ruelle (see Bowen and Ruelle 1975) specifies the necessary conditions for the averages to be independent of the initial conditions of these trajectories. More detailed explanation of the theorem can be found in (Šiška et al. in press). The detailed proof of the theorem will be submitted to a specialized journal.

Systems of differential equations modelling real biological situations may not meet the assumptions of the theorem because the support of the mapping (f,g) may not be compact. As a rule, the considered model is valid only for values of variables from a bounded region, so we can multiply the functions f and g by a suitably chosen bump function λ_a^b (see Hirsch 1976, Šiška et al. (in press) for details). The adapted system describes the same model as the original one in the area of the radius a and moreover has a compact support.

AVERAGING METHOD FOR A RAPIDLY OSCILLATING SYSTEM

If a proper bump function is introduced into system (3) (from the preceding paragraph) the adapted system obeys the assumptions of the above theorem and an approximation $\bar{e}$ of the slow variable e can be obtained. Using formula (5) for the Bowen-Ruelle measure of a limit cycle we have the following equation for $\bar{e}$

$$\frac{d\bar{e}}{dt} = 1 - \bar{e}\left(1 - \frac{1}{P_e} \int_0^{P_e} \frac{p_e^2(t)}{p_e^2(t) + 1}\ dt\right). \tag{6}$$

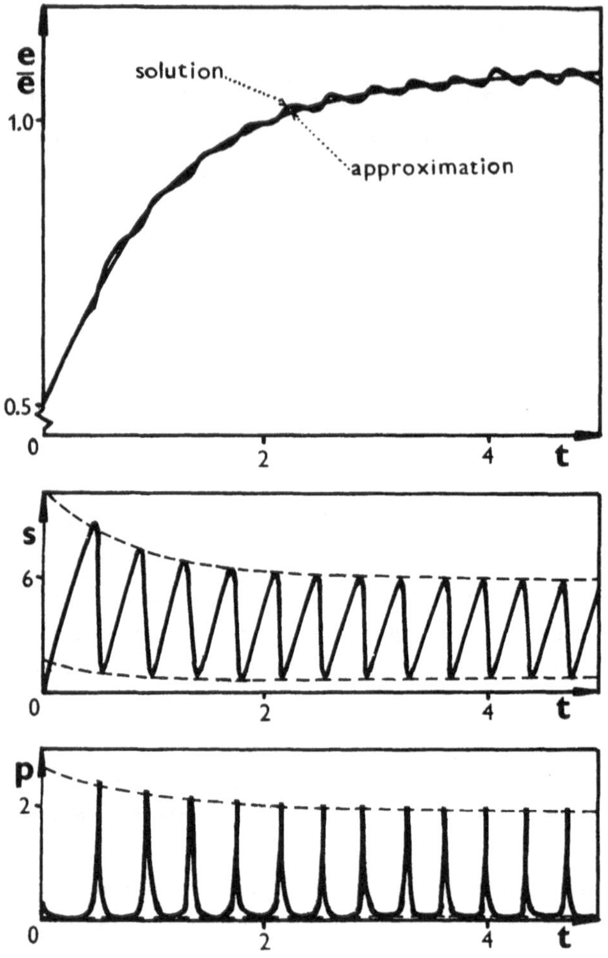

FIGURE 2. Comparison of the averaged solution of system (3), obtained by application of the theorem, with the exact solution, obtained by numerical integration of the whole system. Dashed lines - limits of oscillations of fast variables. Values of the parameters: $v_s = 0.2$, $p_o = 0.05$, $\varepsilon = 0.01$

The integral is over the limit cycle C_e, which is parametrized by the time t. P_e is the period of the cycle C_e for a given value of e. The trajectory $p_e(t)$, $p_e(t) = p_e(t,s_o,p_o)$, $(s_o,p_o) \in C_e$, is the component of the solution of the fast subsystem of system (3) for a fixed value of e.

The solution of equation (6) (calculated by a simple Runge-Kutta method) was compared with the numerical solution of system (3) (calculated by the semi-implicit Michelsen method). The results are presented in Fig. 2. During numerical calculations of formula (6) the extremal values of oscillating fast variables s and p were calculated as well, so they are also depicted in Fig. 2.

Our comparison, together with other numerical studies not presented here, shows that if ε is of the order of 10^{-2} then the difference between both solutions is less then 3 % of the value of the slow variable. Since the accuracy of approximation increases with decreasing value of ε the error of approximation remains below this level. Having in mind the usual inaccuracy of biological measurements we can conclude that the proposed approximation method is reasonably accurate for analysing dynamic models of metabolic systems.

ACKNOWLEDGEMENT

We are grateful to Dr. Petr Karen (Center of Biomathematics, Institute of Physiology, Prague) for help with numerical computations and to Prof. J. G. Reich (Zentralinstitut für Molekularbiologie, Berlin-Buch) for critical revision of the manuscript.

REFERENCES

Bowen, R., Ruelle, D. 1975. The ergodic theory of axiom A flows, *Invent. Math.* 29, 181-202

Dvořák, I., Dainty, J., Janáček, K. 1983. Analysis of time hierarchy in plant cell volume changes, *J. theor. Biol.*, 101, 87-112

Hale, J.K. 1969. *Ordinary Differential Equations*, Wiley-Interscience, New York

Hirsch, M. W. 1976. *Differential Topology*, Springer-Verlag, New York

Reich, J. G., Sel'kov, E. E. 1981. *Energy Metabolism of the Cell. A Theoretical Treatise*, Acad. Press, London

Sel'kov, E. E., Filanovskaya, A. N., Demchenko, L. J. 1966. On the question of kinetic induction (in Russian), in: *Modelling in Biology and Medicine*, 2, 126-141, Naukova Dumka Publishers, Kiev

Sel'kov, E. E., Nazarenko, V. G. 1980. Analysis of a simple open biochemical reaction $\rightarrow S \xrightarrow{E} P \rightarrow$ interacting with the enzyme producing system (in Russian), *Biofizika*, 25, 1006-1011

Sel'kov, E. E., Nazarenko, V. G. 1981. Oscillations and resonance phenomena in a simple enzymic reaction $\rightarrow S \xrightarrow{E} P \rightarrow$ interacting with the enzyme-producing system (in Russian), *Biofizika*, 26, 17-21

Sel'kov, E. E., Dynnik, C. N. 1978. Comparative theoretic analysis of an open reaction $\rightarrow S \xrightarrow{E(R,T)} P \rightarrow$ in which the oligomeric enzyme E(R,T) is isosterically or allosterically activated by the product S_2 (in Russian), *Molec. Biol.*, 12, 1122-1137

Šiška, J., Kubínová, L., Dvořák, I. (in press). Time hierarchy in systems with general attractors, *Proc. Fourth International Conference on Mathematical Modelling*, Pergamon Press, Oxford

Tikhonov, A. N. 1952. Systems of differential equations with small parameters of the derivates (in Russian), *Math. Sb.*, 31, 575-586

MODELING OF THE INTERACTION BETWEEN LIGANDS AND CELL MEMBRANE RECEPTORS IN THE PRESENCE OF EXOGENOUS ELECTRIC FIELDS (*)

M. Grattarola, F. Caratozzolo, R. Viviani and A. Chiabrera
Dipartimento di Ingegneria Biofisica ed Elettronica, Via All'Opera Pia 11a, University of Genoa, 16145 Genoa, Italy

(*) Work supported by the National Research Council (CNR) of Italy

Introduction

Several ligands exist that strongly modify complex cellular functions by means of reversible interactions with cell membrane receptors, which undergo a reorganization of their surface distribution. As typical "in vitro" examples, we recall the action of the nerve growth factor on PC12 cells (1), the action of dimeric IgE on basophils (2), the action of lectins on lymphocytes (3, 4) and the action of parathyroid hormone (PTH) on osteoblast-like cells (5). In the last three cases (as well as in many others), the external ligands have no effect if Ca^{++} is not present in the medium. Moreover, it has recently been shown that a lectin (phytohaemagglutinin (PHA)) induces a rise in free calcium inside lymphocytes a few seconds after it has been added to the culture (6). In accordance with these results, it has been suggested (4, 5) that, as a consequence of the ligand-receptor interactions, changes both in the Ca^{++} binding at the cell surface and in the Ca^{++} fluxes through the cell membrane can occur, that are mainly due to an increase in the clustering of receptors. In recent years, it has been shown that similar phenomena can also be caused by weak exogenous electric fields (7, 8) which can interfere with the action of some ligands too (5, 9).
In order to gain a deeper understanding of such experimental evidence, a quantitative physico-mathematical evaluation of the interplay among receptors, ligands and exogenous electric signals is required.

Kinetics of the interaction between ligands and membrane receptors

We shall adopt a simplified model of the ligand-receptor interaction, considering single receptors and couples of receptors only. The overall picture can be outlined as follows: N structureless objects (receptors) move in a diffusional fashion along the viscous plane formed by the membrane's double layer of lipids. When an object, in approaching another one, exceeds a minimal distance R_E, the couple forms an "encounter" (E), which has physico-chemical properties different from those of single objects. Encounters N_E are reversible entities. Other objects (ligands), which diffuse freely in the outer space near the receptor plane, can interact reversibly with encounters, forming N_A "aggregates" (A) when they exceed a critical distance R_A. The main difference between E and A lies in the mean lifetime spent by receptors near each other. This is longer for A than for E. A detailed discussion of the biological consequences involved is beyond the scope of this paper.

The above schematization yields the following two relationships (10):

$$d(N_E + N_A)/dt = d_+^m (N - 2N_A - 2N_E)^2 - d_-^m N_E$$

$$d(N_A)/dt = - r_- N_A + r_+ C_L N_E$$

where d_+^m and d_-^m are, respectively, the association and dissociation rate constants of encounters N_E ; r_+ and r_- are, respectively, the association and dissociation rate constants of aggregates N_A ; and C_L is the concentration of ligands.

Coupling between ligands and membrane receptors

According to the point of view adopted in this paper, variations in the number of clusters (couples) of membrane receptors do not necessarily result in a biological response, which, on the contrary, is strongly influenced by any variation in the mean lifetime of clusters. The underlying working hypothesis is that couples (and greater clusters) can become effective channels for Ca^{++} ions only if their mean lifetime exceeds some critical duration t_D, which will not be reached without the presence of stabilizing ligands (11). A similar hypothesis can be found in (12). As a consequence of this approach, in the following we shall not be concerned with variations in the number of couples of receptors, but rather with the physico-mathematical characterization of the rate constants of the aggregation process, and with its relationship with cluster lifetimes and experimental observables.

At the microscopic level, the inverse of the rate constant r_- is the time necessary for a ligand and an encounter (which form a couple) to move away at a random walk, travelling a mean square distance equal to R_A^2. We assume that the encounter protrudes outside the cell membrane by a greater length than R_A so that the ligand moves away at a three-dimensional random walk, whereas the encounter random walk takes place on the cell surface. Accordingly, we can write:

$$r_- = (3\, D_L^* + 2\, D_m^*)/R_A^2$$

where the diffusion constants D_L^* and D_m^* are the local values for the ligand and the encounter, respectively, when their interaction occurs, their distance being shorter than R_A. In order to relate these values to the bulk (i. e., no interactions) diffusion constants D_L and D_m, an explicit expression for the coupling is needed. We shall only consider here, in some detail, the limit case of a linear coupling between a ligand and an encounter. Under this simplifying assumption, by neglecting the acceleration term and considering the one-dimensional case, we obtain two Langevin equations coupled by an elastic force:

$$\beta_L \, dx_L/dt = (\lambda/m_L)\,(x_L - x_E) + n_L(t)$$

$$\beta_E \, dx_E/dt = (\lambda/m_E)\,(x_E - x_L) + n_E(t)$$

where $\beta_L = 6\pi a_L \eta / m_L$; $\beta_E = 6\pi a_E \eta_m / m_E$, η being the viscosity of the medium surrounding the cell, η_m the viscosity of the lipid bilayer, a_L, a_E, m_L, m_E the radii and the masses of the ligand and the encounter, respectively; λ is the coupling constant and $n_L(t)$, $n_E(t)$ the white noises acting on the ligand and the encounter, respectively.

Following the approach previously described by Chandrasekar (13), it can be shown that, within the limit of large values of t, the displacements of the encounter and of the ligand in the presence of this coupling are given by:

$$<(x_E - x_E^\circ)^2> \cong <(x_L - x_L^\circ)^2> \cong 2\, D_E^*\, t \cong 2\, D_L^*\, t \cong$$

$$[m_E \beta_E m_L \beta_L / (m_E \beta_E + m_L \beta_L)]^2 \; [\sigma_L^2/(m_E^2 \beta_L^2 \beta_E^2) + \sigma_E^2/(m_L^2 \beta_L^2 \beta_E^2)]\; t$$

where σ_E and σ_L are the variances of n_E and n_L, respectively. Due to the long-range action of the coupling force used in our model, the result is independent of λ. In the absence of any coupling, the free, independent Brownian motions of the encounter and the ligand are governed by the well-known relationship (13):

$$2\, D_E\, t \cong (\sigma_E^2/\beta_E^2)\, t \quad ; \quad 2\, D_L\, t \cong (\sigma_L^2/\beta_L^2)\, t$$

So that we can relate the diffusion coefficients in the presence of such a strong coupling to their free values (for simplicity we assume $m_E = m_L$):

$$D_E^* = D_L^* = D^* = (D_L \beta_L^2 + D_E \beta_E^2)/(\beta_E + \beta_L)^2$$

As a test of consistency of this relationship, if we assume, in accordance with the experimental data, $D_L^* = D_E^* \cong 10^{-11}\ cm^2 s^{-1}$; $D_E \cong 10^{-11} cm^2 s^{-1}$ and $D_L \cong 10^{-7}\ cm^2 s^{-1}$, then we find:

$$\beta_L / \beta_E \cong 10^{-3},$$

which is in reasonable agreement with experimental estimates (14). This result can be regarded as a starting point for more realistic models of weaker couplings; besides, it can justify the simplified assumptions which are utilized in the next section.

Influence of an exogenous electric field on the parameter r_-

We are now in a position to evaluate the effect of an electric field on the values of the rate constant r_-. In order to further simplify this problem we shall only consider the case where a ligand and an encounter are very near each other. Both diffusion constants can be approximated by the limit value D^*, which, from a phenomenological viewpoint, takes into account the effects of the coupling force described in the previous section. On the other hand, we let both L and E to diffuse freely in order to take somehow into account the dropping to zero of the coupling force above some critical mutual distance. In this case, we consider the two vectorial independent equations (15):

$$d^2 x_{j,i}\ /dt^2 + \beta_j\ dx_{j,i}\ /dt = \gamma_{j,i}\ \ E(t-\tau) + n_j(t) \qquad (i=1,\ 2,\ 3\ ;\ \ j=L,\ E)$$

where $E(t) = \sum_{-\infty}^{+\infty}{}_n\ E_n \exp(jn\omega_o t)$ is the exogenous electromagnetic field and γ is the sum of the electric field drift force plus the electric field-induced drag force (16).
By integrating separately the equations for the ligand and the encounter, and by considering the time $1/r_-$ taken by the ligand to reach the critical distance R_A (in the mean square sense) from its encounter partner, we can finally find (15):

$$R_A^2 \cong (5\ D^*)\ (1/r_-) + 2\ \{\sum_1^3{}_i\ [(\gamma_{L,i}/\beta_L) - (\gamma_{E,i}/\beta_E)]^2\} \cdot$$
$$\{(E_o^2/r_-^2) + 4\ \sum_1^{+\infty}{}_n\ [|E_n|/(n\omega_o)]^2\,[1 - \cos(n\omega_o/r_-)]\}$$

where $\gamma_{E,3}=0$.
Letting $r_-^o = 5D^*/R_A^2$ be the zero-field value of r_-, and defining

$$K = \sum_{1}^{3} {}_i \, [(\gamma_{L,i}/\beta_L) - (\gamma_{E,i}/\beta_E)]^2/R_A^2$$

the previous equation can be rewritten in implicit form, as follows:

$$r_- = r_-^o + (K/r_-) \{E_o^2 + 4 \sum_{1}^{+\infty} {}_n [|E_n|/(n\omega_o/r_-)]^2 [1 - \cos(n\omega_o/r_-)]\}$$

which reduces to

$$r_- \cong r_-^o + (K/r_-) E_m^2$$

for a low-frequency band-limited signal, E_m being its r. m. s. value. We observe that, if the average value of the electric field is zero, i. e., $E_o=0$, and if the fundamental signal frequency f_o is equal to r_-^o, then $r_-=r_-^o$.

At this level of approximation, the main effect of the field is a decrease in the mean lifetime $1/r_-$ of a ligand-encounter aggregate, suggesting an inhibitory effect of the field on the biological response induced by the ligand. This effect has already been observed on the action of PTH on osteoblast-like cells (5) and on the action of PHA on human lymphocytes "in vitro" (9, 17).

Conclusions

The interaction process between cell surface receptors and ligands is a fundamental problem in various biomedical areas, like immunology and neurobiology. Recent evidence of possible perturbations induced by weak exogenous electric fields on such interaction suggests a new way of studying and even controlling this phenomenon.

In order to attain this goal, a quantitative assessment of this process at the microscopic level is needed. In this paper, we have proposed a tentative approach in this direction. Even if our model is at present very crude, nevertheless it already allows some predictions, which can be tested by further experiments, and which can provide a link between complex "macroscopic" biological responses and underlying molecular events.

References

(1) A. Levi, Y. Shechter, E. Neufeld, J. Schlessinger: "Mobility, clustering and transport of nerve growth factor in embryonal sensory cells and in a sympathetic neural cell line." Proc. Natl. Acad. Sci. USA, 77, 3469, (1980).

(2) K. Balakrishnan, F. Hsu, A. Cooper, H. McConnell: "Lipid Hapten containing membrane targets can trigger specific immunoglobulin E-dependent degranulation of rat basophils leukemia cells." The Journal of Biochemical Chemistry, 257, 6427, (1982).

(3) G. Edelman: "Surface modulation in cell recognition and cell growth." Science, 192, 218, (1976).

(4) V. Maino, N. Green, M. Crumpton: "The role of calcium ions in initiating transformation of lymphocytes." Nature, 251, 324, (1974).

(5) R. Luben, C. Cain, M. Chen, D. Rosen, W. Adey: "Effects of electromagnetic stimuli on bone and bone cells "in vitro": inhibition of responses to parathyroid hormone by low-energy, low-frequency fields." Proc. Natl. Acad. Sci. USA, 79, 4180, (1982).

(6) R. Tsien, T. Pozzan, T. Rink: "T-cells mitogens caused early changes in cytoplasmic free Ca^{++} and membrane potential in lymphocytes." Nature, 295, 68, (1982).

(7) S. Lin-Liu, W. Adey: "Low-frequency amplitude modulated microwave fields change calcium efflux rates from synaptosomes." Bioelectromagnetics, 3, 309, (1982).

(8) S. Smith, C. Thomas, S. Frasch: "Lanthanum inhibition of electrically induced dedifferentiation in frog erythrocytes." Bioelectrochemistry & Bioenergetics, 5, 117, (1978).

(9) F. Beltrame, A. Chiabrera, A. Gliozzi, M. Grattarola, G. Parodi, D. Vecchio, D. Ponta, G. Vernazza, R. Viviani: "Electromagnetic control of cell reactivation." Proceedings Int. Symp. Electrom. Waves & Biology, 33, (1980).

(10) G. Bell: "Models for the specific adhesion of cells to cells"Science, 200, 618, (1978).

(11) A. Chiabrera, M. Grattarola, R. Viviani, C. Braccini: "Modelling of the perturbation induced by low-frequency electromagnetic fields on the membrane receptors of stimulated human lymphocytes. II. Influence of the field on the mean lifetimes of the aggregation process." Studia Biophysica, 91, 125, (1982).

(12) C. DeLisi: "The biophysics of ligand-receptor interactions." Quart. Rew. Biophysics, 4, 202, (1980).

(13) S. Chandrasekhar: "Stochastic problems in physics and astronomy." Reviews of Modern Physics, 15, 1, (1943).

(14) M. Poo, W. Poo, J. Lam: "Lateral electrophoresis and diffusion of concanavalin A receptors in the membrane of embryonic muscle cell." The Journal of Cell Biology, 76, 483, (1978).

(15) A. Chiabrera, M. Grattarola, R. Viviani: "Interaction between electromagnetic fields and cells: microelectrophoretic effects on ligands and surface receptors." Bioelectromagnetics, In Press, (1984).

(16) S. McLaughin, M. Poo: "The role of electro-osmosis in the electric field-induced movement of charged macromolecules on the surfaces of cells." Biophysical Journal, 34, 85, (1981).

(17) P. Conti, G. E. Gigante, M. G. Cifone, E. Alesse, G. Ianni, M. Reale, P. U. Angeletti: " Reduced mitogenic stimulation of human lymphocytes by extremely low frequency electromagnetic fields"FEBS Letters, 162, 156, (1983).

BALANCE OF GROWTH MODELS OF CELL POPULATIONS: THE SIGNIFICANCE OF SIMPLE MATHEMATICAL CONSIDERATIONS

Zvi Grossman
School of Mathematical Sciences
Tel Aviv University
Tel Aviv 69978, Israel

One purpose of mathematical models in cell biology is to illustrate and support theoretical arguments, e.g. evaluating alternative modes of regulation in complex systems. Since it is not aimed to mimic the natural phenomena in great detail, a highly simplified mathematical description (e.g. by cutting down the system's dimensionality) is quite excusable. At the end of the spectrum of approximations, a model provides just an abstract metaphor of reality.

The "balance of growth" framework incorporates models of variable dimensionality that are amenable to several kinds of elaborations (1,2). The systems under consideration consist of populations of replicating and maturing cells, coupled through lineage relationships (within "clones"), feedback interactions, or cross-activation and cross-suppression (interclonally). It is assumed that for any mitotic cell (a) an activation step must be induced prior to maturation or replication; and (b) the activated cell can either enter regenerative mitosis or undergo maturation, and the relative probabilities are determined by extra-cellular factors. I shall discuss below the local stability of some steady-state solutions of the models' equations. These simple considerations are related to such basic issues as the nature of the coupling between hemopoietic cell compartments; the relationship between chronic leukemia, acute leukemia and the blast crisis; the existence of a natural resistance of normal tissues to tumor growth; the origin of cyclic hemopoiesis; and the basis of immunologic memory.

HEMOPOIESIS

A Model

The model and its theoretical and empirical basis are discussed in (3). In the equations below, x_i and y_i denote densities of resting and activated mitotic cells, respectively, in a given clone within the bone marrow. The index i indicates the maturation stage (the compartment). z is the density of post-mitotic cells. s represents (the density of) a common inducer of activation and maturation, directly related to the density z (feedback) and affecting all compartments. a_i

and a_i' are activation and maturation-rate constants, respectively, and b is the per-cell rate at which activated cells enter regenerative mitosis. A constitutive, non-feedback activation (induced by the microenvironment) is also assumed (the terms $\bar{a}_i x_i$). c is the per-cell rate at which mature cells leave the clone and is presumably geared to the peripheral demand for functional cells. The factor 2 reflects doubling in the numbers of resting cells by mitosis, and p_i are (constitutive) amplification factors associated with the maturation steps. It is not clear whether constitutive replication really exists, but the factors p_i may be required anyway if the heterogeneity of the clone is underestimated, to account (approximately) for compartments which are not included in the model explicitly. (For small i, p_i also reflects a loss of cells via differentiation of pluripotential cells into different lines.) Finally, τ and τ_i' are time delays associated with regeneration and maturation, respectively.

$$\dot{x}_i(t) = -(\bar{a}_i + a_i s(t))x_i(t) + 2by_i(t-\tau) + (1-\delta_{i1})p_{i-1}a'_{i-1}y_{i-1}(t-\tau'_{i-1})s(t-\tau'_{i-1}), \quad (1a)$$

$$\dot{y}_i(t) = (\bar{a}_i + a_i s(t))x_i(t) - a_i' s(t) y_i(t) - by_i(t), \quad (1b)$$

$$\dot{z}(t) = p_n a_n' s(t-\tau_n')y_n(t-\tau_n') - cz(t), \quad (1c)$$

$$\dot{s}(t) = \alpha z(t) - \beta s(t) \; ; \; i = 1, \dots n . \qquad \delta_{i1} = \begin{matrix} 1, i=1 \\ 0, i>1 \end{matrix} \quad (1d)$$

There are several constant steady state solutions characterized by different numbers of non-vanishing compartments. These can be classified by a number j denoting the first non-vanishing population, such that $x_{i0} = y_{i0} = 0$ for $1 \leq i < j$. For $j = 1$ (a fully populated clone) we have (normalizing so that $\alpha = \beta$)

$$z_0 = s_0 = b/a_1', \quad y_{n0} = c/p_n a_n', \quad y_{i-1,0} = [(a_i' - a_1')/p_{i-1}a'_{i-1}]y_{i0},$$
$$x_{i0} = [b(a_i' + a_1')/(ba_i + a_1'\bar{a}_i)]y_{i0} . \quad (2)$$

We note a number of significant consequences. (a) x_{i0} and y_{i0} are proportional to c, for any i. Thus, the system responds to increased peripheral demand by proliferation, as observed, with no direct signals involved. (b) The number of functional cells depends on the maturation-rate constant of the earliest compartment in the series. Hence, although the system does not absolutely depend on primitive stem cells for self-renewal, as is commonly asserted (4), its performance is affected by the coupling to these cells (3). (c) Feasibility requires $y_{i-10} > 0$, and consequently $a_i' > a_1'$ for $i > 1$. This is consistent with the observed hierarchy within hemopoietic clones (4). From Eq. (1b), the probabilities of maturation and regenerative replication for an activated cell are, respectively, $P_i(M) = a_i's/(a_i's + b)$ and $P_i(R) = b/(a_i's + b)$. At steady state, $P_1(M) = P_1(R) = 0.5$, and $P_i(M) > P_i(R)$ for $i > 1$. Hence, the most primitive cells are dynamically adjusted to be self-renewing while later compartments are forced to be transitory (3).

Stability of Population Coupling

If $a_i' > a_2' > a_1'$, $i > 2$, then an uncoupled steady state of the type $j = 2$ is feasible, namely, one in which

$$x_{10} = y_{10} = 0 \quad , \quad z_0 = s_0 = b/a_2' \text{ , etc.} \tag{3}$$

Linearizing Eq. 1 about the steady state, the first two equations become

$$\dot{x}_1 = -(\bar{a}_1 + a_1 z_0)x_1 + 2by_1(t-\tau) \quad , \quad \dot{y}_1 = -(a_1' z_0 + b)y_1 + (\bar{a}_1 + a_1 z_0)x_1 \text{ .} \tag{4}$$

This is a second-order linear system and allows for a straightforward stability analysis. The characteristic equation is

$$\lambda^2 + (A+B)\lambda + AB = 2bBe^{-\lambda\tau} \quad ; \quad A = a_1' z_0 + b \quad , \quad B = a_1 z_0 + \bar{a}_1 \text{ .} \tag{5}$$

Now we note that if initially the densities x_i, y_i, z and s are chosen to be positive, they remain so for all times (as they should): if any one of these variables approaches zero, the corresponding time derivative becomes positive (see Eq. 1). Thus, only real solutions λ of Eq. 5 are of interest, since oscillatory behavior of $x(t)$ and $y(t)$ about zero would lead to negative values. A necessary condition for stability is then $\lambda < 0$, and this is fulfilled, as inspection of Eq. 5 shows, if $AB > 2bB$, or $a_1' z_0 + b > 2b$, namely, if $a_2' < a_1'$. This, however, is not the case (by assumption), and the state of uncoupling of the first population is unstable: it would be destabilized by the presence of a small number of cells belonging to this population. In general, such considerations suggest a dynamic rather than constitutive linkage between cell compartments. The stability of the state $j = 1$ was demonstrated by numerical integration of Eq. 1.

Consider next the possible effect of cell crowding, e.g. by including density-dependent suppression terms in Eq. 1. This elaboration of the model may help to reveal important aspects of hemopoiesis, although density control probably does not play a major role in normal regulation (as suggested by the capacity of the bone-marrow to sustain large and variable numbers of cells). The simplest non-specific feedback suppression is introduced by adding $-x_i f(N)$, $-y_i f(N)$ and $-z_i f(N)$ to the right-hand sides of Eqs. 1a, 1b and 1c, respectively, where f is a positive, monotonically increasing function of the total density N. Instead of Eq. 5, we now have

$$\lambda^2 + (\tilde{A}+\tilde{B})\lambda + \tilde{A}\tilde{B} = 2b\tilde{C}e^{-\lambda\tau} \quad ; \quad \tilde{A} = a_1' z_0 + b + f_0 \quad , \quad \tilde{B} = a_1 z_0 + \bar{a}_1 + f_0 \quad ,$$
$$\tilde{C} = a_1 z_0 + \bar{a}_1 \quad , \quad f_0 \equiv f(N_0) \text{ .} \tag{6}$$

The local stability condition becomes

$$(a_1'z_0+b+f_0)(a_1z_0+\bar{a}_1+f_0) - pb(a_1z_0+\bar{a}_1) > 0 \, . \tag{7}$$

Instead of Eq. 3, z_0 now satisfies the condition

$$(a_2'z_0+b+f_0)(a_2z_0+\bar{a}_2+f_0) - pb(a_2z_0+\bar{a}_2) = 0 \, . \tag{8}$$

Let us assume that the relation $a_2' > a_1'$ also implies $a_2 > a_1$ and $\bar{a}_2 > \bar{a}_1$. Thus, the more primitive cells are less inducible with respect to activation as well (3). Then, if the differences between the rate constants are sufficiently large, Eq. 7 can be fulfilled, as seen by comparing Eqs. 7 and 8. To simplify the argument write $a_i' = k_i\tilde{a}'$, $a_i = k_i\tilde{a}$, $\bar{a}_i = k_i\tilde{\bar{a}}$, $i = 1,2$, and consider the left-hand sides of Eqs. 7 and 8 as a function of the "inductivity" k evaluated at k_1 and k_2, respectively, with $k_2 > k_1$. This function is of the form $F(k) = r_1k^2 - r_2k + r_3$, with positive constants; $F(k_2) = 0$, and $F(k) > 0$ for $k > k_2$ or for k sufficiently smaller than k_2 (since $F(0) > 0$). The conclusion is that although the normal configuration ($j = 1$) is still a stable one, a state of uncoupling ($j \geqslant 2$) may also be stably established. This may account for the conjectured resistance of truncated clones, previously affected by irradiation or chemical agents, to recolonization by migrating primitive stem cells (3).

A similar interpretation can be given to the (conjectured) natural resistance of normal cell populations to the expansion of small numbers of leukemic (or preleukemic) cells. The latter are associated with lower inductivities (see below), and resistance is implied if the established normal clone can keep the other population in a locally stable state of depletion. In Eqs. 7 and 8, z_0 is the density of normal end cells, the index 2 refers to rate constants of the normal stem cells and 1 to those of the transformed cells. It should be noticed that in a reversed situation, where the leukemic clone is established and the normal one depleted, stability is ensured. In the previous notation, $k_1 > k_2$, $F(k_2) = 0$ and $F(k_1) > 0$.

Parametric Singularity and Leukemia

Eq. 2 shows that the numbers of maturing and proliferating cells are inversely related to the maturation rate constants. It is also plausible that activation and maturation constants are interrelated. We may define a common measure of interaction strength, "inductivity", for a cell, a compartment or a clone. For a clone, $a_i = k\tilde{a}_i$, $a_i' = k\tilde{a}_i'$. The model then suggests that chronic leukemia arises from reduced clonal inductivity, k, originating in a transformed early cell and propagating through the clone by proliferation and maturation of the initial cell and its progeny.

z_0 tends to infinity as a_1' approaches zero (Eq. 2). At $a_n' = 0$, however, $z_0 = 0$ (Eq. 1c). Hence, $k = 0$ (zero inductivity) is a singular point in the parameter space. This means that the model, Eq. 1, is no longer meaningful in the vicinity of this point and has to be refined. The singularity is removed if cell density limitations are appropriately taken into account, e.g. by adding the suppression terms specified above (with $f(N)$ increasing linearly or faster for large N), to ensure the boundedness of the steady state(s) for all k. $z_0 = 0$ is then reached as a limit of the closed set $(x_0, y_0, z_0; k)$ as $k \to 0$. (The same is true if a_i' alone are being reduced.)

All the steady state values in Eq. 2 are proportional to $1/k$. If $f(N)$ becomes significant only for large N, then when k is gradually decreased all cell numbers increase at first, but eventually z_0 falls steeply. The precursor populations undergo a process of uncoupling, or selection, which is survived by one dominant compartment approaching a saturation level. In the present model, the dominant population is the one possessing the largest $\bar{a}_i$. A more elaborate model would include also some constitutive induction of maturation, generating certain heterogeneity and imposing an additional (and opposite) requirement, of relative insensitivity to maturation signals, on a dominant compartment. This general course of events parallels closely the progression of chronic leukemia into the blast crisis (5). The analysis offers a unified interpretation of several facets of leukemia (6).

In the present model, refined by density suppression terms but with no constitutive maturation into z, the solution $z_0 = 0$ is feasible and is locally stable for a finite range of small k. Linearizing Eqs. 1c and 1d about $z_0 = s_0 = 0$, the characteristic equation is

$$\lambda^2 + (\beta + c + f_0)\lambda + \beta(c + f_0) = \alpha p_n a_n' y_{n0} e^{-\lambda \tau_n'} . \tag{9}$$

Again, only real λ are of interest. The stability condition is $\beta(c + f_0) > \alpha p_n a_n' y_{n0}$, or $a_n' < \beta(c + f_0)/\alpha p_n y_{n0}$. y_{n0} is bounded, and the r.h.s. of the last inequality is non-zero for any inductivity.

Periodic solutions

Chronic leukemia results in the present approach from a clonal cellular defect that causes reduced probabilities of maturation. This in turn leads to increased transit time (from x_1 to z) and increased amplification. A lowered level of mature cells (e.g. neutropenia) can result, on the contrary, from a (global) defect in which maturation probabilities are increased (e.g. overproduction of the inducer) with a parallelly shorter transit time and decreased amplification. In both forms of abnormal hemopoiesis cyclic (periodic) behavior may be manifested at some stage of the disease (7). Both amplitude and period are considerably larger in cyclic leukemia.

It has frequently been observed that stability of an equilibrium may be lost when delays are increased. Less frequently, it has been seen that further increase in the delay may result in restabilization. In second-order delayed systems there may be arbitrarily many switches as the delay is increased (8). Such behavior is demonstrated here by solving numerically a simplified mathematical model, with a single proliferating compartment. The amplification process is approximated by "exponential delay" (represented as a linear process):

$$\dot{x} = -asx + (p_1/\tau_1)u \quad , \quad \dot{y} = asx - by - a'sy \quad ,$$
$$\dot{u} = by - (1/\tau_1)u \quad , \quad \dot{v} = a'sy - (1/\tau_2)v \quad , \qquad (10)$$
$$\dot{z} = (p_2/\tau_2)v - cz \quad , \quad \dot{s} = \alpha z - \beta s \quad .$$

The delay τ_2 is estimated as a certain function of the parameter which controls the probability of maturation (a' or α), and the amplification factor p_2 is related to τ_2 ; τ_1 and p_1 are associated with mitosis. For a range of the other parameters, simulations manifest, qualitatively, the observed transitions from (stable) under-production of cells through a periodic phase to a normal (stable) steady state, and from that to cyclic and then again stable over-production, with ever-increasing maturation delays.

IMMUNE SYSTEM

A Model

Consider a multiclonal antibody response to antigen. The antigen has n determinants, each of which can bind one type of antibody or B cell. Let x_i , y_i , z_i, Ab_i and c_i denote the concentrations of resting precursor cells, activated precursors, antibody-forming cells, antibody and antigen-antibody complex, respectively, of clone i . Binding of antibody i to its determinant interferes with the binding of antibody j to its respective determinant, and vice-versa. The degradation rate of antigen is proportional to the total amount of complexes. The equations are (see also (2)):

$$\dot{x}_i(t) = s_i - dx_i(t) - a_i\overline{Ag}_i(t)x_i(t) + 2by_i(t-\tau_1) \quad , \qquad (11a)$$
$$\dot{y}_i(t) = a_i\overline{Ag}_i(t)x_i(t) - by_i(t) - a_i'\overline{Ag}_i(t)y_i(t) \quad , \qquad (11b)$$
$$\dot{z}_i(t) = pa_i'\overline{Ag}_i(t-\tau_2)y_i(t-\tau_2) - pa_i'\overline{Ag}_i(t-\tau_3)y_i(t-\tau_3) \quad , \qquad (11c)$$
$$\dot{Ab}_i(t) = \alpha z_i(t) - \beta Ab_i(t) - \gamma Ab_i(t)\overline{Ag}_i(t) + \rho_i c_i(t) \quad , \qquad (11d)$$
$$\dot{c}_i(t) = \gamma Ab_i(t)\overline{Ag}_i(t) - \rho_i c_i(t) - K\sum_{j=1}^{n} c_j(t) \quad , \qquad (11e)$$
$$\dot{Ag}_i(t) = -\sum_{j=1}^{n}\theta_{ij}(\gamma Ab_j(t)\overline{Ag}_j(t) - \rho_j c_j(t)) \quad ; \quad i = 1,\dots n \quad . \qquad (11f)$$

Neglecting the inflow s_i of new lymphocytes, Eqs. 11a and 11b resemble Eqs. 1a and 1b, with $\overline{Ag}_i$ replacing s . However, the feedback is strictly negative — antigen is blocked or removed by the products (Ab_i) of the differentiated cells (z_i). The antibody binds antigen reversibly with association constant γ and dissociation constants ρ_i (2,9). The interaction constants, a_i, a_i', depend on dimensionless "affinities" α_i : $a_i = \alpha_i k$, $a_i' = \alpha_i k'$. Ag_i is the "antigenicity" for clone i. It depends on direct as well as on cross-suppression. Moreover, Ag_i may become negative, reflecting the fact that a finite amount of antibody has to dissociate before this determinant regains its stimulatory capacity. An over-blocked determinant does not interact with B cells or antibody, and thus $\overline{Ag_i}$ was defined as

$$\overline{Ag_i} = \begin{cases} Ag_i, & \text{for } Ag_i \geq 0 \\ 0, & Ag_i < 0. \end{cases} \tag{12}$$

$T = (\theta_{ij})$ is the matrix of cross-suppression. It is assumed that $\theta_{ij} = 1$ for $i = j$, and $\theta_{ij} = \theta_{ji} \leq 1$ for all $i \neq j$.

If the process of complex degradation is a slow one, the unperturbed system can be described by a quasi-steady state approximation. In the limit of no degradation, $K = 0$, integration of Eqs. 11e and 11f yields the relation

$$Ag_i(t) + \sum_{j=1}^{n} \theta_{ij} c_j(t) = Ag_i(0). \tag{13}$$

Multiple Steady States and Memory

If $s_i = K = 0$, a biologically justifiable approximation, then at steady state each clone can either be extinct $(x_{i0} = y_{i0} = z_{i0} = Ab_{i0} = c_{i0} = 0)$, or else Ag_{i0} must assume one of the two values, Ag_i^- or Ag_i^+, where

$$Ag_i^{\pm} = [ba_i - da_i' \pm \sqrt{(ba_i - da_i')^2 - 4bda_ia_i'}]/2a_ia_i'. \tag{14}$$

Ag_i^+ leads to instability and is not of interest here. There are several steady states of the n-clonal system, characterized by the number m, $m \leq n$, of non-extinct clones. From Eq. 13, for a given m we therefore have m linear algebraic equations,

$$Ag_k^- + \sum_{j=1}^{n} \theta_{kj} c_{j0} = Ag_k(0), \quad k = 1, \ldots m. \tag{15}$$

Conditions for the existence of such a steady state are derived from the requirement that all the non-vanishing values c_{k0} be real and positive.

As in previous examples, necessary conditions for stability are derived from the requirement that the state of depletion of the zero-clones be (locally) stable. It can be shown (2), linearizing Eqs. 11a and 11b, that these conditions are simply $Ag_{i0} < Ag_i^-$ for $i \neq k$ (or alternatively $Ag_{i0} > Ag_i^+$, corresponding to the more exotic case of high-dose paralysis). Ag_{i0}, in turn, are given by

$$Ag_{i0} = Ag_i(0) - \sum_{k=1}^{m} \theta_{ik} c_{k0}, \quad i \neq k. \tag{16}$$

The patterns of coexistence which can be maintained were illustrated by detailed analysis of the case $n = 3$, $m = 2$ (9). With $\alpha_1 < \alpha_2 < \alpha_3$ and $Ag_i(0) = Ag(0)$

such that $Ag_1^- < Ag(0) < Ag_3^+$, all three 2-clonal configurations can be stable for suitable choices of the matrix T . Moreover, two of these configurations can be stable for the same set of parameters if

$$\beta_{12}^2 + \beta_{13}^2 + \beta_{23}^2 - \beta_{11}^2 > 2\beta_{12}\beta_{13}\beta_{23}/\beta_{11} \; . \tag{17}$$

In general, the present state of the system would depend on the size and nature of previous manipulations. This suggests a dynamic interpretation of how immunologic memory is maintained (1).

A final comment: the "antigenicity" usually depends also on interactions involving T cells. These affect the previous considerations if they discriminate between clones. In particular, cooperative cell-cell interactions could bring about auto-catalytic effects, enhancing the stability of established configurations with respect to challenging, unexpanded clones.

Conclusion

Unlike many laws in physics, models describing biological systems of interacting cells can seldom be expected to predict numbers. The number "zero" is an important exception (in reality it may mean "small"). The examples considered above show that it may be instructive to derive approximate or even incomplete conditions for the existence and stability of states in which certain subpopulations vanish. While a complete stability analysis of high-dimensional systems may present substantial technical problems, the isolation of zero-components simplifies the analysis considerably and promotes understanding of the structure of the systems. Other phenomena, such as stability switches and periodic solutions, must be studied numerically or evaluated by extrapolation from the analysis of more elementary systems.

References

(1) Z. Grossman, Recognition of self, balance of growth and competition: horizontal networks regulate immune responsiveness. Eur. J. Immunol. 12, 747-756 (1982).

(2) Z. Grossman, R. Asofsky and C. DeLisi, The dynamics of antibody secreting cell production. J. Theor. Biol. 84, 49-92 (1980).

(3) Z. Grossman, The stem cell concept revised, submitted.

(4) L.G. Lajtha, Stem cell concepts. Differentiation 14, 23-33 (1979).

(5) N.L. Warner and D. Metcalf (Eds.), Leukemia. UICC Technical Report Series, 61.

(6) Z. Grossman, Leukemia and the blast crisis: hypothesis, submitted.

(7) A. Morley, Cyclic hemopoiesis and feedback control. Blood cells 5, 283-296 (1979).

(8) K.L. Cooke and Z. Grossman, Discrete delay, distributed delay and stability switches. J. Math. Anal. Applications 86, 592-627 (1982).

(9) S. Novick, Kinetics of affinity maturation. Thesis, Weizmann Inst. (1980).

THE AGE STRUCTURE OF POPULATIONS OF CELLS REPRODUCING BY ASYMMETRIC DIVISION

Mats Gyllenberg
Helsinki University of Technology, Institute of Mechanics
SF-02150 Espoo 15, FINLAND

1. Introduction. The purpose of this paper is to present a new model for the growth and age structure of a population of a species which reproduces by asymmetric division, i.e. cell reproduction where one clearly can distinguish between the mother and the daughter cell after division as opposed to division patterns where each fission produces two equal daughter cells. The cells of such populations are thus naturally divided into different classes according to the number of divisions they have undergone. A typical example of an organism which divides asymmetrically is *Saccharomyces cerevisiae*, which reproduces by budding. After each cell separation a bud scar is left on the wall of the mother cell. These scars can be counted using electron microscopy. We say that a cell with i offsprings and hence i bud scars belongs to the ith scar class. It is supposed that cells of different scar classes behave in a different way, for instance with respect to cell growth and metabolism activity.

When the cell propagates through the cell cycle it passes several specific stages, such as DNA synthesis, premitotic phase, mitosis and so on (cf. [9]). Our model contains a variable τ, which we call cycle age and which describes the phase in which the cell is. Thus our model takes into account that cells behaves differently at different stages of the cell cycle.

2. The model. Let $x_i(t,\tau)$ denote the density of cells, which have entered the ith scar class τ time units ago, at time t. Let $S(t)$ denote the concentration of the growth-limiting substrate at time t. Let $f_i(\tau,S)d\tau$ $[m_i(\tau,S)d\tau]$ denote the fraction of τ-aged organisms in the ith scar class which divide [die] during the infinitesimal age interval $[\tau,\tau+d\tau]$ when the substrate concentration is S. $g_i(\tau,S)$ denotes the rate at which a τ-aged cell in the ith scar class consumes the substrate. We finally let D denote the dilution rate and S_{in} the input substrate concentration. It is clear that the problem is physically meaningful only if the functions f_i, m_i, g_i, D, S_{in} are nonnegative - an assumption we make once and for all.

Population balance yields

$$(2.1)\quad \frac{\partial x(t,\tau)}{\partial t} + \frac{\partial x(t,\tau)}{\partial \tau} + [F(\tau,S)+M(\tau,S)+DI]x(t,\tau) = 0$$

and mass balance of the substrate yields

$$(2.2)\quad S'(t) = -\int_0^\infty G(\tau,S(t))x(t,\tau)d\tau + D[S_{in}-S(t)]$$

(prime denotes differentiation). Here x is an R^{n+1}-valued function the components of which are x_i), F $[M]$ is a diagonal matrix with diagonal elements f_i $[m_i]$, I is the identity matrix and G is a row vector with entries g_i (i = 0,...,n).

The division mechanism is described by the boundary condition

$$(2.3)\quad x(t,0) = \int_0^\infty B(\tau,S(t))x(t,\tau)d\tau ,$$

where

$$B = \begin{pmatrix} f_0 & f_1 & \cdots & & & f_n \\ f_0 & 0 & \cdots & & & 0 \\ 0 & f_1 & \cdots & & & 0 \\ \vdots & & & & & \vdots \\ 0 & \cdots & 0 & f_{n-2} & 0 & 0 \\ 0 & \cdots & 0 & 0 & f_{n-1} & f_n \end{pmatrix}$$

In writing the expression for B we have assumed that when a cell of class n divides the mother cell does not enter class n+1 but returns to the beginning of class n. Thus the fact that x has n+1 components does not mean that a cell can divide at most n times but that cells that have undergone more than n divisions are indistinguishable.

The system (2.1) - (2.3) is supplemented by the initial conditions

$$(2.4)\quad S(0) = S_0 , \qquad x(0,\tau) = x_0(\tau) ,$$

where $S_0 \in R^+$ and $x_0 \in L^1(R^+;R^{n+1})$ are given.

The model (2.1) - (2.4) is a special case of a very general nonlinear extension of the Lotka-von Foerster population model, which has been presented and investigated in [6].

In this paper we shall restrict ourselves to the case $n = 1$, which means that cells that have undergone one or more divisions are indistinguishable. In the case of *S. cerevisiae* this seems to be a fairly good assumption, because although there may be some differences between older scar classes, these are very small compared with the difference between virgin cells and cells which have divided at least once (cf. [1]). The general case $n > 1$ leads to considerations very similar to those of age-dependent multitype branching processes including Perron-Frobenius theory. For the sake of simplicity we shall neglect mortality, i.e. we assume $M = 0$. In the case of *S. cerevisiae* the error made is small (cf. [1]). The results of this paper can easily be reformulated for the case where mortality is taken into account.

<u>3. Batch cultivation with excess of substrate.</u> In this case $D = 0$, the equation (2.2) is superfluous and the matrix valued functions F and B are independent of S. Since we assume $n = 1$, the matrix B has only two rows, which are identical: $[f_0(\tau)\ f_1(\tau)]$. It therefore follows from (2.3) that $x_0(t,0) = x_1(t,0) = y(t)$, where y satisfies the scalar equation

(3.1) $y(t) = \int_0^\infty f_0(\tau)x_0(t,\tau)d\tau + \int_0^\infty f_1(\tau)x_1(t,\tau)d\tau$.

Integration of eq. (2.1) along characteristics yields

(3.2) $x_i(t,\tau) = y(t-\tau)\pi_i(\tau)$, $\tau \leq t$, $x_i(t,\tau) = \phi_i(\tau-t)\pi_i(\tau)$, $\tau \geq t$ $(i = 0,1)$,

where $\pi_i(\tau) = \exp(-\int_0^\tau f_i(\alpha)d\alpha)$, $\phi_i(\tau) = x_{io}(\tau)/\pi_i(\tau)$. If one substitutes (3.2) into (3.1), one finds that

(3.3) $y(t) = \int_0^t p(\tau)y(t-\tau)d\tau + \Phi(t)$,

where $p_i = f_i\pi_i$, $\Phi_i = \int_t^\infty p_i(\tau)\phi_i(\tau-t)d\tau$, $p = p_0 + p_1$, $\Phi = \Phi_0 + \Phi_1$.

The problem is thus reduced to solving the renewal equation (3.3). Once this is done, the age distribution is obtained from (3.2). The sizes of the scar classes are obtained by integration of (3.2) over all ages:

(3.4) $X_i(t) = \int_0^t \pi_i(\tau)y(t-\tau)d\tau + \Psi_i(t)$ $(i = 0,1)$,

where $X_i(t) = \int_0^\infty x_i(t,\tau)d\tau$, $\Psi_i(t) = \int_t^\infty \pi_i(\tau)\phi_i(\tau-t)d\tau$.

Concerning the given functions f_i, x_{io}, we shall make the following assumptions:

(H1) There is a $T \leq \infty$ such that f_i is locally integrable on $[0,T)$, $f_i(\tau) = 0$ for $\tau \geq T$ and $\int_0^\infty f_i(\tau)d\tau = \infty$;

(H2) $\phi_i \in L^\infty(R^+)$.

(H1) implies that $p_i \in L^1(R^+)$ and $\int_0^\infty p_0 d\tau = \int_0^\infty p_1 d\tau = 1$. p_i is the density function of the random variable defined as the time a cell spends in class i. Thus (H1) contains the biological requirement that a cell must complete its cell cycle in the age interval $[0,T)$. If T is finite, f_i and π_i have compact support (we adopt the convention $\exp(-\infty) = 0$) and there is a maximum length of the cell cycle.

$\pi_i(\tau)$ can be interpreted as the probability of a cell staying in class i for τ time units. Since the initial population is selected from real cultures, $x_{io}(\tau)$ cannot be large for such values of τ for which $\pi_i(\tau)$ is small. (H2) is a formalisation of this requirement.

An age distribution is called persistent if it has the form $x_i(t,\tau) = k_i X(t)A_i(\tau)$, where $\int_0^\infty A_0(\tau)d\tau = \int_0^\infty A_1(\tau)d\tau = 1$, k_0 and k_1 are constants with $k_0 + k_1 = 1$. Then $X(t) = X_0(t) + X_1(t)$ stands for the size of the total population. The definition states that the relative age distribution both with respect to different scar classes and within the cell cycle remains unchanged, only the sizes of the scar classes vary.

The following theorem is easily proved using Laplace transform techniques (cf. [7], [8], [11], [12], [15]).

Theorem 3.1: *Suppose* (H1) *and* (H2) *are satisfied. Then all solutions of the system* (3.2), (3.3), *approach persistent age distribution as* $t \to \infty$. *Ultimately both* X_0 *and* X_1 *grow exponentially with exponent* μt, *where* μ *is the unique real root of the equation*

$$(3.5) \quad \hat{p}(\mu) = 1$$

($\hat{}$ *denotes Laplace transform*). *Moreover*

$$(3.6) \quad X_0(t)/X(t) \to \hat{p}_1(\mu), \quad X_1(t)/X(t) \to \hat{p}_0(\mu) \qquad (t \to \infty).$$

Observe that part of the conclusion of Theorem 3.1. is that the root μ of eq. (3.5) is the specific growth rate of the population.

It should be noted that $\hat{p}_1(\mu)$ is the expectation value of the function $\exp(-\mu\tau)$, where τ is the random variable defined as the time a cell spends in class 1. Using different methods, Adams et. al. [1] obtained the same frequency at demographic equilibrium.

Several authors have observed that the relative frequencies of populations of *S. cerevisiae* in batch cultivation with an excess of substrate perform damped oscillations until equilibrium is reached (cf. [1], [2], [10]). Several earlier attempts to model the growth of budding yeast (e.g. the model of Gani and Saunders [3]) have failed to show such oscillations. The model of Adams et. al. [1] does predict oscillations but the model is investigated only for special choices of initial age distributions.

Next we shall show that the model (3.2), (3.3), exhibits damped oscillations and that this is certainly not a consequence of the choice of initial age distribution (of course, some initial distributions may amplify the oscillations) but of the asymmetry of division and the age specific behaviour of the population. In fact, neither the corresponding age independent model nor a model for symmetric division (i.e. $p_0 = p_1$) predict oscillations.

Suppose first that the model is age-independent. Then f_0 and f_1 are constants, integration of (2.1) over all ages (using the fact that $x_i(t,\infty) = 0$ which follows from (H1), (H2) and (3.2)) together with (3.3) yield

$$(3.7) \quad X_0' = f_1 X_1 \,, \qquad X_1' = f_0 X_0 \,.$$

The eigenvalues of the system (3.7) are $\pm\sqrt{f_0 f_1}$ so neither X_0 nor X_0/X perform any oscillations.

Consider now the age-dependent problem and make for simplicity the natural assumption of the existence of a maximal cell cycle length. Then p_i, π_i, Φ_i and Ψ_i have compact support and consequently their Laplace transforms are entire functions, of order $\rho = 1$ (recall that the order of an entire function g is the infimum of all positive numbers λ such that $|g(z)| < \exp(|z|^\lambda)$ for all large enough $|z|$). If the characteristic equation $\hat{p}(z) = 1$ had only a finite number of roots, $z_0 = \mu, z_1, \ldots, z_n$ with multiplicity k_j $(j = 1,\dots,n)$, then $\hat{p}(z)$ would have the form

$$(3.8)\quad \hat{p}(z) = 1 + (z-\mu)(z-z_1)^{k_1}\dots(z-z_n)^{k_n}\exp(Q(z)) ,$$

where $Q(z)$ is a polynomial of degree not greater than $\rho = 1$ (cf. [13, Th. 9.10 p. 267]). The left hand side of (3.8) approaches zero as z tends to infinity along any straight line parallel to the imaginary axis by the Riemann-Lebesgue lemma. The same is certainly not true for the right hand side. This contradiction shows that the characteristic equation has infinitely many roots. Since p is positive the nonreal roots appear as pairs of complex conjugates and have real part less than μ.

By (3.3) $\hat{y} = \hat{\Phi}/(1-\hat{p})$, hence (3.4) implies

$$(3.9)\quad \hat{X}_0 = \hat{\pi}_0\hat{\Phi}/(1-\hat{p}) + \hat{\Psi}_0 , \qquad \hat{X} = \hat{\pi}\hat{\Phi}/(1-\hat{p}) + \hat{\Psi} .$$

$\hat{X}_0$ and $\hat{X}$ are meromorphic functions with poles at the roots of the characteristic equation. An application of the inversion formula and the residue theorem then gives

$$(3.10)\quad \begin{aligned} X_0(t) &= b_0\exp(\mu t) + \Sigma b_i(t)\exp(z_i t) + O(\exp((\gamma-\varepsilon)t)) + \Psi_0(t) , \\ X(t) &= a_0\exp(\mu t) + \Sigma a_i(t)\exp(z_i t) + O(\exp((\gamma-\varepsilon)t)) + \Psi(t) , \end{aligned}$$

for some $\varepsilon > 0$, where $z_i = \gamma + iy_i$ are the complex roots of eq. (3.5) with largest real part ($\gamma < \mu$). a_i and b_i are polynomials of degree at most equal to the multiplicity of the root z_i. Since $\hat{p}(\gamma+iy)$ tends to 0 as $y \to \infty$ there is only a finite number of roots with real part equal to γ, hence the sums in (3.10) are finite. Assume for the sake of simplicity that there are only two roots, $z_1 = \gamma + iy_1$ and $\bar{z}_1 = \gamma - iy_1$, with $\mathrm{Re} z = \gamma$, and that these roots are simple. Then a_i, b_i are constants, $a_i = -\hat{\Phi}(z_i)/z_i\hat{p}(z_i)$, $b_i = \hat{p}_1(z_i)a_i$ $(i = 0,1,\ z_0 = \mu)$. It follows that the deflection of the relative frequency of the zero class from the equilibrium value $\hat{p}_1(\mu)$ is given by

$$(3.11)\quad \frac{X_0(t)}{X(t)} - \hat{p}_1(\mu) =$$

$$\frac{2|a_1|}{a}\exp(-(\mu-\gamma)t)\{|\hat{p}_1(z_1)|\cos(y_1 t+\arg b_1) + \hat{p}_1(\mu)\cos(y_1 t+\arg a_1)\}$$

$$+ O(\exp(-(\mu-\gamma+\varepsilon)t)) .$$

Obviously (3.11) represents damped oscillations.

In the case of symmetric division $\pi_0 = \pi_1$ and hence (3.4) implies that

$$(3.12)\quad \frac{X_0(t)}{X(t)} = \frac{(\pi_0 * y)(t) + \Psi_0(t)}{2(\pi_0 * y)(t) + \Psi_0(t) + \Psi_1(t)} .$$

One observes that all possible oscillations performed by X_0/X are due to asymmetries in the initial age distributions, represented in formula (3.12) by differences between Ψ_0 and Ψ_1. When the effect of the initial state is no longer present, that is, when t is greater than the maximum of the support of Ψ, then the relative frequencies $X_0(t)/X(t)$ and $X_1(t)/X(t)$ are equal to the constant 1/2.

4. Continuous cultivation. We consider the model (2.1) - (2.4) with $D > 0$. Conditions for the existence of a unique, global positive solution have been given in [4] and [5].

By an equilibrium solution we mean a time independent solution $x_i(t,\tau) = x_i^*(\tau)$ ($i = 0,1$), $S(t) = S^*\ \forall (t \in R^+)$. The function y corresponding to this solution must also be constant and is denoted by y^*. For fixed S we let $\pi_i(\tau,S) = \exp(-\int_0^\tau f_i(\alpha,S)d\alpha)$ and $p_i(\tau,S) = f_i(\tau,S)\pi_i(\tau,S)$. $\hat{p}_i(z,S)$ detnotes Laplace transform with respect to the first argument. The following theorem is easily proved using methods analogous to those used in the proof of a similar theorem in [4].

Theorem 4.1.: *Fix $D > 0$. There exists a nontrivial (i.e. $x_i^* \not\equiv 0$) equilibrium solution if and only if the transcendental equation*

$$(4.1)\quad \hat{p}(D,S) = 1$$

has a solution $S = S^* \in (0,S_{in})$. Every root $S^* \in (0,S_{in})$ of Eq. (4.1) is an equilibrium substrate concentration, and conversely, every equilibrium concentration satisfies (4.1). To every root $S^* \in (0,S_{in})$ corresponds a unique equilibrium age distribution x^* given by

$$(4.2)\quad x_i^*(\tau) = y^*\pi_i(\tau,S^*)\exp(-D\tau)\ ,\quad y^* = D[S_{in}-S^*]/\sum_{i=0}^{1}\int_0^\infty g_i(\tau,S^*)\pi_i(\tau,S^*)\exp(-D\tau)d\tau\ .$$

The relative frequencies of the scar classes at equilibrium are given by

$$(4.3)\quad X_0^*/X^* = \hat{p}_1(D,S^*)\ ,\quad X_1^*/X^* = \hat{p}_0(D,S^*)\ .$$

Using the results of Theorem 3.1 we see that the unique positive root $z = \mu(S)$ of the equation $\hat{p}(z,S) = 1$ (if it exists) is the specific growth rate of the population when the substrate concentration is S. Part of the conclusion of Theorem 4.1 is thus that the equilibrium substrate concentrations are exactly those S^* for which the specific growth rate equals the dilution rate.

The question of stability is not a trivial one. For a criterion for local asymptotic stability of equilibria, see [5] and [6].

Beran has observed that during growth of *S. cerevisiae* limited by glucose, the frequency of the virgin cells rises if the dilution rate is decreased [2, p. 165]. We now investigate if our model can explain this phenomenon.

$\mu(S)$ is the unique positive root of $\hat{p}(z,S) = 1$. Since $\frac{\partial\hat{p}}{\partial z} = -\int_0^\infty \tau p(\tau,S)\exp(-z\tau)d\tau < 0$ for all $z \in R^+$, the implicit function theorem states that μ is differentiable and $\mu'(S) = -(\frac{\partial\hat{p}}{\partial S})/(\frac{\partial\hat{p}}{\partial z})$. Let S^* be an equilibrium substrate concentration and suppose $\mu'(S^*) > 0$ i.e. $\frac{\partial\hat{p}}{\partial S}(\mu(S^*),S^*) > 0$. Since $\mu(S^*) = D$, there is by the inverse function theorem a diffeomorphism between a neighborhood of S^* and a neighborhood of D, one has $\frac{dS^*}{dD} = 1/\mu'(S^*)$. The relative frequency of the zero class at equilibrium can therefore locally be considered as a function of D. By Theorem 4.1 and the preceding remarks

$$(4.4)\quad \frac{d}{dD}(X_0^*/X^*) = \frac{d}{dD}\hat{p}_1(D,S^*(D)) = \frac{\partial\hat{p}_1}{\partial D} + \frac{\partial\hat{p}_1}{\partial S}\frac{1}{\mu'(S^*)} = \frac{\partial\hat{p}_1}{\partial D} - \frac{\partial\hat{p}_1}{\partial S}\frac{\frac{\partial\hat{p}}{\partial D}}{\frac{\partial\hat{p}}{\partial S}} =$$

$$= -(1/\tfrac{\partial\hat{p}}{\partial S})\frac{\partial(\hat{p},\hat{p}_1)}{\partial(D,S)} = -(1/\tfrac{\partial\hat{p}}{\partial S})\frac{\partial(\hat{p}_0,\hat{p}_1)}{\partial(D,S)},$$

where the partial derivatives are evaluated at $(\mu(S^*),S^*)$. We thus conclude that under the assumptions made

the frequency of the zero class increases with decreasing dilution rate if and only if the Jacobian determinant $\frac{\partial(\hat{p}_0,\hat{p}_1)}{\partial(D,S)}$ *is positive.*

Von Meyenburg [14] has found that in glucose limited cultivation of *S. cerevisiae* a decrease in $\mu(S)$ implies that the unbudded portion of the cell cycle becomes longer whereas the budded portion is almost unaffected by the growth rate. By Hartwell and Unger [10, p. 423] parent cells produce extremely small daughters during nutrient starvation which implies that the length of the unbudded phase of the virgin cells grows more than the unbudded phase of the other generations. It follows that for *S. cerevisiae* $\hat{p}_0$ is much more sensitive to changes in $\mu(S^*) = D$ than $\hat{p}_1$. The positivity of the Jacobian expresses this in an exact way. Observe that if p_1 is independent of S, then the Jacobian is certainly positive.

References

1. J. Adams, E.D. Rothman and K. Beran, The age structure of populations of *Saccharomyces cerevisiae, Math. Biosci.* 53 (1981), 249-263.
2. K. Beran, Budding of yeast cells, their scars and ageing, *Advan. Microb. Physiol.* 2 (1968), 143-171.
3. J. Gani and I.W. Saunders, Fitting a model to the growth of yeast colonies, *Biometrics* 33 (1977), 113-120.
4. M. Gyllenberg, Age-dependent population dynamics in continuously propagated bacterial cultures, Report No. 8, Helsinki Univ. of Tech. Inst. of Mech., 1981.
5. M. Gyllenberg, Nonlinear age-dependent population dynamics in continuously propagated bacterial cultures, *Math. Biosci.* 62 (1982), 45-74.
6. M. Gyllenberg, Stability of a nonlinear age-dependent population model containing a control variable, *SIAM J. Appl. Math.*, to appear.
7. T. Hamada, S. Kanno, E. Kano, Stationary stage structure of yeast population with stage dependent generation time, *J. Theor. Biol.* 97 (1982) 393-414.
8. T. Hamada, Y. Nakamura, On the oscillatory transient stage structure of yeast population, *J. Theor. Biol.* 99 (1982) 797-805.
9. L.H. Hartwell, *Saccharomyces cerevisiae* cell cycle, *Bacterial. Rev.* 38 (1974) 164-198.
10. L.H. Hartwell and M.W. Unger, Unequal division in *Saccharomyces cerevisiae* and its implications for the control of cell division, *J. Cell. Biol.* 75 (1977) 422-435.

11. M.A. Hjortso, J.E. Bailey, Transient responses of budding yeast populations, *Math. Biosci.* 63 (1983) 121-148.

12. M.A. Hjortso, J.E. Bailey, Steady-state growth of budding yeast populations in well-mixed continuous-flow microbial reactors, *Math. Biosci.* 60 (1982) 235-263.

13. A.I. Markushevich, Theory of functions of a complex variable, Vol. 2, English translation, Prentice-Hall inc., Englewood Cliffs, N.J., 1965.

14. H.K. von Meyenburg, Der Sprossugszyklus von *Saccharomyces cerevisiae*, *Pathol. Microbiol.* 31 (1968) 117-127.

Added im proof:

15. S. Tuljapurkar, Transient dynamics of yeast populations, *Math. Biosci.* 64 (1983) 157-168.

NUMERICAL SIMULATION OF REENTRY OF
THE ACTIVATION WAVE IN THE HEART HISS BUNDLE

Jacques HENRY

INRIA
Domaine de Voluceau
BP 105 - Rocquencourt
78153 LE CHESNAY CEDEX - FRANCE

The contraction of the heart is triggered by an electric signal going throughout the myocardium, the action potential. At the level of one cell, this phenomenon is due to membrane currents carried by various ions whose conductance is time and voltage dependent. For the nerve axon, these currents have been analyzed and modelled by Hodgkin and Huxley. Since, precise data have become available for the cardiac cells. In this paper we present simulations of the propagation of an action potential. We are interested in the conducting tissue between atria and ventricular myocardium ; the Hiss bundle. We use the model of Purkinje cell action potential published by McAllister-Noble-Tsien [4] (MNT).

Ventricular tachycardia is a severe disease whose major cause is a reentry of the activation wave. One possible location for a reentry may be on branches of the Hiss bundle : the activation wave may be stopped by a one way block while being propagated by another branch, through the myocardium, and backwards on the first branch. So it "reenters" a loop.

The target of this paper is to see how data measured at the level of one cell can account for a global phenomenon as a reentry. We also give elements to prove the existence and uniqueness of the mathematical problem and we discuss the numerical methods.

I - MODEL

We aim to model the propagation of an action potential along a branch of the Hiss bundle. From the anatomy of these branches, it is reasonable to consider them as a one-dimensionnal medium. It is assumed that the conduction is due to a low-resistance coupling of the cells at the gap junction. This syncitium is then modelled as an electric cable whose core has a constant resistivity (cytoplasm + gap junction), surrounded by an insulating (membrane), in a bulk solution (extracellular fluid). Then the model is two fold ; the cable equation describing the electric state of the cable ; a model of membrane currents : we use here the model of Purkinje cells published by McAllister-Noble-Tsien [4] which is based on electrophysiological measurements.

- The cable equation

Let V(t,x) be the transmembrane potential, C_M the membrane capacitance by length unit, a the radius, ρ the resistivity of the core, i the transmembrane current, R_M the membrane resistance at rest. V satisfies

$$\tau_M \frac{\partial V}{\partial t} - \lambda^2 \frac{\partial^2 V}{\partial X^2} = R_M i$$

where :

$$\tau_M = R_M \, C_M$$

is the time constant

$$\lambda = \left(\frac{R_M \, a}{2\rho}\right)^{1/2}$$

is the space constant.

Initial and boundary conditions for this partial differential equation are somewhat arbitrary. Usually at time 0, the potential is set to its diastolic resting value V_r (about -80mV). In case of an inhomogeneous cable one waits until a steady state is reached before stimulating. The stimulation is done by depolarizing suddenly one end of the cable. The boundary conditions are :

$$\frac{\partial V}{\partial X}(t,0) = \frac{\partial V}{\partial X}(t,L) = 0$$

meaning that there is no current at each end. One can check that these conditions have no influence outside boundary region of length $O(\lambda)$.

Typical values are :

$\tau_M = 30\text{ms}$ $\lambda = 2\text{mm}$ $R_M = 30\text{k }\Omega\text{-cm}^2$ $L = 1\text{cm}$

- The McAllister-Noble-Tsien model

This model gives the total membrane current as the sum of ionic currents whose conductance is time and voltage dependent. This dependency has been measured by the voltage-clamp method and modelled by the use of Hodgkin-Huxley type gating variables. This model includes : a fast (Na^+) and a slow (Ca^{++}) depolarizing currents, a transient repolarizing (Cl^-) current responsible for the "notch", two K^+ repolarizing currents i_{K_1}, i_{K_2} and two repolarizing currents i_{x_1}, i_{x_2} ending the plateau. The slow inactivation of i_{K_2} is responsible for the diastolic depolarization which causes the automaticity of Purkinje cells. As an example we give the slow current which has a non-linear background part and a linear gated one :

$$i_s = - g_s \, d \, f \, (V - V_s) - i_{s,b}(V)$$

The 9 gating variables (m, h, d, f, q, r, s, x_1, x_2) satisfy at each point a differential equation :

$$\frac{dm}{dt} = \alpha_m(V) \; (1-m) - \beta_m(V) m$$

where α_i and β_i are positive functions of V.

At time 0, these variables are set at their steady state values :

$$m_\infty(V_r) = \frac{\alpha_m(V_r)}{\alpha_m(V_r) + \beta_m(V_r)}$$

II - SHORT MATHEMATICAL STUDY

We just want to indicate how it is possible to prove the existence and uniqueness of a solution of the preceding problem, using the method of Sermange [5]. Let $\vec{y} \in \mathbb{R}^9$ be the vector of the gating variables. The uniqueness follows from the Lipschitz continuity of i(y,V), $\alpha_j(V)$ and $\beta_j(V)$. The existence can be proved by using the invariant domain technique and a fixed point theorem. Let us consider linearization of the non-linear currents using the cord conductance relative to the reversion potential ; for example

$$\tilde{i}_{K_1}(V,\tilde{V}) = g_{K_1}(V) \; (\tilde{V} - V_{K_1})$$

where

$$g_{K_1}(V) = \frac{i_{K_1}(V)}{V - V_{K_1}}$$

It is then possible to define a linear mapping $T : (V,y) \to (\tilde{V},\tilde{y})$. Let V_0 and V_1 be the lowest and greatest reversion potentials (-110mV, +70 mV). The maximum principle is used, knowing that α_j and β_j are positive and that $\tilde{i}(V_0,\tilde{V}) \geq 0$ for $\tilde{V} \geq V_0$; $\tilde{i}(V_1,\tilde{V}) \leq 0$ for $\tilde{V} \leq V_1$. It results that T is a mapping from $A = [V_0,V_1] \times [0,1]^9$ in itself. The existence of a solution and the invariance of A are then a consequence of the contraction mapping fixed point theorem.

The MNT model has too many variables to analyze the corresponding set of ordinary differential equations. It can be seen from the plot of the stationnary membrane current (fig. 1) that there is a stable stationary point at about -40mV but with a small attracting domain. If s is held at 1 (dotted line) there is another stationary point at -85mV (diastolic resting potential) which vanishes with the slow decrease of S.

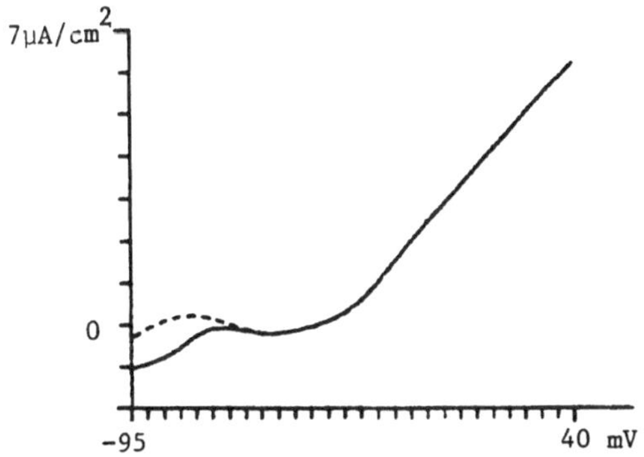

FIGURE 1 : Total steady current of MNT model (dotted line : s held at 1)

III - NUMERICAL METHOD

Sermange [5] has studied the propagation of the Hodgkin-Huxley model of nerve action potential. He proposed a fractionary step scheme which preserves the domain invariance property of the model. Sharp and Joyner [6] used a Crank-Nicholson scheme explicit with respect to the current to compute the propagation of the MNT and the Beeler-Reuter models. Victorri [7], for the same problem, utilized the fact that the equations for y are linear for a fixed V. He uses explicit solutions for y, controlling the time step to have small variations of V. The main difficulty is that the problem is stiff with time constants ranging from 5 μs to 1s. The method used here is a modification of Sermange's one which no more preserves the invariance property but which remains accurate with time steps of the order of the smallest time constant.

This scheme integrates the cable equation at one half step and the MNT model at the other. An implicit Runge-Kunta method of order 2 is used for the MNT model. This scheme has been tested on an explicit solution of a similar problem : we consider a depolarization wave controlled by a unique activation variable m, propagating with velocity w. Let us assume that V(x,t) is given by :

$$V(x,t) = 40\ (\tanh\ [-a(x-wt)] - 1)$$

and that it satisfies

$$\partial V/\partial t - c\ \partial^2 V/\partial x^2 = -\ g_1\ V(V + 40)(V + 80) - g_2\ mV$$

Using the formal computing software MACSYMA, m and then α, β have been calculated such that :

$$dm/dt = -\ (\alpha(V) + \beta(V))\ m + \alpha(V)$$

Numerical values are chosen such that -80 and 0 are stable resting points for the corresponding set of ordinary differential equations, and α, β are positive. Sharp and Joyner's scheme has been compared to ourson this test solution. For $\delta t \simeq \tau_m/7$. Results are given in table 1. :

Sharp-Joyner :	$\delta x = \lambda/6.4$, $w = 0.1004$, $i_m = 40.146$.	$\delta x = \lambda/3.2$, $w = 0.1003$, $i_m = 40.177$
Fractionary Step Method	$w = 0.09997, i_m = 40.28$	$w = 0.09987, i_m = 40.31$

TABLE 1

Computed values of w velocity of propagation (exact value 0.1) and of i_m maximum membrane current (exact value 40.).

The second one is less sensitive to δt. For δt of the order of τ_m the error on w is 2‰, but is is 3% for the first scheme. On the other hand the fractionary step-method gives some local oscillations at the maximum rate of rise, but they do not affect the global error. Both schemes are sensitive to the space step ; large steps have been used by Joyner [3] to simulate the effect of gap functions.

IV - SIMULATIONS

A reentry due to a block of branch is modelled as follows : we consider the branches as two cables stimulated at the same tame at the proximal end, and coupled via the myocardium, at the distal end. On the branch II a one-way block stops the foreward propagation of the action potential but allows it to travel backwards. It is assumed that this block is due to an asymmetric alteration of the tissue : first a zone of ischemia with depolarization allowing only the propagation of slow action potential activated by the slow current i_s alone, then a zone of necrosis with no active current. The action potential going through branch I must be delayed somewhere in order to reach backwards the one-way block after the one going normally on branch II. This delay has been simulated by a zone of slow conduction on branch I. Furthermore one needs to have a short refractory period in order to reactivate the proximal part of branch II after passing through the block. This is done by

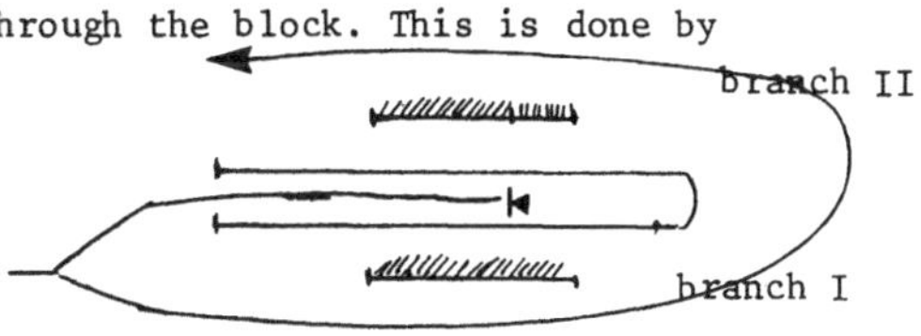

FIGURE 2 : Scheme of a reentry : zone of slow conduction on branch I ; one way block on branch II

decreasing τ_f and τ_{X_1} . Figure 3 shows a situation of reentry, potentials are recorded in the four zone of branch II and at the junction with branch I. Figure 4 shows a reentry with a reflection phenomenon : the summation of the depolarizations of the necrosis zone due to electrotonic currents is sufficient to reexcitate the distal part of branch II. So the backward propagating action potential is transmitted through the block but also reflected on the block in a foreward propagating activation wave. This phenomenon has been studied experimentally by Jalife and Moe [2].

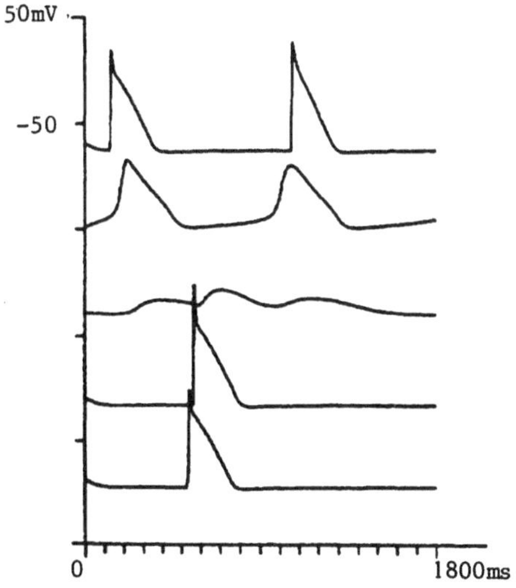

FIGURE 3 : Potential recording along branch II : (normal, slow conduction, no active current zone, normal, branch I junction) : block of the foreward activation backward propagation of A.P. coming from branch I

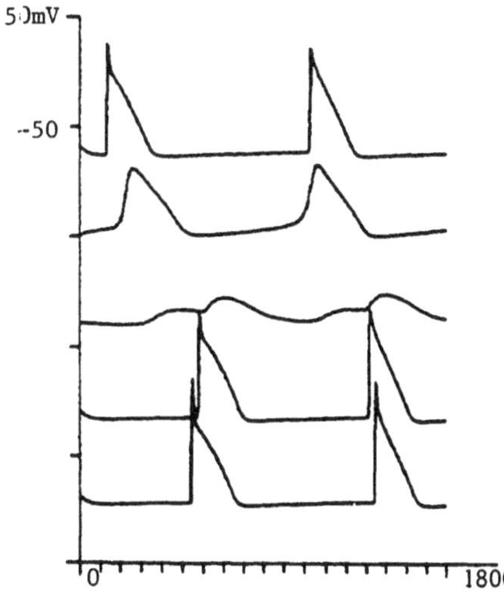

FIGURE 4 : Reflection of the backward propagated A.P. on the block

REFERENCES

[1] E. CORABOEUF. Ionic basis of electrical activity in cardiac tissues. Am. J. Physiol. 234(2), H101-H116, 1978.

[2] J. JALIFE - G. MOE. Excitation, conduction and reflection on impulses in Isolated bovine and canine cardiac purkinje fibers. Circ. Res. 49, pp. 233-247, 1981.

[3] R. JOYNER. Effects of the discrete pattern of electrical coupling on propagation through an electrical syncitium. Circ. Res. 50, pp. 192-200, 1982.

[4] R. Mc ALLISTER-D. NOBLE-R. TSIEN. Reconstruction of the electrical activity of cardiac Purkinje fibers. J. Physiol. 215, pp. 1-59, 1975.

[5] M. SERMANGE. Study of a few equations based on the Hodgkin-Huxley model. Math. Biosc. 36, pp. 45-60, 1977.

[6] G. SHARP-R. JOYNER. Simulated propagation of cardiac action potentials. Biophys. J., Vol 31, pp. 403-424, Sept. 1980.

[7] B. VICTORRI Simulation numérique de potentiels d'action cardiaques. Rapport Technique EP 82-R.23. Ecole Polytechnique, Montréal.

HOPF BIFURCATION AND THE REPETITIVE ACTIVITY OF EXCITABLE CELLS

A.V. Holden
Department of Physiology, University of Leeds,
Leeds LS2 9JT, U.K.

Abstract. Periodic solutions of the Hodgkin-Huxley excitation equations bifurcate from equilibria at super- or sub-critical Hopf bifurcations as the applied current density I, maximal specific K^+-conductance $\bar{g}_K$, the extracellular Ca^{2+} activity $[Ca^{2+}]_o$ and the extracellular K^+ activity $[K^+]_o$ or Nernst potential for K^+, V_K, are varied as bifurcation parameters. Sections through the bifurcation surface produced when $(V_K, \bar{g}_K, I)$ is a bifurcation parameter are presented.

Introduction. The excitable membranes of nerve and muscle cells may be autorhythmic, or may be driven into periodic activity by a variety of physical and chemical agents. These behaviours are the result of a nonlinear current-voltage relation that has voltage-dependent kinetics. It is possible to control the electrical potential across the membrane, and measure the ionic currents: measurements of the amplitudes and kinetics of currents recorded in such voltage-clamp experiments may be used to represent the electrical properties of the membrane by a differential system. The systems of nonlinear o.d.e.s that have been obtained are often of high order, and best approached using numerical methods (Holden, 1980; Holden and Winlow, 1983 a, b).

The first excitation system that was described in detail was the Hodgkin and Huxley (1952) equations for the membrane of the giant axon of the squid. These H-H equations are not generic: other excitable membranes are described by other excitation equations, but the H-H equations have been intensively studied, using qualitative, numerical and analytical methods - see Carpenter (1979), Costen et al. (1975), Hastings (1975), Rinzel (1981). These equations may be represented as

$$\begin{aligned} -dV/dt &= F(V, m, n, h) + I \\ &= I_{Na} + I_K + I_L + I \\ &= I_{Na}(V, m, h) + n^4\bar{g}_K(V-V_K) + I_L(V) + I, \end{aligned} \tag{1}$$

where I_{Na}, I_K, I_L, I are the sodium, potassium, leakage and applied current densities, and the gating variables m, n, h obey first order kinetics, with steady-state values m_∞, n_∞, h_∞ and time constants τ_m, τ_n, τ_h that are functions of voltage. Here I examine bifurcations produced by shifting the voltage-dependence of the gating variables (to simulate changes in extracellular calcium ion activity $[Ca^{2+}]_o$); changing the maximal K^+-conductance $\bar{g}_K$ or its Nernst potential V_K, or changing the applied current density I.

Bifurcations from equilibria. The stationary points of (1) that satisfy

$$F(V, m(V), n(V), h(V)) + I = 0 \quad (2)$$

can, under appropriate parameter ranges, form a curve that exhibits multiple equilibria. The standard H-H system has a single equilibrium solution at any value of I; however, there are multiple equilibria in the presence of hyperpolarising I and low $\bar{g}_K$, or low $[Ca^{2+}]_o$, or high $[K^+]_o$ (Holden et al., 1983; Aihara and Matsumoto, 1983).

If the eigenvalues of the Jacobian matrix evaluated at any stationary point all have negative real parts then the stationary point is linearly stable. Bifurcations between equilibrium points occur where there is a single zero eigenvalue, and a simple Hopf bifurcation occurs when a single complex conjugate pair of eigenvalues cross the imaginary axis. The periodic, small amplitude orbits that emerge are stable at a supercritical bifurcation, and unstable at a subcritical bifurcation: these unstable orbits can provide a transient pathway to other, large amplitude, periodic solutions (Marsden and McCracken, 1976)

Thus it is necessary to evaluate numerically the eigenvalues, and the associated characteristic exponents, of the stationary points of (1) as parameters I, $\bar{g}_K$, V_K, $[Ca^{2+}]_o$ are varied. Methods are given in Hassard et al. (1981).

One-dimensional bifurcation parameters.

I : Sub- and super-critical bifurcations occur at 6.3°C for I ≅ 9.8 and 154 $\mu A\ cm^{-2}$. As the temperature is increased these two points move towards each other, and fuse close to 27.5°C. Just below the sub-critical bifurcation the stable large amplitude periodic solutions (repetitive activity) may be annihilated (Hassard, 1978; Best, 1979; Guttman et al., 1980).

$\bar{g}_K$: Sub-critical bifurcations occur at 6.3°C for $\bar{g}_K \cong 3.8$ and 19.7 $mS\ cm^{-2}$: if autorhythmicity is controlled by K^+-selective conductances, as suggested for embryonic heart muscle (DeHaan, 1980) then the onset of repetitive activity is all-or-none (Holden and Yoda, 1981, 1982).

$[Ca^{2+}]_o$: Sub-critical bifurcations occur at 6.3°C for $[Ca^{2+}]_o \cong$ 3 and 21 $mmol\ dm^{-3}$: the effect of $[Ca^{2+}]_o$ is by binding to surface negative charges, and hence changing the fields experienced by the gating variables (Huxley, 1959; Holden, 1980).

V_K : The Nernst potential $V_K \cong FR/T \log [K^+]_o/[K^+]_i$, and so changes in V_K reflect changes in K^+-concentrations that may result from accumulation or depletion of K^+ in a restricted extracellular space. Hopf bifurcations occur at V_K +14.6 and -1.9 mV.

Two-dimensional bifurcation parameters. When $\bar{g}_K$ or I is a bifurcation parameter, the only bifurcations of (1) are from an equilibrium to periodic orbits at Hopf bifurcations; when ($\bar{g}_K$, I) is a bifurcation parameter there are also bifurcations between equilibria. Bifurcation curves in the $\bar{g}_K$-I, $[Ca^{2+}]_o$ - I and $\bar{g}_K$ - $[Ca^{2+}]_o$ planes are in Holden and Yoda (1982), Holden and Winlow (1983), Holden et al. (1983): Figure 1 shows the bifurcation curves in the V_K - I for (1) at 6.3°C and standard $\bar{g}_K$. The solid lines enclose a region of 3 equilibrium solutions, while the dashed curves are Hopf bifurcations.

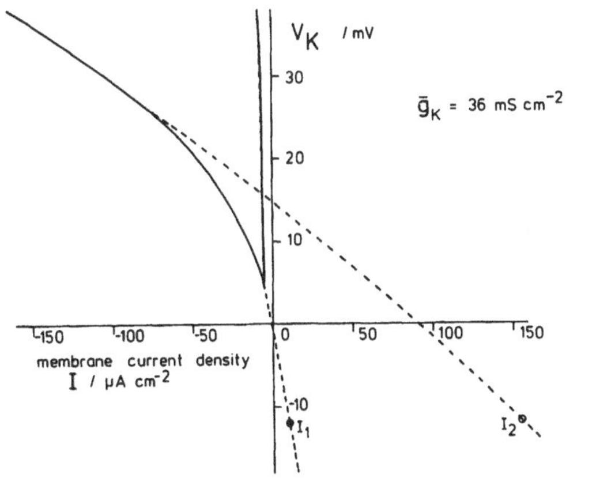

Figure 1. Bifurcations in the V_K - I plane at standard $\bar{g}_K$. The region of 3-equilibria enclosed by the solid line terminates close to I = -6.05 μA cm^{-2}, V_K = 4.48 mV. The dashed lines are Hopf bifurcations, and I_1 and I_2 are the sub- and super-critical Hopf bifurcation points of Hassard (1978).

Three-dimensional parameters. The membrane behaviour is strongly influenced by the repolarizing current ($I_L + I_K$), where $I_K = n^4\bar{g}_K (V - V_K)$, and so a biologically interesting parameter space is ($\bar{g}_K$, V_K, I). $\bar{g}_K$ reflects the number of K^+-selective channels/unit area of membrane, and changes in $\bar{g}_K$ may simulate developmental changes in K^+-channel density, or the effect of K^+-channel blockers. V_K is determined by the intracellular and extracellular K^+ activities: the extracellular volume can be limited, and so outward K^+-current can increase $[K]_o$ and move V_K in the depolarizing direction.

Sections through the ($\bar{g}_K$, V_K, I) bifurcation surface in the (V_K, I) plane are shown in Figure 2: the Hopf bifurcation curve that in-

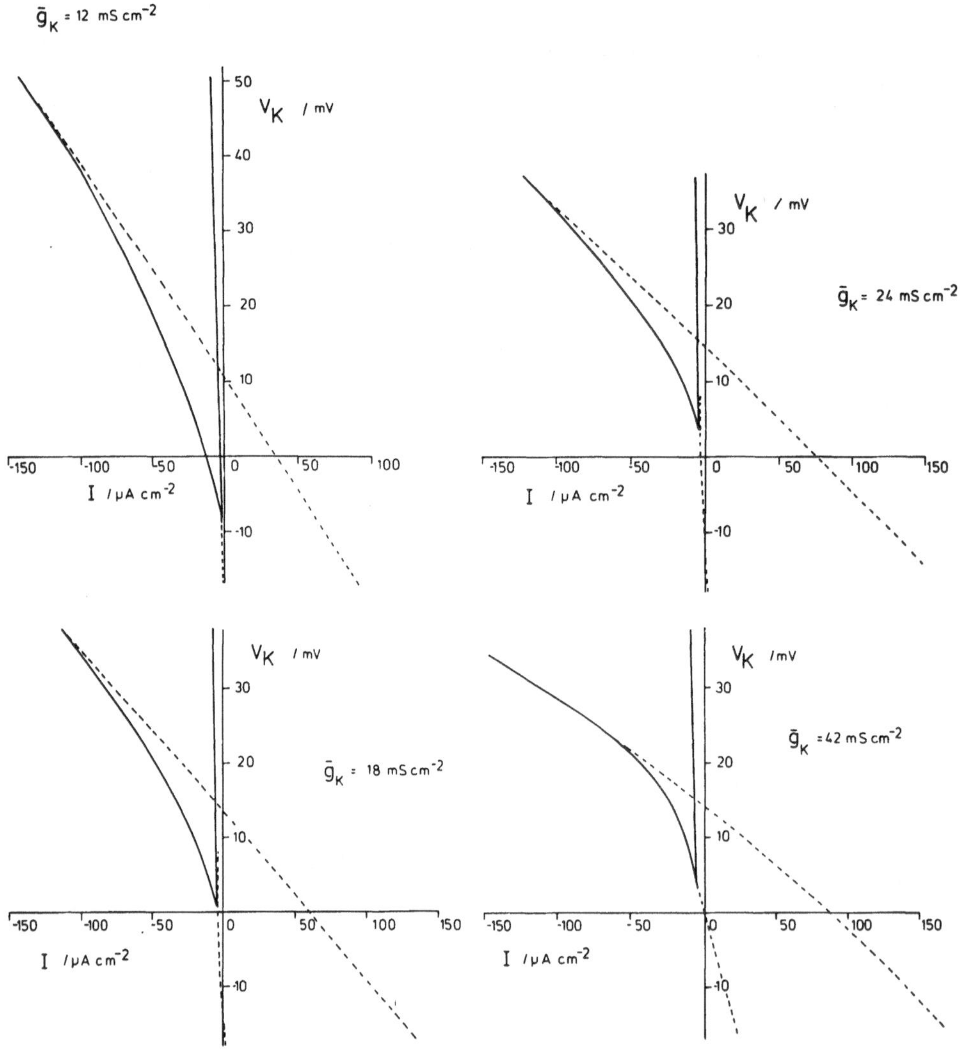

Figure 2. Bifurcation in V_K - I plane with $\bar{g}_K$ = 12, 18, 24, 42 mS cm^{-2}. The solid line encloses the region of 3-equilibria, and the dashed lines are Hopf bifurcations.

cludes the bifurcation I_2 at standard parameter values converges towards the hyperpolarizing edge of the region of multiple equilibria. However, the Hopf bifurcation curve that includes the bifurcation I_1 only runs close to the region of multiple equilibria: this makes numerical tracking of the bifurcation curve difficult.

In spite of these technical problems, it is possible to locate and identify the Hopf bifurcation curves for as complicated a differential system as the Hodgkin-Huxley equations, simply by routine numerical calculations. These bifurcation curves separate equilibria from periodic solutions, and so demonstrate the ways in which simple parameter changes can induce autorhythmicity.

References

Aihara, K., Matsumoto, G.: Two stable steady states in the Hodgkin-Huxley axon. Biophys. J. 41, 87-89 (1983).

Best, E.N.: Null space in the Hodgkin-Huxley equations - a critical test. Biophys. J. 27, 87-104 (1979).

Carpenter, G.: A mathematical analysis of excitable membrane phenomena. Progr. in Cybernetics and Systems Research 3, 505-514 (1979).

Casten, R.G., Cohen, H., Lagerstrom, P.A.: Perturbation analysis of an approximation to the Hodgkin-Huxley theory. Quartl. App. Math. 32, 365-402 (1975).

DeHaan, R.L.: Differentiation of excitable membranes. Current topics in developmental biology 16, 117-164 (1980).

Guttman, R., Lewis, S., Rinzel, J.: Control of repetitive firing in squid axon membrane as a model for a neurone oscillator. J. Physiol. 305, 377-395 (1980).

Hassard, B.: Bifurcations of periodic solutions of the Hodgkin-Huxley model for the squid giant axon. J. Theoret. Biol. 71, 401-420 (1978).

Hassard, B.D., Kazarinoff, N.D., Wan, Y.-H.: Theory and applications of Hopf bifurcation. Cambridge Univ. Press (1981).

Hodgkin, A.L., Huxley, A.F.: A quantitative description of membrane current and its application to conduction and excitation in nerve. J. Physiol. 117, 500-544 (1952).

Holden, A.V.: Autorhythmicity and entrainment in excitable membranes. Biol. Cybernetics 38, 1-8 (1980).

Holden, A.V.: The Mathematics of excitation. Biomathematics in 1980: North Holland Mathematical Studies 58, 15-47. L.M. Ricciardi, A.C. Scott (eds) (1982).

Holden, A.V., Haydon, P.G., Winlow, W.: Multiple equilibria and exotic behaviour in excitable membranes. Biol. Cybernetics 46, 167-172 (1983).

Holden, A.V., Winlow, W.: Neuronal activity as the behaviour of a differential system. I.E.E.E. Trans. SMC (in press) (1983).

Holden, A.V., Winlow, W.: The comparative neurobiology of excitation. In: The Neurobiology of Pain, Holden, A.V., Winlow, W. (eds). Manchester University Press (in press) (1983).

Holden, A.V., Yoda, M.: Ionic channel density of excitable membranes can act as a bifurcation parameter. Biol. Cybernetics 42, 29-38 (1981).

Holden, A.V., Yoda, M.: Bifurcation theory and autorhythmicity of the excitable membrane of nerve cells. Proc. 2nd World Conf. Mathematics at the Service of Mankind, Ballester, A., Cardus, D., Trillas, E. (eds), 355-359 (1982).

Huxley, A.F.: Ion movements during nerve activity. Ann. N.Y. Acad. Sci. 81, 221-246 (1959).

Marsden, J.E., McCracken, M.: The Hopf bifurcation and its applications. Berlin: Springer Verlag (1976).

NONLINEAR EVOLUTION EQUATIONS WITH A CONVOLUTION TERM INVOLVED IN SOME NEUROPHYSIOLOGICAL MODELS.

Michelle SCHATZMAN

Centre de Mathématiques Appliquées
Ecole Polytechnique
91 128 Palaiseau CEDEX - France

1. INTRODUCTION.

The purpose of this paper is to present a family of differential equations yielding a unified scheme which describes in a concise way various phenomena of ontogenetic organization in the central nervous system (C.N.S.).

It has been observed by different authors that patterns are formed during the development of the brain. These patterns are observed by anatomic and physiologic means, and they are usually a result of the interaction with the environment, but this would involve a stochastic process, which will not be considered here.

The equations which are considered here have the form :

(1) $$\frac{\partial u}{\partial t} = (w*u)(x,t) + \text{nonlinear } (u,x,t)$$

or

(2) $$\frac{\partial u}{\partial t} = (w*u)(x,t) \, . \, \text{nonlinear } (u,x,t).$$

Here, x can be

. the location of a neurone in a structure of the C.N.S.

. the location of a couple of neurones in a pair of nervous structures such as for instance, the retina and the tectum

. a parameter of a sensory stimulus, such as a preferred orientation.

Mathematically, x belongs to X, which is either equal to $\mathbb{R}^N$, or to $\mathbb{T}^N = \mathbb{R}^N/\mathbb{Z}^N$ (periodic conditions).

As is natural, t is the time, with a biologically adapted scale. The function u can be :

. synaptic strength ; for instance $u(x_1,x_2)$ is the density of fibres from point x_1 on the retina to the point x_2 on the tectum

. response of a sensory neuron to a given stimulus, defined by x.

We shall assume that u vanishes outside of a given bounded subset of X. If the subset Ω is distinct from X (this is always the case when $X = \mathbb{R}^N$), we shall truncate the functions : define

(3) $$1_\Omega(x) = \begin{cases} 1 \text{ if } x \in \Omega \\ 0 \text{ otherwise ;} \end{cases}$$

then, we shall multiply the right hand side of (1) or (2) by $1_\Omega(x)$.

The function w is an even function of space. We recall that the convolution is defined by

(4) $$(w * u)(x) = \int w(x-y)u(y)\,dy.$$

It can be understood as a weighted averaging process. For instance, if w is constant in a ball of radius r and zero outside, $(w * u)(x)$ is, up to a multiplicative constant, the average of u on a ball of radius r around x.

The nonlinear term in (1) or (2) varies according to the model.

The main principles present in all the models proposed so far are <u>cooperation</u> and <u>competition</u>.

The convolution with the kernel w describes a cooperation (or an anti-cooperation) in the following sense : if, for instance, w is positive in a ball around the origin, then the growth of u is larger at the points where there is a positive average density of u ; for instance, the presence of synapses of a certain kind favors locally the growth of synapses of the same kind. If for instance, w is negative in a ring around the origin, then the growth of u is inhibited in the corresponding translated ring around points with a positive u (lateral inhibition).

This kind of action can be explained by the diffusion of chemical markers which have a very small characteristic time before the time scales considered here ; one of these markers is an activator, the other one is an inhibitor, which diffuses slightly faster, and therefore dominates away from the origin.

The nonlinear term is the competition term : limited resources are available such as space, maximum number of cells, maximum density of synapses,... the competition will be described more precisely in individual models ; it is enough to say at this stage that the competition term is decreasing with respect to u in (1), so that it will prevent some norm of u from blowing up.

The mathematical questions asked are the following :

. The existence and the uniqueness are usually easy.

. Existence of a limit as time increases infinitely is a harder question, usually ; only results on the ω-limit set i.e. the set of limit values of a trajectory for a given topology are given. Observe that the asymptotic states obtained in this fashion are usually less smooth than the initial data : step functions against infinitely differentiable initial data, Dirac measure against continuous initial data.

. Spatial patterns of the possible steady states. Observe that the systems considered below possess usually a large number of steady states.

This work is partially joint with E. BIENENSTOCK and B. MOORE.

Four different types of pattern formation are considered here.

2. DEVELOPMENT OF STIMULUS SELECTIVITY IN INDIVIDUAL NEURONES.

This is the process through an individual neurone tunes to a given stimulus. For instance, the neurone considered is in area 17 of cortex (primary visual cortex), and x is the orientation of a stimulus (a light bar) in the receptive field of the neurone.

The system used to model this phenomenon is as follows : let

(5) $$F(u)(x) = 1_\Omega(x)\,\{(w * u)(x) - [\int_\Omega u(y)\,dy]^2\};$$

the time derivative from the right $\partial^+ u/\partial t$ is given by

(6) $$\frac{\partial^+ u}{\partial t}(x,t) = \begin{cases} F(u)(x,t) \text{ if } u(x,t) > 0, \text{ or if } F(u)(x,t) \geq 0 \\ 0, \text{ otherwise.} \end{cases}$$

with initial condition

(7) $$u(x,0) = u_0(x),$$

where we assume that u_0 is nonnegative almost everywhere in x.

Then, it can be easily checked, even in a very naïve way that, for all t, $u(x,t) \geq 0$ almost everywhere in x.

This model is derived from the stochastic model introduced in [2], after averaging with respect to the distribution of incoming patterns, and some other simplifications. Two essential features remain : stimuli of neighboring orientations cooperate in the modification of synaptic efficacies ; this is natural since such stimuli are overlapping in the space of inputs to the neuron. The second feature is the competition mechanism embodied by the second term of (5). This term acts as a threshold : if $w * u$ is larger than this threshold, u grows ; if $w * u$ is smaller than this threshold, u decays. The nonlinear dependance of this threshold on u, which we assume here, prevents the total mass from growing infinetely, if there is some cooperation ; if w is a Dirac mass centerd at the origin, it can be proved that the total mass is not necessarily bounded uniformly in time. For a discussion of other stochastic models, see [2,4]

It turns out that (5) - (6) can be written as a variational inequality [3,5] : let $K = \{u \in L^2(\Omega) \ / \ u \geq 0 \text{ a.e.}\}$, where $L^2(\Omega)$ is the usual space of square integrable functions. If u is such that

(8) $$u \in C^0([0,T] \ ; \ L^2(\Omega)) \text{ and } \frac{\partial u}{\partial t} \in L^2(0,T \ ; \ L^2(\Omega)),$$

then, u is solution of (5) - (6), if and only if it satisfies

(9) $$\begin{cases} u(.,t) \in K, \ \forall \ t \in [0,T] \\ \int_\Omega \left[\frac{\partial u}{\partial t}(x,t) - F(u)(x,t)\right][v(x) - u(x,t)] \, dx \geq 0, \ \forall \ v \in K. \end{cases}$$

This remark is the key to a very simple existence and uniqueness result.

The main result on the asymptotic behavior of (5) - (6) is the following :

<u>THEOREM 1</u>. Assume that

(10) $$w(x) = z(|x|),$$

with

(11) $$\begin{cases} z \text{ differentiable on } [0,\infty), \\ z' < 0 \text{ on } (0,R), \\ z = 0 \text{ on } [R, \infty). \end{cases}$$

Then, the ω-limit set with respect to the weak topology of the measures, contains, for any trajectory, only finite combinations of Dirac masses of the form

(12) $$u_\infty = \frac{w(0)}{N^2} \sum_{i=1,..,N} \delta(x - x_i),$$

with

(13) $$|x_i - x_j| \geq R, \ \forall \ j, \ \forall \ i \neq j.$$

Moreover, the only stable states contain exactly one Dirac mass.

It can be shown that no function u can be a stable steady state for (5) - (6). On the other hand, it may happen that (5) - (6) has an asymptotic state with N Dirac masses ; we have to only to consider a symmetric initial data, which will retain its symmetry throughout time. If more general w's are considered, we may obtain plateaus in the asymptotic state.

For numerical simulations, see Fig. 1 of [1].

3. SPATIAL ORGANIZATION OF STIMULUS SELECTIVITY.

Orientation selectivity is organized in a fairly regular fashion in primary visual cortex. When cells are recorded along a microelectrode track parallel to the pial surface, preferred orientation progresses regularly, approximately in a piecewise linear manner ; slope reversals or discontinuities are nevertheless not infrequent. After this was first discovered by Huebel and Wiesel, an autoradiography method allowed to observe a fairly regular system of isoorientation stripes - which some authors would rather call patches.

The equation we study now is very close to (5) - (6) ; now, $\Omega = \Omega_x \times \Omega_y$, u is the response of the cell at cortical location y to stimulus x, and it is assumed that the competition mechanism which is responsible for the development of a selective response in cortical neurons acts independantly for each single neuron. However, the convolution operation introduces a general coupling. The new equation reads :

$$F(u)(x,y) = [(w * u)(x,y) - (\int u(x',y)\,dx')^2]\, 1_\Omega(x,y) \tag{14}$$

and

$$\frac{\partial^+ u}{\partial t}(x,y,t) = \begin{cases} F(u(.,t))(x,y) & \text{if } u(x,y,t) \geq 0, \text{ or } F(u(.,t))(x,y) > 0 ; \\ 0 & \text{otherwise.} \end{cases} \tag{15}$$

The theoretical results for (14) - (15) are the following : the existence and uniqueness theory is easy ; the ω-limit set of a trajectory is not empty (use the weak * topology of $L^3(\Omega_y ; M^1(\Omega_x))$, with $M^1(\Omega_x)$ the set of bounded measures on Ω_x), and contains u_∞'s such that

$$(w * u_\infty)(x,y) \leq (\int u_\infty(x',y)\,dx')^2 , \quad \forall\, x,y \tag{16}$$

and

$$\operatorname{supp} u_\infty \subset \{(x,y)/(w * u_\infty)(x,y) = (\int u\ (x',y)\,dx')^2\} . \tag{17}$$

In the numerical experiments, one must choose a w which takes negative values somewhere ; the u is a Dirac mass supported by the graph of a function from Ω_x to Ω_y, whose slope depends on w, but we have presently no precise estimations of the relation between this slope and w. For numerical experiments, see Fig. 2 of [1].

4. DEVELOPMENT OF TOPOGRAPHIC MAPPINGS.

The mapping between two neuronal sheets is a rather general problem, a good example of which is the retinotectal projection.

Let $\Omega = \Omega_x \times \Omega_y$, where x stands for th spatial location on the retina, and y for the spatial location on the tectum, and let u be the density of fibres connecting x to y.

The retina maps accurately onto the tectum if there is a bijection - hopefully bicontinuous between these two structures The principles of interaction which are required are quite analogous to the ones introduced in the previous sections, though this is a phenomenologically very different problem : a short range cooperative interaction is assumed, together with two simultaneaous long range competitions : between $u(x_1,y)$ and $u(x_2,y)$ and between $u(x,y_1)$ and $u(x,y_2)$.

We define a function F by

(18) $$F(u)(x,y) = \{(w * u)(x,y) - [\int u(x',y)\,dx'][\int u(x,y')\,dy']\}\, 1_\Omega(x,y),$$

and the equation we consider is the same as (15), with this new F.

The kernel w is assumed nonnegative.

The existence and uniqueness are known, but, essentially, there are no results on the asymptotic behavior of u, as time goes to infinity.

The behavior observed in numerical experiments is that the asymptotic state is a mass uniformly distributed on the graph of a piecewise continuous bijection between Ω_x and Ω_y. In the experiments which were performed, Ω_x and Ω_y were identical and of dimension 1 ; the slope of the bijection was either +1 or -1 on each continuous piece. The length of the continuous pieces depended on the size of the support of w, and on the initial data. See Fig. 3a and 3b of [1] for these experiments.

For other models of the same problem, see [7,8,13,14,15].

5. SPATIALLY PERIODIC STRUCTURES WITH PURELY LOCAL COMPETITION.

The last two applications we consider are the development of ocularity domains in the primary visual cortex of mammals, and a second model for the spatial organization of stimulus selectivity. In the first case, we have a competition between two populations of neuronal fibers of different origins, which innervate one region. Initially, they are rather uniformly distributed, and in the course of development, they segregate into periodic stripes, which alternately contain cells of the first population, and of the second population. It is known that similar phenomena occur widely in the central nervous system.

In [11], N. Swindale proposes a simple system which formalizes mathematically the appropriate cooperation and competition principles, for the case of ocularity domains. It is

(19) $$\frac{\partial u^L}{\partial t} = (w^{LL} * u^L + w^{LR} * u^R)(N^L - u^L)\, u^L\, 1_\Omega,$$
$$\frac{\partial u^R}{\partial t} = (w^{RL} * u^L + w^{RR} * u^R)(N^R - u^R)\, u^R\, 1_\Omega.$$

Here u^L is the density of fibres from the left eye, and u^R is the density of fibres from the right eye. The kernels w describe the cooperation (negative or positive) between the two populations of fibres, and the nonlinear term $(N^L - u^L)u^L$ (respectively $(N^R - u^R)u^R$) constrains u^L (resp. u^R) to stay between 0 and a threshold N^L (resp. N^R) if the initial data u_0^L (resp. u_0^R) takes its values between 0 and N^L (resp. N^R).

If one makes the following symmetry assumptions :

$$w^{LL} = w^{RR} = - w^{LR} = - w^{RL},$$

$$N^L = N^R = N$$

$$u^L + u^R = N, \text{ almost everywhere,}$$

then, one can show easily that $u_0^L + u_0^R = N$, almost everywhere in x and for all t. After scaling, the equation satisfied by $u = u^L - u^R$ is

$$\frac{\partial u}{\partial t} = (w * u)(1 - u^2)\, 1_\Omega \,. \tag{20}$$

We can modify (20) as follows, in order to have an equation of the form (1) :

$$\frac{\partial^+ u}{\partial t} = \begin{cases} (w * u)\, 1_\Omega & \text{if } -1 < u < 1, \text{ or if } u = 1 \text{ and } w * u \leq 0, \\ & \text{or if } u = -1 \text{ and } w * u > 0, \\ 0 & \text{otherwise.} \end{cases} \tag{21}$$

Global existence and uniqueness for (20) and (21) can be shown easily, because, in both cases, with initial data taking their values between -1 and +1, u remains in the interval $[-1,+1]$ for all time.

To obtain some kind of positive result on the asymptotic behavior of u, one has to make some functional assumptions on w. The assumptions are the following :

(22) If u is such that $w * u$ vanishes on a set of positive measure, then u vanishes identically.

This condiiton is satisfied on a dense subset of the set of integrable functions. Under (22), the only steady states of (21) satisfy either $u = 0$ a.e. or $|u| = 0$ a.e. .

Moreover, a nonzero state is (linearly) stable if $|u| = 1$ a.e. and

$$u\,(w * u) \geq 0, \text{ almost everywhere.} \tag{23}$$

There is a Lyapunov function for both (20) and (22), which is defined by

$$\Phi(u) = - \frac{1}{2} \int_{\Omega\times\Omega} w(x - y)\, u(x)\, u(y)\, dx\, dy, \tag{23}$$

and non-zero stationary stable solutions of (20) are local minima of Φ on the convex set

$$K = \{u / -1 \leq u \leq 1 \text{ a.e.}\} \quad . \tag{24}$$

For a number of theoretical results on (20) and (21), involving a discussion of the properties of the Fourier transform of w in relation with the wave number selection, see [9,10]. Swindale has performed a number of numerical experiments on (19) and (20) ; they are reported in [11] ; there, he suggests that the wave number selected is the one for which the Fourier transform $\hat{w}$ of w has a positive strict maximum ; this is true only in first approximation . in the periodic case, this does not seem to be true.

Another model for the spatial organization of stimulus selectivity has been proposed by Swindale [12]. There, he considers a family of populations dependending on a continuous parameter, the preferred orientation, and models the phenomenon as

$$\frac{\partial u}{\partial t}(x,t) = (w * u)(x,t)\, (1 - |u(x,t)|^2)\, 1_\Omega(x), \tag{25}$$

where u takes its values in $\mathbb{C}$, the field of complex numbers, and initially, $|u_0|$ is lesser than or equal to 1.

The module $|u(x,t)|$ is the synaptic efficiency corresponding to the orientation $\arg(u(x,t))$. Under condition (22) on w, the only steady states satisfy either $u = 0$ a.e., or $|u| = 1$ a.e., amd the linearized stability condition is

$$\mathrm{Re}(\bar{u}\ (w * u)) \geq 0.$$

If $X = \Omega = T^1$, and if the maximum of the Fourier coefficients $\hat{w}(k) = \int w(x)\ e^{2\pi ikx}\ dx$ is positive and attained for $k \neq 0$, then it can be shown that $u(x) = \exp(\pm 2\pi ikx)$ satisfies (26). In some special cases, where the support of w is small, functions of the form

$$u(x) = \begin{cases} e^{2\pi ikx}, & x \in [a,b] \\ e^{-2\pi i(kx+\phi)}, & x \in [a,b] \end{cases}$$

satisfy (26) too. Those can be called solutions with defects, and correspond to observations in the brain, and to the numerical experiments of Swindale [12]. The article [12] contains a number of numerical simulations, and a wonderful color picture of the spatial organization of preferred orientations as obtained with (25).

The main difference between (20),(21), (25) and the previously described models is the purely local nature of the competition between the different species of cells.

REFERENCES.

[1] E. BIENENSTOCK, Cooperation and Competition in the Central Nervous System Development : a unifying approach. To appear in Synergetics, Proceedings of a Symposium held at Elmau, May 83, H.Haken Ed., Springer-Verlag.

[2] E. BIENENSTOCK, L. COOPER, P. MUNRO, Theory for the development of neuron selectivity : orientation specificity and binocular interaction in visual cortex, J. Neuroscience 2(1982)32-48.

[3] H. BREZIS, Opérateurs maximaux monotones et semi-groupes de contractions dans les espaces de Hilbert, North Holland, 1973.

[4] L. COOPER, F. LIEBERMAN, E. OJA, A theory for the acquisition and loss of neuron specificity in visual cortex, Biol. Cybern. 33(1979)9-28.

[5] J.L. LIONS, G.STAMPACCHIA, Variational Inequalities, Comm. Pure Appl. Maths, 20 (1967)493-519.

[6] C. VON DER MALSBURG, Self-organization of orientation sensitive cells in the striate cortex, Kybernetik 14(1973)85-100.

[7] C. VON DER MALSBURG, D.J. WILLSHAW, How to label nerve cells so that they can interconnect in an ordered fashion, Proc. Nat.Acad. Sci. USA 74(1977)5176-5178.

[8] C. VON DER MALSBURG,D.J. WILLSHAW, Differential Equations for the development of topological nerve fibre projections, SIAM-AMS Proccedings 13(1981)39-47.

[9] M. SCHATZMAN, Spatial structuration in a model in neurophysiology, To appear in the proccedings of the applied mathematics seminar of the Collège de France.

[10] M. SCHATZMAN, E. BIENENSTOCK, Neurophysiological models, Spin models, and pattern formation, to appear.

[11] N. V. SWINDALE, A model for the formation of ocular dominance stripes, Proc. R. Soc. Lond. B 208(1980)243-264.

[12] N. V. SWINDALE, A model for the formation of orientation columns, Proc. R. Soc. Lond. B 215(1982)211-230.

[13] D. J. WILLSHAW, C. VON DER MALSBURG, How patterned neural connections can be set up by self organization, Proc. R. Soc. Lond. B 194(1976)431-445.

[14] D. J. WILLSHAW, C. VON DER MALSBURG, A marker induction mechanism for the establishment of ordered neural mappings : its applications to the retinotectal problem. Philosoph. Trans. R. Soc. Lond. B 287(1979)203-243.

[15] A.F. HÄUSSLER, C. VON DER MALSBURG, Development of retinotopic projections, An analytic treatment, J. Theor. Neurobiol. (in press).

INVESTIGATION OF RAPID METABOLIC REACTIONS IN WHOLE ORGANS BY MULTIPLE PULSE LABELLING

Andreas J. Schwab, Adelar Bracht, & Roland Scholz
Institut für Physiologische Chemie, Physikalische Biochemie und Zellbiologie
der Universität München

Goethestrasse 33, D-8000 München 2
Federal Republic of Germany

Flux rates of metabolic and transport processes in whole organs may be determined by conventional compartmental system analysis (1,2). However, this approach is not consistent with the events in real organs characterized by continuous variation of tracer concentrations with space. When the events under study are comparable in rate with transit times or recirculation, their adequate description is only possible by using distributed model systems.

The basis of the distributed systems presented here is the multiple-indicator dilution technique which has been employed mainly by Goresky et al. for the investigation of transport of substances in catheterized or isolated perfused organs (3,4). This method consists in the simultaneous bolus injection of several labelled substances into the arterial perfusate flow, followed by the analysis of the tracer contained in the venous effluent perfusate as a function of time after injection. Various metabolically inert tracers have been used as indicators for the various spaces (vascular, extracellular, intracellular etc.) (5). In addition to the determination of the sizes of these spaces, this technique has been employed for the quantitative investigation of transport processes (3-10).

The mathematical description of the multiple-indicator dilution technique as it has been presented until now can be applied to metabolic processes only in simple cases, when the intracellular metabolism is irreversible and can be described using a single rate constant (3,11). It is not applicable when the metabolism is rapid and involves reversible reactions, or when a labelled metabolic product is released into the effluent, as in hepatic lactate metabolism (12). In the present report, a more general theory will be presented, based on a combination of the principles of the multiple-indicator dilution technique with those of conventional compartmental system analysis, allowing the analysis of complex situations with an arbitrary number of metabolic reactions.

Formulation of the Basic Equations

An organ will be regarded as a collection of capillaries (sinusoids in the case of a liver) with different transit times perfused in parallel, and of great vessels with a uniform transit time. The events in a single capillary will be treated assuming that the tracer equilibrates rapidly across the capillary, whereas lengthwise diffusion will be neglected (3,8-12). By generalizing the equations published previously for simple systems (3,8-12), the events in a single capillary will be described by the following system of partial differential equations:

$$\frac{\partial u(x,t)}{\partial t} + W \frac{\partial u(x,t)}{\partial x} = A\ u(x,t) + \delta(x)\delta(t)e_1 \qquad (1a)$$

$$u(x,t) = 0, \qquad t < 0 \text{ or } x < 0 \qquad (1b)$$

Eq. (1a) is derived from the equation for conventional compartmental systems by adding a term $W\,\partial u(x,t)/\partial x$ describing the longitudinal motion of the tracer due to the convection of the perfusate. t is time; x is the location along a capillary; u(x,t) is a vector-valued function with elements $u_i(x,t)$, $u_i(x,t)$ dt representing the franction of the tracer initially applied to the capillary contained at time t in pool i in the section of the capillary between x and x + dx; A is a compartmental matrix derived from rate constants analogous to that used in compartmental systems (1,2); δ (x)δ(t) is the tensor product of two impulse functions, and $e_1 = (1,0,0...0)^T$ ($\delta(x)\delta(t)e_1$ represents instantaneous injection of tracer into pool 1 at x = 0 and t = 0); W is a diagonal matrix with constant non-negative real elements w_i which represent the mean velocities of motion of the substances in the various pools. Whenever a substance is exchanged rapidly between various spaces including the vascular space ("flow-limited distribution"), it may be regarded to be contained in a single pool with mean velocity

$$w_i = \frac{w_s}{1 + \theta_i}, \qquad (2)$$

where w_s is the velocity of the perfusion fluid and θ_i is the ratio of the volume of the additional distribution space to that of the vascular space.

The impulse response of the whole organ will be formulated in the following way (3,8-12):

$$h(t) = \frac{1}{w_{ref}}\ W \int_0^\infty u(x,t - t_o)\ h_{ref}\left(\frac{x}{w_{ref}} + t_o\right)dx \qquad (3)$$

h(t) is a vector-valued function with elements $h_i(t)$, where $h_i(t)$ dt represents the fraction of the injected radioactivity appearing in the venous perfusate in pool i between the times t and t + dt. $h_{ref}(\cdot)$ is the impulse response of a non-metabolizable reference substance with known space of flow-limited distribtion; it reflects the heterogeneity of the transit times of the capillaries resulting from variations in regional flow (3,5,12). t_o is the combined transit time of the great vessels and of the collecting device; it is determined separately using non-metabolizable tracers (3,5,10).

Note. Because there is no longitudinal diffusion, the label cannot migrate faster than w_{max} (the maximal element of W); mathematically, this is a consequence of the fact that eq. (1a) represents a hyperbolic system of partial differential equations. Hence, $u(x,t) = 0$ for $x > w_{max}t$. The upper limit of the integral in eq. (3) may therefore be replaced by $w_{max}(t - t_o)$.

Solution of the Equations

For an evaluation of the impulse response of the whole organ, a solution u(x,t) of eqs. (1a) and (1b) must be found which is then substituted in eq. (3). Analytical solutions have been published only for some simple cases of eq. (1a) including no more than a single rate constant for metabolism (3,8-12). For the general case, however, a solution in a closed form has not yet been found.

A very efficient method for finding the impulse response of the whole organ is the following: The distribution of the transit times of the capillaries, $h_{ref}(t)$, is approximated by a sum of exponentials:

$$h^*_{ref}(t) = \sum_{k=1}^{N} \alpha_k \, e^{-\beta_k(t - t_o)}, \qquad t \geq t_o \qquad (4)$$

The complex coefficients α_k and β_k may be obtained by the method of moments (13). Because h(t) vanishes for $t \longrightarrow \infty$, the error in the aproximation may be made arbitrarily small by choosing a suitable set of β_k (14). The impulse response of the whole organ may then be approximated by the following equation:

$$h^*(t) = W' \sum_{k=1}^{N} \alpha_k \, e^{(t - t_o)(A - \beta_i W')} e_1, \qquad t \geq t_o, \qquad (5)$$

where $W' = (1/w_{ref})W$. The relative error in approximating h(t) by h*(t) does not exceed the relative error in approximating $h_{ref}(\tau)$ by $h^*_{ref}(\tau)$ in the interval $t_o \leq \tau \leq (t - t_o)w_{max}/w_{ref} + t_o$ (the interval $t_o \leq \tau \leq t$ if $w_{ref} = w_{max}$).

Proof: Let $\bar{u}(s,t) = \int_0^{\infty} u(x,t)\, e^{-sx}\, dx$. Then, by Laplace transformation of eq. (1a) and solution of the resulting system of ordinary differential equations:

$$\bar{u}(s,t) = e^{t(A - sW)} e_1 \qquad (6)$$

Substituting $h_{ref}(\cdot)$ in eq. (3) by $h^*_{ref}(\cdot)$ in eq. (4), one obtains:

$$h^*(t) = W' \sum_{k=1}^{N} \alpha_k \, \bar{u}\left(\frac{\beta_k}{w_{ref}}, t - t_o\right), \qquad t \geq t_o \qquad (7)$$

Substitution of eq. (6) in eq. (7) yields eq. (5).

Let ϵ be the maximum relative error in approximating $h_{ref}(\tau)$ by $h^*_{ref}(\tau)$:

$$h^*_{ref}(\tau) \leq (1 + \epsilon)h_{ref}(\tau), \qquad t_o \leq \tau \leq (t - t_o)w_{max}/w_{ref} + t_o \tag{8}$$

Because of conservation of label, $u(x,\tau)$ and $h_{ref}(\tau)$ are nonnegative for all $x \geq 0$, $\tau \geq 0$. Therefore, ineq. (8) implies (after substituting $\tau = x/w_{ref} + t_o$):

$$\int_0^{w_{max}(t - t_o)} u_i(x,t - t_o)\, h^*_{ref}\left(\frac{x}{w_{ref}} + t_o\right)dx \leq (1 + \epsilon) \int_0^{w_{max}(t - t_o)} u_i(x,t - t_o)\, h_{ref}\left(\frac{x}{w_{ref}} + t_o\right)dx, \quad \forall\ i$$

By comparing with eq. (3) with the upper limit of the integral modified according to the Note in the previous chapter, one obtains $h^*_i(t) \leq (1 + \epsilon)h_i(t)$ and, analogously, $h^*_i(t) \geq (1 - \epsilon)h_i(t)$, i.e., the relative error in approximating $h(t)$ by $h^*(t)$ does not exceed ϵ, q.e.d.

Application to Hepatic Lactate Metabolism

Injection of [2-^{3}H]lactate into the portal vein of a perfused rat liver under non-isotopic steady-state conditions yields the experimental impulse response shown by the data points in the Figure (12). Most of the radioactivity in the effluent is found in non-metabolized lactate; about 20 % of the label represents tritiated water formed metabolically from lactate. The data are interpreted by a system with the structure represented by the digraph in the insert of the figure. Pool 1 represents extracellular lactate, pool 2 total water, pool 3 intracellular lactate, and pool 4 intracellular metabolites exchanging tritium with intracellular lactate. The system is characterized by the following matrices:

$$A = \begin{pmatrix} -a_{31} & 0 & a_{13} & 0 \\ 0 & 0 & a_{23} & a_{24} \\ a_{31} & 0 & -a_{13}-a_{23}-a_{43} & a_{34} \\ 0 & 0 & a_{43} & -a_{24}-a_{34} \end{pmatrix}$$

$$W' = \begin{pmatrix} w'_1 & 0 & 0 & 0 \\ 0 & w'_2 & 0 & 0 \\ 0 & 0 & 0 & 0 \\ 0 & 0 & 0 & 0 \end{pmatrix}$$

The elements of the matrix W' are found as follows: The space for flow-limited distribution and thus the mean velocity of lactate is the same as that of the extracellular reference substance, sucrose; thus, $w'_1 = w_1/w_{ref} = 1$. In contrast, water is exchanging very rapidly between all spaces; its mean velocity w_2 is equal to that of urea, a non-metabolizable substance which behaves like water (3,5,12). From eq. (2) it is deduced that $w'_2 = w_2/w_{ref} = 1/(1 + \theta')$, where $\theta' = (\theta_2 - \theta_{ref})/(1 + \theta_{ref})$,

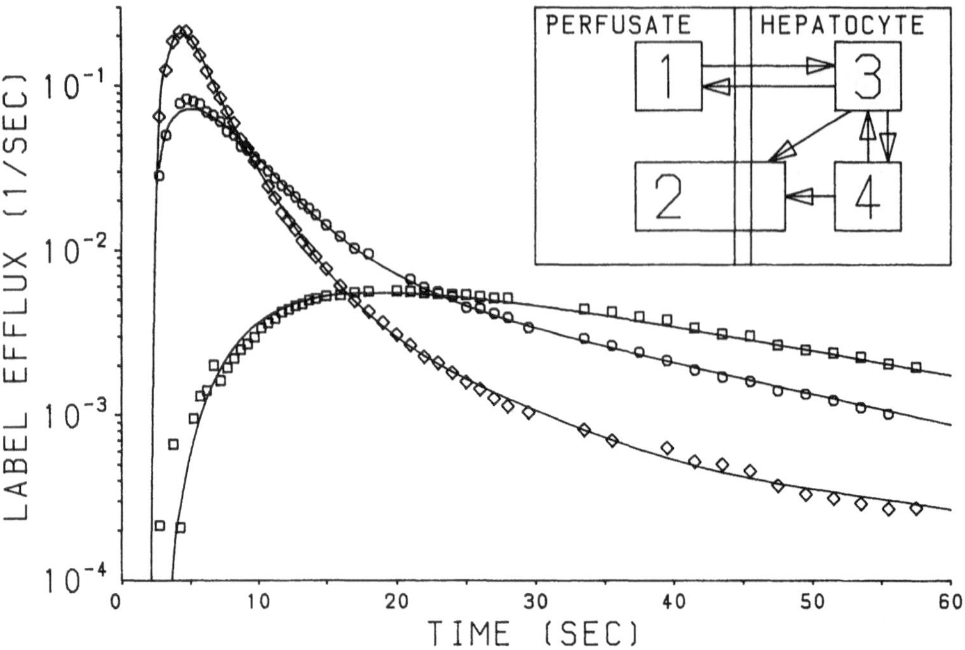

Figure. Impulse response of L-[2-^{3}H]lactate in perfused rat liver. L-[2-^{3}H]lactate, [U-^{14}C]sucrose, and [^{14}C]urea were injected simultaneously into the portal vein of a rat liver perfused in a non-recirculating system. The radioactivity in lactate (○), water as its metabolic product (□), and sucrose (◇), expressed as the fraction of the injected radioactivity appearing per second in the venous effluent, was plotted against the time after injection (data for urea are not shown). The reference curve ([U-^{14}C]sucrose) was approximated by a sum of six complex exponential terms (——). The theoretical impulse responses for lactate and for ^{3}HOH as its metabolic product (——) were calculated using the model discribed in the text and the following parameters: a_{31} = 1.25 s^{-1}, a_{13} = 1.13 s^{-1}, a_{43} =0.14 s^{-1}, a_{34} = 0.032 s^{-1}, a_{23} = 0.0073 s^{-1}, a_{24} = 0.027 s^{-1}, w_2/w_1 = 0.35, t_o = 2.12 s. Insert: Compartmental digraph describing the model system for hepatic lactate metabolism. The double vertical bar denotes the plasma membrane separating the intra- from the extracellular space. Pools are numbered as follows: 1, extracellular lactate pool; 2, total water pool; 3, intracellular lactate pool; 4, additional intracellular metabolite pool

the ratio of the intra- to the extracellular volume, is determined together with t_o from the impulse responses of sumultaneously injected [U-^{14}C]sucrose and [^{14}C]urea, using published procedures (3,5,12). For the experiment shown in the Figure, the following values were determined: t_o = 2.12 s; θ' = 1.83; w_2' = 0.35.

An approximation of the impulse response of the reference substance [U-^{14}C]sucrose by a sum of six exponential terms according to eq. (4) is shown in the Figure. Its maximal relative deviation from the experimental data points is 10 %, and this is also the maximal relative error expected in the approximation of the theoretical impulse resoponse for lactate and water, using eq. (5). In fact, the latter agrees within 1 % with solutions calculated by finite difference or frequency domain methods analogous to those published for simple cases (15,16), but requires far less computer time.

Values for the non-diagonal elements of A (see Figure caption) representing rate constants for transport or metabolism were found by fitting eq. (5) to the data for lactate and water by means of standard least-square procedures. The fairly good agreement between the theoretical curves and the experimental data points indicates that the model is an adequate description of the events in the liver. Like in the case of conventional compartmental systems, the rate constants may be used to calculate fluxes and pool sizes in metabolic steady state (2).

This work was supported by the Deutsche Forschungsgemeinschaft. Computations were performed at the Leibniz Rechenzentrum, Bayerische Akademie der Wissenschaften, using software from International Mathematical and Statistical Libraries (Houston, Tex.). We are grateful to Dipl.-Phys. Cornelia Sedlmeir for carefully reviewing the manuscript.

References

1. Berman, M. & Schoenfeld, R. (1956) Invariants in experimental data on linear kinetics and the formulation of models. J. Appl. Phys. **27**, 1361-1370.
2. Brown, R.F. (1980) Compartmental system analysis: state of the art. I.E.E.E. Trans. Biomed. Engn. **27**, 1-11.
3. Goresky, C.A. & Bach, G.G. (1970) Membrane transport and the hepatic circulation. Ann. N. Y. Acad. Sci. **170**, 18-47.
4. Goresky, C.A. (1982) The processes of cellular uptake and exchange in the liver. Federation Proc. **41**, 3033-3039.
5. Goresky, C.A. (1963) A linear method for determining liver sinusoidal and extravascular volumes. Am. J. Physiol. **204**, 626-640.
6. Schwab, A.J., Bracht, A., & Scholz, R. (1979) Transport of D-lactate in perfused rat liver. Eur. J. Biochem. **102**, 537-547.
7. Bracht, A., Kelmer Bracht, A., Schwab, A.J., & Scholz, R. (1981) Transport of inorganic anions in perfused rat liver. Eur. J. Biochem. **114**, 471-479.
8. Goresky, C.A., Bach, G.G., & Nadeau, B.E. (1973) On the uptake of materials by the intact liver. The concentrative transport of rubidium-86. J. Clin. Invest. **52**, 975-990.
9. Rose, C.P., Goresky, C.A., & Bach, G.G. (1977) The capillary and sarcolemmal barriers in the heart. An exploration of labeled water permeability. Circ. Research **41**, 515-533.
10. Goresky, C.A., Bach, G.G.,& Nadeau, B.E. (1975) Red cell carriage of label. Its limiting effect on the exchange of materials in the liver. Circ. Research **36**, 328-351.
11. Goresky, C.A., Bach, G.G., & Nadeau, B.E. (1973) On the uptake of materials by the intact liver. The transport and net removal of galactose. J. Clin. Invest. **52**, 991-1009.
12. Bracht, A., Schwab, A.J., & Scholz, R. (1980) Studies of flow rates in the isolated perfused rat liver by pulse labelling with radioactive substrates and mathematical analysis of the wash-out kinetics. Hoppe-Seyler's Z. Physiol. Chem. **361**, 357-377.
13. Dyson, R.D. & Isenberg, I. (1971) Analysis of exponential curves by a method of moments, with special attention to sedimentation equilibrium and fluorescence decay. Biochemistry **10**, 3233-3241.
14. Szász, O. (1916) Über die Approximation stetiger Funktionen durch lineare Aggregate von Potenzen. Math. Ann. **77**, 482-496.
15. Bassingthwaighte, J.B. (1974) A concurrent flow model for extraction during transcapillary passage. Circul. Res. **35**, 483-503.
16. Rowlett, R.D. & Harris, T.R. (1976) A comparative study of organ models and numerical techniques for the evaluation of capillary permeability from multiple-indicator data. Math. Biosci. **29**, 273-298.

Cross-linking of Identical Particles by Multiple Ligand-Types

John L. Spouge

Los Alamos National Laboratory
Theoretical Biology, T-10
Mail Stop 465, P.O. Box 1663
Los Alamos, New Mexico 87545
United States of America

Abstract

This paper uses branching processes to model the cross-linking of identical particles by multiple ligand-types. We derive gel points and mole- and weight-average cluster sizes for Binomial and Poisson bonding.

In Immunology, this model might apply to cross-linking by antibodies specific to different antigenic sites. It represents a refinement of the Goldberg-Watson theory of immune complex formation and makes predictions readily tested by experiment.

The model makes the undesirable assumption that no intramolecular bonding occurs. Relaxation of this assumption is mathematically challenging and is of interest to polymer chemists.

1. Introduction

Figure 1 shows several types of ligand bonding identical particles ("units") into a connected cluster. (Note that not all ligands form bonds between units, but every bond-type corresponds to a ligand-type.)

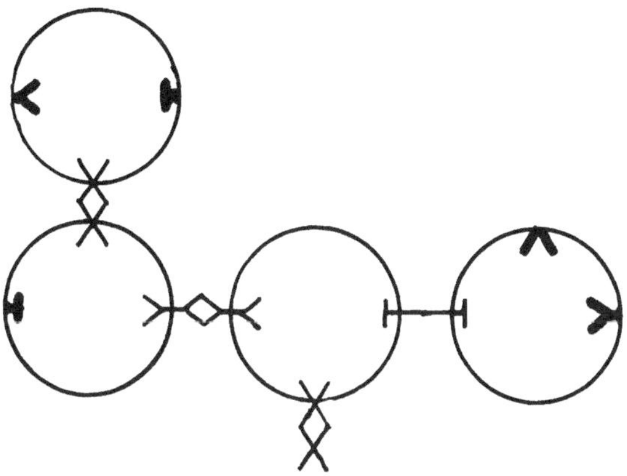

Immunology provides a paradigm for this process. In a mismatched blood transfusion, for example, the donor red blood cells (RBC's) have several antigenic sites foreign to the recipient's immune system. Antibodies specific for each type of site cross-link corresponding sites on RBC's. Similar phenomena occur in solution (e.g. immune complex formation) and on all surfaces (e.g. B-lymphocyte patching, RBC capping, and lectin recpetor clusting). Bell, et. al. (1978) review some of these.

Polymer chemistry has explored models of bonding. These can be kinetic, e.g. Smoluchowski's (1916, 1917) coagulation equation (Dostal and Raff (1936), reviewed by Drake (1972), extended to post-gelation times by Ziff (1980) and Ziff, et. al. (1983)) or equilibrium (Flory (1941), Stockmayer's (1943, 1944) statistical mechanical microcanonical ensembles, Whittle's (1965a,b) detailed balance relations for equilibria). We use the branching process model of Gordon (1962) and Good (1963) because it includes some other models as special cases (Spouge 1983a,b,c,d,e).

Spouge (1983f) examines multiple particle- and bond-types (possibly asymmetric) and generalises this paper.

2. The Model Assumptions

We make two assumptions: (1) No Intramolecular Bonding: clusters contain no cycles, and (2) First Shell Substitution Effect (FSSE): A unit's bonding does not affect its neighbours' bonding.

The second assumption neglects steric effects (e.g., the neighbours of a highly-bonded unit should have fewer bonds) but can be relaxed (Gordon and Scantlebury 1964). The first assumption has also been relaxed (Whittle 1965b, 1980a,b, Gordon and Scantlebury 1966) but the results are not as satisfactory.

We index the bond-types by Roman lower case (a-bonds, b-bonds, etc.) and refer to the number of units in a cluster as cluster "size". Use of brackets, e.g. {a}, indicates an idea relevant to all bond-types a,b,c, etc.

3. Mathematical Method

Pick a unit at random. Let a,b,c... be marker variables for the unit's bonds. The probability generating function (p.g.f.) for the bonding of the unit is

$$P(a,b,c...) = \Sigma P_{qrs...}\, a^q b^r c^s ... \qquad (1)$$

where $P_{qrs...}$ is the proportion of units having q a-bonds, r b-bonds, etc.

Pick at random a unit on the end of an {a}-bond. Because of FSSE, the unit's distribution is $P_{qrs...}$. This *a priori* probability must be weighted by the number of a-bonds, q, coming into the unit (since these increase the probability of the unit's being on the end of an a-bond). Normalizing to get a probability, we divide by

$$r_a = \Sigma P_{qrs...} \cdot q \qquad (2)$$

Hence the p.g.f. for the bonds that this unit gives rise to, excluding the incoming a-bond is

$$A(a,b,c...) = \Sigma \frac{P_{qrs...} \cdot q}{r_a}\, a^{q-1} b^r c^s ... = \frac{1}{r_a} \frac{\partial P}{\partial a} \qquad (3)$$

(and similarly for b,c...).

The cluster containing a random unit can be modelled as a family tree (or branching process). The random unit is a progenitor, giving rise to 1st generation {a}-type individuals of distribution (1). Subsequent generations produce offspring according to (3).

Let ρ be the maximum positive eigenvalue of

$$\underset{\sim}{R} = \left[\frac{1}{r_a}\frac{\partial^2 P}{\partial a \partial b}(1,1,1\ldots)\right] \tag{4}$$

(a and b index rows and columns.) There is an infinite cluster (supercritical branching process, Athreya and Ney (1972)) when $\rho > 1$.

Note: Assumption (1) ensures that the process does branch, i.e. that there are never two lines of descent from the progenitor to a descendant.

In the "Binomial Model"

$$P(a,b,c\ldots) = (1-\alpha+\alpha a)^A (1-\beta+\beta b)^B (1-\gamma+\gamma c)^C \ldots \tag{5}$$

the determinant formula

$$\det[a_j(1-\delta_{ij}b_i)]_{n\times n} = (-1)^n \prod_{i=1}^{n} (a_i b_i).[1-\sum_{i=1}^{n} b_i^{-1}] \tag{6}$$

is useful ($\delta_{ij} = 1, i=j$, and 0 otherwise).

4. Results

In the "Binomial Model", each unit can {a}-bond to most A other units. Each of the A {a}-bonds actually occurs with probability α, independent of other bonds from the unit. (In our example, each RBC has A antigenic sites of one type, B of another, etc., all sites bonding independently.) An infinite cluster forms for (the equivalence zone)

$$\frac{A\alpha}{1+\alpha} + \frac{B\beta}{1+\beta} + \frac{C\gamma}{1+\gamma} + \ldots > 1 \tag{7}$$

If x_k is the proportion of clusters containing k units, the mole-average cluster size is

$$\underset{\sim}{M}_1 = \frac{\Sigma k x_k}{\Sigma x_k} = [1-\tfrac{1}{2}(A\alpha+B\beta+C\gamma+\ldots)]^{-1} \tag{8}$$

and the weight-average cluster size is

$$\underset{\sim}{M}_2 = \frac{\Sigma k^2 x_k}{\Sigma k x_k} = [1 - \frac{A\alpha}{1+\alpha} - \frac{B\beta}{1+\beta} - \frac{C\gamma}{1+\gamma} - \ldots]^{-1} \tag{9}$$

if there is no infinite cluster (specialised from equations (37) and (38) for $\underset{\sim}{M}_1$ and $\underset{\sim}{M}_2$ in Spouge (1983f)).

In the "Poisson Model" the number of possible a-bonds, A, is essentially infinite, but α is so small that the average number of a-bonds from a unit, $A\alpha$, is finite. Substitution of $A\alpha$ and $\alpha = 0$ in (7)-(9) yields the Poisson result.

6. Discussion

Only recently have size distributions become accessible to measurement (Von Schulthess et al. 1980, 1983). On the other hand, experimental polymer gel points (infinite clusters) have long been known to agree well with theoretical predictions (Flory 1941, Spouge 1983a, Stockmayer 1983).

Theoretical success is not as conspicuous in Immunology but classical results (Goldberg 1952, 1953, Watson 1958) predate the experimental refinements allowed by monoclonal antibodies. This paper refines the classical models accordingly. Experimental testing of equation (7) by sequential addition of different monoclonal antibodies would be especially easy. Macken and Perelson (1983) review branching processes in Immunology.

The absence of intramolecular bonding is an undesirable feature of the model under discussion. Allowance for cycles is a difficult graphical enumeration problem (discussion, Spouge 1983f) whose solution would be useful to polymer chemists.

Acknowledgement: This work was supported by a Fellowship from the Medical Research Council of Canada.

Biblography

Athreya, K.B. and Ney, P.E. 1972 Branching Processes (Springer-Verlag, Berlin)

Bell, G.I., Perelson, A.S. and Plimbley, G. 1978 Theoretical Immunology (Marcel Dekker, New York)

Dostal, H. and Raff, R. 1936 Z. Phys. Chem. B32: 117

Drake, R.L. 1972 in Topics in Current Aerosol Research, Hidy and Brock, eds. (Pergamon Press, New York) Vol. 3, Part 2

Flory, P.J. 1941 J. Am. Chem. Soc. 63: 3083, 3091, 3096

Goldberg, R.J. 1952 J. Am. Chem. Soc. 74: 5715

Goldberg, R.J. 1953 J. Am. Chem. Soc. 75: 3127

Good, I.J. 1963 Proc. Roy. Soc. London A272: 54

Gordon, M. 1962 Proc. Roy. Soc. London A268: 240

Gordon, M. and Scantlebury, G.R. 1964 Trans. Faraday Soc. 60: 604

Gordon, M. and Scantlebury, G.R. 1966 Proc. Roy. Soc. London A292: 380

Macken, C.A. and Perelson, A.S. 1983 in Lecture Notes in Biomathematics (Springer-Verlag, Berlin) in press

Schulthess, G.K. von, Benedek, G.B., 1980 Macromolecules 13: 393

Schulthess, G.K. von, Benedek, G.B. and De Blois, R.W. 1983 Macromolecules in press

Smoluchowski, M. von 1916 Physik Z. 17: 585

Smolchowski, M. von 1917 Z. Phys. Chem. 92: 129

Spouge, J.L. 1983a Macromolecules 16: 121

Spouge, J.L. 1983b J. Phys. A.: Math. Gen. 16: 767

Spouge, J.L. 1983c Macromolecules 16: in press

Spouge, J.L. 1983d J. Phys. A.: Math. Gen. 16: in press

Spouge, J.L. 1983e Adv. Appl. Prob. submitted

Spouge, J.L. 1983f Proc. Roy. Soc. London A: in press

Stockmayer, W.H. 1943 J. Chem. Phys. 11: 45

Stockmayer, W.H. 1944 J. Chem. Phys. 12: 125

Stockmayer, W.H. 1983 Personal Communication

Watson, G.S. 1958 J. Immunol. 80: 182

Whittle, P. 1965a Proc. Camb. Phil. Soc. 61: 475

Whittle, P. 1965b Proc. Roy. Soc. London A285: 501

Whittle, P. 1980a Adv. Appl. Prob. 12: 94, 116, 135

Whittle, P. 1980b Theory Prob. Appl. 26: 350

Ziff, R.M. 1980 J. Stat. Phys. 23: 241

Ziff, R.M., Ernst, M.H. and Hendriks, E.M. 1983 J. Phys. A.: Math. Gen.: submitted

ENUMERATION OF RNA SECONDARY STRUCTURES BY COMPLEXITY

G. VIENNOT
Université de Bordeaux I
33405 TALENCE, FRANCE

M. VAUCHAUSSADE de CHAUMONT
Université de Bordeaux II
33076 BORDEAUX, FRANCE

Abstract. - Many investigations in studying primary and secondary structures in Biology require theoretical statistical (that is enumerative) work. We solve one of these problems: enumerate secondary structures of single-stranded nucleic acids (RNA, tRNA, etc...) having a given complexity. This parameter has been introduced for energy computation purpose in order to predict the most stable secondary structure. The method relies on the (non-classical) use of non-commutative variables. Some orthogonal polynomials appear. The final solution shows a relationship between the parameter complexity and another parameter appearing in Hydrography and Botanic.

§ 1. - Introduction.

The primary structure of single-stranded nucleic acids (such as RNA, tRNA, mRNA) is the linear sequence of nucleotides (or bases) linked by phosphodiester bonds. Hydrogen bonds fold the molecule into a planar picture called its secondary structure (see figure 1).

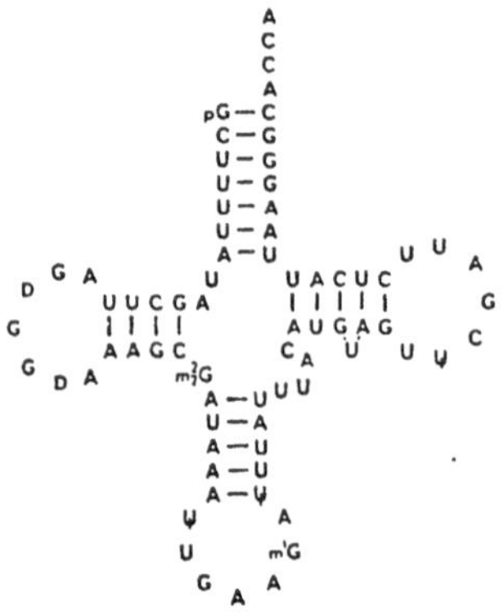

Figure 1 : Secondary structure of $tRNA^{phe}$

Much work has been done by biologists in predicting the most stable secondary structure, once primary structure and rules for evaluation of the Helmhotz free energy are known (see for example Tinoco et al. [11]). Of particular importance are ladders and hairpines, that is parallel hydrogen bonds ending in a loop. Mitiko Gô classified in [5] hairpines by a parameter called order, for energy computation purpose.

Waterman restated rigorously this idea in [11]. He defined mathematically secondary structures as a certain class of planar graphs (containing all known RNAs secondary structures), and the parameter order (or complexity) of the molecules as an integer measuring the complexity of the intricate connexions between hairpines.

Waterman raised the question of finding the number $a_{n,k}$ of secondary structures having n bases and complexity k. In particular for $k=1$, he proves that $a_{n,1}$ is asymptotically of the form λ^n, where λ is the greatest root of the equation $x^3 - 2x^2 - 1 = 0$, that is $\lambda \simeq 2.2055$.

Note that in this context, we are only interested by the planar picture, and not by the particular labelling of the vertices of the graph with the four possible bases A, C, G, U.

Our main result is the following :

THEOREM 1.- The generating function $s_k(t) = \sum_{n\geq 0} a_{n,k} t^n$ of secondary structures of complexity k is

$$s_k(t) = \frac{t^{(5.2^{k-1}-2)}}{(1-t)\, Z_1(t) \dots Z_k(t)} ,$$

where $Z_1(t), \dots, Z_k(t)$ are the polynomials defined by the recurrence

$$Z_1(t) = 1-2t-t^3 , \quad Z_{k+1}(t) = Z_k^2(t) - 2t^{(5.2^{k-1})} .$$

We use a methodology dear to professor M.P. Schützenberger. First, we encode a secondary structure by a word written with letters of an alphabet X. The set of all such words is an algebraic language (a familiar concept in Theoretical Computer Science, called also context-free language or language accepted by a pushdown automaton). Then we introduce the formal sum S_k of all the words coding secondary structures with complexity k. This formal sum satisfied an algebraic system, where the unknowns are formal power series in non-commutative variables X with integer coefficients. Each equation is the traduction of a combinatorial property of the word (and thus of secondary structures classified by complexity).

By replacing all the variables of X by one variable t, we obtain an ordinary algebraic system and S_k becomes $s_k(t)$. Then we solve this system by introducing some one parameter orthogonal polynomials, related to the second kind Tchebycheff polynomials $U_n(x)$, and establishing apparently new identities about these polynomials.

§ 2. - Coding secondary structures with Motzkin words.

If we follow the vertices (or bases) of the graph defined by a secondary structure in the order given by the primary structure, we define a word w in three letters $a, x, \bar{x}$. We write $x, \bar{x}$ or a, according to the fact we meet an hydrogen bond for the first time, second time or the base is not linked by an hydrogen bond.

Notations : The set of words written with the alphabet $X = \{a, x, \bar{x}\}$ is the free monoid denoted by X^*. The concatenation of two words $u = u_1, \ldots, u_p$ and $v = v_1, \ldots, v_q$ is $uv = u_1 \ldots u_p v_1 \ldots v_q$. The empty word is denoted by e. A word f is a factor of the word w if w can be written $w = ufv$. For $x \in X$ and $u \in X^*$, $|u|_x$ denotes the number of occurrences of x in the word u. The length of the word u is $|u| = \sum_{x \in X} |u|_x$.

Waterman's definition for secondary structures is equivalent to say that the associated word w satisfied the three following conditions :

(1) for every factorization $w = uv$, $|u|_x \geqslant |u|_{\bar{x}}$,

(2) $|w|_x = |w|_{\bar{x}}$,

(3) w has no factor $x\bar{x}$.

Remark : The words of $\{a, x, \bar{x}\}^*$ (resp. $\{x, \bar{x}\}^*$) satisfying (1) and (2) are called Motzkin words (resp. Dyck words). The number of such words of length n is classically called the Motzkin number M_n (resp. Catalan number $C_n = \frac{1}{n+1}\binom{2n}{n}$) (see also Stein, Waterman [6]). Condition (3) comes from the fact that two consecutive bases cannot be linked by an hydrogen bond.

The correspondance between secondary structures and words is a coding. The parameter complexity is translated as follows.

If u is a Motzkin word, take $\alpha(u)$ the Dyck word obtained by deleting the letters a. Now, define a pyramid of a Dyck word as to be a factor $f = x^p \bar{x}^p$ $(p \geqslant 1)$ and a maximal pyramid as a pyramid not factor of another pyramid. Then every Dyck word w has a unique factorization $w = u_1 v_1 \ldots u_q v_q u_{q+1}$, where $v_1, \ldots, v_q$ are maximal pyramids and $q \geqslant 1$. We define $\pi(w) = u_1 \ldots u_{q+1}$ by deletion of all maximal pyramids.

The complexity (or order) of the secondary structure coded by the Motzkin word w is the minimum integer k (called order of the word) such that $\pi^k(\alpha(w)) = e$ (empty word).

Example : $w = x\,x\,x\,\bar{x}\,x\,x\,\bar{x}\,\bar{x}\,\bar{x}\,\bar{x}$, $\pi(w) = x\,x\,\bar{x}\,\bar{x}$, $\pi^2(w) = e$, the order is 2 .

For $k \geq 1$, let S_k be the non-commutative (in variables $a, x, \bar{x}$) power serie with integer coefficients, defined as the formal sum of all words coding secondary structures and of order k . Then some combinatorial considerations lead to prove that S_k satisfies the following non-commutative algebraic system (where S_k , $S_{\leqslant k}$, T_k , $T_{\leqslant k}$ are non-commutative series), and this system has only one solution.

(4)
$$\begin{aligned}
S_k &= S_{\leqslant k} - S_{\leqslant k-1} , && (k \geqslant 1)\\
S_{\leqslant k} &= (1 - T_{\leqslant k})^{-1} , && (k \geqslant 0)\\
T_{\leqslant k} &= T_o + T_1 + \ldots + T_k , && (k \geqslant 0)\\
T_k &= x\, S_{\leqslant k-2}\, T_{k-1}\, S_{\leqslant k-2}\, T_{k-1}\, S_{\leqslant k-1}\, \bar{x} + x\, S_{\leqslant k-1}\, T_k\, S_{\leqslant k-1}\, \bar{x} , && (k \geqslant 2)\\
T_o &= a ,\\
T_1 &= x\, S_o\, T_1\, S_o\, \bar{x} + x\, S_o\, T_o\, \bar{x} .
\end{aligned}$$

§ 3. - Proof of theorem 1.

We give here a brief outline of the proof. We introduce the polynomials (with b as parameter) defined by the linear recurrence

(5)
$$\begin{aligned}
F_{n+1}(x,b) &= x\, F_n(x,b) - \lambda_n\, F_{n-1}(x,b) ,\\
&\text{with} \quad \lambda_n = 1 \text{ if } n \text{ odd} ,\\
&\qquad\qquad\quad = b \text{ if } n \text{ pair},\\
F_o &= 1 , \quad F_1 = x .
\end{aligned}$$

These polynomials are orthogonal with respect to a linear form $P(x) \to \int_\alpha^\beta P(x)\, d\psi$ defined by its moments $\mu_n(b)$

(6)
$$\mu_n(b) = \int_\alpha^\beta x^n\, d\psi = \sum_{k \geqslant 1} \frac{1}{n} \binom{n}{k} \binom{n}{k-1}\, x^b .$$

Define $G_n = G_n(x,b)$ as to be the reciprocal of the polynomial $F_n(x,b)$, that is $G_n = x^n F_n(\frac{1}{x}, b)$. By combinatorial arguments using recent studies on orthogonal polynomials, we prove the two following identities, for every $1 \leqslant q < p$,

(7)
$$G_{2p} = G_p^2 - b t^2\, G_{p-1}^2 ,$$

(8) $$G_{2p+2}\, G_{2q+1} - G_{2p+1}\, G_{2q+2} = b^{p+1}\, t^{4p+4}\, G_{2q-2p-1} \,.$$

Let $H_n(x,b)$ be the reciprocal of the polynomial $F_n(x-1,b)$ and $V_n(x) = H_n(x,x)$. Replacing in (4) all the variables $a, x, \bar{x}$ by t gives an algebraic system with unknowns the generating function $s_k(t)$ (resp. $s_{\leqslant k}(t)$) of words (or secondary structures) of order k (resp. order $\leqslant k$).

Using (7) and (8) we prove by recurrence on k that the solution is given by

(9) $$s_k(t) = \frac{t^{(5.2^{k-1}-2)}}{V_{(2^{k+1}-1)}(t)} \quad , \quad s_{\leqslant k}(t) = \frac{V_{(2^{k+1}-2)}(t)}{V_{(2^{k+1}-1)}(t)} \,.$$

Using other combinatorial arguments, we deduce theorem 1.

§ 4. - Concluding remarks.

If we delete the letter a in the system (4), then everything becomes simpler. We obtain the generating function $d_{\leqslant k}(t)$ for Dyck words of order $\leqslant k$:

(10) $$d_{\leqslant k}(t) = \frac{G_{(2^{k+1}-2)}(t,1)}{G_{(2^{k+1}-1)}(t,1)} \,.$$

Note that for $b = 1$, $F_n(x,1) = U_n(x/2)$ where $U_n(x)$ is the Tchebycheff polynomial of second kind defined by $\sin(n+1)\theta = \sin\theta\, U_n(\cos\theta)$. It is surprising that the generating function (10) has already appeared in Computer Science as describing a parameter over trees [2]. This parameter is known as the Strahler number of a tree [3]. It was introduced by Strahler [9] in fluvial Hydrography, and appears also in Anatomy and Botanic [7]. Thus, the parameters "order of a Dyck word" and "Strahler number of a tree" have the same distribution.

Another interesting application of coding secondary structures with words would be algorithmic for computing the most stable secondary structure. One of the most important methods, due to Tinoco et al. [10] uses the base pairing matrix. This is also related to Waterman algorithm using the adjacency matrix of the corresponding graph. It is much more convenient to manipulate words of length n rather than $n \times n$ matrices, especially if n is more than 1000 (as for example the Escherichia Coli 16-S ribosomal RNA). In fact, we need now an alphabet with twelve letters formed by a pair (z, B) where z is a, x or $\bar{x}$, as in § 2 and B is a base A, C, G, U . This word can also be generated by an algebraic (context-free) grammar. Also, particularly important are algorithms comparing several analoguous secondary structures at the same time (as for example in [8]).

Other combinatorial problems, using the same kind of ideas (coding a planar picture with words of an algebraic language) has recently been solved, as for example the number of certain polymers in dilute solutions [1] [4] (or the so-called polyominoes or animals of the cell growth problem) .

REFERENCES

[1] M. DELEST and G. VIENNOT, Algebraic languages and polyominoes enumeration. To be published in Lecture Notes in Computer Science, Proceedings 10th ICALP, Barcelona, July 1983.

[2] P. FLAJOLET, J.C. RAOULT, J. VUILLEMIN, The number of registers required for evaluating arithmetic expressions. Theor. Comp. Sc. (1979), 99-125.

[3] J. FRANÇON, Sur le nombre de registres nécessaires à l'évaluation d'une expression arithmétique. A paraître R.A.I.R.O.

[4] D. GOUYOU-BEAUCHAMPS and G. VIENNOT, Equivalence of the two-dimensional directed animal problem to a one-dimensional path problem, submitted to J. of Physics A.

[5] MITIKO GÔ , Statistical mechanics of biopolymers and its application to the melting transition of polynucleotides. J. Phys. Soc. Japan, 23 (1967).

[6] P.R. STEIN and M.S. WATERMAN, On some new sequences generalizing the Catalan and Motzkin numbers. Discrete Maths 26 (1979), 261-272.

[7] P.S. STEVENS, Patterns in Nature, Little Brown and Co. (1974).

[8] P. STIEGLER, P. CARBON, J.P. EBEL and C. EHRESMANN, A general secondary-structure model for procaryotic and eucaryotic RNAs of the small ribosomal subunits. Eur. J. Biochem. 120 (1981) 487-495.

[9] A.N. STRAHLER, Hypsometric (area-altitude) analysis of erosional topology. Bull. Geological Soc. Amer., 63 (1952), 1117-1142.

[10] I. TINOCO, O.C. UHLENBECK and M.D. LEVINE, Estimation of secondary structure in ribonucleic acids, Nature 230 (1971), 362-367.

[11] M.S. WATERMAN, Secondary structure of single-stranded nucleic acids, in "Studies in Foundations and Combinatorics", Advances in Maths. Suppl. Studies, 1 (1978), 167-212.

Growth of Cell Populations

Eberhard O. Voit and Georg Dick

Zoologisches Institut der Universität zu Köln
Weyertal 119, 5000 Köln 41
Bundesrepublik Deutschland

Introduction

Growth is one of the most evident features of life. We know that it has been studied mathematically for at least several thousand years (in Savageau 1979). Nevertheless, there are still more open questions than answered ones. The growth of cell populations can be examined under many different aspects (reviewed in Voit and Dick 1983 a). The study of growth in one or more dimensions brings up questions about allometric growth and thus the formation of patterns in a very general sense. Both are phenomena that are tightly connected to differentiation and all kinds of control mechanisms. A cell population can be considered as an entity - for instance a tissue or an organism - that increases in volume or length, or as a set of proliferating individuals such as protozoa. Cell population growth obviously depends on the cell cycle whose many detáils are subject to intensive research. The cell cycle in turn - and therefore cell population growth - is based on genetical, biochemical, and physiological mechanisms. Particular motivation in researching cell population growth stems from the hope to understand cancer and to provide a more effective cancer therapy.

In our paper, we will present a mathematical model (cf. Voit and Dick 1983 a,b) that deals with just one aspect of cell population growth. We address the question how a cell population increases in number if its cells have different cycle durations. As an example, we have analyzed the experimental data shown in Figure 1. Prescott (1959) studied the growth of ciliates of the species Tetrahymena pyriformis. He also determined the frequency distribution of generation times as shown in Fig. 1. Our model tries to demonstrate the connections between the frequency distribution and the growth curve. We will show that the frequency distribution contains enough information to produce a growth curve like that observed. But it will turn out that the curve fit is considerably improved if one assumes that the generation times of a ciliate is correlated to that of its mother.

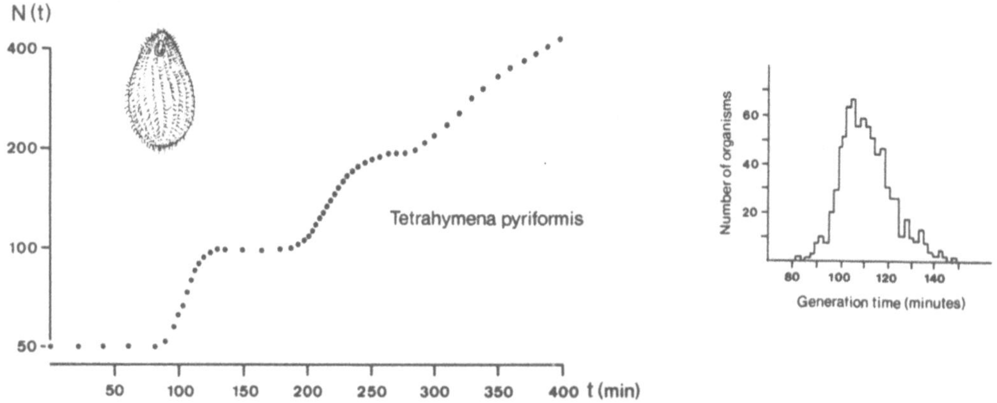

Figure 1: Growth of a population of ciliates (Tetrahymena pyriformis) and distribution of generation times (Prescott 1959).

The Model

Let us first outline the general concept of the model. We assume that each cell grows for a while and then divides into two daughter cells. We confine our considerations to doubling because this is the most common mode of cell proliferation. However, doubling can easily be replaced by other constant proliferation rates if required. At each cell division, each daughter cell is assigned an individual cycle duration, that is the length of the time period after which this cell will divide again. All cells are assumed to survive and to stay within the population. That means, we do not consider cell death, emigration, immigration, differentiation and so forth.

We use a discrete time scale, because cell divisions correspond to discrete steps in the growth process. However, we will also show how a continuous function approximates the discrete model, if the population grows larger.

Our basic terms are the expected number $N(t)$ of cells at time t and a discrete frequency distribution of cycle durations with at most m different cycle durations. This frequency distribution is used to estimate the probability that a particular cycle duration is assigned to a cell at a cell division.

We will now derive a formula for the expected number of cells, $N(t)$. We define $A_i(t)$ as the number of cells with age i at time t. The number of cells with age 1 at time t, $A_1(t)$, is twice the number of new cells, $N(t)-N(t-1)$, because all new cells are age 1 and for each new cell there is a sister which is also age 1.

(1) $$A_1(t) = 2\cdot(N(t) - N(t-1))$$

The fraction of cells with age 1 and cycle duration k is $p_k\cdot A_1(t)$. All cells with age 2 at time t were age 1 at time $t-1$ and those with age i were age 1 at time $t-i+1$. Consequently, we have to expect $p_k\cdot A_1(t-i+1)$ cells with cycle duration k and age i at time t. When we collect the cells of all ages and all cycle durations, we obtain as the expected number of cells at time t

(2) $$N(t) = \sum_{k=1}^{m} \sum_{i=0}^{k-1} p_k\cdot A_1(t-i)$$

Similar to Eq.(1), $A_1(t-i)$ can also be expressed in terms of $N(t-i)$ and $N(t-i-1)$. Using these substitutions, we obtain as the expected number of cells at time t

(3) $$N(t) = 2\cdot\sum_{k=1}^{m} p_k\cdot N(t-k)$$

Note that in Eq.(3) only the probabilities p_k and certain earlier cell numbers appear.

$N(t)$ is a linear combination of $N(t-1),\dots,N(t-m)$. Since these earlier cell numbers are again linear combinations of again earlier cell numbers, $N(t)$ can finally be written as a linear combination of the first m cell numbers $N(1),\dots,N(m)$. Therefore, we can write $N(t)$ in its most general form as a linear combination with real coefficients $x_i(t)$ that depend on the p's but not on any cell numbers.

(4) $$N(t) = \sum_{i=1}^{m} x_i(t)\cdot N(i) \qquad N(t) = \vec{x}(t)^{tr}\cdot\vec{N}$$ (5)

Eq.(4) can equivalently be written in vector notation; the coefficients x_i are then entries of a column vector $\vec{x}(t)$, the entries of the column vector $\vec{N}$ are $N(1),\dots,N(m)$.

Because the coefficients $x_i(t)$ are independent of any cell numbers, the expected number of cells at time t+1, N(t+1), can be represented in two different ways: either as linear combination of N(1),...,N(m)

$$(6) \qquad N(t+1) = \sum_{i=1}^{m} x_i(t+1)\; N(i)$$

or as linear combination of N(2),...,N(m+1)

$$(7) \qquad N(t+1) = \sum_{i=2}^{m+1} x_{i-1}(t)\; N(i)$$

These two representations allow us to compare the coefficients $x_i(t)$ and $x_i(t+1)$. N(m+1) is substituted by the m previous cell numbers as shown in Eq.(3).

The comparison of coefficients yields a simple matrix equation for the dependence of the coefficient vector $\vec{x}(t+1)$ on the previous coefficient vector $\vec{x}(t)$ and, hence, on the initial vector $\vec{x}(1)$, which must be the first unit vector. The expected number of cells, N(t), can now be calculated from the initial cell numbers N(1),...,N(m) by iterative multiplication with the matrix M:

$$(8) \qquad M = \begin{bmatrix} 0 & \cdots & 0 & 2p_m \\ 1 & \ddots & \vdots & \vdots \\ 0 & \ddots & 0 & \vdots \\ \vdots & \ddots & \ddots & \vdots \\ 0 & \cdots\, 0 & 1 & 2p_1 \end{bmatrix}$$

An Example

Let us consider a population with the five different cycle durations 8,...,12. Each of these cycle durations may have the same expectations 0.2. Shorter cycle durations are not possible in our population, i.e. $p_1,\ldots,p_7$ are equal to zero. We assume that in the beginning there is a single cell with age 1 and cycle duration 10 (Fig. 2).

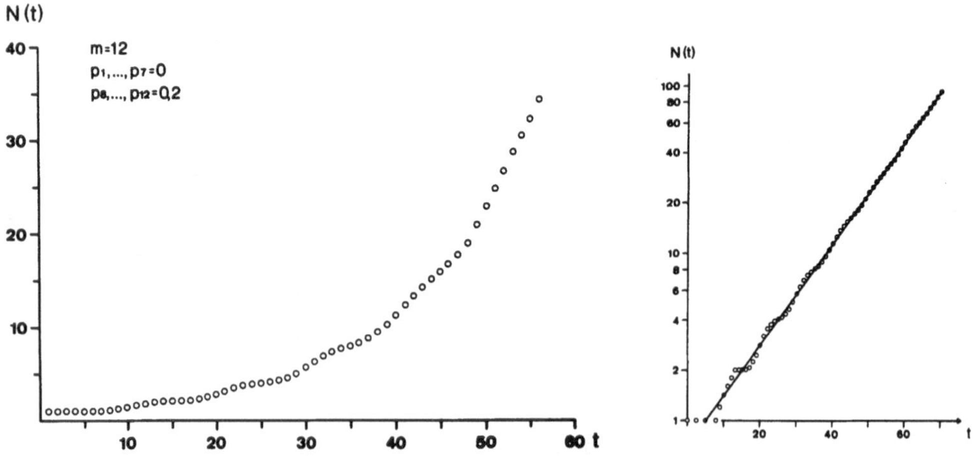

Figure 2: Growth of a cell population; metric and logarithmic scaling. See text for further details.

The growth begins rather slow but becomes faster as the cell number increases. Although there are different cycle durations, one can surmise that the population eventually will grow exponentially. This hypothesis is supported by logarithmic scaling (see Fig. 2, right panel), where the growth curve increasingly approximates a straight line. The intuitive impression of exponential growth of a large population can be proven mathematically within the model (Voit and Dick 1983 a). We will skip the proof here and only make some remarks.

The population growth is tightly connected to the matrix M, and, in particular, to its convergence. It can be shown that the matrix multiplied to a vector converges to a positive eigenvector that corresponds to the dominant real eigenvalue, the so-called spectralradius (Pullman 1976, Voit and Dick 1983 a). This spectral radius is closely related to the average cycle duration during the exponential growth. The average cycle duration is the time needed to double the number of cells. Equation (9) shows a possible representation of the exponential increase in the cell number.

$$N(t) \approx \alpha \cdot 2^{t/\tau} \qquad (9)$$

It doesn't matter mathematically which base we choose. We took base 2 in order to indicate cell doubling. The parameter α is a positive constant and can be considered as some kind of an initial value. τ is the average cycle duration. As mentioned before, τ is connected to the matrix M and can be calculated from the characteristic polynomial. However, it is much easier to assume equality in equation (9) and to express N(t) in terms of N(t-1) (Eq.(10)) and in terms of earlier cell numbers (Eq.(11)).

$$(10) \qquad N(t-1)\; 2^{1/\tau} = N(t) = 2 \sum_{k=1}^{m} p_k\, N(t-k) \qquad (11)$$

From Eqs.(10) and (11) we obtain

$$1 = \sum_{k=1}^{m} p_k\, 2^{(\tau-k)/\tau} \qquad (12)$$

Eq.(12) is an implicit representation of the average cycle duration τ with dependency on the probability distribution $p_1,\ldots,p_m$. The equation in general has no known analytical solution, but is easily solved by numerical methods. It should be mentioned that τ is none of the famous average values like arithmetic or geometric mean, although its numerical value might not differ much from the arithmetic mean. In our constructed example, the arithmetic mean is 10, whereas τ is about 9.93.

We will now return to our original example concerning the growth of a synchronized population of the ciliate Tetrahymena (cf. Fig. 1). We simplified the distribution of generation times from experiments to only eight generation times 80,90,...,150 minutes. The relative generation times were read off Prescott's distribution. Although this method may seem to be rather crude, it turned out not to be very sensitive and yielded, for instance, almost identical results when slightly different frequencies were used or even, when only six or seven generation times were distinguished. Figure 3 shows the curve fit with the model.

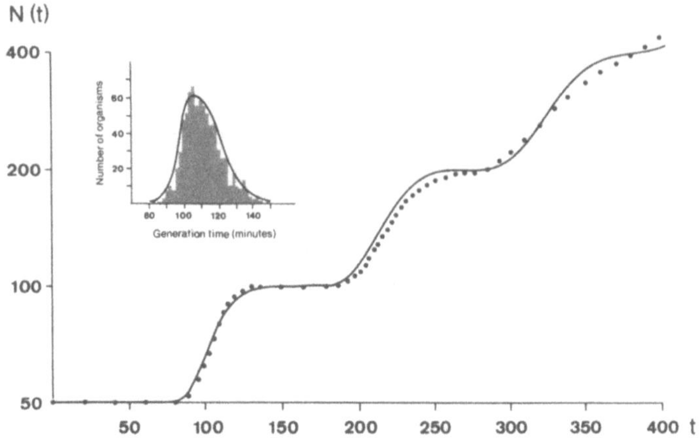

Figure 3: Approximation to Prescott's frequency distribution and observed (•) and calculated (line) growth curve of Tetrahymena.

It should be emphazised that in the present model only Prescott's data are used and no further assumptions are made. The curve fit is good in the beginning and clearly shows the stepwise growth of the population. However, later the synchrony does not fade in the model as fast as it does in the ciliate population. This discrepancy cannot be due to the probability distribution we used nor to poor synchronization of the ciliates. This fact leads us to the question how our basic model can be extended. In particular, there is some discussion among the cell cycle specialists about correlations between mother and daughter cells (references in Voit and Dick 1983 b). All of their hypotheses are based on experimental data. We will not participate in this discussion but study how our model must be altered in order to account for possible mother-daughter correlations, that could not be described with our basic model.

The probabilities p_i have to be replaced by double-indexed probabilities p_{ij} that reconsider the mother's cycle duration when the daughter cell is assigned its cycle duration. For convenience, we define $d_{ij}(t)$ as the number of cells with age i and cycle duration j. Since i cannot be greater than j, there are $m \cdot (m+1)/2$ d's and we call the number of d's μ. The d's can be collected in a time-dependent vector $\vec{d}(t)$. The expected number of cells, N(t), is equal to the sum of all d's at time t. It can be shown that the vector $\vec{d}(t)$ can again be calculated from $\vec{d}(t-1)$ by multiplication with the square correlation matrix M_c with μ rows:

$$
M_c = \begin{bmatrix}
2p_{11} & \cdot & 2p_{21} & \cdot & \cdot & 2p_{31} & & \cdots & 2p_{m1} \\
2p_{12} & \cdot & 2p_{22} & \cdot & \cdot & 2p_{32} & & \cdots & 2p_{m2} \\
\cdot & 1 & \cdot & \cdot & \cdot & \cdot & & \cdots & \cdot \\
2p_{13} & \cdot & 2p_{23} & \cdot & \cdot & 2p_{33} & & \cdots & 2p_{m3} \\
\cdot & \cdot & \cdot & 1 & \cdot & \cdot & & \cdots & \cdot \\
\cdot & \cdot & \cdot & \cdot & 1 & \cdot & & \cdots & \cdot \\
2p_{14} & \cdot & 2p_{24} & \cdot & \cdot & 2p_{34} & & \cdots & 2p_{m4} \\
\cdot & \cdot & \cdot & \cdot & \cdot & \cdot & 1 & \cdots & \cdot \\
 & & & & & & & \ddots & \vdots
\end{bmatrix}
$$

It can be shown that M_c with appropriate interpretation includes the simple matrix M as a special case (Voit and Dick 1983 b).

We analyzed again the growth of the ciliate population. In the basic model, the single-indexed probabilities p_i could directly be taken from the published experimental data. In contrast, the double-indexed p_{ij} now needed are not at hand. They, therefore, had to be estimated with a least-square method that minimizes the differences between the experimental data and the model calculation which depends on the set of probabilities that is being optimized. The data and calculated growth curve are shown in Figure 4. In contrast to the basic model, the present data fit not only describes the growth behavior of the population at the beginning but also mimics the increasing decay of synchrony.

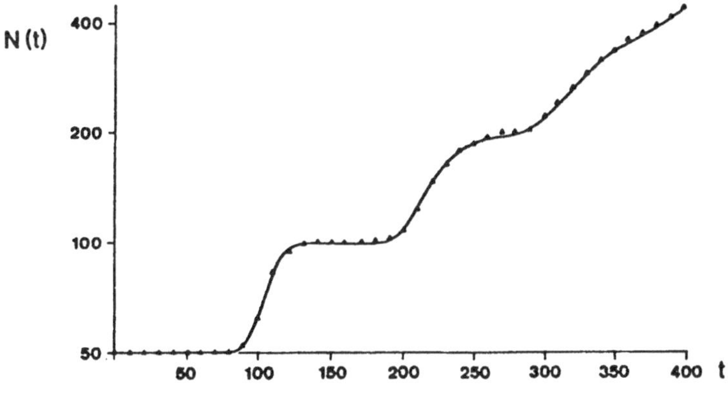

Figure 4: Curve fit (▲) to Prescott's data (line) with the extended model. See text for further information.

Conclusions

The proposed models describe the growth of cell populations. In contrast to earlier models, our models allow the cell cycle durations to be arbitrarily distributed. The basic model assumes that the cycle durations of mother and daughter cells are independent of each other, whereas the extended model accounts for arbitrary correlations. In the analysis of experimental data with our basic model, no assumptions or additional data were required except those found experimentally. In contrast, the extended model needed the specification of the correlations. These data have not yet been found experimentally but people are now beginning to study cell lineages, from which those specifications may some day be derived.

The basic and the extended model include correlations between two sister cells, a well-known phenomenon which was not discussed here (but in Voit and Dick 1983 b). We also did not show here, how both models can be used to calculate the age distributions of cells within the population and under which circumstances such age distributions do not change in time.

In the future, we will address questions on the influence of time-dependent or population-size-dependent proliferation rates, and we will study labelling experiments and problems concerning cancer therapy.

References

Prescott, D.M.: Variations in the individual generation times of Tetrahymena geleii HS. Exp.Cell Res. 16, 279-284 (1959)

Pullman, N.J.: Matrix Theory and its Applications. Marcel Dekker, Inc., New York, Basel (1976)

Savageau, M.A.: Growth of complex systems can be related to the properties of their underlying determinants. Proc.Natl.Acad.Sci.U.S.A. 76,11, 5413-5417 (1979)

Voit, E.O. and Dick, G.: Growth of cell populations with arbitrarily distributed cycle durations, I: Basic model. Math. Biosci. (in press 1983 a)

Voit, E.O. and Dick, G.: Growth of cell populations with arbitrarily distributed cycle durations, II: Extended model for correlated cycle durations of mother and daughter cells. Math. Biosci. (in press 1983 b)

PART V

COMPARTMENTAL ANALYSIS

A STAGING PROCESS WITH APPLICATIONS IN BIOLOGY AND MEDICINE

C. L. Chiang
University of California
Berkeley, California 94720 USA

1. Introduction

The staging process is not a new concept. The process existed even before the creation of man. Formation of the heavenly body, evolution of living things, advancement of civilization and of sciences, as well as the development of social structure, religious faith, political systems, etc. are all by stages. At the micro level, metamorphosis in biology and development of a fetus are good examples, each of which follows a definite staging process. This concept became eminent in recent years in survival analysis when it was recognized that development of many chronic conditions is by stages and patients in different stages are subject to different chances of dying. It was in the study of survival analysis of such patients that a stochastic model of the staging process was formulated. We shall briefly review the staging process in terms of survival and death of chronic patients, and present its applications in carcinogenesis, fertility and epidemics.

2. The Staging Process in Survival Analysis

Suppose that a morbidity process consists of k stages, $s_1 \dots s_k$ and a death state R. For a patient in stage s_i at time τ, for $\tau \varepsilon [0,\infty)$, let

$$\nu_i \theta(\tau) d\tau = \Pr\{\text{the patient will enter } s_{i+1} \text{ in } (\tau, \tau+d\tau)\} \tag{1}$$

and

$$\mu_i \theta(\tau) d\tau = \Pr\{\text{the patient will enter R in } (\tau, \tau+d\tau)\} \tag{2}$$

and let

$$\nu_{ii} = -(\nu_i + \mu_i) \quad i=1,\dots,k-1. \tag{3}$$

For a patient in the final stage s_k at time τ, $\nu_k=0$,

$$\nu_{kk} = -\mu_k \; . \tag{4}$$

Each of the intensity functions in (1) and (2) is a product of two factors; ν_i or μ_i is a function of the stage s_i from which a patient exits and enters the next stage s_{i+1} or enters the absorbing state R. The quantity $\theta(\tau)$ is a function of time at which the transition takes place. We assume that the integral

$$\beta(0,t) = \int_0^t \theta(\tau) d\tau \to \infty \qquad \text{as } t \to \infty \quad . \tag{5}$$

While a disease develops continuously, the time of transition is somewhat restricted. Suppose the transition from stage s_i to stage s_{i+1} takes place during the time interval $(\tau_i, \tau_i + d\tau_i)$, for $i=1,2,\ldots,k-1$. Then by the order of occurrences, we have $0 \leq \tau_1 \leq \tau_2 \leq \cdots \leq \tau_{k-1}$.

For a patient who is in s_1 at the initial time 0, we let T be his survival time. By definition, the density function of T is

$$f_T(t)dt = \Pr\{t<T\leq t+dt \mid T>0\} \quad . \tag{6}$$

Since he must be in one of the k stages at time t, $f_T(t)dt$ is the sum of k terms:

$$f_T(t)dt = \sum_{j=1}^{k} f_j(t)dt \quad . \tag{7}$$

Each $f_j(t)$ in (7) corresponds to the realization of the sequence of transitions $s_1 \to s_2 \to \cdots \to s_j \to R$ and is given by

$$f_j(t)dt = \prod_{i=1}^{j-1} \nu_i \mu_j \theta(t) \left[\sum_{i=1}^{j} \frac{1}{\prod_{\substack{\ell=1 \\ \ell \neq i}}^{j} (\nu_{ii} - \nu_{\ell\ell})} e^{\nu_{ii}\beta(0,t)} dt \right] . \tag{8}$$

Substituting (8) in (7) yields the explicit formula for the density function of T,

$$f_T(t) = \sum_{j=1}^{k} \left\{ \prod_{i=1}^{j-1} \nu_i \mu_j \theta(t) \left[\sum_{i=1}^{j} \frac{1}{\prod_{\substack{\ell=1 \\ \ell \neq i}}^{j} (\nu_{ii} - \nu_{\ell\ell})} e^{\nu_{ii}\beta(0,t)} \right] \right\} \quad . \tag{9}$$

From the relation

$$F_T(t) = \int_0^t f_T(t)dt \tag{10}$$

we find the formula for the distribution function

$$F_T(t) = \sum_{j=1}^{k} \left\{ \prod_{i=1}^{j-1} \nu_i \mu_j \sum_{i=1}^{j} \frac{1}{\prod_{\substack{\ell=1 \\ \ell \neq i}}^{j} (\nu_{ii} - \nu_{\ell\ell})\nu_{ii}} [e^{\nu_{ii}\beta(0,t)} - 1] \right\} \quad . \tag{11}$$

As $t \to \infty$, the integral $\beta(0,t) \to \infty$, the exponential functions vanish. As a result

$$F_T(\infty) = - \sum_{j=1}^{k} \left\{ \prod_{i=1}^{j-1} \nu_i \mu_j \left[\sum_{i=1}^{j} \frac{1}{\prod_{\substack{\ell=1 \\ \ell \neq i}}^{j} (\nu_{ii} - \nu_{\ell\ell})\nu_{ii}} \right] \right\}$$

$$= 1 \quad . \tag{12}$$

The algebraic proof of the last equality in (12) is straightforward.

For a detailed treatise on the staging process and the formulas for the expectation E[T] and the variance V[T] of T, refer to Chiang [5] and [6, p. 319-332].

3. An Application in Carcinogenesis

It has been well established that cancerous cells in many tissues develop by stages. A normal cell must undergo k mutations in order to become a neoplastic cell. It is the neopolastic cell whose growth may escape the control mechanism inherent in the normal tissue and whose uncontrolled proliferation gives rise to a tumor.

Various statistical and stochastic models for carcinogenesis have been proposed according to biological interpretations of the process. For example, N. Arley and S. Iverson (1952) suggest that a normal cell can become a cancerous cell through a single mutation. P. Armitage and R. Doll (1957) explain carcinogenesis by means of a two-stage mutation theory. D. G. Kendall (1960), J. Neyman and E. L. Scott (1967) and H. Tucker (1961) all use the birth-death process to derive stochastic models for the two-stage mutation phenomenon. C. P. Nordling in 1953, recognizing the improbability that a single mutation causes the first cancerous cell, proposes a multi-stage theory of the cancer-inducing mechanism, a view which is shared by cancer biologists at least for certain tissues in the human body as well as in experimental animals.

Consider then a time interval $(0,t)$ and a normal cell $t=0$, denoted by c_0. During the interval $(0,t)$, the cell c_0 is subject to k mutations to become a neoplastic cell, $c_0 \rightarrow c_1 \rightarrow \dots \rightarrow c_k$. For each time element $(t,t+dt)$, we assume that an i-th order mutant c_i is subject to an additional mutation with an intensity function $\nu_i\theta(t)$ and to death with an intensity function $\mu_i\theta(t)$, for $i=0,1,\dots k-1$. After k mutations, the cell c_k is no longer subject to further mutation but will experience a high rate of division. Each function $\nu_i\theta(t)$ (or $\mu_i\theta(t)$) is a product of two quantities ν_i (or μ_i) and $\theta(t)$, where ν_i (or μ_i) is specific for the i-th order mutation and $\theta(t)$ is an integrable function of time (or age) t at which the transition takes place and is subject to the condition in (5). According to the above description, carcinogenesis is in effect a staging process.

The paramount question facing a cancer biologist is how long it takes a normal cell to become a neoplastic cell, or more generally, the length of time needed for j mutations to take place, for $j=1,\dots,k$. Starting from time $t=0$ with a normal cell c_0, let t_j be the time period required for a sequence of j mutations to occur, for $j=1,\dots,k$. The density function of t_j is given by

$$f_j(t) = \prod_{i=0}^{j-1} \nu_i \theta(t) \left[\sum_{i=0}^{j-1} \frac{1}{\prod_{\substack{\alpha=0 \\ \alpha \neq i}}^{j-1} (\nu_{ii} - \nu_{\alpha\alpha})} e^{\nu_{ii}\beta(0,t)} \right], \quad j=1,\ldots,k, \tag{13}$$

The distribution function of t_j is

$$F_j(t) = \prod_{i=0}^{j-1} \nu_i \left[\sum_{i=0}^{j-1} \frac{1}{\prod_{\substack{\alpha=0 \\ \alpha \neq i}}^{j-1} (\nu_{ii} - \nu_{\alpha\alpha})\nu_{ii}} \left(e^{\nu_{ii}\beta(0,t)} - 1\right) \right], \quad j=1,\ldots,k . \tag{14}$$

As $t \to \infty$,

$$F_j(\infty) = - \prod_{i=0}^{j-1} \nu_i \left[\sum_{i=0}^{j-1} \frac{1}{\prod_{\substack{\alpha=0 \\ \alpha \neq i}}^{j-1} (\nu_{ii} - \nu_{\alpha\alpha})\nu_{ii}} \right]$$

$$= \prod_{i=0}^{j-1} \left(\frac{-\nu_i}{\nu_{ii}}\right) < 1 , \quad j=1,\ldots,k . \tag{15}$$

Therefore, for each j, t_j is an improper random variable.

Since for j=k,

$$F_k(\infty) = \Pr\{t_k < \infty\} < 1 , \tag{16}$$

there is a positive probability

$$1-F_k(\infty) = 1 - \prod_{i=0}^{k-1} \frac{-\nu_i}{\nu_{ii}}$$

$$= 1 - \prod_{i=0}^{k-1} \frac{\nu_i}{\nu_i + \mu_i} \tag{17}$$

that neoplastic cells will never develop. This probability increases with the number of mutations k needed to convert a normal cell to a neoplastic cell and is also dependent upon the relative magnitude of the mutation coefficient ν_i and the mortality intensity μ_i. If ν_i is much smaller than μ_i, then the i-th mutant is more likely to die than undergo a next mutation. If this is true for most of the i=1,...,k, then the probability is overwhelming that neoplastic cells will never develop. The probability of developing neoplastic cells in a given time interval and the problem of estimation have been discussed in [8].

4. An Application in Fertility Analysis

From a stochastic viewpoint human reproduction is a staging process; each stage is defined by birth of a child. The process advances from one stage to the next until a family is completed. A woman of parity 0 (with no children) may live through her reproductive years without producing any child, or she may have a baby and then stop reproducing, or she may continue to give birth to a second child and then stop, etc. Once a woman has given birth, the reproductive mechanism starts anew. The process may stop at any stage (parity). The parameters that govern human reproduction are a set of fertility intensity functions $[\lambda_0(x), \lambda_1(x),\ldots]$ prevailing in a study population over the reproductive period.

Each $\lambda_i(x)$ is a function of parity i and age x within the reproductive period of women and is defined as follows:

$$\lambda_i(x)\Delta x + o(\Delta x) = \Pr\{\text{a woman of age } x \text{ and parity } i \text{ will give birth to an infant in the time element } (x,x+\Delta x)\} \tag{18}$$

for $i=0,1,\ldots,$ and $x\varepsilon(x_i,x_w)$. Here x_i is the age of women of parity i and x_w is the age of women at the end of the reproductive period. It is easy to show that the difference

$$1 - \exp\{-\int_{x_i}^{x_w} \lambda_i(t)dt\} = p_i \tag{19}$$

is the probability that a woman of age x_i and parity i will have an (i+1)th child. This probabiity p_i is also known as the parity progression ratio for parity i.

The fertility rate specific for parity i for a woman of age x_i, denoted by r_i, is defined as the ratio of the expected number of births of order i+1 to the expected length of exposure to the risk of having an (i+1)th child, or in formula:

$$r_i = \frac{\text{expected number of (i+1)th births during interval } (x_i,x_w)}{\text{expected length of exposure to the risk of having an (i+1)th birth in interval } (x_i,x_w)} \qquad i=0,1,\ldots \tag{20}$$

A woman will have a child with a probability p_i, and will not have a child with a probability $1-p_i$. Therefore, the expected number of children of birth order (i+1) that she will have is p_i, which is the numerator in (20). For the denominator we let a random variable τ_i be the interval between the i-th and the (i+1)th birth. This is also the length of time that a woman of age x_i and parity i waits for the birth of the (i+1)th child. The density function of τ_i denoted by $h_{\tau_i}(t)$ is given by

$$h_{\tau_i}(t)dt = \Pr\{t < \tau_i \le t+dt | \tau_i > x_i\}$$

$$= \frac{1}{p_i} \exp\{-\int_{x_i}^{x_i+t} \lambda_i(\xi)d\xi\}\lambda_i(x_i+t)dt \ . \tag{21}$$

The expectation of τ_i, $E(\tau_i)$, can be computed from (21).

While a woman's period of exposure to the risk of having an (i+1)th child extends from x_i to x_w, her exposure ends as soon as she has an (i+1)th child. If a woman is to reproduce (with a probability p_i) her expected length of exposure is $E(\tau_i)$; if she is not to reproduce (with a probability $1-p_i$) her exposure time extends from x_i to x_w. Therefore, her total length of exposure to the risk of having an (i+1)th child is

$$p_i E(\tau_i) + (1-p_i)(x_w - x_i) \ . \tag{22}$$

And the parity specific fertility rate defined in (20) has an explicit formula:

$$r_i = \frac{p_i}{p_i E(\tau_i) + (1-p_i)(x_w - x_i)} \ , \qquad i=0,1,\ldots \ . \tag{23}$$

From (23), we can also express the parity progression ratio p_i in terms of the fertility rate r_i. Namely

$$p_i = \frac{(x_w - x_i)r_i}{1 + [(x_w - x_i) - E(\tau_i)]r_i} \ , \qquad i=0,1,\ldots \tag{24}$$

Equations (23) and (24) are used in two distinctly different situations. In a cohort study of the reproductive experience of a female population, the probability p_i and age x_i for each parity are determined from the study population. The fertility rate r_i is then estimated from equation (23). In a reproductive analysis of a current population, the fertility rate r_i and age x_i are first computed directly from vital statistics and census data. The probability p_i is estimated from equation (24). The following paragraphs illustrate the two types of studies.

Suppose that in a <u>cohort study</u>, ℓ_0 women of parity 0 (with no children) of an average age of x_0 are observed for reproductive experience during their life time. Each of the women is subject to the sequence of probabilities of reproduction $\{p_0, p_1, \ldots\}$. At the end of their reproductive period, there will be d_i women who have i (live) births in their life, and

$$\ell_i = d_i + d_{i+1} + \ldots \ , \qquad i = 0,1,\ldots \tag{25}$$

women who have i or more children. It is clear that the sequence $(d_0, d_1, \ldots)$ has a multinomial distribution with the probability

$$\frac{\ell_0!}{\prod_{i=0} d_i!} \prod_{i=0} [p_{0i}(1-p_i)]^{d_i} , \qquad d_i = 0,1,\ldots; \; d_0+d_1+\ldots = \ell_0 , \tag{26}$$

where $p_{0i} = p_0 p_1 \cdots p_{i-1}$. The sequence $(\ell_1,\ell_2,\ldots)$ form a chain of binomial distributions with the probability

$$\prod_{i=0} \binom{\ell_i}{\ell_{i+1}} p_i^{\ell_{i+1}} (1-p_i)^{\ell_i-\ell_{i+1}} , \qquad \ell_{i+1} = 0,\ldots,\ell_i . \tag{27}$$

From equation (27), we find the maximum-likelihood estimate of p_i

$$\hat{p}_i = \frac{\ell_{i+1}}{\ell_i} , \tag{28}$$

with the expectation

$$E[\hat{p}_i] = E[\frac{1}{\ell_i} E(\ell_{i+1}|\ell_i)] = p_i , \quad i=0,1,\ldots . \tag{29}$$

Thus $\hat{p}_i = \ell_{i+1}/\ell_i$ is the most efficient unbiased estimator of p_i.

Estimation of the expectation $E(\tau_i)$ in (23) is slightly involved. According to the structure of the fertility process, intensity functions $[\lambda_0(x),\lambda_1(x),\ldots]$ are dependent upon both age and parity of women; the length of exposure for a woman of parity i is related to her future reproduction. This relationship must be taken into account in the estimation of $E(\tau_i)$ and of the expected age at the i-th birth. Let us introduce for the α-th woman in the cohort a vector

$$(\varepsilon_{0\alpha}, \varepsilon_{1\alpha}, \ldots)' , \tag{30}$$

such that

$$\begin{aligned} \varepsilon_{i\alpha} &= 1 \qquad \text{if she has i children in her life} \\ &= 0 \qquad \text{otherwise} , \end{aligned} \tag{31}$$

and

$$\varepsilon_{0\alpha} + \varepsilon_{1\alpha} + \ldots = 1 , \tag{32}$$

for $i=0,1,\ldots; \alpha=1,\ldots,\ell_0$. Clearly

$$\sum_{\alpha=1}^{\ell_0} \varepsilon_{i\alpha} = d_i , \qquad i=0,1,\ldots , \tag{33}$$

and

$$\sum_{j=i} \sum_{\alpha=1}^{\ell_0} \varepsilon_{j\alpha} = \sum_{j=i} d_j = \ell_i , \qquad i=0,1,\ldots . \tag{34}$$

Let $x_{i\alpha}$ be the age of the α-th woman at the birth of the i-th child; let

$$\frac{1}{d_j}\sum_{\alpha=1}^{\ell_0}\varepsilon_{j\alpha}x_{i\alpha}=x_{i\cdot j},\quad j=i,i+1,\ldots \tag{35}$$

the mean age at the birth of the i-th child of the d_j women, and let

$$\frac{1}{\ell_i}\sum_{j=i}\sum_{\alpha=1}^{\ell_0}\varepsilon_{j\alpha}x_{i\alpha}=\frac{1}{\ell_i}\sum_{j=i}d_jx_{i\cdot j}=x_i \tag{36}$$

be the mean age at the birth of the i-th child of the ℓ_i women.

The best estimators of the expectation $E(\tau_i)$ and of the difference (x_w-x_i) in formula (23) are the corresponding sample values,

$$\text{Est of } E(\tau_i)=\frac{1}{\ell_{i+1}}\sum_{j=i+1}\sum_{\alpha=1}^{\ell_0}\varepsilon_{j\alpha}(x_{i+1,\alpha}-x_{i\alpha}) \tag{37}$$

$$=\frac{1}{\ell_{i+1}}\sum_{j=i+1}d_j\,(x_{i+1\cdot j}-x_{i\cdot j}),$$

which is a weighted sample mean difference of $x_{i+1.j}-x_{i\cdot j}$, $j=i+1,\ldots$, and

$$\text{Est of } (x_w-x_i)=\frac{1}{d_i}\sum_{\alpha=1}^{\ell_0}\varepsilon_{i\alpha}(x_{w\alpha}-x_{i\alpha}). \tag{38}$$

$$=(x_{w\cdot i}-x_{i\cdot i}).$$

Both estimators in (37) and (38) possess the optimality properties of an estimator. Substituting (37) and (38) in (23) and simplifying yields the estimate of the fertility rate

$$\hat{r}_i=\frac{\hat{p}_i}{\hat{p}_i(x_{i+1}-x_i)+(1-\hat{p}_i)(x_{w\cdot i}-x_i)} \tag{39}$$

where $\hat{p}_i$, x_i, and $x_{w.i}$ are given in (28), (36) and (35), respectively. Note that formula (39) does not explicitly depend on the mean ages $x_{i.j}$.

In a <u>fertility study of a current population,</u> the basic data are the number of births by order of birth and by age of mother, the number of women by parity and age, and the mean age of mothers at delivery for each birth in a current population under study. From these quantities the parity specific fertility rates are computed. Let b_{ix} be the number of (live) births for order i for women of age x, and P_{ix} the number of women of parity i and age x. Then the sum

$$\sum_x b_{ix} = b_i, \qquad i=1,2,\dots ,$$

is the total number of births of order i and the sum

$$\sum_x P_{ix} = P_i , \qquad i=0,1,\dots , \tag{40}$$

is the total number of women of parity i. The ratio

$$\frac{b_{i+1,x}}{P_{ix}} = \hat{r}_{ix} , \qquad i=0,1,\dots,$$

is the fertility rate for women of parity i and age x, and the ratio

$$\frac{b_{i+1}}{P_i} = \hat{r}_i, \qquad i=0,1,\dots, \tag{41}$$

is the fertility rate for women of parity i and of all ages. Clearly, $\hat{r}_i$ is the weighted mean of $\hat{r}_{ix}$ with the proportion of women P_{ix}/P_i in each age used as weights.

Let x_i be the mean age of women at birth of the i-th child in a current population. Using similar reasoning as in the cohort study, we can show that the probability p_i in formula (24) is estimated from

$$\hat{p}_i = \frac{(x_{w\cdot i} - x_i)\hat{r}_i}{1 + (x_{w\cdot i} - x_{i+1})\hat{r}_i} \tag{42}$$

where $x_{w.i}$ is the age at the end of the reproductive period for women whose final family size is i. This age may be taken as $x_{w.i}$ = 45 or 50 years for all i. On the basis of the observed mean ages x_i and the computed values of $\hat{r}_i$ and $\hat{p}_i$, and by means of life table methodology, a fertility table has been constructed for the U.S. white female population, 1978, and is given in Chiang and van den Berg (1982) and in [9, pp. 259-272].

5. An Application in Stochastic Epidemics

W. O. Kermack and A. G. McKenrick in a series of papers published between 1926-1939 discussed the mathematical theory of epidemics and provided several mathematical models. A simple epidemic model may be described as follows:

A population consists of two groups of individuals: susceptibles and infectives. A susceptible becomes infected through contact with infectives. There are no removals, no deaths, no immunes, and no recoveries from infection. Suppose that at the initial time t=0, there are N susceptibles and one infective. For every $t\in[0,\infty)$, let the number of infectives be denoted by X(t) and the number of susceptibles by N+1-X(t), so that the total population size remains N+1. The primary purpose is to derive an explicit formula for the probability distribution of the random variable X(t):

$$p_{1,n}(0,t) = \Pr\{X(t)=n|X(0)=1\} \quad , \ n=1,\ldots,N+1 \ . \tag{43}$$

For each interval (τ,t), $0 \leq \tau \leq t < \infty$, and for each k, there is a non-negative continuous function $\lambda_k(\tau)$ such that

$$\frac{\partial}{\partial t} p_{k,n}(\tau,t)\Big|_{t=\tau} = \begin{cases} -\lambda_k(\tau) & \text{for } n=k \\ \lambda_k(\tau) & \text{for } n=k+1 \\ 0 & \text{otherwise .} \end{cases} \tag{44}$$

Assuming homogeneous mixing of the population, we let

$$\lambda_k(\tau) = k(N+1-k)\theta(\tau) = a_k\theta(\tau) \tag{45}$$

where

$$a_k = k(N+1-k) = a_{N+1-k} \ , \quad k=1,\ldots,N \ . \tag{46}$$

Thus $a_1,a_2,\ldots,a_n$ are not all distinct for $n > (N+1)/2$. The quantity $\theta(\tau)$, known as the infection rate, is a function of τ and the integral

$$\beta(0,t) = \int_0^t \theta(\tau)d\tau \quad \underset{\text{as } t\to\infty}{\to} \quad \infty \ .$$

as in Section 2.

In this model, stage is defined as the occurrence of a new infection, the number of stages is the population size N+1. The model had been studied by N. T. J. Bailey (1963), M. S. Bartlett (1956), and D. G. Kendall (1956), among others. An explicit solution for the transient probability $p_{1,n}(0,t)$ in (43) was obtained by G. L. Yang and C. L. Chiang (1971). The solution depends on the value of n. For $n \leq (N+1)/2$,

$$p_{1,n}(0,t) = (-1)^{n-1} a_1\ldots a_{n-1} \sum_{i=1}^{n} \frac{e^{-a_i\beta(0,t)}}{\prod_{\substack{\alpha=1\\ \alpha\neq i}}^{n} (a_i-a_\alpha)} \ , \tag{47}$$

for $n=1,\ldots,(N+1)/2$.

For $(N+1)/2 < n \leq N+1$, the above formula does not apply since the a_k's are not all distinct. In this case, we use the relationship

$$p_{1,n}(0,t) = \int_0^t p_{1,k}(0,\tau)a_k\theta(\tau)p_{k+1,n}(\tau,t)d\tau \tag{48}$$

to obtain the required formula

$$p_{1,n}(0,t) = (-1)^{n-1}\, a_1 \dots a_{n-1} \left[-\sum_{j=k+1}^{n} \frac{\beta(0,t)\exp\{-a_j\beta(0,t)\}}{\prod_{\delta=1}^{n}(a_j-a_\delta)} \right.$$

$$\left. + \sum_{i=1}^{k} \sum_{\substack{j=k+1 \\ a_i \neq a_j}}^{n} \frac{\exp\{-a_i\beta(0,t)\}-\exp\{-a_j\beta(0,t)\}}{(a_i-a_j)\prod_{\substack{\alpha=1 \\ \alpha\neq i}}^{k}(a_i-a_\alpha)\prod_{\substack{\delta=k+1 \\ \delta\neq j}}^{n}(a_j-a_\delta)} \right] \tag{49}$$

where $k = N/2$ when N is even, and $k = (N+1)/2$ when N is odd, and $n=k+1,\dots,N+1$.

The density function and the distribution function of the length of time up to the nth infection (T_n) or of the duration of an epidemic (T_{N+1}) can be derived from the probabilities in (47) and (49) using the following relationships:

$$f_n(t)dt = p_{1,n-1}(0,t)a_{n-1}\theta(t)dt \tag{50}$$

and

$$F_n(t) = \int_0^t p_{1,n-1}(0,\tau)a_{n-1}\theta(\tau)d\tau\ , \qquad \text{for } n=2,\dots,N+1\ . \tag{51}$$

For $1<n<(N+1)/2$, for example, the density function and the distribution function of T_n are given respectively by

$$f_n(t) = (-1)^n\, a_1 \dots a_{n-1}\theta(t) \sum_{i=1}^{n-1} \frac{e^{-a_i\beta(0,t)}}{\sum_{\substack{\alpha=1 \\ \alpha\neq i}}^{n-1}(a_i-a_\alpha)} \tag{52}$$

and

$$F_n(t) = (-1)^n\, a_1 \dots a_{n-1} \sum_{i=1}^{n-1} \frac{i}{\sum_{\substack{\alpha=1 \\ \alpha\neq i}}^{n-1}(a_i-a_\delta)a_i}\,[1-e^{-a_i\beta(0,t)}] \tag{53}$$

The expected value $E[T_n]$ and the variance $V[T_n]$ of T_n can be determined from (50) for a particular function $\theta(t)$. For $\theta(t) = 1$, we find from (50) the expected value

$$E[T_n] = \sum_{i=1}^{n-1} a_i^{-1}\ , \qquad n=2,3,\dots,N+1\ , \tag{54}$$

and the variance

$$V[T_n] = \sum_{i=1}^{n-1} a_i^{-2}\ , \qquad n=2,3,\dots,N+1\ . \tag{55}$$

References

[1] Arley, N. and S. Iversen (1952). "On the mechanism of experimental carcingenesis. III. Further developments of the bit theory of carcinogenesis," Acta. Path. Microb. Scand., Vol. 30, pp. 21-53.

[2] Armitage, P. and R. Doll (1957). "A two-stage theory of carcinogenesis in relation to the age distribution of human cancer," Brit. J. Cancer, Vol. 11, pp. 161-169.

[3] Bailey, N. T. J. (1963). "The simple stochastic epidemic: A complete solution of terms of known functions," Biometrika, Vol. 50, pp. 235-240.

[4] Bartlett, M. S. (1956), "Deterministic and stochastic models of recurrent epidemics," Proc. Third Berkeley Symp. Math. Stat. and Prob., Univ. of California Press, Berkeley and Los Angeles, Vol. 4, pp. 81-109.

[5] Chiang, C. L. (1979). "Survival and stages of disease," Math. Biosci. Vol. 43, pp. 159-171.

[6] Chiang, C. L. (1980). An Introduction to Stochastic Processes and Their Applications, Robert E. Krieger, Melbourne, FL.

[7] Chiang, C. L. and B. J. van den Berg (1982). "A fertility table for the analysis of human reproduction," Math. Biosci., Vol. 62, pp. 237-251.

[8] Chiang, C. L. (1983). "The theory of multistage carcinogenesis," Math. Biosci., Vol. 67, pp. 33-40.

[9] Chiang, C. L. (1984). The Life Table and Its Applications, Robert E. Krieger, Melbourne, FL.

[10] Kendall, D. G. (1956). "Deterministic and stochastic epidemics in a closed population," Proc. Third Berkeley Symp. Math. Stat. and Prob. Univ. of California Press, Berkeley and Los Angeles, Vol. 4, pp. 149-165.

[11] Kendall, D. G. (1960). "Birth-and-death processes, and the theory of carcinogenesis," Biometrika, Vol. 47, pp. 13-21.

[12] Neyman, J. and E. L. Scott (1967). "Two-stage mutation theory of carcinogenesis," Proc. Fifth Berkeley Symp. Math. Stat and Prob., Univ. of California Press, Berkeley and Los Angeles, Vol. IV, pp. 745-776.

[13] Nordling, C. O. (1953). "A new theory on the cancer-inducing mechanism." Brit. J. Cancer, Vol. 7, pp. 68-72.

[14] Tucker, H. G. (1961). "A stochastic model for a two-stage theory of carcinogenesis," Proc. Fourth Berkeley Symp. Math. Stat. and Prob., Univ. of California Press, Berkeley and Los Angeles, Vol. IV, pp. 387-403.

[15] Yang, G. L. and Chiang, C. L. (1971). "A time dependent simple stochastic epidemic," Proc. Sixth Berkeley Symp. Math. Stat and Prob., Univ. of California Press, Berkeley and Los Angeles, Vol. IV, pp. 147-158.

ON THE USE OF RESIDENCE TIME MOMENTS IN THE STATISTICAL ANALYSIS OF AGE-DEPENDENT STOCHASTIC COMPARTMENTAL SYSTEMS

J. H. Matis and T. E. Wehrly
Institute of Statistics
Texas A&M University
College Station, TX 77843

1. Introduction

Most of the present compartmental modeling and analysis found in the applied literature is based on a classical deterministic formulation (1-4). An alternative stochastic formulation has developed rapidly in the mathematical modeling literature (5-7), but the practical application of this formulation to experimental data analysis has developed at a much slower pace. Two recent reviews have now shown that the stochastic model may be useful in practical applications for several reasons related to the statistical analysis of data (8,9).

This paper uses the stochastic model primarily to motivate and then generate residence time moments which are shown to be helpful in data analysis. Section 2 outlines the derivation of such moments for a standard compartment model with age-invariant rates, or equivalently with exponential lifetimes. Section 3 reviews the practical application and statistical utility of these moments. Section 4 then extends the concepts to a generalized model where the rates are age-dependent and hence are the resultant of non-exponential lifetimes. It is shown that under certain specified conditions the solutions of two-compartment models with gamma lifetimes have damped oscillations. The mean residence time for such models is derived and illustrated with an example.

2. Residence Time Moments in a Standard Compartment Model

Rescigno et al. (10) use the following operational definition of a (deterministic) compartment: "A variable $X(t)$ of a system is called a compartment if it is governed by the differential equation:

$$dX/dt = -KX + f(t)$$

with K constant." Their analogous definition in a stochastic context is as follows: "A compartment can be considered as being made up of an ensemble of particles, molecules, or parts of molecules which have the same probability of passing from that state to other possible states." For the purpose of this paper, we will denote a system with constant rates (K), or analogously, with particles having identical probabilities, as a "standard" compartmental system.

In order to investigate some of the implications of a standard stochastic model, consider the following definitions relating to a stochastic particle model:

1. Let $x_{ij}(t)$, $i,j = 1, \ldots, n$; denote the random count of particles which originated in i and are in j at time $t \geq 0$.
2. Let $\underset{\sim}{X}(t) = [x_{ij}(t)]$ denote the random matrix of counts.
3. Let k_{ij}, $i=1, \ldots, n$; $j=0, 1, \ldots, n$; with $i \neq j$; denote the probability intensity coefficient, in units of $time^{-1}$. The k_{ij}'s are defined through the assumption: Prob {any given particle in i at time t will be in j at $t + \Delta t\} = k_{ij}\Delta t + o(\Delta t)$ (1)
4. Let $k_{ii} = - \sum_{\substack{j=0 \\ j \neq i}}^{n} k_{ij}$
5. Let $\underset{\sim}{K} = [k_{ij}]$ be the transition intensity matrix.
6. Let $\underset{\sim}{\Lambda} = \text{diag}(\lambda_i)$ and $\underset{\sim}{T} = [\underset{\sim}{T}_1, \ldots, \underset{\sim}{T}_n]$ be respectively the diagonal matrix of eigenvalues of $\underset{\sim}{K}$, and the corresponding matrix of right eigenvectors of $\underset{\sim}{K}$. By definition, on has $\underset{\sim}{K}\underset{\sim}{T} = \underset{\sim}{T}\underset{\sim}{\Lambda}$.

For present simplicity, we will consider only models with pulse input, i.e. where there is no input (or immigration) of particles to the system after $t = 0$.

The constant probability intensity coefficients in Eq. 1 are analogous to the constant flow rates of the deterministic model. Eq. 1 defines the basic chance mechanism of the model, which implies an "uncertain" count, $x_{ij}(t)$, for $t>0$. Eq. 1 also implies that the transfer probability is independent of the elapsed time, t, and of the arrival time of a particle into the compartment. Such compartments are said to "lack a memory" of the elapsed and arrival times.

Consider now the following definition:

7. Let R_{ij} denote the random time in i for the population of all particles in i whose next transfer is to j.

It can be shown that an alternative definition of k_{ij} in Eq. 1 is provided by the following equivalent assumption:

8. Let k_{ij}, $i \neq j$, denote the parameter of the exponential distribution of R_{ij}, i.e. let R_{ij} have probability density function

$$f_{R_{ij}}(a; k_{ij}) = k_{ij} \exp\{-k_{ij}\, a\}, \quad a \geq 0. \tag{2}$$

It is easy to show that Eq. 2 implies Eq. 1, and it is well known that Eq. 1 characterizes Eq. 2 [see e.g. (11)]. Therefore, one can define a standard stochastic compartment as one with exponentially distributed R_{ij} retention times, and a standard system as one with "exponential" compartments.

The following basic results hold for the standard compartmental system:

9. Let $p_{ij}(t)$, $i, j=1, \ldots, n$; denote the probability that a particle originally in i at $t=0$ will be in j at time t.
10. Let $\underset{\sim}{P}(t) = [p_{ij}(t)]$ be the matrix of probabilities.

The following may now be proven:

<u>Result 1</u>: In a system the usual regularity conditions (e.g. openness, particle independence, and distinct eigenvalues),

$$\underset{\sim}{P}(t) = e^{\underset{\sim}{K}t} = \underset{\sim}{T}e^{\underset{\sim}{\Lambda}t}\underset{\sim}{T}^{-1} . \tag{3}$$

This is a well-known result which is proven, e.g., in (12, p.425) and is related to compartmental modeling in (5).

Often in modeling, the structure of the system is known a priori. In such cases, the eigenvalues are usually assumed to be real and the resulting individual $p_{ij}(t)$ sum-of-exponentials models in Eq. 3 are written as explicit functions of the k_{ij} rate parameters. The k_{ij}'s are typically estimated by fitting the model to data using nonlinear least squares.

Consider now the following definitions:

11. Let V_{ij} denote the total number of visits to j prior to leaving the system of a particle which originated in i. By definition, $V_{ii} \geq 1$.
12. Let S_{ij} denote the total residence (or sojourn) time in j (accumulated during the V_{ij} visitations) of a particle originating in i.
13. Let $\theta_{ij} = E[S_{ij}]$ and $\gamma_{ij} = Var[S_{ij}]$ be the mean and variance of S_{ij}, respectively.
14. Let $\underset{\sim}{\theta} = [\theta_{ij}]$ and $\underset{\sim}{\Gamma} = [\gamma_{ij}]$ be the matrices of means and variances of residence times, respectively.
15. Let $\theta'_{ij} = E[V_{ij}]$ and $\gamma'_{ij} = Var[V_{ij}]$ be the mean and variance of V_{ij}, respectively.
16. Let $\underset{\sim}{\theta}' = [\theta'_{ij}]$ and $\underset{\sim}{\Gamma}' = [\gamma'_{ij}]$ be the matrices of mean and variances of the number of visitations, respectively.
17. Let $\underset{\sim}{K}^* = (k_{ij}^*)$ be the normalized $\underset{\sim}{K}$ intensity matrix, such that $k_{ii}^* = 0$, and $k_{ij}^* = -k_{ij}/k_{ii}$.

One can then prove the following result:

Result 2: For the stochastic particle model in Result 1,

i) $\underset{\sim}{\theta} = -\underset{\sim}{K}^{-1}$ (4)

ii) $\underset{\sim}{\Gamma} = 2\underset{\sim}{\theta}\underset{\sim}{\theta}_{(d)} - \underset{\sim}{\theta}_{(2)}$ (5)

iii) $\underset{\sim}{\theta}' = (\underset{\sim}{I} - \underset{\sim}{K}^*)^{-1}$ (6)

iv) $\underset{\sim}{\Gamma}' = 2\underset{\sim}{\theta}'\underset{\sim}{\theta}'_{(d)} - \underset{\sim}{\theta}'_{(2)}$ (7)

where $\underset{\sim}{\theta}_{(d)}$ and $\underset{\sim}{\theta}_{(2)}$ denote respectively the matrices of diagonal elements and of squared elements of $\underset{\sim}{\theta}$ and $\underset{\sim}{I}$ is the identity matrix.

Eq. 4 is a result which is well known and has been used by many experimenters. Eqs. 6 and 7 are obtainable using standard Markov chain theory. They are proven and applied in (13), among others. Eq. 5 may be obtained as a limiting form of Eq. 7, but to our knowledge, it is proven for the first time in (14). It can be shown subsequently that each may be useful in its own right.

Clearly, all of the moments in Result 2 may be obtained directly if the $\underset{\sim}{K}$ intensity matrix is known. In practice, one has only the estimated matrix, say $\hat{\underset{\sim}{K}}$, which is consistent. It has been shown (14) that substitution of the consistent $\hat{\underset{\sim}{K}}$ into Result 2 yields consistent estimates of the various moments.

In many applications, much of the structure of the system is unknown. In many such cases, Rescigno (15,16) has shown how one may estimate certain residence time moments directly from the estimated a_{ijk} and λ_k parameters of the sum-of-exponentials model

$$p_{ij}(t) = \sum_{k=1}^{n} a_{ijk} \exp\{\lambda_k t\}. \qquad (8)$$

3. The Practical Application of Residence Time Moments

Estimated residence time moments are finding increased use in the statistical analysis of kinetic data for one or more of the following reasons (8): i) they are of inherent biological interest; ii) they are convenient to use in comparing results of different models; and iii) they have been found to be relatively powerful in detecting treatment differences in designed experiments. Warner (17) points out several advantages of the use of mean residence times as compared to the traditional half-life measures, particularly when applied to multi-compartment systems. Among these advantages are the biological interpretability of residence time moments and their relative invariance over different structural assumptions of the model. To our knowledge, the relative power of the estimated residence time moments has only recently been noted. A few studies (14, 18) have analyzed data using both estimated rate parameters and mean residence times, and they have found the latter to be more powerful in detecting treatment effects. These two cases were followed by a simulation study (8) of certain completely specified two-compartment open models illustrated in Figure 1.

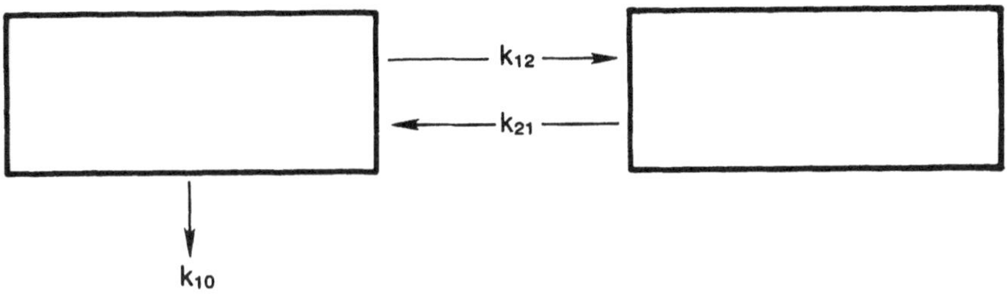

Figure 1. Two Compartment Open Model

This study simulated an experiment with two populations, treated and control groups where the treatment effect consisted of a specified difference in one or more k_{ij} rate parameters. The data from simulated replicate observations over time were analyzed, using different response variables to determine which variables detected the treatment effect most often. The study found that, as a group, the estimated k_{ij} rate coefficients were not as powerful in detecting the treatment effect as were the estimated θ_{ij} mean residence times.

Clearly, the statistical power of an experiment is a function of many variables, including the size of the effect, the number of observations, the structural model, and the precise specification of the parameter space. Therefore, much further study of the problem seems warranted. However, all of our experience to date suggests that the estimated residence time moments are powerful variables for the statistical analysis of designed experiments, as well as being inherently attractive as useful descriptors of kinetic systems.

4. Residence Time Moments for an Age-Dependent Compartmental Model

4.1 Introduction to Age-Dependent Models

Let a generalized compartmental model be defined as one where one or more of the k_{ij} rates are functions of either the absolute time, t, at which the particle entered the system, or of the age, a,of the particle in a specified compartment. Generalized models with time-varying rates have been studied, using time-dependent Markov theory (12,19), but they are not of present interest.

Models with age-varying rates have been investigated with increasing interest recently for, among others, the following two reasons:

i) Many processes have rates which are inherently age-dependent, e.g., various digestion processes and other enzyme-kinetic activity.

ii) Some complex standard models with many compartments may be simplified by using approximate, age-dependent models with fewer parameters, and thus often with superior subsequent statistical analysis. One such application is the description of mixing in passage models.

These models with age-varying rates do not have a deterministic analog which is expressible as a set of differential equations. We call them generalized compartmental models since they satisfy Eq. 1 with k_{ij} being a function of age, a.

Marcus (20) discusses several examples relating to ecological succession, and also to the uptake, distribution, and retention of trace contaminants in mammals. Both Marcus (20) and Purdue (21) review the use of semi-Markov theory from which, in principle, the requisite $p_{ij}(t)$ regression function may be determined for arbitrary (non-exponential) residence time distributions. However, the solutions are given in general terms involving an infinite sum of convolutions, and the complexity generally rules out an analytical solution for the $p_{ij}(t)$ function.

A more limited approach which has been applied in practice involves the assumption of some specific non-exponential parametric family, rich in its possible forms. A commonly assumed family is the integral gamma (or Erlang) residence time distribution. Analytical forms of the $p_{ij}(t)$ regression functions for two-compartment irreversible models have been developed in (22) for one "gamma" compartment, and in (23) for two "gamma" compartments. These regression functions are obtained by representing each gamma variable as a convolution of exponential variables. The corresponding $\underset{\sim}{K}$ matrix is diagonal, with resulting multiple eigenvalues; hence, $p_{ij}(t)$ involves powers of t in addition to exponential terms. This simple, irreversible model has been found to provide a useful description of passage of digestion in ruminants (24).

The solution for the generalized, two-compartment open (and reversible) model with gamma lifetimes is a natural extension of the problem. The two-compartment open model occurs widely in kinetic work (25); hence, incorporation of gamma age-dependency would be of general interest. Also, the model has been suggested specifically for passage in the gastro-intestinal tract. We are not aware of any analytical $p_{ij}(t)$ solution for this model, or indeed, for any other reversible, age-dependent model in the current literature.

4.2 A Generalized Model with Gamma Lifetimes

Consider now the following special case of a generalized, two-compartment open model, using the notation of Purdue (7,21).

i) The probability density function of the residence time, R_i, of a particle in compartment i during a single visit is the integral gamma distribution:
$f_i(t;\ \phi_i,\ n_i) = (\phi_i)^{n_i}\ (t)^{n_i-1}\ \exp\{-\phi_i t\}/(n_i-1)!$ where $t \geq 0$, $\phi_i > 0$,
$n_i = 1, 2, \ldots,$ for $i=1, 2$.
The distribution functions are $F_i(t) = \int_0^t f_i(u)du$, $i = 1, 2$.

ii) The conditional probabilities, α_{ij}, of entering j upon departure from i are

$$\alpha_{12} = \rho \qquad \alpha_{10} = 1 - \rho \qquad \alpha_{21} = 1 \qquad \text{with } 0 < \rho < 1 .$$

The cumulative hazard (or transfer) rate from i is the age-dependent function:

$$k_i(t) = \frac{(\phi_i)^{n_i}(t)^{n_i-1}/(n_i-1)!}{\sum_{\ell=0}^{n_i-1} (\phi_i t)^{\ell}/\ell!} \tag{10}$$

as given in (26) and illustrated for compartmental models in (8).

As noted previously, the hazard rate $k_i(t)$ is equated to the passage probability from the last of a sequence of n_i exponential compartments, each with rate ϕ_i. Thus, one can simulate the ith gamma compartment through the mathematical artifice of n_i exponential sub-compartments. Clearly the sub-compartments have no real existence and are not separately measurable; hence, they are called pseudo-compartments. After employing this relationship, and also after partitioning passage from the last exponential pseudo-compartment according to the conditional probabilities given in assumption (ii), one may represent the present generalized, two-compartment model by the equivalent standard compartment model given in Figure 2.

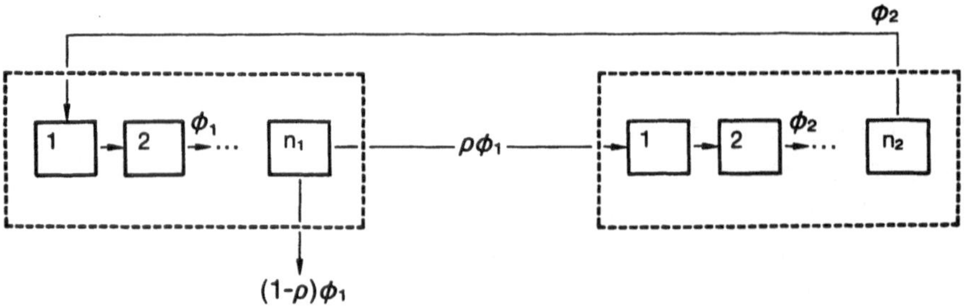

Figure 2. Standard (Exponential) Representation of Two Compartment Model with Gamma (n_i, ϕ_i) Lifetimes.

Consider the following additional notation :

18. Let $\pi_{u,v}(t)$, $u,v = 1, \ldots, n_1 + n_2$; denote the probability that a particle originally in pseudo-compartment u at t=0 will be in pseudo-compartment v at time t.

19. One may immediately relate the two systems by noting correspondences such as:

$$p_{11}(t) = \sum_{v=1}^{n_1} \pi_{1v}(t) \qquad p_{12}(t) = \sum_{v=n_1+1}^{n_1+n_2} \pi_{1v}(t) \tag{11}$$

It follows that a generalized system with gamma lifetimes is easily transformed to an expanded standard system provided $\underset{\sim}{K}$ is specified. One can solve for the eigenvalues of $\underset{\sim}{K}$, which could be either real or complex, then solve for the $\pi_{uv}(t)$'s using Result 1, and finally obtain the $p_{ij}(t)$'s for the generalized model from Eq. 11. The $p_{ij}(t)$'s are useful in some applications for prediction and/or control of the process.

However, in most pharmacokinetic applications one has the inverse problem where $\underset{\sim}{K}$ is unknown. In such cases it is virtually always estimated based on the implicit assumption that the eigenvalues are real and distinct, and hence that $p_{ij}(t)$ has the familiar sum of exponentials form in Eq. 8. It will be shown subsequently that such assumption is usually untenable in the present model, which fact complicates the inverse (estimation) problem for gamma compartments.

In order to investigate the intensity matrix $\underset{\sim}{K}$ of the standard system in Figure 2, let it first be scaled, with no loss of generality, by dividing the rates by ϕ_2. Letting $\phi = \phi_1/\phi_2$, the non-zero entries of the scaled $\underset{\sim}{K}$ matrix are:

$$\tilde{K} = \left[\begin{array}{cccccccc} -\phi & \phi & & & & & & \\ & -\phi & \phi & & & & & \\ & & \ddots & & & & & \\ & & & -\phi & \rho\phi & & & \\ & & & & -1 & 1 & & \\ & & & & & -1 & 1 & \\ & & & & & & \ddots & \\ 1 & & & & & & & -1 \end{array}\right] \begin{array}{l} \left.\begin{array}{l} \\ \\ \\ \end{array}\right\} n_1 \text{ rows} \\ \left.\begin{array}{l} \\ \\ \\ \end{array}\right\} n_2 \text{ rows} \end{array} \tag{12}$$

Now let $n=n_1+n_2$. The following theorem pertains to the $\tilde{K}$ matrix of the specified two-compartment open model.

<u>Theorem 1</u>: The $\tilde{K}$ matrix in Eq. 12 has at least 2 complex eigenvalues for $n\geq 3$ except under the following conditions:

i) $n_1=n_2=2$ with $\rho\leq(\phi-1)^4/16\phi^2$,

ii) $n_1=2$, $n_1=1$ with $\rho\leq 4(1-\phi)^3/27\phi^2$,

iii) $n_1=1$, $n_2=2$ with $\rho\leq 4(\phi-1)^3/27\phi$.

The theorem is proven in the Appendix. Its practical significance is that for two-compartment gamma models with large n or large ρ, the $p_{ij}(t)$'s do not have the simple, commonly-used sum-of-exponentials form in Eq. 8. Instead, when the roots are distinct, the $p_{ij}(t)$'s will be of the form:

$$p_{ij}(t) = \sum_{\ell=1}^{r} a_{ij\ell}e^{\lambda_\ell t} + \sum_{k=1}^{s} e^{\nu_k t} [b_{ijk}\cos(\zeta_k t) - c_{ijk}\sin(\zeta_k t)] \tag{13}$$

where λ_ℓ denotes a real root and the complex roots have form $\nu_k \pm i\zeta_k$. Since $\nu_k < 0$ for all k (27), the $p_{ij}(t)$'s contain damped oscillations.

The models with oscillations may be fitted to data as before to estimate the parameters and the $p_{ij}(t)$'s. If the structure (with n_1 and n_2) is known, it is possible in principle to solve for the $p_{ij}(t)$'s as a function of ϕ_1, ϕ_2 and ρ. The parameters could then be estimated from data, whereupon the $\hat{\tilde{K}}$ matrix is immediate. However the explicit solutions for the $p_{ij}(t)$'s for large n are difficult. The alternative procedure is to fit models of the form in Eq. 13 with a parsimonious number of real <u>and complex</u> roots to the data. One could then find estimated $p_{ij}(t)$'s directly from the estimated a, b, c, λ, ν and ζ parameters.

4.3 Mean Residence Times (MRT) for Compartments in the Generalized Model

Clearly, if the n_i, ϕ_i and ρ parameters are known (and/or estimated) one may use Theorem 2 to obtain the moments (or their estimates) for the pseudo-compartments from the $\tilde{K}$ (or $\hat{\tilde{K}}$) matrix. These moments may then be combined to give various residence time moments for the gamma compartments.

As indicated before, some estimated moments may be obtained from Rescigno's results (15,16) for a standard model through estimates of various $p_{ij}(t)$'s even though the

k_{ij} estimates might not be available. However, the age-dependent gamma model in Figure 2 does not meet the conditions for Rescigno's results. The following result is useful in such models.

Result 3: The mean residence time, θ_{ii}, for any compartment in a gamma age-dependent model is

$$\theta_{ii} = -\sum_{\ell=1}^{r} \frac{a_{ii\ell}}{\lambda_\ell} - \sum_{k=1}^{s} \frac{b_{iik}\nu_k + c_{iik}\zeta_k}{(\nu_k^2 + \zeta_k^2)} \tag{14}$$

where the a, b, c, λ, ν and ζ's are parameters of $p_{ii}(t)$ in Eq. 13.

The proof is immediate since the integral of the $p_{ii}(t)$ "survival" function for any generalized model yields the MRT (28). In this case, integrating Eq. 13 from 0 to ∞ gives the desired result.

The practical significance of Result 3 is that one can find estimated MRT for gamma compartments even though the estimation of the ϕ_i and ρ parameters might be intractable. One instead fits parsimonious models of the form in Eq. 13 to the data, and then finds the MRT through Eq. 14. Result 3 may be generalized to find the MRT in any set of compartments which contains the site of introduction summary. Result 3 makes it possible to extend the statistical analysis based on MRT from standard models to age-dependent gamma models. It is expected that such analysis for gamma models will be as successful in general, and as statistically powerful in particular, as it has been for standard models.

4.4 An Illustration

Often in passage modeling, compartment 1, the site of introduction, is considered gamma age-dependent but compartment 2 is not. Consider now such a model with additional specifications that $E[R_1] = E[R_2] = 1$, and $\alpha_{12} = \frac{1}{2}$. For the gamma distribution, $E[R_i] = \frac{n_i}{\phi_i}$ hence ϕ_1 is integer n_1 and $\phi_2 = 1$. For simplicity, let $n_1 = k$ whereupon the $\underset{\sim}{K}$ matrix under the present assumptions is:

$$\underset{\sim}{K} = \begin{bmatrix} -k & k & & & \\ & -k & k & & \\ & & \ddots & & \\ & & & -k & k/2 \\ 1 & & & & -1 \end{bmatrix} \tag{15}$$

The roots may be obtained through analytical formulas for k = 2 and 3, and numerically for any k. Some specific solutions for small k are given in Table 1.

Certain patterns are apparent. $\underset{\sim}{K}$ has either 1 or 2 real roots; the rest are distinct pairs of complex conjugates. The dominant real root, λ_1, approaches a limit which may

Table 1. Roots of K matrix in Eq. 5 for various values of k.
(λ_i denote real roots; ν_i, ζ_i denote real and imaginary parts of complex roots)

k	2	3	4	5	6	7
λ_1	-0.3044	-0.3080	-0.3098	-0.3108	-0.3115	-0.3120
λ_2	-	-4.559	-	-7.953	-	-11.529
ν_1	-2.357	-2.566	-2.633	-2.645	-2.636	-2.619
ζ_1	1.029	1.740	2.223	2.566	2.820	3.015
ν_2			-5.712	-6.222	-6.403	-6.420
ζ_2			1.471	2.794	3.877	4.746
ν_3					-9.304	-10.040
ζ_3					1.758	3.432

be determined as follows. The characteristic polynomial of $\underset{\sim}{K}$ is

$$Q(\lambda) = (1 + \lambda/k)^k(1 + \lambda) - \rho = 0$$

which has the limit

$$\lim_{k\to\infty} Q(\lambda) = e^{\lambda}(1 + \lambda) - \rho = 0$$

For $\rho = \frac{1}{2}$, the real root of the above limit is approximately $\lambda = -0.3149$.

The eigenvectors may also be obtained numerically, for example by using the IMSL computer program package. As an illustration, for k=2 the eigenvector matrix is

$$T = \begin{bmatrix} 1 & 1 & 1 \\ 0.85 & -0.17 + 0.51i & -0.17 - 0.51i \\ 1.44 & -0.47 - 0.36i & -0.47 + 0.36i \end{bmatrix}$$

One may then solve for the T^{-1} matrix and for the $\pi(t)$'s using Eq. 3. The solutions for k=2 are

$$\pi_{11}(t) = .225e^{-.304t} + e^{-2.35t}[0.775\cos(1.03t) - 0.110\sin(1.03t)]$$

$$\pi_{12}(t) = .266e^{-.304t} + e^{-2.35t}[-0.266\cos(1.03t) + 1.416\sin(1.03t)]$$

$$\pi_{13}(t) = .382e^{-.304t} + e^{-2.35t}[-0.382\cos(1.03t) - 0.759\sin(1.03t)],$$

and from Eq. 11 one has, for example,

$$p_{11}(t) = .491e^{-.304t} + e^{-2.35t}[.509\cos(1.03t) + 1.306\sin(1.03t)]$$

The MRT, θ_{11}, for the first gamma compartment may be found from Result 3 to be $\theta_{11} = 2.00$. Similarly, by integrating $p_{11}(t) + p_{12}(t)$, the MRT in the system is $\theta_{11} + \theta_{12} = 3.00$. Since the parameters are known, these may be verified by finding the MRT for the exponential, pseudo-compartments from Result 2 and then summing. In usual practice one would estimate $p_{ii}(t)$ from data and then use Result 3 to find the estimated MRT.

Appendix

Proof of Theorem 1:

The characteristic equation of $\underset{\sim}{K}$ is

$$|\underset{\sim}{K} - \lambda \underset{\sim}{I}| = (-1)^n(\phi + \lambda)^{n_1}(1 + \lambda)^{n_2} + (-1)^{n+1}\phi^{n_1}\rho = 0$$

Let
$$Q(\lambda) = (1 + \lambda/\phi)^{n_1}(1 + \lambda)^{n_2} - \rho$$

The eigenvalues of $\underset{\sim}{K}$ are the roots of $Q(\lambda) = 0$.

Note that $Q'(\lambda) = (1 + \lambda/\phi)^{n_1-1}(1 + \lambda)^{n_2-1}(n_1 + n_2\phi + n\lambda)/\phi$

It is clear from Rolle's Theorem that if $Q'(\lambda)$ has at most m real roots, $\lambda_1', \lambda_2', \ldots, \lambda_m'$, then $Q(\lambda)$ can have at most m+1 real roots, and that the real roots of $Q'(\lambda)$ separate the real roots of $Q(\lambda)$. Clearly if $Q(\lambda_k) \neq 0$, $k = 1, \ldots, n$, then the real roots of $Q(\lambda)$ are simple.

Consider now the following special cases:

a) $n \geq 5$ with $n_1 \geq 2$ and $n_2 \geq 2$. $Q'(\lambda)$ has at most 3 roots, $\lambda_1' = -\phi$, $\lambda_2' = -1$, and $\lambda_3' = -(n_1 + n_2\phi)/n = (n_1\lambda_1' + n_2\lambda_2')/n$. Thus λ_3' is in the closed interval with endpoints λ_1' and λ_2'. If $\lambda_1' = \lambda_2'$, then $Q'(\lambda)$ has only 1 real root, λ', and $Q(\lambda') = -\rho \neq 0$. In this case, $Q(\lambda)$ has at most 2 real roots and they are simple. If $\lambda_1' \neq \lambda_2'$ then $Q(\lambda_1') = Q(\lambda_2') = -\rho \neq 0$. $Q(\lambda_3') = 0$ would imply a root of $Q(\lambda)$ in between λ_1' and λ_3', which is false. Hence $Q(\lambda_3') \neq 0$ which implies that $Q(\lambda)$ has at most 4 real roots, all of which are simple. However K has at least 5 roots, therefore at least 1 is complex.

b) $n_1 = 1$, $n_2 \geq 3$. The roots of $Q'(\lambda) = 0$ are $\lambda_1' = -1$ and $\lambda_2' = -(1 + n_2\phi)/(1 + n_2)$. Clearly $Q(-1) < 0$. If $Q(\lambda_2') \neq 0$, the real roots are distinct and number at most 3. Suppose $Q(\lambda_2') = 0$. $Q''(\lambda) = 0$ has roots $\lambda_1'' = -1$ and $\lambda_2'' = -(2 + n_2\phi + \phi)/(1 + n_2)$. Note that $\lambda_2' = \lambda_2''$ only if $\phi=1$ which implies $\lambda_2' = -1$. Since $Q(-1) < 0$, λ_2' could be at most a double root which implies $Q(\lambda) = 0$ has at most 3 real roots. However K has at least 4 roots, therefore at least 1 is complex.

c) $n_1 \geq 3$, $n_2 = 1$. Proof is similar to case b.

d) $n_1 = n_2 = 2$. The roots of $Q'(\lambda) = 0$ are $\lambda_1' = -1$, $\lambda_2' = -\phi$, $\lambda_3' = -(1+\phi)/2$. There exist at most 3 real roots iff $Q(\lambda_3') = (\phi-1)^4/16\phi^2 - \rho < 0$ which implies $\rho > (\phi-1)^4/16\phi^2$.

e) $n_1 = 1$, $n_2 = 2$. The roots satisfy $Q(\lambda) = \lambda^3 + (2 + \phi)\lambda^2 + (1 + 2\phi)\lambda + \phi(1-\rho) = 0$. Let $w = \lambda + (2 + \phi)/3$, whereupon $Q(w) = w^3 + aw + b = 0$ where $a = -(\phi-1)^2/3$, $b = 2(\phi-1)^3/27 - \rho\phi$. The cubic equation has 2 complex roots iff $c = b^2/4 + a^3/27 = (\rho\phi)^2/4 - \rho\phi(\phi-1)^3/27 > 0$ which implies $\rho > 4(\phi-1)^3/27\phi$

f) $n_1=2$, $n_2=1$. Proof is similar to case e.

In each case above, conditions have been established for the existence of at least 1 complex root. However it is well known that complex roots occur in conjugate pairs (27), therefore under the identical conditions above, K has at least 2 complex roots.

References:

1. Atkins GL: Multicompartment Models in Biological Systems. London, Methuen, 1969

2. Gibaldi M, Perrier D: Pharmacokinetics. New York, Marcel Dekker, 1975

3. Jacquez JA: Compartmental Analysis in Biology and Medicine. New York, Elsevier, 1972

4. Rescigno A, Segre G: Drug and Tracer Kinetics. Waltham MD, Blaisdell, 1966

5. Matis JH, Hartley HO: Stochastic compartmental analysis: Model and least squares estimation from time series data. Biometrics 27: 77-102, 1971

6. Matis JH, Wehrly TE: Stochastic models of compartmental systems. Biometrics 35: 199-220, 1979

7. Purdue P: Stochastic compartmental models: A review of the mathematical theory with ecological applications. In Matis JH, Patten BC, White GC, Eds. Compartmental Analysis of Ecosystem Models. Burtonsville MD, Int. Co-op. Publ. House, 1979

8. Matis JH, Wehrly TE, Gerald KB: The statistical analysis of pharmacokinetic data. In Lambrecht RM and Rescigno A, Eds. Tracer Kinetics and Physiologic Modeling - Lecture Notes in Biomath. #48, New York, Springer - Verlag, 1983

9. Thakur AK: Some statistical principles in compartmental analysis. In Robertson J, Ed. Compartmental Distribution of Radio Tracers, Boca Raton, FL, CRC Press, to appear, 1983

10. Rescigno A, Lambrecht RM, Duncan CC: Mathematical methods in the formulation of pharmacokinetic models. In Lambrecht RM and Rescigno A, Eds. Tracer Kinetics and Physiologic Modeling - Lecture Notes in Biomath. #48, New York, Springer-Verlag, 1983

11. Parzen E: Stochastic Processes. San Francisco, Holden-Day, 1962

12. Chiang CL: An Introduction to Stochastic Processes and Their Applications. Huntington, NY, Krieger, 1980

13. Matis JH, Patten BC: Environ analysis of linear compartmental systems: The static, time invariant case. Proc. 42nd Session Int. Stat. Inst. Manila, Philippines, NEDA-APO, 1979

14. Matis JH, Wehrly TE, Metzler CM: On some stochastic formulations and related statistical moments of pharmacokinetic models. J Pharmacokin. Biopharm 11: 77-92, 1983

15. Rescigno A: On transfer times in tracer experiments. J. Theo. Biol. 39: 9- , 1973

16. Rescigno A, Gurpide E: Estimation of average times of residence, recycle and interconversion of blood-borne compounds using tracer methods. J Clin Endocrinal Metab 36: 263-276, 1973

17. Warner ACI: Rate of passage of digesta through the gut of mammals and birds. Nutr. Abstr. Rev. B 51: 789-820, 1981

18. Saffer SI, Mize CE, Bhat UN, Szygenda SA: Use of nonlinear programming and stochastic modeling in the medical evaluation of normal-abnormal liver function. IEEE Trans Biomed Eng BME-23: 200-207, 1976

19. Epperson JO, Matis JH: On distribution of the general irreversible n-compartmental model having time-dependent transition probabilities, Bull Math Biol 41: 737-749, 1979

20. Marcus AH: Semi-Markov compartmental models in ecology and environmental health. In Matis JH, Patten, BC, White GC, Eds. Compartmental Analysis of Ecosystem Models. Burtonsville, MD, Int. Co-op Publ. House, 1979

21. Purdue P: Stochastic theory of compartments: An open, two-compartment, reversible system with independent Poisson arrivals. B. Math. Biol. 37: 269-275, 1975

22. Matis JH: Gamma time-dependency in Blaxter's compartmental model. Biometrics 28: 597-602, 1972

23. Hughes TH, Matis JH: An irreversible two-compartment model with age-dependent turnover rates. Biometrics, to appear, 1983

24. Pond KR, Ellis WC, Matis JH: Compartmental models for estimating various parameters of digesta flow. Manuscript.

25. Metzler CM: Usefulness of the two-compartment open model in pharmacokinetics. J Am Stat Assoc 66: 49-53, 1971

26. Gross AJ, Clark VA: Survival Distributions: Reliability Applications in the Biomedical Sciences. New York, Wiley, 1975

27. Hearon JZ: Theorems on linear systems. Ann N Y Acad Sc 108: 36-68, 1963

28. Morse PM: Queues, Inventories and Maintenance. New York, Wiley, 1967

Acknowledgements

We are indebted to J. Eisenfeld who reviewed this paper and made many helpful suggestions, among them the use of Rolle's Theorem in the proof of our Theorem 1.

AN ALGORITHM FOR RECONSTRUCTION OF COUNT RATE CURVES FROM TOTAL COUNTS

Aldo Rescigno[1,2] and Richard M. Lambrecht[1]
[1]Department of Chemistry
Brookhaven National Laboratory
Upton, New York 11973
and
[2]Section of Neurological Surgery
Yale University School of Medicine
New Haven, Connecticut 06520

Dynamic studies with radioactive tracers require measurements of instantaneous tracer activity $X(t)$ as a function of time; this is usually done either with a counter set for a sequence of short intervals of time, or with a ratemeter.

A counter set for a finite interval of time does not measure $X(t)$, but its average value

$$\frac{1}{t_2-t_1}\int_{t_1}^{t_2} X(\tau)d\tau;$$

an ordinary scanner for instance has an acquisition time of the order of minutes, therefore it does not give accurate readings when the activity is not a smooth function of time.

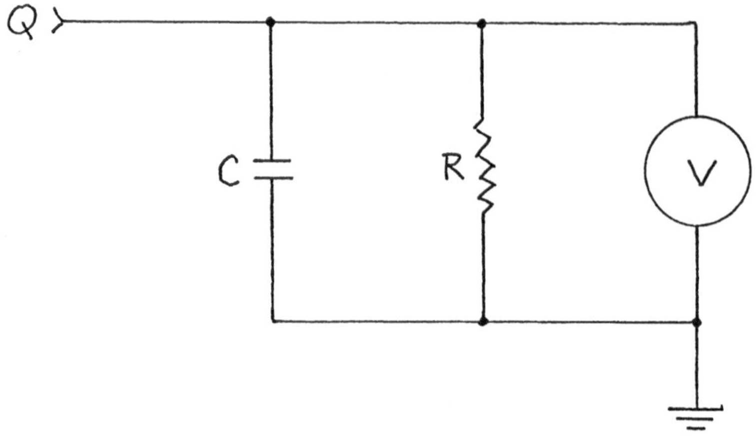

In its basic form a ratemeter generates a pulse of fixed charge Q for each pulse received from a detector; those standard pulses are fed into a condenser of capacity C, while a voltmeter measures the voltage V across a resistor of resistance R in parallel to the condenser. If the pulses are generated at a rate X(t), the charges received by the condenser per unit time are X(t)·Q, and those lost through the resistor are V/R; it follows that

$$C \cdot dV/dt = X(t) \cdot Q - V/R,$$

or

$$dV/dt = -\frac{1}{RC} V + X(t) \cdot Q/C;$$

this is the equation of a compartment with turnover rate 1/RC and feeding function X(t)·Q/C; thus V(t) is a measure of X(t) as if in series with a compartment with turnover rate 1/RC.

As an alternative to the two methods above, consider recording the time of arrival of a pulse generated by the detector, raising that value to the power i, where i=0,1,2,..., and finally accumulating those powers in separate registers.

We can now introduce a few definitions:

$N_i(t)$ = sum of the i powers of the time of arrival of the pulses generated by a detector in the interval 0,t;

$p_i(r,t)dr$ = probability that $N_i(t)$ is a number in the interval r,r+dr;

$X(t)dt$ = probability that a pulse is generated by a detector in the interval t,t+dt.

Obviously $N_o(t)$ is equal to the number of pulses generated in the interval 0,t; the probability of two or more pulses being generated in the same interval is an infinitesimal of order higher than dt and can be neglected.

If in the interval t,t+dt a pulse is generated, the random variable N_i increases by t^i; if no pulse is generated in that interval, the random variable does not change its value; therefore

$$p_i(r,t+dt)dr = [1-X(t)dt]\cdot p_i(r,t)dr \quad \text{for } r < t^i$$

$$= X(t)dt\cdot p_i(r-t^i,t)dr + [1-X(t)dt]\cdot p_i(r,t)dr$$

$$\text{for } r \geqq t^i.$$

Rearranging,

$$(1) \qquad \frac{\partial p_i}{\partial t} dr = - X(t)\cdot p_i(r,t)dr \qquad \text{for } r < t^i$$

$$(2) \qquad \frac{\partial p_i}{\partial t} dr = - X(t)\cdot [p_i(r,t)dr - p_i(r-t^i,t)dr] \qquad \text{for } r \geqq t^i.$$

Define the generating function

$$G_i(s,t) = \int_o^\infty s^r p_i(r,t)dr;$$

multiply the two equations above by s^r, then integrate equation (1) from 0 to t^i and equation (2) from t^i to ∞; add them together,

$$(3) \qquad \frac{\partial G_i}{\partial t} = - X(t)(1-s^{t^i})G_i(s,t);$$

at time 0 no counts have been recorded, therefore

$$G_i(s,0) = 1,$$

by integration equation (3) gives

$$G_i(s,t) = \exp\left[-\int_0^t (1-s^{t^i})X(\tau)d\tau\right], \quad i=0,1,2,\ldots .$$

For i=0 this is the generating function of a Poisson random variable; in general

$$\frac{\partial G_i}{\partial s} = \int_0^t \tau^i s^{\tau^i-1} X(\tau)d\tau \cdot G_i(s,t)$$

$$\frac{\partial^2 G_i}{\partial s^2} = \int_0^t \tau^i(\tau^i-1)s^{\tau^i-2} X(\tau)d\tau \cdot G_i(s,t) +$$

$$+ \int_0^t \tau^i s^{\tau^i-1} X(\tau)d\tau \cdot \frac{\partial G_i}{\partial s}$$

$$\frac{\partial^3 G_i}{\partial s^3} = \int_0^t \tau^i(\tau^i-1)(\tau^i-2)s^{\tau^i-3} X(\tau)d\tau \cdot G_i(s,t) +$$

$$+ 2\int_0^t \tau^i(\tau^i-1)s^{\tau^i-2} X(\tau)d\tau \cdot \frac{\partial G_i}{\partial s} +$$

$$+ \int_0^t \tau^i s^{\tau^i - 1} \; X(\tau)d\tau \cdot \frac{\partial^2 G_i}{\partial s^2}$$

and from these derivatives we compute the expected value, the variance and the third moment around the mean of the random variable $N_i(t)$,

$$E[N_i(t)] = \int_0^t \tau^i \, X(\tau)d\tau,$$

$$Var[N_i(t)] = \int_0^t \tau^{2i} \, X(\tau)d\tau,$$

$$M_3[N_i(t)] = \int_0^t \tau^{3i} \, X(\tau)d\tau.$$

Separate registers can be used to compute the sums

$$\hat{N}_i(t) = \sum_{j=1}^{N} t_j^i; \; i=0,1,2,\ldots; \quad 0 < t_1 < t_2 < \ldots t_N < t$$

of the successive values $t_1, t_2, \ldots, t_N$ of the times of counts generated by the radioactive source. Of course

$$N = \hat{N}_0(t) = \sum_{j=1}^{N} 1$$

and $\hat{N}_i(t)$ is the best estimator of $E[N_i(t)]$, while $\hat{N}_{2i}(t)$ is the best estimator of $Var[N_i(t)]$ and $\hat{N}_{3i}(t)$ the best estimator of $M_3[N_i(t)]$.

The unknown function $X(t)$ can be reconstructed to any degree of accuracy in the interval $0,t$ from a sufficient number of estimators $\hat{N}_i(t)$.

For instance put

$$X(t) = a_0 + a_1 t + a_2 t^2 + \dots ;$$

it follows

$$N_0(t) = a_0 t + \frac{1}{2} a_1 t^2 + \frac{1}{3} a_2 t^3 + \dots$$

$$N_1(t) = \frac{1}{2} a_0 t^2 + \frac{1}{3} a_1 t^3 + \frac{1}{4} a_2 t^4 + \dots$$

$$N_2(t) = \frac{1}{3} a_0 t^3 + \frac{1}{4} a_1 t^4 + \frac{1}{5} a_2 t^5 + \dots$$

----

or more compactly

$$\underset{\sim}{N} = \underset{\sim}{A} \cdot \underset{\sim}{T}$$

where $\underset{\sim}{N}$ and $\underset{\sim}{A}$ are the row vectors of the $N_i(t)$ and the a_i, and $\underset{\sim}{T}$ is the persymmetric matrix

$$\begin{bmatrix} t & t^2/2 & t^3/3 & --- \\ t^2/2 & t^3/3 & t^4/4 & --- \\ t^3/3 & t^4/4 & t^5/5 & --- \\ \multicolumn{4}{c}{-----------------------} \end{bmatrix} ;$$

since $\underset{\sim}{N}$ and $\underset{\sim}{T}$ are known, $\underset{\sim}{A}$ can be easily computed.

When t is sufficiently large so that X(t) is approximately zero, the random variables $N_i(t)$ have some useful properties, as shown elsewhere.[1-7] For instance if X(t) and Y(t) are the probability density functions of the counting rate of a precursor and of its successor, respectively, then

$$\int_0^t X(\tau)G(t-\tau)d\tau = Y(t), \tag{4}$$

where G(t) is the transfer function from the precursor to the successor; define $M_i(t)$ and $N_i(t)$ the random variables associated to X(t) and Y(t), respectively. If $E[M_i(t)]$ and $E[N_i(t)]$ are evaluated, then function G(t) can be reconstructed.

In fact, from equation (4),

$$\int_0^t \tau^i Y(\tau)d\tau = \int_0^t \tau^i \int_0^\tau X(\sigma)G(\tau-\sigma)d\sigma \cdot d\tau$$

$$= \int_0^t X(\sigma)\int_\sigma^t \tau^i G(\tau-\sigma)d\tau \cdot d\sigma$$

$$= \int_0^t X(\sigma)\int_0^{t-\sigma} (\tau+\sigma)^i G(\tau)d\tau \cdot d\sigma$$

$$= \sum_{j=0}^{i} \binom{i}{j}\int_0^t \sigma^j X(\sigma)\int_0^{t-\sigma} \tau^{i-j}\, G(\tau)d\tau \cdot d\sigma$$

and for t sufficiently large, as stated above,

$$E[N_i(\infty)] = \sum_{j=0}^{i} \binom{i}{j}\, E[M_j(\infty)] \cdot \int_0^\infty \tau^{i-j} G(\tau)d\tau .$$

We can write

$$G_i = \int_0^\infty \tau^i G(\tau)d\tau ,$$

so that the equations above become

$$E[N_0(\infty)] = E[M_0(\infty)]\cdot G_0$$

$$E[N_1(\infty)] = E[M_1(\infty)]\cdot G_0+E[M_0(\infty)]\cdot G_1$$

$$E[N_2(\infty)] = E[M_2(\infty)]\cdot G_0+2E[M_1(\infty)]\cdot G_1+E[M_0(\infty)]\cdot G_2$$

and so forth. In this way the successive moments of the transfer function G(t) can be computed from the expected values of $M_i(\infty)$ and $N_i(\infty)$.

It is a well known fact[3] that in the case of a precursor of order one,

$$G_2/G_0 = 2(G_1/G_0)^2,$$

$$G_3/G_0 = 3!(G_1/G_0)^3,$$

and in general

$$G_i/G_0 = i!(G_1/G_0)^i;$$

it has also been shown that[6] the rank m of the persymmetric matrix

$$\begin{bmatrix} G_0 & G_1 & G_2/2! & G_3/3! & \cdots \\ G_1 & G_2/2! & G_3/3! & G_4/4! & \cdots \\ G_2/2! & G_3/3! & G_4/4! & G_5/5! & \cdots \\ \hline \end{bmatrix}$$

is equal to the minimum number of compartments that are consistent with the transfer function G(t).

Another problem connected with the measurement of fast changing radioactive sources is the correction for radioactive decay.[7] When measurements are made over a finite interval of time and the decay is very fast, the exact time for the correction depends on the shape of the radioactivity curve, which of course is unknown. If X(t) is a measure of the activity present at time t, as though no decay were present, and if γ is the disintegration constant, then the actual activity that can be measured is

$$e^{-\gamma t}X(t);$$

calling $\tilde{N}_i(t)$ the random variables associated with this function, then

$$E[\tilde{N}_i(t)] = \int_0^t \tau^i e^{-\gamma\tau} X(\tau)d\tau$$

$$= \int_0^t \tau^i (1-\gamma\tau+\gamma^2\tau^2/2!-\ldots)X(\tau)d\tau$$

$$= E[N_i(t)]-\gamma\cdot E[N_{i+1}(t)]+\gamma^2/2!\cdot E[N_{i+2}(t)]-\ldots$$

Thus if the $\tilde{N}_i$ are measured, the N_i can be computed.

Acknowledgements

This research was performed at Brookhaven National Laboratory under contract with the U. S. Department of Energy and supported by its Office of Health and Environmental Research. Partial support from the National Institute of Neurological Communicative Disease and Stroke Grant No. NS-16801-02 is also acknowledged.

References

1. J. Eisenfeld, et al. System Identification Problems and the Method of Moments. Math. Biosc. 36, 199-211 (1977).

2. J. H. Matis, et al. On Some Stochastic Formulations and Related Statistical Moments of Pharmacokinetic Models. J. Pharmacokin. Biopharm. (to appear).

3. A. Rescigno and L. D. Michels. On Dispersion in Tracer Experiments. J. Theor. Biol. 41, 451-460 (1973).

4. A. Rescigno. Multiple Compartmental Localization and the Method of Moments. In Proceedings of the Second Portonovo Conference on Biomathematics, page 569. Università-degli Studi, Ancona, 1978.

5. A. Rescigno. Multiple Compartmental Localization by Diffusion. In Principles of Radiopharmacology (L. Colombetti, Ed.), Vol. 3, pages 35-42. C. R. C. Press, Boca Raton, Florida, 1979.

6. A. Rescigno, et al. Mathematical Methods in the Formulation of Pharmacokinetic Models. In Tracer Kinetics and Physiologic Modeling (R. M. Lambrecht and A. Rescigno, Eds.), pages 59-119. Springer-Verlag, Berlin, 1983.

7. A. Rescigno, et al. Stochastic Modelling of Physiologic Processes with Radiotracers and Positron Emission Tomography. In Proceedings of the International Conference on Application of Physics to Medicine and Biology (G. Alberi, Z. Bajzer, P. Baxa, Eds.). World Scientific Publ. Co., Singapore, 1983.

ON IDENTIFIABILITY OF LINEAR COMPARTMENTAL SYSTEMS: SOME RESULTS OBTAINED BY MEANS OF STRUCTURAL PROPERTIES OF THE ASSOCIATED GRAPH

S. Audoly - Istituto di Matematica per Ingegneri, Cagliari

L. D'Angio - Istituto Matematico, Facoltà di Scienze, Cagliari

A. Introduction - This paper is concerned with the problem of a priori identifiability of linear time-invariant compartmental systems for experimental configurations with two separate input and two output. We will check for unique identifiability for a particular class of models which often occur in both biomedical and ecological studies /1/.

Let us consider the dynamic equations of a compartmental model

$$(1) \qquad S \equiv \begin{cases} \dot{\underline{x}} = A\underline{x} + B\underline{u} \\ \underline{x}(\emptyset) = \emptyset \\ \underline{y} = C\underline{x} \end{cases}$$

where $\underline{x},\underline{u},\underline{y}$ are the state, input and output vectors respectively, A is a constant matrix and B,C are defined as follows:

$$(1') \qquad B = \{e_i e_j\} \qquad C = B^T$$

In the sequel we summarise some notions of a method introduced in /2/ and we deduce, by structural properties, information about the identifiability of a class of systems which verifies Eqs (1').

B. Outline of the general method - Let G_N with $N \equiv \{1,..,n\}$ be a digraph associated with the matrix A: to each cycle and each path are associated respectively the functions:

$$(2) \quad c_{i_1 i_2 .. i_k} = (-1)^{k+1} a_{i_2 i_1} .. a_{i_1 i_k} ; \quad p_{i_1 i_2 .. i_k} = (-1)^{k+1} a_{i_2 i_1} .. a_{i_k i_{k-1}}$$

A vector f is associated with G_N, its rth component $f^{(r)}$ being the sum of all the cycles and cycle products not having common nodes, of length r, r=1..n; a vector Φ_{ij} is associated to each pair of nodes i,j its rth component being the sum of all the paths from i to j, each multiplied by the cycles or cycle product, without common nodes, which contains r arcs for r=1..n-1. In addition we assume the convention that $f^{(o)} = 1$ and $f^{(r)} = 0$ if $r > n$ or $r < 0$.

Given a decomposition of the graph G_N into two subgraphs G_K and G_{N-K} let us call "connecting cycles" the cycles which link the subgraphs, and "complementary vector" the vector $V_{K/N-K}$ defined as follows:

$$V_{K/N-K} = \{v_j\} \qquad v_j = \begin{cases} 0 & j=1..i-1 \\ \sum c_{k_1 k_2 .. k_l} \tilde{f}^{(j-l)}_{N-\{k_1 k_2 .. k_l\}} & j=i..n \end{cases}$$

where i is the length of the connecting cycles with minimum length, the summation is extended to all the connecting cycles having length $l \leq j$; $\tilde{f}^{(r)}_{N-\{k_1 k_2 .. k_l\}}$ is the form $f^{(r)}_{N-\{k_1 k_2 .. k_l\}}$ expropriated from all the cycles already considered in the sum. We have:

$$f_N^{(i)} = \sum_{j=0}^{i} f_K^{(i-j)} \cdot f_{N-K}^{(j)} + v_i \tag{3}$$

The overall complex of Eq(3) for i=1,2..n will be indicated as

$$f_N = f_K \oplus f_{N-K} + V_{K/N-K}$$

The transfer function for a strongly connected model can be written:

$$h_{ij}(s) = \frac{N_{ij}(s)}{D(s)} = \frac{\sum_{k=0}^{n-1} b_k^{(ij)} s^k}{\sum_{k=0}^{n} a_k s^k} \tag{4}$$

where

$$a_k = (-1)^{n-k} f_N^{(n-k)} \qquad k=0,..,n-1; \qquad a_n = 1$$

$$b_k^{(ij)} = \begin{cases} (-1)^{n-k-1} f^{(n-k-1)}_{N-\{i\}} & \text{if } i=j \\ (-1)^{n-k} \phi_{ij}^{(n-k-1)} & \text{if } i \neq j \end{cases} \qquad k=0,..,n-2; \qquad b_{n-1}^{(ij)} = \begin{cases} 1 & \text{if } i=j \\ 0 & \text{if } i \neq j \end{cases}$$

In the following we will furnish a scheme as a starting point to study the identifiability of a strongly connected model if Eqs(1') hold. The experiments provide f_N, $f_{N-\{i\}}$, $f_{N-\{j\}}$, Φ_{ij}, Φ_{ji}. From Eqs (3) we get

(5) $$f_N = f_i \oplus f_{N-\{i\}} + V_{i/N-\{i\}}$$

(6) $$f_{N-\{i\}} = f_j \oplus f_{N-\{ij\}} + V_{j/N-\{ij\}}$$
$$f_{N-\{j\}} = f_i \oplus f_{N-\{ij\}} + V_{i/N-\{ij\}}$$

hence f_i, f_j, $V_{i/n-\{i\}}$ can be uniquely calculated.

The vector $V_{i/N-\{i\}}$ can be decomposed as follows:

(7) $$v^{(r)}_{i/N-\{i\}} = x_r + y^{(i)}_r + f_j y^{(i)}_{r-1} + z_r \quad r=2..n; \quad v^{(1)}_{i/N-\{i\}}=0$$ where

$X = \{x_r\}$; $x_r = C_{ij}\, f^{(r-2)}_{N-\{i,j\}}$ $r=1..n$; hence $x_1 = 0$

$Y^{(i)} = \{y^{(i)}_r\}$ the vector generated by all the cycles connecting the node i to N-{i,j}, hence $y^{(i)}_1 = y^{(i)}_n = 0$. Thus $y^{(i)}_r$ $r=1..n-1$ is identical to $v^{(r)}_{i/N-\{ij\}}$; analogously $v^{(r)}_{j/N-\{ij\}} = y^{(j)}_r$; $y^{(j)}_1 = y^{(j)}_n = 0$

$Z = \{z_r\}$ the vector generated by the cycles of cycle products of length greater than 2, containing both node i and j; hence $z_1 = z_2 = 0$.

Let us suppose first a_{ij} and a_{ji} are different from zero. From Φ_{ij} and Φ_{ji}, p_{ij} and p_{ji} can be calculated, hence C_{ij} and f_{ij} are known. Then we have:

(8) $$-\phi^{(1)}_{ij}\,\phi^{(r)}_{ji} = x_{r+1} + z^{(i,j)}_{r+1} \quad r=1..n-1 \quad z^{(i,j)}_1 = z^{(i,j)}_2 = 0$$
$$-\phi^{(1)}_{ji}\,\phi^{(r)}_{ij} = x_{r+1} + z^{(j,i)}_{r+1} \quad r=1..n-1 \quad z^{(j,i)}_1 = z^{(j,i)}_2 = 0$$

where $Z^{(i,j)}$ and $Z^{(j,i)}$ are the part of the vector Z generated by the cycles containing the arc (i,j) or the arc (j,i) respectively. Hence

$$Z = Z^{(i,j)} + Z^{(j,i)} + \hat{Z}$$

denoting for convenience with $\hat{Z}$ the part of the vector Z related with the cycles and cycle products containing nodes i and j, but not the arc ij, or ji; hence $\hat{z}_1 = \hat{z}_2 = \hat{z}_3 = 0$.

This scheme allows to individuate four fundamental vectors which appear in the equations linearly and are related respectively to: X, the direct arcs between i and j; $Y^{(i)}$, $Y^{(j)}$, the connections between i <u>or</u> j and the remaining subgraph; Z, between i <u>and</u> j and the remaining subgraph. If these vectors can be calculated we can more easily investigate if the system is identifiable.

If $\hat{z}_r = 0$ r=1..k but $z_{k+1} \neq 0$, or if the first k components of the vector Z are in some way known, by exploiting each time one of Eqs(8),(7), (6) the first k components of the vectors $Y^{(i)}$, $Y^{(j)}$ and the first k+2 of the vectors X, $Z^{(ij)}$, $Z^{(ji)}$ can be uniquely calculated; moreover if k=n-2 the vectors as a whole are known: this always happens if $n \leq 5$.

If a_{ij} or a_{ji} are equal to zero, only one set of Eqs(8) is useful, none if useful if both a_{ij} and a_{ji} vanish, but since $X \equiv 0$ it is still possible to calculate k components of the vectors $Y^{(i)}$, $Y^{(j)}$ if the first k components of the vectors Z are in some way known, for example by utilizing the vectors Φ.

C. New results on a class of strongly connected models - With reference to the n-compartment model of Fig.1 with I/O in two consecutive nodes, e.g. i=1, j=2, we will prove the unique structural identifiability.

For such a system the knowledge of the decomposition vectors and consequently of $f_{N-\{12\}}$ is sufficient to state that the system is uniquely identifiable. In fact $f_{N-\{1\}}$ $f_{N-\{1,2\}}$ allows to construct a catenary sequence which, as the $f_1, f_2, p_{12}, p_{21}, C_{1n}$ are known, leads to uniquely

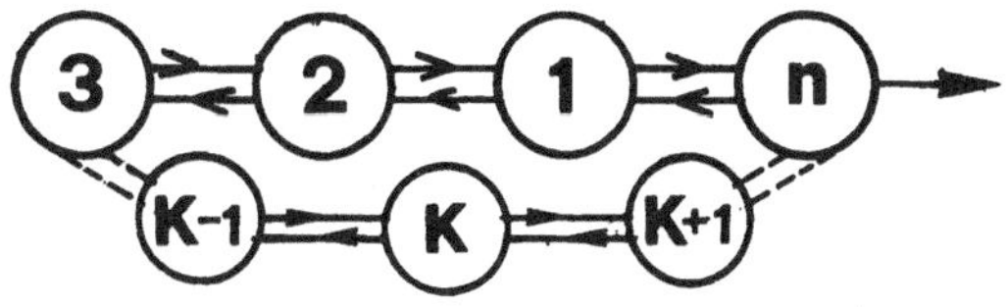

Fig.1 - Graph of an n-compartmental model

calculate all the a_{ij} /2/.

Because of the particular structure of the system, we get:

$z_r^{(12)} = z_r^{(21)} = 0$ and hence $z_r = \hat{z}_r$ for r=3..n-1.

Due to the absence of cycles from the 3rd to the n-1st order,

$\hat{Z} = C_{1n}\ C_{23}\ f_{N-\{123n\}};\quad Y^{(1)} = C_{1n}\ f_{N-\{12n\}};\quad Y^{(2)} = C_{23}\ f_{N-\{123\}}.$

Eqs (5),(7),(6),(8) provide:

$$f_1,\ f_2,\ C_{12};\quad f^{(k)}_{N-\{12\}},\ Y_k^{(1)},\ Y_k^{(2)},\ z_k \quad k=1..n-3$$

On the strength of the general results we need only calculate z_{n-2}, hence $f^{(n-6)}_{4..n-1}$.

By using as a starting point the known components of $f_{N-\{12\}}$ and $f_{N-\{12n\}}$ a partial catenary sequence can be constructed. This sequence, although incomplete, allows to calculate:

$$f_{n-r},\ f^{(k)}_{3..n-r-2}\quad k=1..n-7-2r,\quad r=0..\left[\frac{n-6}{2}\right];\ C_{n-r-1,n-r}\quad r=0..\left[\frac{n-7}{2}\right]$$

$f^{(n-6)}_{4..n-1}$ can be calculated from the following linear system:

$$f^{(k)}_{4..n-r} = f^{(k)}_{4..n-r-1} + f_{n-r}\cdot f^{(k-1)}_{4..n-r-1} + C_{n-r-1,n-r}\ f^{(k-2)}_{4..n-r-2}$$

$$r=0..\left[\frac{n-7}{2}\right]\quad k=1..n-2r-6$$

The system is structurally uniquely identifiable.

Let us now suppose the inputs are not consecutive but any whatsoever, say i=1, $j=3..\left[\frac{n+3}{2}\right]$. On the strength of the general results, taking into account that $z_1 = z_2 = z_3 = 0$ and $z_4 = y_2^{(1)} \cdot y_2^{(j)}$ if $j=4..\left[\frac{n+3}{2}\right]$ or $z_4 = y_2^{(1)} y_2^{(j)} - \phi_{1j}^{(1)} \phi_{j1}^{(1)}$ if j=3, the first four components of the vectors $Y^{(1)} Y^{(2)} f_{N-\{1j\}}$ are uniquely known; it ensues that if $n \leq 6$ the whole vector $f_{N-\{1j\}}$ hence $Y^{(1)} Y^{(2)} Z$ can be calculated. A catenary sequence can be performed by comparing $f_{N-\{1\}}$ or $f_{N-\{j\}}$ with $f_{N-\{1j\}}$; it furnishes $\binom{n-1}{j-1}$ solutions for the binary cycles and f_k. It is now possible to construct $\hat{z}_6$ thus the sum of the two n-order cycles, C', C" is known, hence C' and C" can be calculated. According to each set of cycles one matrix A is obtained /3/; the multiplicity can be removed by comparing the a_{ij} with Φ_{ij}.

The n-compartment system $n \leq 6$ is uniquely structurally identifiable.

D. Conclusion - For models with two perturbed and observed compartments a standard decomposition is proposed that suggests a starting point for identifiability check. For a particular structure of the matrix A, this decomposition provides a resolutive scheme which leads to state the unique identifiability for a class of n-compartment systems.

REFERENCES

/1/ C. Cobelli, R. Nosadini, G. Toffolo, A. McCulloch, A. Avogaro, A. Tiengo, G.K.M.M. Alberti: Model of the kinetics of ketone bodies in humans, Am.J.Physiol, 243, R7-R17 (1982).

/2/ S. Audoly, L. D'Angiò: On the identifiability of linear compartmental systems: a revisited transfer function approach based on topological properties, Math.Biosci., 66,2, 201-228(1983).

/3/ L. D'Angiò: On some structural properties of compartmental matrices, RSFSAK, 53/1 (1983).

COMPARTMENTAL VS NONCOMPARTMENTAL MODELING OF KETONE BODY KINETICS

Claudio Cobelli and Gianna Toffolo
Istituto di Elettrotecnica e di Elettronica, Università di Padova
and LADSEB-CNR, Padova, Italy

1. INTRODUCTION

Compartmental and noncompartmental linear models are widely used in physiological and clinical studies (e.g. endocrinology, metabolism, pharmacokinetics) for the interpretation of kinetic data obtained from linearising input-output experiments /1/. Recently some bases for choice between these two approaches have been discussed for the case where only one compartment is accessible for test input and output /2/. Both modeling strategies have been applied for studying the individual kinetics of ketone bodies (KB), i.e. of acetoacetate (AcAc) and 3hydroxybutyrate (βOHB), in man /3,4/. This is a situation where two pools are accessible and a two input-four output tracer experiment has been proved to provide an adequate data base. In this paper we discuss the relative merits of compartmental vs noncompartmental modeling by using the ketone body system as a prototype. The example allows us to evidence the structural (i.e. in terms of system connectivity) conditions for which noncompartmental modeling is inappropriate and to discuss some (superimposed) ambiguities which arise from computational problems when a noncompartmental model is numerically quantified from the data.

2. COMPARTMENTAL MODEL

Compartmental modeling of KB system has involved the assumption of a specific model structure, its numerical identification from experimental data and validation /1,3, 4/. The four compartment model of Fig.1 resulted to be the best representation of KB kinetics in normal humans. It evidences AcAc and βOHB in blood (compartment 1 and 2 respectively) and assumes the existence of two other compartments (liver and extrahepatic tissues, compartment 3 and 4 respectively) where the two ketone bodies interconvert very rapidly. All transfer rate parameters between compartments (k_{ij}, min^{-1} in Fig.1) can be estimated with accuracy from discrete time measurements of labelled AcAc and βOHB concentrations in blood following the two impulsive injections of labelled ketones by using nonlinear least squares estimation techniques /3/. A number of important parameters of tracee metabolism can then be quantified. Of particular concern here in view of the comparison with noncompartmental analysis of the data are the rates of appearance in blood of AcAc and βOHB due to interconversion (Ra_1^i, Ra_2^i) and recycling (Ra_1^r, Ra_2^r) which can be split in their liver and extrahepatic tissue components; the rates of irreversible disposal from blood (R_{01}, R_{02}) of AcAc and βOHB for subsequent utilization in extrahepatic tissues; and their rates of appearance in blood (Ra_1^n, Ra_2^N) from endogenous liver production. It is also possible to describe the system in terms of residence times in various compartments, via the mean residence time matrix /5/ which is related to the state matrix A of the model by

$$\Theta = -A^{-1} \qquad (A: \quad a_{ij} = k_{ij},\ i \neq j;\ a_{ii} = -\sum_{j=0}^{4} k_{ji}) \tag{1}$$

The generic element θ_{ij} represents the mean residence time that a particle introduced into compartment j is expected to stay in compartment i. As a result the sum of all the elements of the i-th column gives the mean total residence time in the system when input is in the i-th compartment:

$$MRT_i = \sum_{j=1}^{4} \theta_{ji} \tag{2}$$

Since endogenous production $R_{30}(=Ra_1^n + Ra_2^n)$ takes place in the liver (compartment 3) MRT_3 represents the mean time that an endogenously produced particle will spend in the system, and thus MRT_3 times R_{30} will give the total KB mass Q_D:

$$Q_D = MRT_3\ R_{30} \tag{3}$$

Finally the steady state total equivalent distribution volume is:

$$V_D = Q_D/(c_1 + c_2) \tag{4}$$

where c_1 and c_2 are steady state concentrations of AcAc and OHB in blood. Numerical values of relevant parameters for five normal subjects are given in Tab.1.

3. NONCOMPARTMENTAL MODEL

Noncompartmental analysis of kinetic data was carried out as proposed in /6/, and the system schematization is shown in Fig.2. It focuses on the accessible compartments only (AcAc and βOHB in blood) and assumes them to be embedded in a network of an undetermined number of connected compartments. As a consequence, only parameters related to the two accessible compartments can be estimated, in particular Ra_1^r, Ra_2^r, Ra_1^i, Ra_2^i (without evidencing components due to the liver and extrahepatic tissues) and also $R_{01}, R_{02}, Ra_1^n, Ra_2^n$, together with four residence times $T_1^1, T_2^1, T_1^2, T_2^2$ representing the expected time a particle introduced into compartment 1 or 2, indicated by the superscript leaves the system forever from compartment 1 or 2, indicated by the subscript. From these parameters, the mean total residence times in the system after input in compartment 1 and 2, MRT_1^* and MRT_2^* can be estimated:

$$MRT_1^* = T_1^1 \frac{R_{01}}{R_{01}+Ra_2^i R_{02}/(R_{02}+Ra_1^i)} + T_2^1 \left[1 - \frac{R_{01}}{R_{01}+Ra_2^i R_{02}/(R_{02}+Ra_1^i)} \right] \tag{5}$$

$$MRT_2^* = T_2^2 \frac{R_{02}}{R_{02}+Ra_1^i R_{01}/(R_{01}+Ra_2^i)} + T_1^2 \left[1 - \frac{R_{02}}{R_{02}+Ra_1^i R_{01}/(R_{01}+Ra_2^i)} \right] \tag{6}$$

(by $*$ we label in the following noncompartmental estimates) where terms multiplying T_1^1 and T_2^1 in eq. (5) can be interpreted as probabilities for a particle introduced into pool 1 to leave the system irreversibly from pool 1 and 2 respectively, a similar interpretation holds for terms in eq.(6). Finally the whole KB mass is given by:

$$Q_D^* = MRT_1^* Ra_1^n + MRT_2^* Ra_2^n \tag{7}$$

and the steady state equivalent distribution volume V_D^* can be estimated by dividing Q_D^* by (c_1+c_2) (cf. Eq.(4)).

The quantification of these parameters from experimental data requires only computation and algebraic manipulation of: i) initial values and slopes of the two decaying curves; ii) areas under the four experimental curves, from t=0 to ∞, and iii) areas under the four experimental curves multiplied by time, from t=0 to ∞. Recommendations given in /6/ were followed, and three computational approaches were considered: A) extrapolation to t=0 by an exponential function determined from the first three data points; extrapolation to $t\to\infty$ by an exponential function determined from the last four points; linear interpolation between data points. B) extrapolation to t=0 by assuming blood (80 ml/kgbw) as initial distribution volume and evaluating the initial slope assuming an exponential fall between the theoric sample at t=0 and first experimental point; extrapolation to $t\to\infty$ and interpolation between points as in A). C) A sum of two exponentials was used to describe the four discrete time curves, and areas, initial values and slopes were thus computed analytically. Mean percentual difference and range between noncompartmental estimates with approaches A,B,C and corresponding compartmental ones are shown in Table 2 for the five subjects.

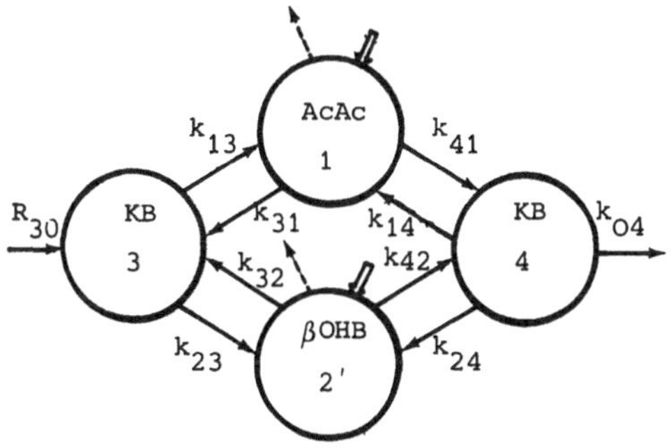

Fig. 1. Compartmental model of KB kinetics and input-output experiment.

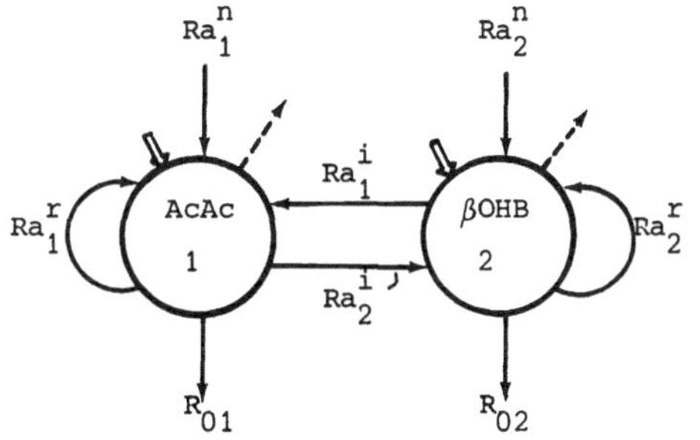

Fig. 2. "Two accessible pool" noncompartmental model and input-output experiment.

4. ERROR ANALYSIS IN NONCOMPARTMENTAL MODELING

Relevant and systematic differences exist in noncompartmental parameter estimates in comparison with the true compartmental ones. If the causes of this discrepancy are analyzed, the most important source of error in noncompartmental analysis appears to be an intrinsic, structural one: some parameters are incorrectly recovered as the allowed system connectivity is in general an oversimplified one. In addition, computational errors superimpose which arise from ambiguities in calculations. These components are discussed now. Structural errors affect the mean residence times in the system, MRT_1^* and MRT_2^*, the total body mass, Q_D^*, and the total equivalent distribution volume, V_D^*. More precisely, it follows from their definitions (eq. 5 and 6) that MRT_1^* and MRT_2^* will recover the true residence times MRT_1 and MRT_2 if and only if all particles exit the system from accessible compartments 1 and 2; from eq. 7 Q_D^* and thus V_D^* will equal the true Q_D and V_D if, in addition, all endogenous particles enter the accessible pools directly. Examining the physiologically based model of KB metabolism (Fig.1), neither conditions are met, since irreversible loss takes place in extrahepatic tissues (compartment 4) and endogenous production enter the system in the liver (compartment 3).

The structural errors in noncompartmental estimates of mean residence times and total mass/volume can be quantified by transforming the four compartment model of Fig. 1 into one which is equivalent in terms of input-output relationships and has the same connections between compartments but irreversible loss and production associated with compartments 1 and 2 only. Definitions (1) (2) (3) are then applied to both the original (Fig.1) and the transformed model and the following relations are obtained for the errors:

$$MRT_1^* - MRT_1 = MRT_2^* - MRT_2 = -\frac{1}{k_{04}+k_{14}+k_{24}} \quad (8)$$

$$Q_D^* - Q_D = -\frac{R_{30}}{k_{13}+k_{23}} - \frac{k_{04}Q_4}{k_{14}+k_{24}+k_{04}}$$

The structural erros can be computed from the data reported in Table 1. The error in Q_D^* is in mean 54±12% (SD) (range 41-69%); it is made up of two terms, the first one due to endogenous input R_{30} entering a compartment different from the accessible ones (44±12%), the second due to irreversible loss taking place from a compartment different from accessible ones (10±3%). Errors in MRT_1^* and MRT_2^* arise from this last source only and are respectively -25±9% (range -18,-38%) and -18±5% (range -15,-27%).

Table 1. Parameters estimated from the four compartment model.

Subjects	Ra_1^r	Ra_1^i	Ra_1^n	R_{01}	Ra_2^r	Ra_2^i	Ra_2^n	R_{02}	MRT_1	MRT_2	Q_D
				(μmol/min m²)					(min)		(mmol/m²)
1	29	35	42	71	28	6	86	57	9.0	19.6	5.30
2	110	156	104	133	199	127	164	135	13.7	17.2	7.28
3	60	83	172	175	100	80	165	162	11.4	13.7	8.70
4	105	191	59	147	319	103	310	222	14.7	17.7	9.04
5	89	159	65	125	162	89	253	183	14.2	17.9	8.25
Mean	79	125	86	130	162	81	196	152	12.6	17.2	7.71
±SD	±34	±64	±53	±38	±109	±46	±87	±62	±2.4	±2.2	±1.50

Table 2. Difference between noncompartmental and compartmental estimates. Values for computational approaches A,B,C are shown in percentage (Mean ±SD and range).

Method	Ra_1^r	Ra_1^i	Ra_1^n	R_{01}	Ra_2^r	Ra_2^i	Ra_2^n	R_{02}	MRT_1	MRT_2	Q_D
A	-43±15 (-62÷24)	19±7 (9÷27)	25±29 (-5÷68)	28±20 (7÷53)	-8±35 (-50÷29)	23±18 (0÷44)	-20±30 (-61÷ 20)	-9±21 (-25÷18)	-9±18 (-33÷6)	-11±19 (-42÷6)	-45±16 (-62÷-26)
B	73±48 (28÷145)	-33±14 (-50÷-18)	-3±15 (-24÷10)	-17±11 (-29÷-5)	37±51 (-40÷99)	-31±15 (-51÷-9)	-5±2 (-7÷-3)	-1±13 (-22÷9)	-20±18 (-41÷-3)	-21±6 (-28÷-14)	-59±14 (-73÷-42)
C	-9±34 (-43÷31)	-4±17 (-23÷13)	67±37 (29÷119)	25±17 (10÷51)	21±38 (-32÷65)	19±24 (-6÷53)	-14±17 (-36÷41)	1±21 (-18÷30)	-10±22 (-38÷5)	-33±5 (-39÷25)	-55±10 (-70÷46)

Computational errors arise when the noncompartmental model is quantified from experimental data, i.e. when areas under the curve, extrapolations to zero and infinity are numerically evaluated by resorting to some numerical procedure. They will obviously affect all the parameters. In order to evaluate their role per se, i.e. independently of any structural errors, compartmental and noncompartmental estimates can be compared for those parameters which are exactly recovered, i.e. no structural error, by the two approaches. For these parameters, i.e. rates of appearance in blood due to recycling, interconversion; disposal and synthesis, the observed discrepancy is due to numerical sources only. Table 2 shows that approaches A), B), C) lead to different estimates, the computational errors being always relevant. With method A) recycling rates are systematically underestimated while interconversion are overestimated; the opposite is true with method B) where recycling rates are overestimated while interconversion and disposal rates are underestimated. With method C) errors are somewhat less evident when expressed as mean values but large deviations are present in individual results as it clearly emerges from reported ranges. Also Ra_1^n and Ra_2^n are not exactly recovered, the errors being different with procedures A), B) and C), but their sum is able to approximate liver production reasonably well (error +7%, -9%, +11% with method A), B) and C) respectively).
As concerns mean residence times and total KB mass/volume it should be noted that the differences in Table 2 reflect the total, i.e. structural plus computational errors: as expected noncompartmental parameters always underestimate the true value. Carrying out the noncompartmental analysis with different computational procedures, has evidenced that the implicit models which are built into the method, i.e. one exponential decay in the early phase of disappearance, or a two exponential description for each curve, strongly bias the final results. As an example, method A) and B) which differ only in the extrapolation to time zero, i.e. the time of injections, lead to two different quantitative descriptions of the system, even if the two assumed models of initial distribution are reasonable and commonly adopted. The use of an analytical description of the experimental curves (method C) seems a more straightforward method, however the precision of the results is not always satisfactory. It should be noted that noncompartmental estimation has been performed here in a situation where great attention was devoted to the experimental design: frequent samples were collected, also near the time of injections, measurement errors were low, the presence of a steady state condition during the experiment has been ascertained. The numerical results of Table 2 show that this analysis leads to relevant ambiguities in describing KB metabolism and one may easily speculate on the consequences of a less careful experimental design.

5. CONCLUSIONS

Noncompartmental analysis constitutes, when compared to compartmental modeling, an appealing approach for kinetic analysis of data, both in terms of modeling and computational demands. However some difficulties related to its use which are likely to cause errors seem not to be fully appreciated. Noncompartmental modeling is often referred to as model-independent or structure-free: we have shown that a specific structure is assumed by noncompartmental analysis in order to be correct, i.e. endogenous sources and irreversible losses must take place in the accessible compartments only.
For the ketone body system these assumptions do not hold and the resulting structural errors cause an underestimation for mean residence times of about 20% and for total body mass/equivalent distribution volume of about 50%. Over and above these structural errors, the approach does not appear robust in terms of needed computational techniques. In the evaluation of areas / extrapolations some models are implicitely assumed and we have shown that the final results of the analysis are strongly sensitive to these assumptions. In our study on ketone body metabolism, better use is made of the informational content of kinetic data by using compartmental modeling: while more demanding in terms of modeling and identification techniques it has also revealed greater potentiality in terms of structural insight. While different, the compartmental and noncompartmental approaches should however complement each other

and as some results obtained with the two approaches should agree one approach may be used to validate the other.

Acknowledgment

This research was supported by a grant of the Ministero della Pubblica Istruzione.

REFERENCES

/1/ CARSON E.R., COBELLI C. and FINKELSTEIN L. Mathematical Modeling of Metabolic and Endocrine Systems. Model Formulation, Identification and Validation. New York, John Wiley 1983.

/2/ DI STEFANO J.J. III. Noncompartmental vs. compartmental analysis: some bases for choice. Am. J. Physiol. 243, R1-R6, 1982.

/3/ COBELLI C., TOFFOLO G. and NOSADINI R. Mathematical models of ketone bodies kinetics in the human. Experiment design and their identification / validation. In Preprints 6th IFAC Symp. on Identification and System Parameter Estimation. Ed. G.A. Bekey and G.M. Saridis, Oxford: Pergamon 1982, vol.1, p.222-227.

/4/ COBELLI C., NOSADINI R., TOFFOLO G., McCULLOCH A., AVOGARO A., TIENGO A. and ALBERTI K.G.M.M. Model of the kinetics of ketone bodies in humans. Am. J. Physiol. 247, R7-R17, 1982.

/5/ EISENFELD J. Relationship between stochastic and differential models of compartmental systems. Math. Biosci. 43, 289-305, 1979.

/6/ RESCIGNO A. and GURPIDE E. Estimation of average times of residence, recycles and interconversion of blood-borne compounds using tracer methods. J. Clin. Endocrinol. Metab. 36, 263-276, 1973.

IDENTIFIABILITY OF COMPARTMENTAL MODELS : ON CHARACTERIZATION OF TWO EXTREMAL TYPES OF INPUT-OUTPUT EXPERIMENTS

Arezki Mohammedi

Ecole Polytechnique Fédérale de Lausanne
Département de Mathématiques
MA Ecublens

CH-1015 Lausanne (Switzerland)

INTRODUCTION

Bellman-Åström (1970) [2] have stated the question of identifiability for time-invariant linear compartmental systems in general input-output configurations. Di Stefano (1977) [8], Eisenfeld (1979) [9], Cobelli and all (1979) [3] have shown that the controllability, the observability and the minimality are not in general necessary and/or sufficient conditions for the identifiability of such systems.

In this contribution, we give a statement relating the controllability and the identifiability of a structured linear system. Then for the compartmental systems we study two extremal situations for which the controllability or the observability are necessary and sufficient for their identifiability.

PRELIMINARY CONSIDERATIONS

Describing a real process S of transformations, we consider the time-invariant linear system (A,B,C) of order n

$$\begin{aligned} \dot{x}(t) &= A\,x(t) + B\,u(t) \qquad t > 0 \\ x(0) &= 0 \\ y(t) &= C\,x(t) \end{aligned} \tag{1}$$

where $x \in R^n$, $A \in R^{n\times n}$, $B \in R^{n\times p}$, $C \in R^{q\times n}$. We suppose that

$U_\Omega = \{u;\ R^+ \to \Omega;\ \Omega \subset R^p,\ \Omega \text{ open}\}$ and $Y = \{y : R^+ \to R^q\}$

satisfy the ordinary conditions for dynamical systems (Kalman and al. (1969) [15]). For the system (1) we have as input-output relation

$$y(t;u) = C \int_0^t e^{(t-\tau)A}\, B\, u(\tau)\, d\tau, \quad t > 0 \tag{2}$$

and we then say that the input-output experiment on S is E(p;q).

Now, from a priori knowledge of S some of the model parameters, i.e. those a priori null, are known. In order to determine the ℓ unspecified parameters , unknown coefficients of A,B,C, defining a vector parameter $\theta \in P \subset R^{\ell}$, we plan on S some E(p;q). The identifiability is the qualitative question of existence and unicity of θ, i.e. of a triple (A,B,C) from the potential data of this experiment (Jacquez (1982) [14], Nguyen-Wood (1982) [17], Cobelli-Di Stefano (1980) [4], Walter (1982) [24]).

For the notion of structured system, structural controllability and observability (SC and SO), digraph associated with (A,B,C), form (I) or form (II), generic rank, input or output or input-output reachability (i.r. or o.r. or i.o.r.) and canonical decomposition relatively to the reachability, we refer to Davison (1977) [5], Siljak (1978) [21], Cobelli et al. [3], Glover-Silverman (1976) [11], ShieldsPearson (1976) [19], Walter [24]. Moreover we suppose here that our systems cannot be a priori decomposable into isolated subsystems (compare to Walter [24] p. 30).

We recall here

Proposition 1. Let $M^c = (B, AB, \ldots, A^{n-1}B)$ and $M^o = \begin{bmatrix} C \\ CA \\ \vdots \\ C A^{n-1} \end{bmatrix}$ the $n \times np$ and the $nq \times n$ controllability matrix and observability matrix. Then rank $M^c = n(g)$ iff (A,B) is SC; rank $M^o = n(g)$ iff (C,A) is SO, where $n(g)$ denotes that the rank is n almost everywhere in P. (We will always use in this paper the term rank in the above sense).

Proposition 2 (Davison [5]). If $A \in R^{n \times n}$ possesses at least (n-1) diagonal unspecified coefficients, then (A,B) is SC iff the system is i.r.

CONTROLLABILITY AND IDENTIFIABILITY OF A LINEAR STRUCTURED SYSTEM

THEOREM 1. Let (A,B,I) where I is the $n \times n$ identity matrix and suppose that Ω contains a basis $\{\omega_i\}$ of R^p.

(a) The input-output relation is characterized by

$$x^{(k)}(0;\omega_i), \quad k = 1,\ldots, n, \ i = 1,\ldots, p \tag{3}$$

(b) These data with $x^{(n+1)}(0;\omega_i)$, $i = 1,\ldots, p$, determine A,B, uniquely if (A,B) is SC (sufficient condition of identifiability)

(c) If A possesses at least n-1 diagonal unspecified coefficients, the SC is also necessary condition for identifiability of A .

Proof

(a) After some classical manipulations we define the $n \times p$ matrix

$$H_k \triangleq (x^{(k)}(0;\omega_1),\ldots, x^{(k)}(0;\omega_p)) = (A^{k-1}B\omega_1,\ldots, A^{k-1}B\omega_p) = A^{k-1}BE \quad (4)$$

where the $p \times p$ constant matrix $E = (\omega_1,\ldots, \omega_p)$ is non singular and $k = 0,1,\ldots,n,\ldots$. In view of the Cayley-Hamilton theorem applied to A we stop at $k = n$ and this is why we say that the input-output relation is characterized by (4), i.e. by (3). In fact the $q \times p$ impulse response $Ce^{At}B$ of (A,B,C) is expressed in terms of the Markov parameters $M_j = CA^j B$, $j = 0,1,\ldots,n-1$ (Eisenfeld [9], Jacquez [14]).

(b) For (A,B,I) we have $M_k = A^{k-1} B = H_k E^{-1}$, $k = 1,\ldots,n$, and then

$$B = H_1 E^{-1} \quad (5)$$

Since $A(M_0,M_1,\ldots,M_{n-1}) = (M_1,M_2,\ldots, M_n) \triangleq M$ we get

$$A M^c = M \quad (6)$$

where M^c is the $n \times np$ controllability matrix observed and adjusted via (3) and (4). M is defined by (4) and $x^{(n+1)}(0;\omega_i)$, $i = 1,\ldots,p$. If (A,B) is SC i.e. if rank $M^c = n(g)$, then (6), which is a system of linear algebraic equations in a_{ij}, has a unique solution A almost everywhere in P, i.e. (A,B,I) is identifiable.

(c) It is almost evident (Cobelli and al. [3]) that (A,B,I) must be i.r. for identifiability since then it is i.o.r. But the condition (c), via the proposition 2, implies that the i.r. is equivalent to SC.

LINEAR TIME INVARIANT COMPARTMENTAL SYSTEMS

The linear time independant compartmental model is frequently encountered in the study of transformation systems (metabolism, exchange, ...) in biomedicine and biology, especially when the method of radioactive tracer is used (Jacquez (1972) [13]). This model is given in (1) where $A = (a_{ij})$ is a compartmental matrix that is

$$a_{ij} > 0,\ i \neq j\ ;\quad a_{ii} = \sum_{j=0}^{n}{}' a_{ji} < 0\ ,\quad i,j = 1,\ldots,n \quad (7)$$

The a_{ij} are the coefficient of exchange from compartment j to compartment i ; a_{0j} $(j = 1,\ldots, n)$ is the excretion coefficient from compartment j to the environment. When $a_{ii} = 0$, the column i is null and one says that the compartment i is a trap.

Here the use of tracer implies that we inject instantaneously at $t = 0$ (the input is supposed impulsional) p compartments of the system initially at rest and that we observe later q compartments, individually (neither fractioned nor additioned inputs or outputs).

1. Identifiability and controllability of (A, B, I)

As inputs we have $\omega_j = e_j\,\delta(t)$ where $\{e_j\}$ is the canonical basis of R^p and $\delta(t)$ is the "delta function". Then we put $x(0-) = 0$ in (1).

THEOREM 2. Let a compartmental system (A,B,I)

(a) The input-output relation is characterized by
$x^{(k)}(0+;\omega_i)$, $k = 0, 1,\ldots, n-1$, $i = 1,\ldots,p$ (8)

(b) These data with $x^{(n)}(0+;\omega_i)$, $i = 1,\ldots, p$ determine A,B uniquely if (A,B) is SC (sufficient condition for identifiability)

(c) If rank $A > n-1$, then the SC is also necessary for identifiability of A.

Proof

It is similar to that of theorem 1. The condition stated in (c) is an interpretation of the condition (c) of theorem 1 and states that the system contains at most one trap (Fife (1972) [10]).

Example 1

The condition (c) excludes the configurations like the following for which we may have identifiability without SC

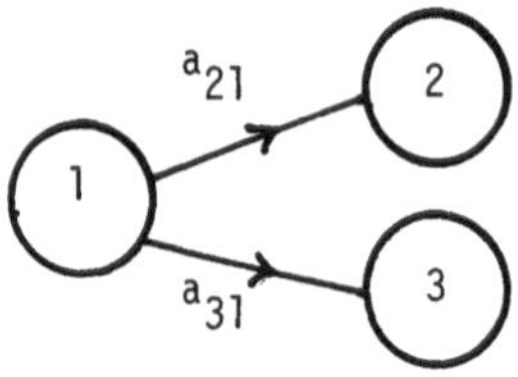

$$A = \begin{bmatrix} a_{11} & 0 & 0 \\ a_{21} & 0 & 0 \\ a_{31} & 0 & 0 \end{bmatrix}, \text{ rank } A = n-2 = 1 ;$$

for $B = (1\ \ 0\ \ 0)'$, (A,B) is not SC but (A,B,I) is identifiable.

If we add an excretion a_{03}, we get rank $A = n-1 = 2$; then if $B = (1\ \ 0\ \ 0)'$ the theorem 2 implies that A is identifiable.

2. Identifiability and observability of (A, I, C)

Let $\omega_j = e_j\,\delta(t)$ where $\{e_j\}$ is the canonical basis of R^n. In the same way we have done in theorems 1 and 2, let the $q \times n$ matrix

$$K_j \triangleq (y^{(j)}(0+;\ \omega_1),\dots,\ y^{(j)}(0+;\ \omega_n)) = CA^j\ ,\ j = 0,1,\dots,\ n-1 \qquad (9)$$

and

$$M^o\ A = \begin{bmatrix} K_1 \\ K_2 \\ \vdots \\ K_n \end{bmatrix} \triangleq K \qquad (10)$$

M^o is the $nq \times n$ observability matrix observed and adjusted by (9). If A is non singular, (A,I,C) is i.o.r. iff (C,A) is SO, since rank $\binom{C}{A} = n(g)$ (Cobelli et al. [3]). Thus we state

THEOREM 3. Let a compartmental system (A,I,C)

(a) The input-output relation is characterized by
$$y^{(j)}(0+;\omega_i),\ j = 0,1,\dots,n-1,\ i = 1,\dots,n \qquad (11)$$

(b) These values with $y^{(n)}(0+;\omega_i)$, $i = 1,\dots,n$, gives A,B uniquely if (C,A) is SO

(c) If A is non singular, i.e. if the system has no traps then the SO is also a necessary condition for identifiability of A.

Example 2

$$A = \begin{bmatrix} -a_{31} & 0 & 0 \\ 0 & -a_{32} & 0 \\ a_{31} & a_{32} & -a_{03} \end{bmatrix};\ \text{rank}\ A = 3$$

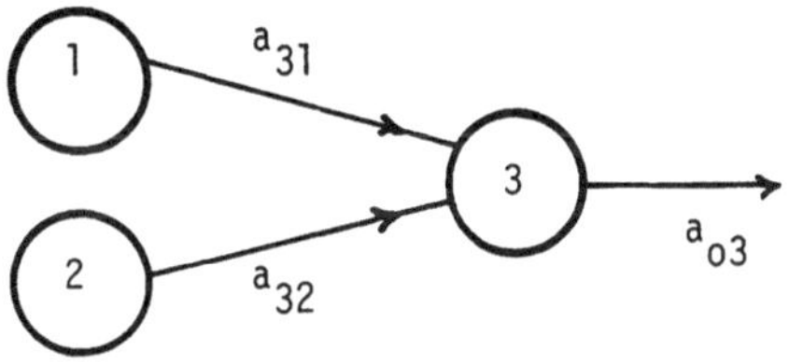

(a) $C = \begin{pmatrix} 1 & 0 & 0 \\ 0 & 1 & 0 \end{pmatrix}$; (C,A) not SO implies (A,I,C) not identifiable

(b) $C = (0\ \ 0\ \ 1)$; (C,A) is SO implies $A = M^{o-1}\ K$.

COROLLARY (two compartmental system)

(1) (A,B) is SC iff (A,B,I) is i.r. iff A,B identifiable

(2) (C,A) is SO iff (A,I,C) is o.r., and if A is non singular (C,A) is SO iff A,C identifiable.

(3) For (A,I,I) we have always identifiability.

CONCLUSIONS

The theorem 1 may be of some interest when one considers tracer experiments via constant infusions (Shipley-Clarck (1972) [20]).

The sufficiency of the SC for identifiability of (A,B,I) is mentioned by Bellman-Åström [2], Eisenfeld [9], Travis-Haddock (1982) [22] and at least implicitly by Rubinow-Winzer (1971) [18]); see also Walter [24].

In these theorems we have made no hypothesis on the spectrum (distinct or not distinct) of A . (Compare to Delforge (1981) [7], Anderson (1982) [1], Le Cardinal and al. (1976) [16]).

The condition (c) introduces structural constraints (Walter [24]). The equations we have to study for determining A from the observed M^c and M or M^o and K are linear and are in principle easier to handle than those non linear we obtain from the transfer matrix method (Delforge (1977) [6]).

The drawback is that the rank conditions result in general in ill-posed numerical problems (Siljak [21], Glover-Willems (1974) [12]) compared to the purely structural manipulations via the boolean matrices of interconnections which are used in the notion of reachability (Vidyasagar (1981) [23]).

Nevertheless it is shown that when one tries to exhibit situations where the SC and/or the SO are necessary and sufficient for identifiability of compartmental models one observes that such situations are not so frequent even for the systems (A,B,I) and (A,I,C).

REFERENCES

1. ANDERSON D.H. : Mathematical Biosciences 58, 61-81 (1982)
2. BELLMAN R. , ÅSTROM K.J. : Mathematical Biosciences 7, 329-339 (1970)
3. COBELLI C., LEPSHY A., ROMANIN-JACUR G. : Mathematical Biosciences 44, 1-18 (1979)
4. COBELLI C., DI STEFANO J.J. : IEEE AC-25, 4, 830-833 (1980)
5. DAVISON E.G. : Automatica 13, 109-123 (1977)
6. DELFORGE J. : Mathematical Biosciences 36, 119-125 (1977)
7. DELFORGE J. : Mathematical Biosciences 54, 159-180 (1981)
8. DE STEFANO J.J. : IEEE AC-22, 652 and comments (1977)
9. EISENFELD J. : Mathematical Biosciences 47, 15-33 (1979)
10. FIFE D. : Mathematical Biosciences 14, 311-315 (1972)
11. GLOVER K., SILVERMAN L.M. : IEEE AC-21, 534-537 (1976)
12. GLOVER K., WILLEMS J.C. : IEEE AC-19, 640-645 (1974)
13. JACQUEZ J.A. : Compartmental Analysis in Biology and Medicin, Elsevier Publ. Co., Amsterdam (1972)
14. JACQUEZ J.A. : Mathematics and Computers in Simulation XXIV, 452-459 (1982)
15. KALMAN R.E., FALB L.E., ARBIB M.E., Topics in Mathematical Systems Theory, Mc Graw Hill (1969)
16. LE CARDINAL G., BERTRAND P., WALTER E. : Mathematical Biosciences 31,131-141 (1976)
17. NGUYEN V.V., WOOD E.F. : SIAM review 24, 1, 34-51 (1982)
18. RUBINOW S.I., WINZER A. : Mathematical Biosciences 11, 203-247 (1971)
19. SHIELDS R.W., PEARSON J.B. : IEEE AC-21, 2, 203-212 (1976)
20. SHIPLEY R.A., CLARCK R.E. : Tracer methods for in vivo kinetics, Theory and applications, Academic Press (1972)
21. SILJAK D.D. : Large scale dynamic systems, North-Holland (1978)
22. TRAVIS C.E., HADDOCK G. : Mathematical Biosciences 57, 157-173 (1981)
23. VIDYASAGAR M. : Input-output analysis of large-scale interconnected systems, Springer-Verlag (1981)
24. WALTER E. : Itendifiability of state space models, Springer-Verlag (1982)

PART VI

GENERAL MATHEMATICAL METHODS

ESTIMATION TECHNIQUES FOR TRANSPORT EQUATIONS

H. T. BANKS
Div. of Applied Math
Brown University
Providence, RI 02912

P. KAREIVA
Div. of Biology and Medicine
Brown University
Providence, RI 02912

P. K. LAMM
Dept. of Mathematics
Southern Methodist Univ.
Dallas, Texas 75275

Abstract. We present convergence arguments for algorithms developed to estimate spatially and/or time dependent coefficients and boundary parameters in general transport (diffusion, advection, sink/source) models in a bounded domain $\Omega \subset R^2$. A brief summary of numerical results obtained using the algorithms is given.

I. Introduction. In this note we present theoretical results for estimation of function space (i.e., time and spatially varying) parameters in general transport equations. The presentation here is motivated by our own efforts on problems in transport of labeled substances in brain tissue [3], population dispersal (in particular insect movement--see [2], [3], [7]), and bioturbation [6], among other applications in the biological sciences. Due to limitations in space, we shall not discuss here any of those particular efforts. Rather we provide an outline of a general convergence theory for a class of approximation schemes that we have used and are continuing to use successfully in a number of biological applications. In the first two sections we present for the first time general theoretical arguments underlying these approximation schemes; our use of the methods in specific problems is discussed elsewhere ([2], [3]). In a final section we summarize briefly one aspect of the numerical performance of our methods that is pertinent to any convergence theory that one might develop.

As our fundamental state system we consider the scalar equation

$$(1)\qquad \frac{\partial u}{\partial t} + \nabla\cdot(\mathfrak{V}u) = \nabla\cdot(\mathfrak{D}\otimes\nabla u) + \alpha u + f, \qquad t\in(0,T],$$

on the bounded domain $\Omega \subset R^2$ with boundary conditions $u(t,\cdot)|_{\partial\Omega} = 0$ and given initial conditions $u(0,\cdot)|_{\Omega} = u_0(\gamma)$. Here we assume that $\mathfrak{V}$ and $\mathfrak{D}$ are functions depending on (t,x,y), $t \geq 0$, $(x,y)\in\Omega$ and $f = f(\beta,t,x,y)$, $\beta = \beta(t,x,y)$, $\alpha = \alpha(t,x,y)$, $\gamma = \gamma(x,y)$. While we treat only trivial Dirichlet boundary conditions in this note, our ideas are sufficiently general to allow nontrivial boundary conditions depending possibly on unknown parameters. We assume that any such boundary conditions have been transformed in the usual manner so that the unknown boundary parameters are included in the vector parameters β and γ in f and u_0 above. We have also simplified our exposition in treating only a scalar equation even though our methods are applicable to (and have been used with) vector systems.

Along with the state equation (1) we assume that we have observations $\hat{u}_i \in H^0(\Omega)$ for $u(t_i,\cdot)$, or $\hat{u}_{ijk} \in R^1$ for $u(t_i,x_j,y_k)$ and that we wish to choose the parameter functions $\mathfrak{v}$, $\mathfrak{D}$, α, β, γ so that the corresponding solution of (1) best fits the observations. For our discussions here we shall assume that this problem is posed in terms of an optimization problem for a least squares fit-to-data criterion. Letting $q = (\mathfrak{D},\mathfrak{v},\alpha,\beta,\gamma)$ represent the set of unknown parameters and Q represent the class of admissible parameter functions, we denote by $q \to J(q)$ the least squares criterion function. For the observations mentioned above, this function is given by

$$(2) \quad J(q) = \sum_i |u(t_i,q) - \hat{u}_i|^2$$

in the case of distributed data $\hat{u}_i \in H^0(\Omega)$, and

$$(3) \quad J(q) = \sum_{i,j,k} |u(t_i,x_j,y_k,q) - \hat{u}_{ijk}|^2$$

in the case of pointwise data $\hat{u}_{ijk} \in R^1$, where in both cases $u(\cdot,q)$ is the solution of (1) for a given $q = (\mathfrak{D},\mathfrak{v},\alpha,\beta,\gamma)$. In either case our basic problem consists of minimizing J over Q.

This problem is difficult in part because it is, in general, infinite dimensional in both the state u and the parameters q, each of which lies in a function space. Therefore algorithms for its solution will, in most cases, involve two separate and often unrelated approximation ideas, one for the state space and one for the parameter set. In the next section we consider these approximations and outline convergence arguments that indicate that the schemes can yield useful computational results.

II. Convergence Results. We first rewrite equation (1) in its weak form in the state space $H = H^0(\Omega)$ with the usual inner product. We have, dropping the Kronecker product sign $\otimes$ for ease in notation, that for all $\phi \in H_0^1(\Omega)$, the weak solution u must satisfy

$$(4) \quad \begin{aligned} &\langle u_t,\phi\rangle + \langle \mathfrak{D}\nabla u(t),\nabla\phi\rangle - \langle \mathfrak{v}u(t),\nabla\phi\rangle - \langle \alpha u(t) + f,\phi\rangle = 0, \qquad t \in (0,T], \\ &u(0) = u_0(\gamma). \end{aligned}$$

Retaining the notation $q = (\mathfrak{D},\mathfrak{v},\alpha,\beta,\gamma)$ we make the standing assumptions on the parameter set Q:

[A] The set Q is a bounded subset of $L_\infty([0,T] \times \Omega)$.

[B] There exists a positive constant m such that for every $q = (\mathcal{D},\mathcal{V},\alpha,\beta,\gamma)$ in Q, $\mathcal{D}_i(t,x,y) \geq m$ for i = 1, 2, and (t,x,y) in $[0,T] \times \Omega$.

We also assume throughout that the perturbation function f is C^1 in all of its arguments and that $\gamma \to u_0(\gamma)$ is continuous from $H^0(\Omega)$ to $H^0(\Omega)$. This will suffice for most of the results stated below, but the regularity of solutions u assumed later can be guaranteed only under more stringent smoothness assumptions on f along with conditions relating f to the initial data u_0.

Under our standing assumptions on Q and f, we can, further assuming that $u_0 \in H^0(\Omega)$, use rather standard arguments (e.g., see [10, p. 104]) to guarantee existence and uniqueness of solutions u to (4) with $u(t,q) \in H_0^1(\Omega)$.

For the state space approximation of (4), we consider a Galerkin scheme on finite dimensional subspaces H^N of $H^0(\Omega)$. We assume $H^N \subset H_0^1(\Omega)$ for each N = 1, 2, ..., and define the Galerkin approximation for a given $q \in Q$ as the solution u^N, $u^N(t) \in H^N$, of the equations

$$(5)\qquad \begin{aligned} &\langle u_t^N,\psi\rangle + \langle \mathcal{D}\nabla u^N(t),\nabla\psi\rangle - \langle \mathcal{V}u^N(t), \nabla\psi\rangle - \langle \alpha u^N(t) + f,\psi\rangle = 0, \qquad \psi \in H^N, \\ &u^N(0) = P^N u_0(\gamma), \end{aligned}$$

where P^N is the orthogonal projection of $H^0(\Omega)$ onto H^N.

It is useful to define the bilinear form $\mathcal{L}(q) : H_0^1(\Omega) \times H_0^1(\Omega) \to R^1$ by

$$\begin{aligned} \mathcal{L}(q)(\psi,\phi) &= \langle \mathcal{D}\nabla\psi,\nabla\phi\rangle - \langle \mathcal{V}\psi,\nabla\phi\rangle - \langle \alpha\psi,\phi\rangle \\ &\equiv a(q)(\psi,\phi) - b(q)(\psi,\phi) - \langle \alpha\psi,\phi\rangle . \end{aligned}$$

Then we may rewrite the original equation and its Galerkin approximation in H^N as

$$(6)\qquad \begin{aligned} &\langle u_t,\phi\rangle + \mathcal{L}(q)(u,\phi) = \langle f,\phi\rangle, \qquad \forall\phi \in H_0^1(\Omega), \\ &u(0) = u_0(q) \end{aligned}$$

and

$$(7)\qquad \begin{aligned} &\langle u_t^N,\psi\rangle + \mathcal{L}(q)(u^N,\psi) = \langle f,\psi\rangle, \qquad \forall\psi \in H^N, \\ &u^N(0) = P^N u_0(q). \end{aligned}$$

Defining, for solutions u^N of (7), the approximate fit-to-data function J^N (corresponding to (2)) by

(8) $J^N(q) = \sum_i |u^N(t_i,q) - \hat{u}_i|^2$,

we find that our estimation problems with approximate states (which are still optimization problems over an infinite dimensional function set Q) consist of minimizing J^N over Q. Before turning to a second level of approximation (for the parameter set) we give some convergence results for these approximate problems that will prove useful in discussing the state-and-parameter approximation problems. We make the following standing assumptions on the approximation properties of H^N relative to H.

[C] Let P^N denote the orthogonal projection of $H^0(\Omega)$ onto H^N. Then for any $\psi \in C^2(\Omega) \cap H_0^1(\Omega)$ the following estimates hold: $|P^N\psi - \psi|^2, |(P^N\psi - \psi)_x|^2$, $|(P^N\psi - \psi)_y|^2$ are each dominated by some functional $g(N,\psi)$ satisfying $|g(N,\psi)| \leq \varepsilon(N) \{|\psi_{xx}|^2 + |\psi_{yy}|^2\}$ where $\varepsilon(N) \to 0$ as $N \to \infty$.

We remark that for $\Omega = (0,1) \times (0,1)$, tensor products of the subspaces $L_0(\Delta^N)$ and $S_0(\Delta^N)$ of linear and cubic splines (corresponding to a grid size $\Delta^N = 1/N$) modified to satisfy homogeneous boundary conditions, are readily seen to satisfy the condition C (see [11, Chap. 6] for further discussion and details). The following fundamental convergence statement (e.g., see [2], [3]) is most helpful in establishing the desired approximation theorems.

Theorem 1. Suppose $q^N, q^* \in Q$ with $q^N \to q^*$ in $H^0((0,T) \times \Omega)$. Suppose further that $u(q^*) \in C^2(\overline{(0,T) \times \Omega})$. Then $u^N(t,q^N) \to u(t,q^*)$ in $H^0(\Omega)$ for each $t \in [0,T]$, where u, u^N are solutions of (6), (7), respectively.

Since this theorem is fundamental to our discussions in this note, we shall outline the essential steps of its proof. First, we note that under assumptions A and B above, it is easy to establish the following Gärding inequality (e.g., see also [9, p. 144], [4, p. 34]): there are positive constants c_0, c_1 depending on Ω, the bounds for Q, and m such that $\mathcal{L}(q)(\phi,\phi) \geq c_1|\phi|_1^2 - c_0|\phi|^2$ for all $\phi \in H_0^1(\Omega)$ and all $q \in Q$.

Let u^N, u^* be the solutions of (7), (6) corresponding to q^N, q^*, respectively. We wish to argue that $|u^N(t) - u^*(t)| \to 0$. But from the inequality $|u^N - u^*| \leq |u^N - P^N u^*| + |P^N u^* - u^*|$ (to simplify notation here and throughout we suppress the dependence on t) and the approximation results of condition C, it suffices to argue $|u^N - P^N u^*| \to 0$.

Let f^N denote f at β^N of q^N and note that the convergence hypotheses on q^N and the smoothness of f imply $|f^N - f^*| \to 0$ where f^* corresponds to q^*. We have that u^N, u^* satisfy the equations (see (6), (7))

$$(9)\qquad \begin{aligned} &\langle u_t^N,\psi\rangle + \mathcal{L}(q^N)(u^N,\psi) = \langle f^N,\psi\rangle \qquad \text{for all } \psi \in H^N , \\ &u^N(0) = P^N u_0(q^N) \end{aligned}$$

and

$$(10)\qquad \begin{aligned} &\langle u_t^*,\phi\rangle + \mathcal{L}(q^*)(u^*,\phi) = \langle f^*,\phi\rangle \qquad \text{for all } \phi \in H_0^1(\Omega) , \\ &u^*(0) = u_0(q^*). \end{aligned}$$

Since $H^N \subset H_0^1(\Omega)$, letting D_t denote $\frac{\partial}{\partial t}$, we have for all $\psi \in H^N$

$$\begin{aligned} &\langle D_t(u^N - P^N u^*),\psi\rangle + \mathcal{L}(q^N)(u^N - P^N u^*,\psi) \\ &\quad = - \langle D_t P^N u^*,\psi\rangle - \mathcal{L}(q^N)(P^N u^*,\psi) + \langle f^N,\psi\rangle \\ &\quad = \langle D_t(u^* - P^N u^*),\psi\rangle + \mathcal{L}(q^*)(u^*,\psi) - \mathcal{L}(q^N)(P^N u^*,\psi) + \langle f^N - f^*,\psi\rangle \end{aligned}$$

where we have used (9) and (10). Choosing $\psi = z^N = u^N - P^N u^*$ in H^N we thus find using the above identity

$$\begin{aligned} \tfrac{1}{2} D_t|z^N|^2 + \mathcal{L}(q^N)(z^N,z^N) &= \langle D_t(u^* - P^N u^*),z^N\rangle \\ &\quad + \mathcal{L}(q^*)(u^*,z^N) - \mathcal{L}(q^N)(P^N u^*,z^N) + \langle f^N - f^*,z^N\rangle . \end{aligned}$$

Use of the Gärding inequality then yields

$$\begin{aligned} \tfrac{1}{2} D_t|z^N|^2 + c_1|z^N|_1^2 - c_0|z^N|^2 &\le \langle D_t(u^* - P^N u^*),z^N\rangle \\ &\quad + \mathcal{L}(q^*)(u^*,z^N) - \mathcal{L}(q^N)(P^N u^*,z^N) + \langle f^N - f^*,z^N\rangle \\ (11)\qquad &\le \tfrac{1}{2} |u_t^* - P^N u_t^*|^2 + \tfrac{1}{2} |z^N|^2 + a(q^*)(u^*,z^N) - a(q^N)(P^N u^*,z^N) \\ &\quad + b(q^N)(P^N u^*,z^N) - b(q^*)(u^*,z^N) \\ &\quad + \langle \alpha^N P^N u^* - \alpha^* u^*,z^N\rangle + \langle f^N - f^*,z^N\rangle . \end{aligned}$$

But

$$a(q^*)(u^*,z^N) - a(q^N)(P^N u^*,z^N) = \langle \mathcal{D}^*\nabla u^* - \mathcal{D}^N\nabla P^N u^*,\nabla z^N\rangle$$

$$\leq \frac{1}{2}\,\frac{2}{c_1}\,|\mathcal{D}^*\nabla u^* - \mathcal{D}^N\nabla P^N u^*|^2 + \frac{1}{2}\,\frac{c_1}{2}\,|\nabla z^N|^2$$

while

$$b(q^N)(P^N u^*,z^N) - b(q^*)(u^*,z^N) = \langle \mathfrak{v}^N P^N u^* - \mathfrak{v}^* u^*,\nabla z^N\rangle$$

$$\leq \frac{1}{2}\,\frac{2}{c_1}\,|\mathfrak{v}^N P^N u^* - \mathfrak{v}^* u^*|^2 + \frac{1}{2}\,\frac{c_1}{2}\,|\nabla z^N|^2 .$$

Thus, using these estimates in (11) we find

$$\frac{1}{2} D_t|z^N|^2 + c_1|z^N|_1^2 - c_0|z^N|^2$$

$$\leq \frac{1}{2}\,|u_t^* - P^N u_t^*|^2 + \frac{1}{2}|z^N|^2 + \frac{1}{c_1}\,|\mathcal{D}^*\nabla u^* - \mathcal{D}^N\nabla P^N u^*|^2$$

$$+ \frac{1}{c_1}\,|\mathfrak{v}^N P^N u^* - \mathfrak{v}^* u^*|^2 + \frac{c_1}{2}\,|\nabla z^N|^2$$

$$+ \frac{1}{2}\,|\alpha^N P^N u^* - \alpha^* u^*|^2 + \frac{1}{2}\,|z^N|^2 + \frac{1}{2}\,|f^N - f^*|^2 + \frac{1}{2}\,|z^N|^2 ,$$

or, since $|z^N|_1^2 \geq |\nabla z^N|^2$,

$$(12)\quad \frac{1}{2} D_t|z^N|^2 + \{-\frac{3}{2} - c_0\}\,|z^N|^2 \leq h^N$$

where

$$h^N \equiv \frac{1}{2}\,|u_t^* - P^N u_t^*|^2 + \frac{1}{c_1}\,|\mathcal{D}^*\nabla u^* - \mathcal{D}^N\nabla P^N u^*|^2$$

$$+ \frac{1}{c_1}\,|\mathfrak{v}^N P^N u^* - \mathfrak{v}^* u^*|^2 + \frac{1}{2}\,|\alpha^N P^N u^* - \alpha^* u^*|^2 + \frac{1}{2}\,|f^N - f^*|^2 .$$

Using the Gronwall inequality and defining $\varepsilon^N = |u^N(0) - P^N u^*(0)|^2$, we may obtain from (12) the estimate

$$|u^N(t) - P^N u^*(t)|^2 \leq \varepsilon^N e^{2\delta t} + e^{2\delta t}\int_0^t e^{-2\delta s}\, 2h^N(s)ds$$

$$(13)\qquad \leq \varepsilon^N e^{2\delta t} + 2e^{2\delta t}\,(T\int_0^T |h^N(s)|^2 ds)^{1/2}$$

where $\delta = c_0 + \frac{3}{2}$ is independent of N. Since $\varepsilon^N \to 0$ follows from the continuity assumption on u_0, for the desired convergence it suffices to argue that $h^N \to 0$ in $H^0(0,T)$. However, using the convergence $q^N \to q^*$, the assumption that u^* is in $C^2(\overline{(0,T) \times \Omega})$, the estimates of condition C, and the bounds from condition A, this convergence is readily established. Thus is the statement of Theorem 1 proved.

We remark that the regularity required of $u(q^*)$ in Theorem 1 can be guaranteed by rather standard smoothness theorems (e.g., see [4, p. 141]). It is a straightforward exercise to verify that such theorems require sufficient smoothness of the coefficients as well as of the perturbing function f (this also involves the initial data u_0). An alternate approach, which permits relaxation of the smoothness of the coefficients (and requiring this smoothness only on $(\varepsilon,T] \times \Omega$) could be taken (see [1], [5]) at the expense of some technical tedium. However, we shall be approximating the parameter functions q below on $[0,T]$, not $(\varepsilon,T]$, and hence the stronger smoothness assumptions are more appropriate here.

To use the statements in Theorem 1 to obtain a convergence theory for approximate parameters, we shall need a continuous dependence result for solutions u of (4). To state this result, we define $C^1_B(\Omega)$ as the set of C^1 functions with bounded derivatives on Ω.

<u>Theorem 2</u>. For any solution u of (4) such that $u(t,q) \in C^1_B(\Omega)$, we have that $q \to u(t,q)$ is continuous on Q in the $H^0((0,T) \times \Omega)$ topology.

The arguments for this theorem, which involve estimates for $|u(t,q) - u(t,\tilde{q})|$, make use of conditions A and B above. They are very similar to the arguments outlined for Theorem 1 and so we shall not give them here. Instead, we explain how these results are used to obtain a parameter convergence theorem.

<u>Theorem 3</u>. Suppose Q is compact in the $H^0((0,T) \times \Omega)$ topology. Then a solution $\bar{q}^N$ to the problem of minimizing J^N over Q exists, N = 1,2,... . Let $\{\bar{q}^{N_k}\}$ be any convergent subsequence, $\bar{q}^{N_k} \to q^*$ in $H^0((0,T) \times \Omega)$. If $u(q) \in C^2(\overline{(0,T) \times \Omega})$ for each $q \in Q$, then q^* is a solution to the problem of minimizing J over Q.

Since the arguments are similar to those we have presented elsewhere (see [1, pp. 28-29]), we only sketch them here. The existence statement follows once one establishes continuity of $q \to J^N(q)$ on the compact set Q (the continuity arguments are similar to those behind Theorem 2). From Theorem 1, (2), and (8) we have

$$J(q^*) = \lim_{N_k\to\infty} J^{N_k}(\bar{q}^{N_k}) \leq \lim_{N_k\to\infty} J^{N_k}(q)$$

for any $q \in Q$. But Theorem 1 (with the constant sequence $\{q\}$) also guarantees $J^{N_k}(q) \to J(q)$. Thus $J(q^*) \leq J(q)$ for any $q \in Q$.

As we have indicated previously, the state approximation results of Theorem 3 are only first level results that are not satisfactory from a computational point of view since the approximate problems still involve minimization over the infinite dimensional set Q. We turn next to a second level of approximation where we combine the state approximation ideas outlined above with ideas for approximation of the parameter set Q.

We suppose that Q^M, $M = 1,2,\ldots$, are subsets of $H^0((0,T) \times \Omega)$ defined by $Q^M = i_M(Q)$ where i_M is a mapping from $Q \subset H^0((0,T) \times \Omega)$ into $H^0((0,T) \times \Omega)$. The approximation properties for the Q^M are given in terms of the mappings i_M. Specifically we assume

D (a) The mapping $i_M : Q \to H^0$ is continuous;

(b) For each $q \in Q$, $i_M(q) \to q$ as $M \to \infty$ and the convergence is, moreover, uniform in $q \in Q$.

We note that we do not require that $Q^M \subset Q$. Furthermore, in the event $\Omega = (0,1) \times (0,1)$, there are several useful special cases of approximation sets that satisfy the assumptions of condition D. Under sufficient regularity assumptions on Q, we may choose $i_M = I^M$ = the linear spline (or cubic spline) interpolatory map--for precise definitions and details, see [1], [11]. As a second example, we (again for sufficient regularity on Q) may verify that condition D is satisfied when we choose $i_M = P^M$ = the orthogonal projection mapping (in H^0) onto the subspace $L(\Delta^M)$ of linear B-splines (or the subspace $S(\Delta^M)$ of cubic B-splines)--see [11].

To see that this approximation idea does indeed fulfill our needs theoretically, we first observe that if Q is compact in H^0, then part (a) of condition D guarantees that $Q^M = i_M(Q)$ is compact. Hence the problem of minimizing J^N of (8) over Q^M has a solution $\bar{q}_M^N$. From the compactness of Q^M, we have a convergent subsequence $\bar{q}_M^{N_k} \to \bar{q}_M$ in Q^M. It follows (under sufficient regularity) from Theorem 3 that $\bar{q}_M$ is a solution of the problem of minimizing J over Q^M. Let $\hat{q}_M \in Q$ be chosen such that $\bar{q}_M = i_M(\hat{q}_M)$. Then $\{\hat{q}_M\} \subset Q$ and the compactness of Q guarantees existence of a subsequential limit $q^* = \lim \hat{q}_{M_j}$. From (b) of condition D and the fact that $\bar{q}_{M_j} = i_{M_j}(\hat{q}_{M_j})$, it is easily seen that $\bar{q}_{M_j}$ also converges to q^*. We finally observe that $J(\bar{q}_{M_j}) \leq J(q)$ for all $q \in Q^{M_j}$. But since $Q^{M_j} = i_{M_j}(Q)$, we actually have $J(\bar{q}_{M_j}) \leq J(i_{M_j}(q))$ for all $q \in Q$. Taking the limit as $M_j \to \infty$ in this inequality and using the continuity of J and part (b) of condition D, we find $J(q^*) \leq J(q)$ for all $q \in Q$.

For details of the above double limit results in the case of cubic spline state approximations and linear or cubic interpolatary splines for the parameter approximations, the reader may consult [1]. We further observe that a careful consideration of the detailed arguments in [1] and those sketched above will reveal that the order of the limits in the double limit procedure is immaterial. Summarizing we have:

Theorem 4. Let Q^M be as given above where Q is compact and let $\bar{q}_M^N$ be a solution of the problem of minimizing J^N over Q^M. Then for any convergent subsequence $\bar{q}_{M_j}^{N_k} \to q^*$, the limit q^* is a solution of minimizing J over Q.

III. Numerical Findings. We have carried out extensive numerical tests of the methods described in this note on problems involving estimation of both constant and time and/or spatially varying coefficients in parabolic equations. Computations (involving linear and cubic splines) for both test examples and inverse problems using experimental data have been performed. See, for example [1], [2], [3], where descriptions of the algorithms and software packages employed can also be found. We report here briefly on one aspect of our numerical findings which concerns the possible difference in performance of the algorithms when one employs a pointwise (in Ω--see (3)) fit-to-data criterion as opposed to an integral criterion (e.g., as in (2), where the $H^0(\Omega)$ norm is used).

We first observe that the theory presented in this note (for $H^0(\Omega)$ convergence of the approximating states) promises adequate performance of the methods when a distributed criterion such as (2) (and the analogous form of J^N--see (8)) is used in computations. In fact the arguments given above can, under reasonable assumptions, be readily extended to establish pointwise (in Ω) convergence of the states. (The additional arguments involve a choice of $\psi = z_t^N$ in the basic identity derived from (9) and (10).) Hence a convergence theory for a pointwise criterion such as (3) can be easily developed (see also [8] for a different approach to pointwise state convergence). This raises the natural question as to which, if either, criterion is preferable from a computational viewpoint. Our preliminary numerical investigations suggest that whenever one is estimating constant or temporally varying parameters (the ones of interest in the insect dispersal studies of [2], [3]), it does not matter whether the problem is posed using (2) or (3).

To be more specific, consider the Example 4.4 of [1], which is a test example where both the true parameters and true solution are known. The equation involved has the form (1) with $\alpha = 0$, f known, $\Omega = (0,1)$, $\mathfrak{v} = 100(4-t)(.5-x)$ and $\mathfrak{D} = 20$. Tests to estimate the temporal part of $\mathfrak{v}$ (i.e., the term $4-t$) or $\mathfrak{D}$ were performed (see [1] for details). The methods yield essentially the same parameter values and functions (to 3 decimal places) regardless of whether the integral form (i.e., (2)) or the pointwise form (see (3)) of the criterion function J (and the corresponding J^N) are used. In general the computer time to carry out the estimation with distributed criterion was equal to (or less than) that needed for the same example using a pointwise criterion.

We also raised this "criterion" question in some of our extensive uses of the methods with field data from insect dispersal experiments (see [2], [3] for a full report). Again we compared performance of the algorithm using a pointwise criterion with that employing a distributed (in Ω) criterion. Results very similar to those reported above for the test examples were obtained (i.e., same parameters with comparable computational efficiency). For a more complete discussion see [2].

Finally we note that we have successfully used the methods discussed in this paper for test examples and experimental data (again for insect dispersal) in the case of two-dimensional spatial domains Ω. These results will be reported in a manuscript currently in preparation.

Acknowledgements. This research was supported in part by the National Science Foundation under NSF grants MCS-8205355 (to H.T.B), MCS-8200883 (to P.K.L.), and DEB-8207117 (to P.K.), and by the U.S. Air Force Office of Scientific Research under contract AFOSR 81-0198. Part of the research was carried out while the first two authors were visitors at the Institute for Computer Applications in Science and Engineering (ICASE), NASA Langley Research Center, Hampton, VA. which is operated under NASA contracts NAS1-17130 and NAS1-16394. Part of the research was carried out while the first author was a Visiting Professor in the Department of Mathematics, Southern Methodist University.

References

[1] H.T. Banks and P. Daniel (Lamm), Estimation of variable coefficients in parabolic distributed systems, LCDS #82-22, Brown University, Sept. 1982; IEEE Trans. Auto. Control, to appear.

[2] H.T. Banks, P. Daniel (Lamm) and P. Kareiva, Estimation of temporally and spatially varying coefficients in models for insect dispersal, LCDS #83-14, Brown University, June, 1983; J. Math. Biology, to be submitted.

[3] H.T. Banks and P. Kareiva, Parameter estimation techniques for transport equations with application to population dispersal and tissue bulk flow models, LCDS #82-13, Brown University, July, 1982; J. Math. Biology, 17 (1983), 253-273.

[4] A. Friedman, Partial Differential Equations, Holt, Rinehart and Winston, 1969.

[5] A. Friedman, Partial Differential Equations of Parabolic Type, Prentice-Hall, 1964.

[6] N.L. Guinasso and D.R. Schink, Quantitative estimates of biological mixing rates in abyssal sediments, J. Geophys. Res. 80 (1975), 3032-3043.

[7] P. Kareiva, Experimental and mathematical analyses of herbivore movement: quantifying the influence of plant spacing and quality of foraging discrimination, Ecological Monographs, 52(3), (1982), 261-282.

[8] K. Kunisch and L. White, The parameter estimation problem for parabolic equations in multidimensional domains in the presence of point evaluations, preprint, 1983.

[9] O.A. Ladyzhenskaya and N.N. Ural'tseva, Linear and Quasilinear Elliptic Equations, Academic Press, 1968.

[10] J.L. Lions, Optimal Control of Systems Governed by Partial Differential Equations, Springer-Verlag, 1971.

[11] M.H. Schultz, Spline Analysis, Prentice-Hall, 1973.

STABILITY OF DELAY DIFFERENTIAL EQUATIONS WITH APPLICATIONS IN BIOLOGY AND MEDICINE

Kenneth L. Cooke
Pomona College
Claremont, California, 91711, U.S.A.

1. Introduction

Delay differential or differential difference or functional differential equations arise in models of biological phenomena when the time delays occurring in these phenomena are taken into account. This will be illustrated below with some predator-prey models. If there are equilibrium states in these equations, it is important to determine sets of parameter values for which these states are stable (locally or globally), and critical parameter values at which stability may be lost. Often the loss of stability may be associated with Hopf bifurcation and the onset of oscillatory behavior. The local stability problem is usually analyzed by linearization around the equilibrium point. For an equation with a single discrete time lag T this could lead to a delay differential equation of retarded type of the form

$$\sum_{k=0}^{n} a_k \frac{d^k u(t)}{dt^k} + \sum_{k=0}^{m} b_k \frac{d^k u(t-T)}{dt^k} = 0 \qquad (n > m). \tag{1}$$

Associated with this equation is the characteristic equation

$$P(z) + Q(z)e^{-Tz} = 0 \tag{2}$$

where

$$P(z) = \sum_{k=0}^{n} a_k z^k, \quad Q(z) = \sum_{k=0}^{m} b_k z^k.$$

It is well-known that the necessary and sufficient condition for uniform asymptotic stability is that all roots of the characteristic equation lie in the left half-plane, $\mathcal{Re}\, z < 0$. However, it is a difficult mathematical problem to determine when this condition will be satisfied. The primary purpose of this report is to survey several recently suggested methods for this problem, for Eq. [2], for equations with several lags, and also for the more difficult transcendental equations arising when there are "distributed delays". Because of lack of space, we shall describe, in some detail, only a few of these methods, and give references to papers that explain several others.

The methods described here will be illustrated by a few biological examples. A second major purpose of this article is to raise some general questions about the effects of various kinds of time delays in such models. We first consider a prey-predator model that has been analyzed by Hastings (7), of the form

$$\begin{aligned} H'(t) &= rH(t-T) - dH(t) - f(H(t))P(t) \\ P'(t) &= P(t)f(H(t)) - P(t). \end{aligned} \tag{3}$$

Here H denotes the adult prey or host population and P denotes the predator population. The parameter r is the natural birth rate and d the natural death rate of the prey, and T is the length of the juvenile period or period of immaturity. The term $Pf(H(t))$ represents the effect of the predation, and a scaling has already been performed so that the coefficients in the second equation are both one. It is assumed that predation is entirely on adults, and only adults can reproduce; therefore it suffices to consider only the adult prey population H. (A number of other models were discussed by Hastings.) The function $f(H)$ is the so-called functional response of prey and predator. Various possible analytical forms have been suggested for f, but no particular assumption about f is needed to state the results here.

A different class of prey-predator models has been analyzed by Freedman and Rao (5). For purposes of comparison with Eq. [3], we shall examine a special case of their model, of the form

$$\begin{aligned} H'(t) &= rH(t) - dH(t) - f(H(t))P(t) \\ P'(t) &= P(t)f(H(t-\tau)) - P(t) \end{aligned} \qquad [4]$$

The interpretation is similar, except that no juvenile period is included. Instead, a delay τ is postulated in the response of the predator to predation. A model incorporating both kinds of delay is

$$\begin{aligned} H'(t) &= rH(t-T) - dH(t) - f(H(t))P(t) \\ P'(t) &= P(t)f(H(t-\tau)) - P(t) \end{aligned} \qquad [5]$$

We want to raise the following questions about these (and other) models. Is it true that increasing delay is always destabilizing? Is the equilibrium always unstable for sufficiently large delay? As we shall point out, if the delay is taken as a parameter and is increased continuously, there can be, in some cases, alternating intervals of stability and instability. We refer to this phenomenon as stability "switching" and ask when it can occur and when not. If it does occur, how can the stability intervals and switch points be computed, and how many switches can occur? In connection with the models, we think it is important to ask whether there is an essential difference between the two kinds of delay appearing in [3] and [4]. Similar questions arise, for example, in relation to epidemiological models containing different delays.

2. Stability as a function of delay

The "classical" method of determining stability for Eq. [2] is due to Pontryagin and is described and applied in Bellman and Cooke (1). Since the method is venerable and well-known, we shall not describe it here. Instead, we shall outline a method that emphasizes the delay time as a parameter and looks for stability switches. The method was used by Lee and Hsu (11), in different form by Cooke and Grossman (3), and additional results were obtained by Cooke and van den Driessche (4).

Consider the equation [2] under the following assumption:

(H_1) P and Q are analytic functions in $\mathcal{Re}\, z > -\delta$ ($\delta > 0$), with no common imaginary zero. Also $P(0) + Q(0) \neq 0$, and for real y

$$\overline{P(-iy)} = P(iy), \qquad \overline{Q(-iy)} = Q(iy)$$

where the bar denotes complex conjugation.

In particular, (H_1) is satisfied if P and Q are polynomials with real coefficients. As T varies, a change in stability can occur only when a root $z(T)$ of [2] crosses the imaginary axis. Therefore, we look for imaginary roots $z = iy$, and because of (H_1) it suffices to assume $y > 0$. Substituting $z = iy$ in [2] we get

$$|P(iy)| = |Q(iy)| \quad . \qquad [6]$$

This shows that crossing points y are independent of the time T at which they occur. The points y can be computed easily from [6] when P and Q are polynomials. From [2], the crossing times T (if any) are determined from

$$\begin{aligned} \sin Ty &= (Q_R P_I - P_R Q_I)/(Q_R^2 + Q_I^2) \\ \cos Ty &= -(P_R Q_R + P_I Q_I)/(Q_R^2 + Q_I^2) \end{aligned} \qquad [7]$$

where $P(iy) = P_R(y) + iP_I(y)$, $Q(iy) = Q_R(y) + iQ_I(y)$. It can be proved that the direction of crossing at a simple root $z = iy$ is given by $s = \operatorname{sign} F'(y)$ where

$$F(y) = P_R^2 + P_I^2 - Q_R^2 - Q_I^2 = |P(iy)|^2 - |Q(iy)|^2 , \qquad [8]$$

provided $s \neq 0$. More information can be obtained under the following additional hypothesis.

(H_2) The equation $P(z) + Q(z) = 0$ has a finite number of roots in $\mathcal{Re}\, z \geqq 0$. $F(y)$ has a finite number of real zeros, and the positive roots are simple.

For each positive root y_j of $F(y)$, there are infinitely many crossing times T determined by [7], occurring at time intervals $2\pi/y_j$. The following general result, proved in (4), answers some of the questions raised in Section 1.

<u>Theorem 1</u>. Assume hypotheses (H_1) and (H_2).

(a) If $F(y)$ has no positive zeros, then no stability switches occur. Either the equation is stable for all $T \geqq 0$ or it is unstable for all $T \geqq 0$.

(b) If $F(y)$ has positive zeros, stability switches may occur. At most a finite number of switches are possible. There exists T^* such that the equation is unstable for all $T > T^*$.

This method can often be used to compute more specific results. For example, Cooke and Grossman (3) considered the general second degree equation

$$z^2 + az + bze^{-Tz} + c + de^{-Tz} = 0 \qquad (a + b \neq 0,\ c + d \neq 0)$$

and found specific stability conditions in terms of the parameters a, b, c, d, and T.

Some other methods for the stability problem have been presented by Stepan (17), Mahaffy (14), and Thowsen (18). Certain others will be described with some detail in Sections 4-6 below.

3. Application to prey-predator models

Now we shall apply the above method to Eq. [3]. The system has a unique non-trivial equilibrium $H^* = f^{-1}(1)$, $P^* = (r-d)H^*$. Linearizing around this equilibrium one obtains a linear variational system with characteristic equation $z^2 - Mz - rze^{-Tz} + S = 0$, where $M = -d - (r-d)k$, $S = (r-d)k$, $k = H^*f'(H^*)$. The parameter k indicates whether the functional response is "stabilizing" or "destabilizing". The characteristic equation is of the form [2] with $P(z) = z^2 - Mz + S$, $Q(z) = -rz$. Consequently, we find from [8] that

$$F(y) = y^4 + (M^2-r^2-2S)y^2 + S^2 .$$

Note that the system is stable at $T = 0$ if and only if, $S > 0$ and $M+r < 0$. Assume that $r > d$. Then it is stable if $k > 1$ and unstable if $0 < k < 1$. If stable at $T = 0$, then $F(y)$ has no positive roots and the system remains stable for all $T \geq 0$. If it is unstable at $T = 0$, Hastings shows that there are cases in which stability switching occurs. For example, if $r = 0.2$, $d = 0.1$, $k = 0.8$, the system is stable for T in the intervals (1.86,17.7), (27.8,36.7), (53.7,55.8), and otherwise unstable.

Now consider system [4], which has the same equilibrium. We now find that $F(y) = y^4 + (M+r)^2y^2 - S^2$, which has exactly one positive root. Therefore, all root crossings of the imaginary axis are from left to right. If the equation is stable at $\tau = 0$, it switches to unstable at some τ^* and is unstable for $\tau > \tau^*$; if it is unstable at $\tau = 0$, it is unstable for $\tau \geqq 0$. For example, for $r = 0.2$, $d = 0.1$, $k = 1.2$, this model is stable only for $0 \leq \tau < 0.173$, whereas Eq. [3] is stable for all $T \geqq 0$. Thus, it appears that delay in the response of the predator is more destabilizing than maturation delay in the prey, so far as these particular models are concerned.

For the two-lag model [5], the appropriate characteristic equation is $z^2 - Mz - rze^{-Tz} + Se^{-\tau z} = 0$. It is very hard to find the stability regions for this equation but in the special case $T = \tau$ it can be shown that all root crossings are from left to right and so switching back to stability is impossible. This again suggests that delay in the predator response is strongly destabilizing. A full analysis for arbitrary T and τ remains an open problem.

4. Stability for all delays

For equations with several lags, the available methods of analysis are usually not adequate to determine the exact stability regions. Consequently, several authors have proposed the idea of looking for the set of coefficients such that the equation is stable for all values of the lags. Lewis and Anderson (13), Zivotovskii (19), Silkowski (16), and Hale *et al* (6), have obtained useful criteria for this strong form of stability. We shall describe some of the results of Hale *et al*. Consider the system

$$x'(t) = A_o x(t) + \sum_{k=1}^{N} A_k x(t-r_k) \qquad [9]$$

where A_o, A_k are matrices, and the $r_k > 0$ are the lags. The characteristic function is

$$f(z,r,A) = \det[zI - A_o - \sum_{k=1}^{N} A_k e^{-zr_k}] \qquad [10]$$

where I is the identity matrix, $r = (r_1,\ldots,r_N)$, $A = (A_o,A_1,\ldots,A_N)$. Call the equation hyperbolic at (r,A) if $f(z,r,A) = 0$ implies $\mathcal{Re}\, z \neq 0$, and asymptotically stable at (r,A) if $f(z,r,A) = 0$ implies $\mathcal{Re}\, z < 0$. Now consider the effect of varying r along a ray. Thus, given $r^o = (r_1^o,\ldots,r_N^o)$, the hyperbolic cone at r^o is the set, denoted H_{r^o}, of A for which the system is hyperbolic at $(\alpha r^o,A)$ for all $\alpha \geqq 0$. S_{r^o} is defined similarly, and $H = \cap H_{r^o}$, $S = \cap S_{r^o}$ where the intersection is over all r^o with $r_j^o > 0$. Utilizing these concepts, these authors prove a number of general Theorems, including the following.

<u>Theorem 2</u>. A belongs to H if, and only if,

(i) $\det \sum_{k=0}^{N} A_k \neq 0$

(ii) $\det [iy - A_o - \sum_{k=1}^{N} A_k\omega_k] \neq 0$ for all real y, $y \neq 0$, and all complex ω_k of modulus one.

A is in S if and only if (ii) holds and all characteristic roots of $\sum_{k=0}^{N} A_k$ have negative real parts.

More specific results are then obtained for scalar equations

$$y^{(n)}(t) + \sum_{j=1}^{n} a_{jo} y^{(n-j)}(t) + \sum_{k=1}^{N} \sum_{j=1}^{n} a_{jk} y^{(n-j)}(t-\omega_k) = 0 \qquad [11]$$

The necessary and sufficient condition for the coefficients to be in the set S is that

$$\text{(i)} \quad \sum_{j=0}^{N} a_{nj} \neq 0$$

$$\text{(ii)} \quad |p_o(iy,a)| > \sum_{k=1}^{N} |p_k(iy,a)| \quad \text{for } y \neq 0,\ y \text{ real}, \qquad [12]$$

$$\text{(iii)} \quad p_o(z,a) = 0 \text{ implies } \mathcal{Re}\, z < 0 \text{ where}$$

$$p_o(z,a) = z^n + \sum_{j=1}^{n} a_{jo} z^{n-j}, \quad p_k(z,a) = \sum_{j=1}^{n} a_{jk} z^{n-j}.$$

Condition [12] shows that the non-delayed terms must be dominant, in a sense, in order to have this type of stability independent of the delays ω_k. These conditions can be used to obtain explicit conditions on the coefficients if, for example, $n = 2$ and $N = 1$. However, they are difficult to apply to equations with several lags if $n > 1$.

5. Retarded functional differential equations

Cooke and Ferreira (2) have extended some of the ideas of Section 4 to general linear functional differential equations of the form

$$x'(t) = \int_{-1}^{0} [d\eta(\theta)] \; x(t-r(\theta)) \qquad [13]$$

where $x \in \mathbf{R}^n$, r is in the set C^+ of nonnegative continuous real functions on $[-1,0]$, and $\eta \in BV$, the space of matrix functions of bounded variation. The characteristic function is

$$F(z,\eta,r) = \det[zI - \int_{-1}^{0} e^{-zr(\theta)} d\eta(\theta)]$$

and we let $Z(\eta,r)$ denote the set of zeros of F. It can be proved that $f(\eta,r) = \sup Re\, Z(\eta,r)$ is well defined and is continuous for (η,r) in $BV \times C^+$, and also $f(\alpha\eta,r) = \alpha f(\eta,\alpha r)$ for positive α. Define $S(r)$ to be the set of η in BV for which Eq. [13] is stable at (η,r), and let S be the intersection of $S(r)$ for r in C^+. S is the "region of stability globally in the delays." Characterizations of the sets $S(r)$ and S are obtained, and applied to special cases. For example, the scalar equation

$$\dot{x}(t) = a_0 x(t) + a_1 \int_{-1}^{0} x(t-r(\theta))d\theta$$

is stable for every r in C^+ if and only if either $a_1 \geq 0$, $a_0 + a_1 < 0$, or else $a_0 \leq a_1 < 0$. For the case $r(\theta) = -\theta$, explicit sufficient conditions for stability are given in terms of a_0 and a_1.

6. Commensurate delays -- a computational method

As we have suggested above, the problem of finding stability regions for an equation with more than one lag is analytically intractable, in general. However, if the lags are commensurate, more complete information can be obtained, using methods proposed by Kamen (10), and Jury and Mansour (9). In this case, the characteristic function may be written in the form $P(z,e^{-hz})$ where h is the basic lag, and

$$P(z,w) = z^n + \sum_{p=0}^{n-1} \sum_{q=0}^{m} a_{pq} z^p w^q \qquad [14]$$

is a polynomial in two variables. Kamen has proved that the equation is asymptotically stable for all $h \geq 0$ and $P(0,w) \neq 0$ for $|w| = 1$, if, and only if,

$$P(z,w) \neq 0 \text{ for all } Re\, z \geq 0 \text{ and all } |w| = 1. \qquad [15]$$

If we let $w = e^{i\omega}$, [15] is equivalent to

all roots of $P(z,e^{i\omega}) = 0$ lie in $Re\, z < 0$, for all $\omega \in [0,2\pi]$. [16]

Equation [16] is an equation in z, with complex coefficients that are polynomials in $e^{i\omega}$. A condition that all its roots lie in $Re\, z < 0$ is that the Hermite matrix $D(e^{i\omega})$ obtained from its coefficients be positive definite. This, in turn, is equivalent to

$D(1)$ is positive definite and $\det D(e^{i\omega}) > 0$ for $\omega \in [0,2\pi]$. [17]

Condition [17] can be used to obtain analytic results if the order of the equation is not too great. For higher order, there is a recursive scheme to check [17] in a finite number of steps; see Jury and Mansour and Siljak (15).

7. Conclusions

We have given a survey of some of the recent work on the stability of linear delay equations. Despite the large amount of work on this problem, no easily used and universally applicable criteria are known. This makes it difficult to deduce general principles concerning the effect of time delays, due to different biological causes, in models such as the prey-predator models described above. More exploratory work is needed to seek out such principles as may exist.

Correction

It has been pointed out that there is an error in Sections 6 and 9 of (3). For corrections see "Stability switches in distributed delay models" by S.P. Blythe, R.M. Nisbet, W.S.C. Gurney, N. MacDonald, to appear in J. Math. Anal. Appl.

References

(1) R. Bellman and K.L. Cooke, Differential Difference Equations, New York, Academic Press, 1963.

(2) K.L. Cooke and J.M. Ferreira, "Stability conditions for linear retarded functional differential equations", J. Math. Anal. Appl., 96(1983), 480-504.

(3) K.L. Cooke and Z. Grossman, "Discrete delay, distributed delay and stability switches", J. Math. Anal. Appl. 86 (1982), 592-627.

(4) K.L. Cooke and P. van den Driessche, "On zeroes of some transcendental equations", in preparation.

(5) H.I. Freedman and V.S.H. Rao, "The tradeoff between mutual interference and time lags in predator-prey systems", preprint.

(6) J.K. Hale, E.F. Infante, and F-S.P. Tsen, "Stability in linear delay equations", J. Math. Anal. Appl., to appear.

(7) A. Hastings, "Age dependent predation is not a simple process", Theor. Pop. Biol. 23(1983), 347-362.

(8) E.F. Infante, "A note on the stability in retarded delay equations for all delays", in Dynamical Systems II, A.R. Bednarek and L. Cesari (eds.), New York, Academic Press, 1982.

(9) E.I. Jury and M. Mansour, "Stability conditions for a class of delay differential systems", Int. J. Control 35 (1982), 689-699.

(10a) E.W. Kamen, "On the relationship between zero criteria for two-variable polynomials and asymptotic stability of delay differential equations", IEEE Trans. Automat. Control, AC-25 (1980), 983-984.

(10b) E.W. Kamen, 'Correction to "Linear systems with commensurate time delays: stability and stabilization independent of delay"', IEEE Trans. Automat. Control, AC-28(1983), 248-249.

(11) M.S. Lee and C.S. Hsu, "On the τ-decomposition method of stability analysis for retarded dynamical systems", SIAM J. Control 7 (1969), 242-259.

(12) S.H. Lehnigk, Stability Theorems for Linear Motions, Englewood Cliffs, N.J., Prentice-Hall, 1966.

(13) R.M. Lewis and B. Anderson, "Necessary and sufficient conditions for delay-independent stability of linear autonomous systems", IEEE Trans. Automat. Control AC-25 (1980), 735-739.

(14) J.M. Mahaffy, "A test for stability of linear differential delay equations", Quart. App. Math. 40 (1982), 193-202.

(15) D.D. Siljak, "Stability criteria for two-variable polynomials", IEEE Trans. Circuits Syst., CAS-22 (1975), 185-189.

(16) R. Silkowski, "A star shaped condition for stability of linear retarded functional differential equations", Proc. Roy. Soc. Edinburgh, Ser. A, 83 (1979), 189-198.

(17) B. Stepan, "Stability investigation of retarded differential equations", Acta. Tech. Acad. Sci. Hungaricae 90 (1980), 109-132.

(18) A. Thowsen, "The Routh-Hurwitz method for stability determination of linear differential-difference systems", Int. J. Control 33 (1981), 991-995.

(19) L.A. Zivotovskii, "Absolute stability for the solutions of differential equations with retarded arguments", Trudy Sem. Diff. Urav. Otkly. Arg. 7 (1969), 82-91.

EXISTENCE OF QUASI-SOLUTIONS OF SYSTEMS OF NONLINEAR ELLIPTIC BVP'S SUGGESTED BY BIOCHEMICAL REACTIONS

V. Lakshmikantham
Department of Mathematics
University of Texas at Arlington
Arlington, Texas

1. Introduction

In all biochemical processes of practical interest such as fermentation of industries, pharmaceuticals and waste treatment, the reaction rates are influenced by diffusion and the equations are nonlinear [11,12]. Several interacting populations related to ecology [10] as well as problems of chemical engineering [14] also yield nonlinear elliptic equations as a by-product. Since constructive proofs of existence, which can also provide numerical procedures for the computation of solutions, are of greater value than nonconstructive proofs, the method of upper and lower solutions coupled with the monotóne iterative method has been, in recent years, employed to prove existence of multiple solutions of a variety of nonlinear problems [8].

Keller [6] has studied a class of nonlinear elliptic BVP's that are suggested by steady state diffusion processes of interest, has developed iterative schemes that converge monotonically to extremal solutions and has illustrated the results for the diffusion kinetics equation governing the steady state concentration of some substate in an enzyme-catalyzed reaction. For a formulation of a variety of steady state diffusion problems for enzyme-catalyzed reactions see [11]. Sattinger [13] has considered monotone methods for both time dependent diffusion equations and the corresponding steady state problems and has pointed out the importance of extending this method to systems of equations when nonlinear terms possess a mixed monotone property as well as contain gradient terms. See also Amann [1]. Recently Amann and Crandall [2] and Bernfeld and Chandra [3] have extended the monotone technique for the scalar case in a general form. Assuming that nonlinear terms possess a quasimonotone nondecreasing property, Chandra, Lakshmikantham and Leela [4] extended this method to countable systems of second order BVP's. Also, Zygourakis and Aris [14] have discussed monotone technique for elliptic systems (with no gradient terms) when the nonlinear terms enjoy either quasimonotone increasing or decreasing property.

In many problems of interest, the functions involved in the systems do not possess quasimonotone property but do enjoy a mixed monotone property which makes the constructive schemes difficult. For example, when we consider two interacting populations, one prey u_1 and one predator u_2, we obtain the following BVP

$$\begin{cases} -\Delta u_1 = au_1 - bu_1^2 - cu_1u_2, \\ -\Delta u_2 = eu_2 - gu_2^2 + fu_1u_2 \quad \text{in} \quad \Omega. \\ u_1 = u_2 = 0 \quad \text{on} \quad \partial\Omega, \end{cases} \qquad [1.1]$$

where a,b,c,e,g and f are positive constants. Since only positive solutions are of physical interest, it is enough to consider $(u_1,u_2) \in R_+^2$. Then it is easy to see that the system possesses a mixed monotone property. The notion of quasisolutions introduced in [5,9] is very helpful in dealing with mixed monotone systems.

In this paper we discuss general systems of nonlinear elliptic BVP when nonlinear terms possess a mixed quasi-monotone property, obtain extremal quasisolutions as limits of monotone sequences and indicate how one step cyclic monotone schemes can be generated which yield accelerated rate of convergence of iterates. We then consider the relation between coupled quasisolutions and actual solutions. Furthermore, we show that the monotone iterates actually converge to the unique solution. It then follows that the given coupled quasi upper and lower solutions provide bounds for the solution which can be improved by the iterative procedure. One step cyclic iteration that is discussed accelerates the rate of convergence of iterates. For details see [7].

2. Definitions and Assumptions

Consider the second order nonlinear elliptic BVP

$$\begin{cases} -Lu_k = f_k(x,u,Du_k) \quad \text{in} \quad \Omega, \\ Bu_k = \phi_k \quad \text{on} \quad \partial\Omega, \quad \text{for all} \quad k = 1,2,\dots,n, \end{cases} \qquad [2.1]$$

where $u \in R^n$ and Ω is a bounded domain in R^m. Here, L is a second order differential operator given by

$$L = \sum_{i,j=1}^{m} a_{ij}(x) \frac{\partial^2}{\partial x_i \partial x_j} + \sum_{i=1}^{m} b_i(x) \frac{\partial}{\partial x_i} + C(x), \qquad [2.2]$$

and B is a boundary operator defined by

$$Bu_k = p(x)u_k + q(x) \frac{du_k}{d\nu}, \qquad [2.3]$$

where $\frac{du_k}{d\nu}$ denotes the normal derivative of u_k and $\nu(x) = (\nu_1(x),\nu_2(x),\dots,\nu_m(x))$ is the unit normal vector field on $\partial\Omega$.

To define quasisolutions of [2.1], for each $k \in I$, let b_k, d_k be two non-negative integers such that $b_k + d_k = n - 1$. By splitting $u \in R^n$ into $u = (u_k, [u]_{b_k}, [u]_{d_k})$ we rewrite [2.1] in the form

$$\begin{cases} -Lu_k = f_k(x, u_k, [u]_{b_k}, [u]_{d_k}, Du_k), \\ Bu_k = \phi_k. \end{cases} \qquad [2.4]$$

Also, for any $u, v \in R^n$, we let $[u,v]_k$ denote an element of R^n with the description $[u,v]_k = (u_k, [u]_{b_k}, [v]_{d_k})$. Without further mention, we assume that $K \in I$ and all inequalities between vectors are component-wise.

Definition 2.1.

The functions $u, v \in C^2[\bar{\Omega}, R^n]$ are said to be coupled quasisolutions of [2.1] if

$$\begin{cases} -Lu_k = f_k(x, u_k, [u]_{b_k}, [v]_{d_k}, Du_k) & \text{in } \Omega, \\ -Lv_k = f_k(x, v_k, [v]_{b_k}, [u]_{d_k}, Dv_k) & \text{in } \Omega, \\ Bu_k = \phi_k = Bv_k & \text{on } \partial\Omega. \end{cases} \qquad [2.5]$$

The functions $v, w \in C^2[\bar{\Omega}, R^n]$ with $v(x) \leq w(x)$ on $\bar{\Omega}$ are said to be coupled lower and upper quasisolutions, respectively, if

$$\begin{cases} -Lv_k \leq f_k(x, v_k, [v]_{b_k}, [w]_{d_k}, Dv_k) & \text{in } \Omega, \\ Bv_k \leq \phi_k & \text{on } \partial\Omega, \end{cases} \qquad [2.6]$$

and

$$\begin{cases} -Lw_k \geq f_k(x, w_k, [w]_{b_k}, [v]_{d_k}, Dw_k) & \text{in } \Omega, \\ Bw_k \geq \phi_k & \text{on } \partial\Omega. \end{cases}$$

Definition 2.2.

The function f is said to possess a mixed quasimonotone property (mqmp for short) if for each $k \in I$, $f_k(x, u_k, [u]_{b_k}, [u]_{d_k}, Du_k)$ is monotone non-decreasing in $[u]_{b_k}$ and monotone non-increasing in $[u]_{d_k}$.

Let us list the following assumptions for convenience:

(A_1): (i) For each $i,j = 1,2,\dots,m$, a_{ij}, b_i and $c \in C^\alpha[\bar{\Omega}, R]$, $c(x) \leq 0$, and L is strictly uniformly elliptic in $\bar{\Omega}$;

(ii) $p, q \in C^{1,\alpha}[\partial\Omega, R]$, p and q are non-negative functions which do not vanish simultaneously;

(iii) $\partial\Omega$ belongs to class $C^{2,\alpha}$;

(iv) $f \in C^{\alpha}[\bar{\Omega}\times R^n \times R^m, R^n]$;

(v) f satisfies a Nagumo condition, that is , there exists an increasing function $\Psi_k : R_+ \to R_+$ such that for $(x,u,y) \in \bar{\Omega} \times R^n \times R^m$, $|f_k(x,u,y)| \le \Psi_k(\|u\|)(1+\|y\|^2)$;

(vi) $\phi \in C^{1,\alpha}[\partial\Omega, R^n]$.

(A_2) v,w are coupled lower and upper quasisolutions relative to [2.4];

(A_3) for some $M \ge 0$, $f_k(x,u,h(y)) - f_k(x,\bar{u},h(y)) \ge -M(u_k-\bar{u}_k)$ whenever $u_i = \bar{u}_i$ for $i \neq k$, $v(x) \le \bar{u} \le u \le w(x)$, $x \in \bar{\Omega}$, and $y \in R^m$.

In (A_3), the function $h \in H = [h \in C^1[R^m,R^m]: h(y) = y$ for $\|y\| < N$, $\|h(y)\| \le \lambda\|y\|$ for $y \in R^n$, and $h(R^m)$, $h_y(R^m)$ are bounded], where h_y denotes the Jacobian matrix function of h, $\lambda > 1$; N is defined by

$$N > \max\{\bar{N}, \max_{x\in\bar{\Omega}} \|v_x(x)\|, \max_{x\in\bar{\Omega}} \|w_x(x)\|\},$$

$\bar{N}$ being the Nagumo constant relative to v,w and $\Psi_k(\mu)\lambda^2(1+\|y\|^2)$, $\mu = \max\{\|\underline{v}\|,\|\bar{w}\|\}$, $\underline{v}_k = \min_{x\in\bar{\Omega}} \{v_k(x)\}$, $\bar{w}_k = \max_{x\in\bar{\Omega}} \{w_k(x)\}$, $\underline{v} = (\underline{v}_1,\underline{v}_2,\ldots,\underline{v}_n)$, $\bar{w} = (\bar{w}_1,\bar{w}_2,\ldots,\bar{w}_n)$, $k = 1,2,\ldots,n$.

3. Auxiliary Results

Consider the modified second order nonlinear elliptic BVP

$$\begin{cases} -Lu_k = G_k(x,u_k,Du_k) - Mu_k & \text{in } \Omega, \\ Bu_k = \phi_k & \text{on } \partial\Omega, \text{ for } k \in I \end{cases} \tag{3.1}$$

where

$$G_k(x,u_k,Du_k) = f_k(x,\eta_{1k},[\eta_1]_{b_k},[\eta_2]_{d_k},h(Du_k)) + M\eta_{1k} \tag{3.2}$$

and $\eta_1, \eta_2 \in C^{1,\alpha}[\bar{\Omega},R^n]$ such that $v(x) \le \eta_1(x),\eta_2(x) \le w(x)$ on $\bar{\Omega}$.

Our first objective is to show that the BVP [3.1] has a unique solution. The proof of existence and uniqueness of solutions of [3.1] is equivalent to the existence and uniqueness of solutions of

$$\begin{cases} -\tilde{L}u_k = G_k(x,u_k,Du_k) & \text{in } \Omega, \\ Bu_k = \phi_k & \text{on } \partial\Omega, \text{ for } k \in I \end{cases} \tag{3.3}$$

where $\tilde{L} = L - M$.

We note that [3.3] is completely decoupled system. Moreover, G_k is independent of

u_k. To prove the existence of solutions of [3.3], we need to verify the hypotheses of the existence theorem with respect to each component of [3.3]. In fact, we can prove the following result.

Theorem 3.1.

Let assumptions (A_1), (A_2) and (A_3) hold. Assume further that f in [2.1] possesses mqmp. Then there exists a unique solution $u \in C^{2,\alpha}[\bar{\Omega},R^n]$ to the BVP [3.3] such that $v(x) \leq u(x) \leq w(x)$ on $\bar{\Omega}$. Moreover, there exists an $N_0 > 0$ such that $\|u_x(x)\| \leq N_0$ on $\bar{\Omega}$, where N_0 depends on ψ,h,f,v and w.

For each $\eta_1, \eta_2 \in C^{1,\alpha}[\bar{\Omega},R^n]$ such that $v(x) \leq \eta_1(x), \eta_2(x) \leq w(x)$ on $\bar{\Omega}$, we define a mapping A by

$$A(\eta_1,\eta_2) = u$$

where $u \in C^{2,\alpha}[\bar{\Omega},R^n]$ is the unique solution of the BVP [3.3]. The properties of A are given by the following result.

Lemma 3.1.

Under the assumptions of Theorem 3.1, the mapping defined by [3.4] possesses the following properties:

(i) $v \leq A(v,w)$, $w \geq A(w,v)$;

(ii) A is a mixed monotone operator on the segment $\langle v,w \rangle$ where

$\langle v,w \rangle = [u \in C^2[\bar{\Omega},R^n]: v(x) \leq u(x) \leq w(x),\ x \in \bar{\Omega}]$.

4. Monotone Iterative Technique

Because of Lemma 3.1, we define the sequences

$$v_i = A(v_{i-1},w_{i-1}),\quad w_i = A(w_{i-1},v_{i-1}) \qquad [4.1]$$

with $v_0 = v$, $w_0 = w$ and conclude that

$$v \leq v_i \leq v_{i+1} \leq w,\quad v \leq w_{i+1} \leq w_i \leq w,$$

and $v_i, w_i \in C^{2,\alpha}[\bar{\Omega},R^n]$ for $i = 1,2,\dots$. Concerning these sequences, we have the following main result.

Theorem 4.1.

Suppose that assumptions (A_1), (A_2) and (A_3) hold. Further assume that f in [2.1] possesses mqmp. Then the sequences $\{v_i\}$ and $\{w_i\}$ defined by [4.1] converge in $C^2[\bar{\Omega},R^n]$. Moreover, the limits $\rho(x)$ and $r(x)$ of $\{v_i\}$ and $\{w_i\}$, respectively, are the coupled minimal and maximal quasisolutions of [2.1], and $\rho,\ r \in \langle v,w\rangle$.

Let us next discuss an iteration scheme which accelerates the rate of convergence of the sequence of iterates. A method that is useful for the solution of BVP [2.1], is the one-step cyclic iterative procedure. We note that the procedure [4.1] is equivalent to

$$v_{ki} = A_k(v_{i-1},w_{i-1}),\ w_{ki} = A_k(w_{i-1},v_{i-1}),\ i = 1,2,\ldots$$

with $v_0 = v = (v_{10},v_{20},\ldots,v_{n0})$ and $w_0 = w = (w_{10},w_{20},\ldots,w_{n0})$ where v_{ki} and w_{ki} are the unique solutions of [3.3] with respect to $[v_{i-1},w_{i-1}]_k$ and $[w_{i-1},v_{i-1}]_k$ for $k \in I$, respectively. In this case, this procedure is modified as follows:

$$\overset{*}{v}_{ki} = A_k(\overset{*k-1}{v}_{i-1},\overset{*k-1}{w}_{i-1}),\ \overset{*}{w}_{ki} = A_k(\overset{*k-1}{w}_{i-1},\overset{*k-1}{v}_{i-1}),\quad i = 1,2,\ldots \qquad [4.2]$$

with $\overset{*0}{v}_0 = v = (v_{10},v_{20},\ldots,v_{n0})$ and $\overset{*0}{w}_0 = w = (w_{10},w_{20},\ldots,w_{n0})$ where $\overset{*}{v}_{ki}$ and $\overset{*}{w}_{ki}$ are the unique solutions of [3.3] relative to $[\overset{*k-1}{v}_{i-1},\overset{*k-1}{w}_{i-1}]_k$ and $[\overset{*k-1}{w}_{i-1},\overset{*k-1}{v}_{i-1}]_k$ for $k \in I$, repsectively. For fixed $i = 1,2,\ldots,$ vectors $\overset{*k-1}{v}_{i-1}$, $\overset{*k-1}{w}_{i-1}$ are defined by

$$\overset{*k-1}{v}_{i-1} = [\overset{*}{v}_{1i},\overset{*}{v}_{2i},\ldots,\overset{*}{v}_{k-1i},\overset{*}{v}_{ki-1},\overset{*}{v}_{k+1i-1},\ldots,\overset{*}{v}_{ni-1}],$$

$$\overset{*k-1}{w}_{i-1} = [\overset{*}{w}_{1i},\overset{*}{w}_{2i},\ldots,\overset{*}{w}_{k-1i},\overset{*}{w}_{ki-1},\overset{*}{w}_{k+1i-1},\ldots,\overset{*}{w}_{ni-1}],\quad \text{for } K \in I,$$

with

$$\overset{*0}{v}_{i-1} = \overset{*}{v}_{i-1} = [\overset{*}{v}_{1i-1},\overset{*}{v}_{2i-1},\ldots,\overset{*}{v}_{ni-1}],\quad \text{and}$$

$$\overset{*0}{w}_{i-1} = \overset{*}{w}_{i-1} = [\overset{*}{w}_{1i-1},\overset{*}{w}_{2i-1},\ldots,\overset{*}{w}_{ni-1}].$$

Theorem 4.2.

Assume that the hypotheses of Theorem 4.1 hold. Then the sequences $\{\overset{*}{v}_i\}$ and $\{\overset{*}{w}_i\}$ defined by [4.2] converge in $C^2[\bar{\Omega},R^n]$. Moreover, the limits $\rho(x)$ and $r(x)$ of $\{\overset{*}{v}_i\}$ and $\{\overset{*}{w}_i\}$, respectively, are coupled quasi-minimal and maximal solutions of [2.1], and $\rho,\ r \in \langle v,w\rangle$.

Having obtained coupled extremal quasisolutions of [2.1], let us first observe that if $d_k = 0$ so that $u = (u_k,[u]_{b_k})$, $b_k = n-1$, then f possesses quasimonotone nondecreasing property. In this special case, the coupled quasiextremal solutions ρ, r reduce to extremal solutions of [2.1]. If f satisfies further a uniqueness condition, one can show that $\rho = r$ is the unique solution of [2.1].

In the general case also, one can show $\rho = r$ so that coupled quasisolutions become actual solutions. For this purpose let us assume that

(A_4) $[u_1-u_2, f(x,u_1,Du_{1k}) - f(x,u_2,Du_{2k})]$

$$\leq \tilde{C}(x)\|u_1-u_2\|^2 + \sum_{i=1}^{m} d_i(x)(u_1-u_2)(Du_1-Du_2),$$

whenever $u_1, u_2 \in \langle v,w\rangle$ and $\|Du_1\|, \|Du_2\| \leq \bar{N}$ on $\bar{\Omega}$ where $\bar{N}$ is the Nagumo constant, $\tilde{C}, d_i \in C[\bar{\Omega},R]$ and $C + \tilde{C} \leq 0$ on $\bar{\Omega}$. We then have the following result.

Theorem 4.3.

Let the assumptions of Theorem 4.1 hold. Suppose further that (A_4) is satisfied. Then there exists a unique solution u of [2.1] such that $v \leq u \leq w$ on $\bar{\Omega}$ which can be obtained as a limit of monotone sequences $\{v_i\}$ and $\{w_i\}$.

Proof.

It is enough to show that $\rho \equiv r$ on $\bar{\Omega}$. Set $p = \|r-\rho\|^2$. Then, using (A_4), we can easily obtain

$$- Lp \leq Cp + 2[\tilde{C}p + \sum_{i=1}^{m} d_i p_{x_i}] \quad \text{in } \Omega,$$

and $Bp \leq 0$ on $\partial\Omega$. This implies that

$$- \tilde{L}p \equiv -[\sum_{i,j}^{m} a_{ij}p_{x_i x_j} + \sum_{i=1}^{m} (b_i+d_i)p_{x_i} + 2(C+\tilde{C})p]$$

$$\leq 0 \quad \text{in } \Omega \quad \text{and} \quad Bp \leq C \quad \text{on } \partial\Omega.$$

Consequently, by the maximum principle, we can conclude that $\rho \equiv r$ on $\bar{\Omega}$ and the proof is complete.

Let us illustrate the results relative to example [1.1]. Suppose that $a, e > \lambda$ where $\lambda > 0$ is the principle eigenvalue of $-\Delta\omega = \lambda\omega$ in Ω and $\omega = 0$ on $\partial\Omega$. Let $\omega(x)$ be the eigen function. Assume that

$$gb - cf > 0 \quad \text{and} \quad (gb-cf)a > (g\lambda+ce)b.$$

We set $v_1 = \varepsilon w$, $v_2 = \varepsilon w$, $w_1 = \frac{a}{b}$, $w_2 = \frac{be+af}{bg}$, where $\varepsilon > 0$ is sufficiently small. Then it is easy to check that, v,w are coupled quasi lower and upper solutions of [1.1]. Assumption (A_3) is satisfied with $M_1 = gab + c[be+af]/gb$ and $M_2 = \frac{be+2fa}{b}$. Assumption (A_4) is satisfied with $\tilde{C} = a - 2\varepsilon\, b\omega + \frac{ca}{b} + c(\frac{be+af}{bg}) > 0$. Hence by Theorem 4.1, we conclude that there exist positive solutions for the problem [1.1] that can be obtained from monotone sequences.

References

[1] H. Amann, "Fixed point equations and nonlinear eigenvalue problems in ordered Banach spaces," SIAM Rev., 18 (1976) 620-709.

[2] H. Amann and M. G. Crandall, "On some existence theorems for semi-linear elliptic equations, Indiana Univ. Math J., 27 (1978) 779-790.

[3] S. Bernfeld and J. Chandra, "Minimal and maximal solutions of nonlinear boundary value problems," Pacific J. Math., 71 (1977) 13-20.

[4] J. Chandra, V. Lakshmikantham and S. Leela, "A monotone method for infinite system of nonlinear boundary value problems," Arch. Rat. Mech. Anal., 68 (1978) 179-190.

[5] K. Deimling and V. Lakshmikantham, "Quasi-solutions and their role in the qualitative theory of differential equations," Nonlinear Analysis, 4 (1980) 657-663.

[6] H. B. Keller, "Elliptic boundary value problems suggested by nonlinear diffusion processes," Archs. Ration. Mech. Analysis, 35 (1969) 363-381.

[7] G. S. Ladde, V. Lakshmikantham and A. S. Vatsala, "Existence of coupled quasi-solutions of systems of nonlinear boundary value problems," To appear in Nonlinear Analysis.

[8] V. Lakshmikantham, G. S. Ladde and A. S. Vatsala, "Monotone technology for nonlinear problems," (unpublished manuscript).

[9] V. Lakshmikantham and A. S. Vatsala, "Quasi-solutions and monotone method for systems of nonlinear boundary value problems," J. of Math. Anal. and Appl., 79 38-47.

[10] A. Leung, "Monotone schemes for semilinear elliptic systems related to ecology," Math. Meth. in the Appl. Sciences, 4 (1982) 272-285.

[11] J. D. Murray, "A simple method for obtaining approximate solutions for a class of diffusion kinetics enzyme problems," Math. Biosciences, 2 (1968) 379-411.

[12] L. W. Ross, "Perturbation analysis of diffusion coupled biochemical reaction kinetics," SIAM J. Appl. Math., 19 (1970) 323-329.

[13] D. H. Sattinger, "Monotone methods in nonlinear elliptic and parabolic boundary value problems," Indiana Univ. Math. J., 21 (1972) 979-1000.

[14] K. Zygourakis and R. Aris, "Weakly coupled systems of nonlinear elliptic boundary value problems," Nonlinear Analysis, 5 (1982) 1-15.

DIFFUSION APPROXIMATIONS AND FIRST PASSAGE TIME PROBLEMS IN POPULATION BIOLOGY AND NEUROBIOLOGY (*)

L. M. RICCIARDI

Dipartimento di Matematica e Applicazioni

Università di Napoli

Introduction

Everyone would certainly agree that mathematics is becoming an indispensable tool of investigation in the biological sciences. What is probably less evident is that in turn mathematicians have been often guided in their investigations by needs and indications deeply rooted in the realm of biology, to the point that a number of by now well developed fields of mathematics would have probably remained unexplored without the pressing demands posed by their colleagues engaged in the biological discovery. The theory of reaction-diffusion equations is an example of a branch of mathematics whose origin relies on questions and problems of biological nature. A second example is provided by the theory of (stochastic) singular diffusion equations developed by W. Feller in order to account for some new and somewhat "strange" features exhibited by simple equations derived to describe the growth of a population subject to random effects. A third example is offered by the revamped interest of mathematicians towards the so-called first passage time problems for random processes with a view to working out algorithms and effective numerical procedures for the evaluation of neuronal firing distributions and population extinction probabilities.

In the sequel we shall briefly describe some of the endeavours in these latter two directions setting the emphasis on the neurobiological and population dynamics applications. As a starting point, let us consider a population consisting of similar individuals and assume that each of them, independently of all others, is able to generate a number 0,1,2.... of offsprings with probabilities Π_0, Π_1, Π_2,...Let us initially consider <u>one</u> individual, defining generation zero. The offspring of this individual will define generation one, their offspring will be taken to form

---

(*) Work performed within a scientific cooperation programme between the Italian National Research Council (CNR) and the Japanese Society for Promotion of Science (JSPS), contract no. 83.00032.01 and partly financed by the Italian Ministry of Education (MPI).

generation two, etc. In general, generation r consists of the offspring of the individuals in generation r-1. We have thus defined a branching process. Clearly, if no individual in a certain generation produces offspring the population goes extinct. We shall assume $\Pi_0 > 0$ and $\Pi_0 + \Pi_1 < 1$ to make extinction possible and to avoid the case where each generating consists of at most one individual. Let

$$G(z) = \sum_{i=0}^{\infty} \Pi_i z^i \tag{1}$$

be the probability generation function (p.g.f.) of the number of offspring per individual and let X_n denote the number of individuals in generation n.

We then have

$$P(X_{n+1} = k \mid X_n = j) = P\left(\sum_{i=1}^{j} Z_i = k\right), \tag{2}$$

Z_i's being i.i.d. random variables with probabilities $\Pi_0, \Pi_1, \Pi_2, \ldots$. Upon setting

$$H_j(z) = \sum_{k=0}^{\infty} P\left(\sum_{i=1}^{j} Z_i = k\right) z^k \equiv [G(z)]^j \tag{3}$$

we see that $P(X_{n+1} = k \mid X_n = j)$ is the coefficent of z^k in the power series expansion of $[G(z)]^j$. Denoting by $p_r^{(s)}$ the probability of having r individuals in generation s, we have:

$$p_k^{(n)} = \sum_{j=0}^{\infty} p_j^{(n-1)} \{ \text{coefficent of } z^k \text{ in } \sum_{j=0}^{\infty} p_j^{(n-1)} [G(z)]^j. \tag{4}$$

Using this relation, the p.g.f. $F_n(z)$ of the population size in generation n reads:

$$F_n(z) \equiv \sum_{k=0}^{\infty} p_k^{(n)} z^k = F_{n-1}[G(z)] = G[F_{n-1}(z)]. \tag{5}$$

Let us now define the cumulant generating functions

$$\begin{aligned} K(\alpha) &\equiv \ln G(e^{-\alpha}) = \ln \sum_{i=0}^{\infty} \Pi_i e^{-i\alpha} \\ K_n(\alpha) &\equiv \ln F_n(e^{-\alpha}) = \ln \sum_{k=0}^{\infty} p_k^{(n)} z^k. \end{aligned} \tag{6}$$

Denoting by μ and σ^2 the mean value and the variance of the population size in generation zero, and by μ_n and σ_n^2 the corresponding quantities in generation n, by direct differentiation of (6) we obtain

$$\begin{aligned} \mu &= -\left(\frac{dK}{d\alpha}\right)_{\alpha=0}, & \sigma^2 &= \left(\frac{d^2K}{d\alpha^2}\right)_{\alpha=0} \\ \mu_n &= -\left(\frac{dK_n}{d\alpha}\right)_{\alpha=0}, & \sigma_n^2 &= \left(\frac{d^2K_n}{d\alpha^2}\right)_{\alpha=0} \end{aligned} \tag{7}$$

To determine μ_n and σ_n^2 in terms of μ and σ^2, we make use of (5) and (6a) to rewrite (6b) as

$$K_n(\alpha) = K[\,-K_{n-1}(\alpha)\,] \tag{8}$$

so that from (7) we find

$$\mu_n = \mu^n$$
$$\sigma_n^2 = \frac{\sigma^2}{\mu(\mu-1)}\,\mu^n\,(\mu^n-1). \tag{9}$$

Let us now remove the assumption that in generation zero there is just one individual and instead let the number of individuals be k_0. Since we have assumed that individuals reproduce independently of one another, after denoting by X_n the total number of individuals in generation n from (9) we obtain

$$E\,[X_n] = k_o\,\mu^n$$
$$V\,[X_n] = \frac{k_o\,\sigma^2}{\mu(\mu-1)}\,\mu^n\,(\mu^n-1), \tag{10}$$

whereas from one generation to the next we find:

$$E[X_{n+}\,|X_n = k\,] \equiv E[X_1 \,|\, X_o = k\,] = k\mu$$
$$V[X_{n+1}\,|X_n = k\,] \equiv V[X_1 \,|\, X_o = k\,]\; k\sigma^2. \tag{11}$$

Following Feller (1951a) we now look into a limit diffusion process X(t) for X_n. To this purpose we make a change of scale and denote by τ the time interval between two consecutive generations. Then setting

$$t = n\tau, \qquad \sigma^2 = 2r\tau, \qquad \mu = 1+p\tau, \qquad k = x \tag{12}$$

from (11) we obtain

$$\tau^{-1}\,E\,[\,X_{(n+1)\tau} - X_{n\tau}\,|X_{n\tau} = x\,] = \beta x$$
$$\tau^{-1}\,E\{[\,X_{(n+1)\tau} - X_{n\tau}]^2\,|X_{n\tau} = x\} = \alpha\,x + O(\tau) \tag{13}$$

indicating that in the limit as $\tau \to 0$, $X_{n\tau}$ converges to the diffusion process X(t) having drift and infinitesimal variance given by

$$A_1(x) = px$$
$$A_2(x) = 2rx. \tag{14}$$

This is a singular diffusion process defined over the interval $[\,0,\infty\,)$, with x=0 an exit boundary. Its transition p.d.f. $f(x,t|y) = \frac{\partial}{\partial x}\, P\{X(t) \le x\,|\,X(0) = y\}$ can be shown to be given by

$$f(x,t\,|y) = \frac{p}{r(e^{pt}-1)}\,\exp\left\{-\,\frac{p(x+ye^{pt})}{r(e^{pt}-1)}\right\}\left[\,e^{-pt}\,\frac{x}{y}\,\right]^{(q-r)/2r} I_{1-q/r}\left\{\frac{2p}{r(1-e^{-pt})}\,[e^{-pt}\,xy]^{\frac{1}{2}}\right\} \tag{15}$$

where $I_k(x)$ is the Bessel function defined as

$$I_k(x) = \sum_{n=0}^{\infty} \frac{(x/2)^{2n+k}}{n!\,\Gamma(n+k+1)} \tag{16}$$

It is worth pointing out that the limit process X(t) thus obtained is characterized by vanishing drift and infinitesimal variance at the zero boundary, which makes the integration problem of the associated Fokker-Planck equation quite challenging. An interesting generalization consists of adding a constant q to the drift term. The boundary $x = 0$ then changes its nature according to whether it is $q \le 0$ (exit), $0 < q < r$ (regular) or $q \ge r$ (entrance). This strange behavior was the starting point of Feller's celebrated work on the mentioned integration theory of singular diffusion equations (cf. Feller 1951b; 1952; 1954), which can be taken as a paradigm of the non trivial role of biology in mathematical discovery.

2. Extinction Probability

We now make use of (16) to calculate the probability

$$\delta(t|x_0) = P\{X(t) < 0 \mid X(0) = x_0\} \tag{17}$$

that extinction occured before time t for a population initially consisting of x_0 individuals. After a straightforward integration we obtain:

$$\delta(t|x_0) = \exp\left(-\frac{2px_0 e^{pt}}{r(e^{pt}-1)}\right) \tag{18}$$

Hence, for the probability $P(x_0)$ of ultimate extinction it follows:

$$P(x_0) \lim_{t\to\infty} \delta(t|x_0) = \begin{cases} 1, & p \le 0 \\ e^{-2px_0/r}, & p > 0 \end{cases} \tag{19}$$

Can the same considerations be extended to the case when x=0 is not an accessible boundary? Assume, for instance, that the population size X(t) is modeled as a diffusion process defined in $(0, \infty)$, with x=0 a natural boundary. Then:

$$\delta(t|x_0) \equiv 1 - \int_0^{\infty} dx\, f(x,t|x_0) = 0 \quad \text{for all } t \text{ and } x_0 \tag{20}$$

and:

$$P(x_0) = 0 \qquad \text{for all } x_0. \tag{21}$$

One may wonder whether such conclusions are satisfactory. The answer is clearly in the negative, as the following example discloses. Let X(t) be generated by the stochastic differential equation

$$\frac{dx}{dt} = [\alpha + \Lambda(t)]\, X - \beta X^2$$
$$P\{X(0) = x_0\} = 1 \tag{22}$$

obtained by parameterizing the logistic equation, i.e. by assuming that the intrinsic fertility α is changed into $\alpha + \Lambda(t)$, where $\Lambda(t)$ is a stationary normal process with zero mean and delta-type correlation function (white noise):

$$E[\Lambda(t)] = 0 \qquad E[\Lambda(t)\Lambda(t')] = \sigma^2 \delta(t - t'). \tag{23}$$

As is well known, eq. (22) implies that X(t) is a diffusion process having drift $A_1(x)$ and infinitesimal variance $A_2(x)$ given by

$$A_1(x) = \left(\alpha + \frac{\sigma^2}{2}\right)x - \beta x^2 \qquad A_2(x) = \sigma^2 x^2. \tag{24}$$

Alternatively, the logistic equation can be parameterized in the form

$$dY(t) = [\alpha\, dt + dW(t)]X - \beta X^2 dt \qquad P\{Y(0) = x_0\} = 1 \tag{25}$$

where W(t) is the standard Wiener process with

$$E\ W(t) = 0 \qquad E\ W(t)\, W(t') = \sigma^2 \min(t,t'). \tag{26}$$

Then, Y(t) is a diffusion process whose infinitesimal moments are

$$B_1(x) = \alpha x - \beta x^2 \qquad B_2(x) \equiv A_2(x) = \sigma^2 x^2 \tag{27}$$

Both X(t) and Y(t) are defined over the interval $(0,\infty)$, with x=0 a natural boundary so that, according to (20), $\delta(t|x_0) = 0$ for all t and x_0 and $P(x_0) = 0$ for all x_0. Still, the processes X(t) and Y(t) exhibit quite different behaviors in that the steady state p.d.f.

$$W(x) = \lim_{t\to\infty} f(x,t|x_0) \tag{28}$$

always exists for X(t) whereas it exists in the case of Y(t) only if $k \equiv \alpha/\beta > \sigma^2/(2\beta)$. Furthermore, let $0<\varepsilon<x_0$ and $x_0< \eta <\infty$ and let Z(t) be a diffusion process with infinitesimal moments $C_1(z)$ and $C_2(z)$. The probability $P(\varepsilon|x_0)$ that $\{Z(t)|Z(0) = x_0\}$ eventually reaches level ε for the first time is (Ricciardi, 1977):

$$P(\varepsilon|x_0) = 1 - \frac{H(\varepsilon,x_0)}{K(\varepsilon)} \tag{29}$$

where:

$$H(\varepsilon, x_0) = \int_\varepsilon^{x_0} dy \exp\left[-2\int^y \frac{C_1(z)}{C_2(z)} dz\right] < \infty \text{ for all } \varepsilon \text{ and } x_0$$

and

$$K(\varepsilon) = \int_\varepsilon^\infty dy \exp\left[-2\int^y \frac{C_1(z)}{C_2(z)} dz\right]. \tag{30}$$

Hence, $P(\varepsilon|x_o) = 1$ iff $K(\varepsilon) = \infty$. Furthermore, for $x_o \leqslant \eta < \infty$ one has

$$P(\eta|x_o) = 1 - \frac{H(x_o,\eta)}{L(\eta)} \tag{31}$$

with

$$L(\eta) = \int_0^{\eta} dy \exp\left[-2\int^{y} \frac{C_1(z)}{C_2(z)}\, dz\right] \tag{32}$$

so that $P(\eta|x_o) = 1$ iff $L(\eta) = \infty$. By identifying $Z(t)$ with the process $X(t)$ and $Y(t)$ respectively defined in the foregoing, one can easily prove that for the case of $X(t)$ it is $P(\varepsilon|x_o) = P(\eta|x_o) = 1$ for all ε and η such that $0<\varepsilon<x_o\leqslant\eta<\infty$. Instead, for $Y(t)$ one has $P(\varepsilon|x_o) = 1$ always, whereas $P(\eta|x_o) = 1$ only if $\alpha \geqslant \sigma^2/2$. Hence, $X(t)$ is recurrent whatever α,β and σ^2, while $Y(t)$ is recurrent only if $\alpha \geqslant \sigma^2/2$ (i.e. for a "small" noise). If $\alpha<\sigma^2/2$ ("large" noise) any state below x_o will be reached with certainty by $Y(t)$ whereas it is $P(\eta|x_o) < 1$ no matter how small is the difference $\eta - x_o$. Hence, the origin appears to be "attracting" for $Y(t)$ even though it is a natural (and thus unaccessible) boundary. Still, with a positive probability any large value η can be attained by the population. One concludes that it would not be judicious to assume that for $Y(t)$ extinction is a sure event even when $\alpha<\sigma^2/2$, i.e. even when $x=0$ is an attracting point. Of course, similar features are shared by a variety of more interesting population growth models all clearly indicating the need for a different definition of extinction probabilities in the case of singular diffusion population models (see, for istance, Nobile and Ricciardi, 1983 and 1984).

The discrepancy between the process $X(t)$ and $Y(t)$ earlier discussed can be eliminated if the probability of extinction is identified with the probability $P(S|x_o)$ that the population size eventually attains some small threshold value S (or, more in general, $S(t)$). More precisely, in place of the functions (17) and (19) let us set:

$$\delta(t\,|x_o) = g[S(t),\, t|x_o] \equiv \frac{\partial}{\partial t} P\left\{\inf_{0<\tau<\infty}[\tau:\, X(\tau) > S(\tau)] < t|X(0) = x_o\right\} \tag{33}$$

and

$$P(x_o) = \int_0^{\infty} dt\; g\,[S(t),\, t|x_o] \tag{34}$$

Hence, the extinction probability problem becomes a problem of first passage time p.d.f. calculation, which in general is quite a formidable task. Before mentioning some recent results in this direction, a brief discussion will be given in the sequel to show that, <u>mutatis mutandis</u>, similar needs also arise within the context of neuronal firing in theoretical neurobiology.

3. The Neuronal Firing Problem.

The interest of diffusion models for the spontaneous activity of single neurons belonging to complex networks has very frequently emerged in the neurophysiological literature. Particulary relevant from an historical perspective is the contribution by Gerstein and Mandelbrot (1964) in which the neuron's membrane potential is depicted as a Wiener process with drift in order to provide an interpretation for a variety of experimentally recorded spike trains. The firing time distribution thus identifies with the first passage time distribution for the Wiener process through a boundary representing the neuron's threshold. Apart from some serious problems of interpretation, this model nowadays appears to be too oversimplifed in that an infinitely large membrane time constant is considered in order to overcome the difficulty of spontaneous exponentially varying membrane potentials. While referring to Ricciardi and Sacerdote (1979) and to the references therein for some complementary remarks, here we shall sketch a procedure to obtain a sensible diffusion approximation for the neuron's membrane potential under the effect of very many uncorrelated exicitatory and inhibitory inputs.

Let x denote the variation of the potential difference across the neuronal membrane. In the absence of inputs, x exponentially decays toward the (zero) resting potential with a time constant θ. We shall now assume that the neuron's input consists of p+q sequences of approximately zero-width pulses Poisson distributed in time with rates $\alpha_1, \alpha_2, \ldots\ldots \alpha_p$ and $\beta_1, \beta_2, \ldots\ldots \beta_q$, respectively. The pulses characterized by rates α_k (k=1,2,...,p) are taken as excitatory while the others are inhibitory. Denoting by $e_k > 0$ (k=1,2,...,p) and by $i_k < 0$ (k=1,2,...,q) the corresponding EPSP's and IPSP's, the istantaneous transition $x \to x + e_k$ (k=1,2,...,p) represents the effect of an excitatory input pulse belonging to the sequence characterized by the rate α_k whereas $x \to x + i_k$ (k=1,2,...,q) is the instantaneous transition occurring when the neuron is hit by a pulse belonging to the sequence that has rate β_k. The membrane potential X(t) is therefore a Markov process whose transition p.d.f.

$$f(x,t|y,\tau) = \frac{\partial}{\partial x} P\{X(t) < x \mid X(\tau) = y\} \tag{35}$$

for any $t+\Delta t > t > 0$ satisfies the Kolmogorov equation

$$f(x,t+\Delta t|x_0) = \int_{-\infty}^{\infty} dz\, f(x,\Delta t \mid z)\, f(z,t \mid x_0) \tag{36}$$

where we have set $X(0) = x_0$ and have omitted the specification of the initial time. We now note that, apart from infinitesimal quantities $o(\Delta t)$, it is:

$$f(x,\Delta t\,|z) = \Big\{ 1 - \Delta t \Big[\sum_{k=1}^{p} \alpha_k + \sum_{k=1}^{q} \beta_k \Big] \Big\} \delta[\, x - (z - z\,\Delta t/\theta)]$$
$$+ \Delta t \sum_{k=1}^{p} \alpha_k \, \delta[\, x - (z - z\,\Delta t/\theta + e_k)]$$
$$+ \Delta t \sum_{k=1}^{q} \beta_k \, \delta[\, x - (z - z\,\Delta t/\theta + i_k)], \tag{37}$$

where $\delta(\cdot)$ is the Dirac delta-function. Substituting (37) in (36), after some straightforward calculations, and in the limit when $\Delta t \to 0$, we obtain:

$$\frac{\partial f}{\partial t} = \frac{\partial}{\partial x}\left(\frac{x}{\theta} f\right) + \sum_{k=1}^{p} \alpha_k [\, f(x-e_k, t\,|\,x_o) - f(x, t\,|\,x_o)]$$
$$+ \sum_{k=1}^{q} \beta_k [\, f(x-i_k, t\,|\,x_o) - f(x, t\,|\,x_o)] \,. \tag{38}$$

In order to smooth down the process, we first expand the functions on the r.h.s. of (38) as Taylor series about x and then apply a suitable limit procedure to the magnitude of PSP's and of input arrival rates. After the mentioned expansion, eq. (38) yields:

$$\frac{\partial f}{\partial t} = -\frac{\partial}{\partial x}\left[\left(-\frac{x}{\theta} + \mu_1\right) f\right] + \sum_{j=2}^{\infty} \frac{(-1)^j}{j!} \mu_j \frac{\partial^j f}{\partial x^j} \tag{39}$$

where we have set:

$$\mu_j = \sum_{k=1}^{p} \alpha_k e_k^j + \sum_{k=1}^{q} \beta_k i_k^j \qquad (j=1,2,\ldots\ldots,n). \tag{40}$$

We then proceed with the smoothing and take the parameters appearing in (39) and (40) as indicated hereafter (note that without loss of generality we have assumed $p > q$):

$$\left.\begin{aligned}
\alpha_k &= \lim_{y\to 0} a_k/y^2 && (a_k > 0)\\
\beta_k &= \lim_{y\to 0} b_k/y^2 && (b_k > 0)\\
i_k &= \lim_{y\to 0} d_k\, y && (d_k < 0)\\
e_k &= \lim_{y\to 0} c_k\, y && \left(c_k = \frac{|d_k|\, b_k}{a_k}\right)
\end{aligned}\right\} \quad (k=1,2....,q) \tag{41}$$

$$\left.\begin{array}{lll}\alpha_r = \lim_{y\to 0} a_r / y & & (a_r > 0) \\ \beta_r = \lim_{y\to 0} b_r / y & & (b_r > 0) \\ i_r = \lim_{y\to 0} d_r\ y & & (d_r < 0) \\ e_r = \lim_{y\to 0} c_r\ y & & (0 < c_r \neq \frac{|d_r|\ b_r}{a_r}) \end{array}\right\} (r = q+1, q+2, \ldots, p)$$

Here e_k's, b_k's, c_k's and d_k's are otherways arbitrary constants. From (40) and (41) we are thus lead to the following result:

$$\mu_1 + \lim_{y\to 0} [y^{-1} \sum_{k=1}^{q} (a_k c_k - b_k |d_k|) + \sum_{r=q+1}^{p} (a_r c_r - b_r |d_r|) \equiv \delta$$

$$\mu_2 + \sum_{k=1}^{q} b_k d_k^2 (1 + \frac{b_k}{a_k}) + \lim_{y\to 0} [y \sum_{r=q+1}^{p} (a_r c_r^2 + b_r d_r^2)] \equiv \mu > 0$$

$$\mu_j + \lim_{y\to 0} \{y^{j-2} \sum_{k=1}^{q} b_k d_k [1 + (-1)^j (\frac{b_k}{a_k})^{j-1}]$$

$$+ y^{j-1} \sum_{r=q+1}^{p} [a_r c_r^j + (-1)^j b_r |d_r|^j]\} = 0 \qquad (j = 3,4,\ldots\ldots).$$

We have thus proved that eq. (39) tends to the Fokker-Planck equation

$$\frac{\partial f}{\partial t} = - \frac{\partial}{\partial x} (- \frac{x}{\theta} + \delta) f] + \frac{\mu}{2} \frac{\partial^2 f}{\partial x^2}, \tag{43}$$

where μ denotes the infinitesimal variance. The quantity δ in the drift is the net rate of excitation. Note that in the limit $\theta\to\infty$ eq. (43) yields the model postulated by Gerstein and Mandelbrot (1964) whereas for $\delta \to 0$ it identifies with that derived in Capocelli and Ricciardi (1971).

The limit procedure carried out in the foregoing models the neuron's membrane potential as an Ornstein-Uhlenbeck diffusion process. Hence, firing p.d.f. coincides with the first passage time p.d.f. this process through a boundary representing the neuron's threshold. Particulary in the present context, such boundary would be time dependent in several situations of neurobiological interest.

4. Evaluation of First Passage Time Densities.

In Section 3 and 4 we have sketched a few diffusion approximations arising in population dynamics and in neurobiology and we have pointed out the interest of the evaluation of first passage time densities to gather information on population extinction probabilities and on neuronal firing distribution. The common feature of the models discussed so far is that they lead to continuous Markov processes. However, non Markov processes would naturally arise if the assumption that $\Lambda(t)$ is of the white noise type is relaxed or if the input processes in the neuronal model are assumed to be correlated. As far as the first passage time distribution is concerned, it is essential to consider separately the Markov and non Markov cases.

Markov processes. Analytical results are scarce and fragmentary, so that it is necessary to make use of computational methods. These belong to essentially two categories. The first consists of approximating the boundary by a suitable function (Durbin, 1973; Ricciardi and Sato, 1983; Ricciardi et al, 1983) while the second rests on numerical procedures to solve integral equations of the Volterra type (Anderson et al, 1973; Favella and De Griffi, 1981; Favella et al, 1982; Balossino et al, 1983; Ricciardi et al, 1983). For the description of these methods we refer to the quoted papers.

Non Markov processes. Consider, as an example, a Malthusian growth process in random environment and let $x(t)$, the population size at time t, satisfy

$$\frac{dX}{dt} = X[\alpha + \xi(t)] \qquad P\{X(0) = x_0\} = 1 \tag{44}$$

Here, more realistically than in the diffusion approximation case, we assume that the fluctuating part $\xi(t)$ of the intrinsic growth rate is a normal process with zero mean and correlation function

$$E[\xi(t)\xi(t+\tau)] = g(\tau) \tag{45}$$

of a non delta-function. This implies that we are dealing with a correlated environment. The n-dimensional joint p.d.f. of $\xi(t)$ is then given by

$$f_n(x_1, t_1; x_2, t_2; \dots; x_n, t_n) = \frac{1}{(2\pi)^{n/2}\sqrt{\Delta}} \exp\{-\tfrac{1}{2} X^T K^{-1} X\} \tag{46}$$

where $X^T \equiv (x_1, x_2, \dots, x_n)$, K is the correlation matrix and Δ is the determinant of K. In particular, for $n=1$ we have:

$$f_1(\xi, t) \equiv f(\xi) = \frac{1}{\sqrt{2\pi\, g(0)}} \exp\{-\frac{\xi^2}{2g(0)}\} \tag{47}$$

which yields the white noise case in the limit as $g(\tau) \to \delta(\tau)$ (infinite variance). Using the substitution

$$Y = \ln X - \alpha t \tag{48}$$

from (45) we get:

$$\frac{dY}{dt} = \xi(t)$$
$$P\{Y(o) = \ln x_o\} = 1 \qquad (49)$$

Hence

$$Y(t) = Y(0) + \int_0^t d\tau\, \xi(\tau) \qquad (50)$$

showing that Y(t) is in turn a normal process with mean $\ln x_o$ and covariance

$$h(t, t') = \int_0^t d\tau \int_0^{t'} d\eta\; g(\eta-\tau). \qquad (51)$$

The conditional density $f(x, t|x_o)$ of X(t) can therefore be expressed as:

$$f(x, t|x_o) = \frac{1}{\sqrt{2\pi h(t,t)}} \exp\left\{-\frac{(\ln\frac{x}{x_o} - \alpha t)^2}{2h(t,t)}\right\} \qquad (52)$$

in terms of the covariance of Y(t). Note, however, that X(t) is non Markov. According to the remarks of Section 2, the extinction p.d.f. of X(t) can be identified with the first passage time p.d.f. of X(t) through some threshold value $S < x_o$. Due to transformation (48), this can be calculated in terms of the first passage time p.d.f. of the process Y(t) through the time dependent boundary $S'(t) = \ln S - \alpha t$ which is a constant only if $\alpha = 0$. Therefore, the real problem one has ultimately to solve is that of calculating a first passage time p.d.f. through some time dependent boundary for a non Markov Gaussian process. As is easily understood , this is a very complicated task. However, some preliminary results are already available, as sketched in the sequel.

Let $\{X(t);\ t > 0\}$ be a one-dimensional non singular stationary Gaussian process with

$$E[X(t)] = 0$$
$$E[X(t), X(t')] = \gamma(t - t'), \quad \gamma(0) = 1 \qquad (53)$$

and let $\ddot{\gamma}(0) = \left(\frac{d^2\gamma(t)}{dt^2}\right)_{t=0}$ exist and be finite. Then $\dot{X}(t) = dX/dt$ exists in the mean square sense. Furthermore, let $X(0) = x_o$ and let S(t) be a function possessing a continuous derivative and such that $S(0) > x_o$ (the case $S(0) < x_o$ can be treated in a similar way). Then (see Ricciardi and Sato, 1983) the first passage time p.d.f. $g[S(t), t|x_o]$ of X(t) through S(t) can be expressed as

$$g[S(t), t|x_o] = W_1(t|x_o) - \int_0^t dt_1\, W_2(t_1, t|x_o)$$
$$+ \ldots + (-1)^n \int_0^t dt_1 \int_{t_1}^t dt_2 \ldots \int_{t_{n-1}}^t dt_n\, W_{n+1}(t_1, t_2, \ldots, t_n, t|x_o) + \ldots$$

where the functions W_1, W_2,..... have the following meaning:

$W_n(t_1, t_2, \ldots, t_n | x_o) dt_1\, dt_2 \ldots dt_n$ = {joint probability that X(t) crosses S(t) from below in $(t_1, t_1 + dt_1)$, $(t_2, t_2 + dt_2)$,...., $(t_n, t_n + dt_n)$ given that $X(0) = x_o$ }.

Let now Δ_{2n+1} be the covariance matrix whose entries δ_{ij} are defined as follows:

$$\delta_{i+1,\, j+1} = E[X(t_i)\, X(t_j)] \equiv \gamma(t_i - t_j), \qquad i,j=0,1,\ldots,n$$

$$\delta_{n+i+1,\, n+j+1} = E[\dot{X}(t_i)\, \dot{X}(t_j)] \equiv -\ddot{\gamma}(t_i - t_j), \qquad i,j=1,2,\ldots,n \tag{55}$$

$$\delta_{i+1,\, n+j+1} = E[X(t_i)\, \dot{X}(t_j)] \equiv -\dot{\gamma}(t_i - t_j), \qquad i=0,1,\ldots,n,\; j=1,2,\ldots,n$$

and let L_{ij} be the (i,j)-element of $(\Delta_{2n+1})^{-1}$. Then:

$$W_n(t_1, t_2, \ldots, t_n | x_o) = (2\pi)^{-n} |\Delta_{2n+1}|^{-1/2} \prod_{i=1}^{n} \int_{\dot{S}_i}^{\infty} d\dot{x}_i\, (\dot{x}_i - \dot{S}_i)$$
$$\exp\{-\frac{1}{2} \sum_{i,j=1}^{2n} L_{i+1,j+1} [x_i - \gamma(t_i) x_o]\, [x_j - \gamma(t_j) x_o]\} \tag{56}$$

where we have set:

$$x_i = x(t_i),\; x_{n+i} = \dot{x}(t_i),\; \gamma_i = \gamma(t_i),\; \gamma_{n+i} = \dot{\gamma}(t_i) \tag{57}$$

By means of (54) and (56) the first passage time p.d.f. can in principle be evaluated to an arbitrary degree of accuracy. It should however be explicitly mentioned that the involved calculations are outrageously complicated so that an ad hoc numerical procedure should be worked out. While their description will be the object of future reports, here we limit ourselves to providing the first order approximation to g:

$$g\,[S(t), t | x_o] \approx \frac{|\Delta_3|^{1/2}}{2\pi\,(1-\gamma^2)} \exp\,[-\frac{1}{2}(\omega^2 + \frac{s^2}{1-\gamma^2})]$$
$$\cdot\{1 + \sum_{n=0}^{\infty} \frac{\sigma^{2n+2}}{(2n+1)!!} - \sqrt{\frac{\pi}{2}} \sum_{n=0}^{\infty} \frac{\sigma^{2n+1}}{(2n)!!}\}, \tag{58}$$

where we have set:

$$\Delta_3 = \begin{vmatrix} 1 & \gamma(t) & \dot{\gamma}(t) \\ \gamma(t) & 1 & 0 \\ \dot{\gamma}(t) & 0 & -\ddot{\gamma}(0) \end{vmatrix} \tag{59}$$

$\gamma = \gamma(t)$, $s = S(t) - \gamma(t)x_0$, $\dot{s} = \frac{ds}{dt}$, $\sigma = \{\frac{1-\gamma^2(t)}{|\Delta_3|}\}^{1/2}$ $\{\dot{s} + \frac{\gamma(t)\dot{\gamma}(t)s}{1-\gamma^2(t)}\}$

Note that in the special case when

$$\lim_{t\to\infty} s \equiv \lim_{t\to\infty} [S(t) - \gamma(t)x_0] = c \qquad \text{(c a constant)} \tag{60}$$

one has

$$\sigma \to 0, \ |\Delta_3| \to - \ddot{\gamma}(0) \tag{61}$$

so that (58) yields

$$g[S(t), t|x_0] \approx \frac{\sqrt{-\ddot{\gamma}(0)}}{2\pi} \exp\left(-\frac{c^2}{2}\right) \tag{62}$$

in agreement with the well known result on the c-level crossing provided by Kac (1943). Finally, it should be pointed out that the approximations (58) and (62) become progressively better as t decreases.

References

[1] Anderssen, R.S., De Hoog, F.R. and Weiss, R. (1973). On the numerical solutions of Brownian motion process, J. Appl. Prob. 10, 409-418.

[2] Balossino, N.; Ricciardi, L.M. and Sacerdote, L. (preprint 1983). On the evaluation of first passage time densities for diffusion processes.

[3] Capocelli, R.M. and Ricciardi, L.M. (1971). Diffusion approximations and first passage time problem for a model neuron, Kybernetik 8, 214-223.

[4] Durbin, J. (1971). Boundary crossing probabilities for Brownian motion and Poisson processes and techniques for computing the power of the Kolmogorov--Smirnov test, J. Appl. Prob. 8, 431-453.

[5] Favella, L.F. and De Griffi, E.M. (1981). On a weakly singular Volterra integral equation, Calcolo Vol. XVIII, 153-195.

[6] Feller, W. (1951a). Diffusion processes in genetics, Proc. 2nd Berkeley Symp. Math. Statistics and Probability, 227-246.

[7] Feller, W. (1951b). Two singular diffusion processes, Ann. Math. 54, 173-182.

[8] Feller, W. (1952). Parabolic differential equations and associated semigroup transformations, Ann. Math. 55, 468-518.

[9] Feller, W. (1954). Diffusion processes in one dimension, Trans. Amer. Math. Soc. 77, 1-31.

[10] Gerstein, G.L. and Mandelbrot, E. (1964). Random walk models for the spike activity of a single neuron, Biophys. J. 4, 41-68.

[11] Kac, M. (1943). On the distribution of values of trigonometric sums with linearly independent frequencies, Amer. J. Math. Vol. LXV, 609-615.

[12] Nobile, A.G. and Ricciardi, L.M. (1984). Growth with regulation in fluctuating environments. I. Alternative logistic-like models, Biol. Cybernetics 49, 179-188.

[13] Nobile, A.G. and Ricciardi, L.M. (preprint 1983). Growth with regulation in

fluctuating environments. II. Intrinsic lower bounds to population size, Biol. Cybernetics (in press).

[14] Ricciardi, L.M. (1977). Diffusion Precesses and Related Topics in Biology. Lecture Notes in Biomathematics, Vol. 14, Springer-Verlag, Heidelberg-New York.

[15] Ricciardi, L.M. and Sacerdote, L. (1979). The Ornstein-Uhlenbeck process as a model for neuronal activity, Biol. Cybernetics 35, 1-9.

[16] Ricciardi, L.M. Sacerdote, L. and Sato, S. (1983). Diffusion approximation and first passage time problem for a model neuron. II. Outline of a computation method, Math. Biosciences 64, 29-44.

[17] Ricciardi, L.M., Sacerdote, L. and Sato, S. (preprint 1983). On an integral equation for first passage time probability densities, J. Appl. Prob. (in press).

[18] Ricciardi, L.M. and Sato, S. (1983). A note on the evaluation of first passage time probability densities, J. Appl. Prob. 20, 197-201.

[19] Ricciardi, L.M. and Sato, S. (1983). A note on first passage time problems for Gaussian processes and varying boundaries, IEEE Trans. Inform. Theory, Vol IT-29, 454-457.

THE ROLE OF EXCHANGEABILITY IN BIOLOGY, GENETICS AND MEDICINE

R. Ahmad
Department of Mathematics, University of Strathclyde, Glasgow, Scotland.

ABSTRACT

In the context of various models appearing in biology, genetics and medicine, the concepts of exchangeability and its extension partial exchangeability are discussed. It is pointed out that these concepts appear either as a basic assumption in some models or are inherent in the composition of a population (or subpopulations). That is, when the population is homogeneous as a whole or can be partitioned to several subgroups which are homogeneous within themselves. Also extensions of some of the existing results in the literature are carried out.

1. INTRODUCTION

At the occasion of the International Congress of Mathematicians, Toronto, Jules Haag (1924) discussed exchangeable sequences of random variables. He dealt with only two-valued random variables. For this case he hints at, but does not rigorously state or prove, the famous representation theorem which has many applications in various fields including biology, genetics and medicine. The concept of exchangeability, in its present generality, is essentially due to De Finetti (1931). A probability distribution on R^p is defined as exchangeable if almost surely

$$F(x_1,\ldots,x_p) = F(\pi(x)) = F(x_{\pi(1)},\ldots,x_{\pi(p)}) \tag{1}$$

for all π in the permutation symmetric group S_p. That is, the distribution is invariant under permutation of all its components. This idea, as we shall see later, can be extended to partial exchangeability, by considering appropriate subgroups of S_p. For the latest development and a comprehensive list of references see Koch and Spizzichino (1982). The hypotheses testing structure is given in Ahmad and Peterson (1978).

Exchangeability is applicable to a large class of models which arise in genetics, biology, ecology, intelligence whether artificial or not, medicine, and elsewhere, when the main interest centers on the evolution of one or more characteristics in a large and finite population. Here, the concept of exchangeability occurs naturally when the members of a population are not specially labelled or are indistinguishable. Such situations arise in random genetic drift, in the genealogy of large populations, the so-called N-coalescent problems, the common ancestor problems, and the Ewens sampling formula; just to mention a few cases which are being actively pursued in current literature. However, for more refined analysis and somewhat more sophisticated problems a population can be thought of as composed of several homogeneous subpopulations, then the restricted exchangeability will be the natural candidate for evolutionary characteristics. In this connection one can recast (which we hope to do elsewhere) and slightly extend some ideas of Kingman (1982).

2. EXCHANGEABILITY STRUCTURE AND ITS IMPLICATIONS

Consider a population of size N with k subpopulations (homogeneous within themselves) of sizes $N_1, N_2, \ldots, N_k$ with $N = \sum_{j=1}^{k} N_j$.. Suppose that the evolutionary processes within various subpopulations are behaving independently. In practice in many experimental situations a sample of size n_j is drawn at random from the jth subpopulation at a particular time, and that either univariate or multivariate measurement gives values $X_{1j}, X_{2j}, \ldots, X_{n_j j}$ for the n_j objects of the sample. Any evolutionary model of such a population leads to a joint probability distribution $P_{N_1, \ldots N_k;\, n_1, \ldots, n_k} \equiv P^*$, say, for these random variables, which will depend on N_j's and other relevant parameters like mutation rates, migration rates etc.. Because of the sampling scheme, P^* must be partially exchangeable. A distribution F is said to be partially exchangeable if for x_i in R^{k_i} and π in S_{k_i}, one has:

$$F(x_1, \ldots, x_q) = F(\pi_1 x_1, \ldots, \pi_q x_q) \tag{2}$$

In some situations it is reasonable to assume that for each j, N_j is large, and the underlying parameters in jth sub-population are of order N_j^{-1}. Assuming that the respective parameters when multiplied by N_j converge to finite limits, the limiting behaviour of P^* can be investigated as $N_j \to \infty$ for $j = 1,2,\ldots,k$. In some genetical models, it turns out that for each $\{n_j, j=1,2,\ldots,k\}$, P^* converges to a limit $P_{n_1, \ldots, n_k} \equiv P_\infty$, say. Each P_∞ is a partially exchangeable joint distribution for random variables $\{X_{1j}, X_{2j}, \ldots, X_{n_j j};\ j = 1,2,\ldots,k\}$.

Since each P_∞ is partially exchangeable, the extended version of Finetti's representation theorem applies to the infinite sequence $\{X_{n_j j}\}$. Consequently, there is a probability distribution P such that P_∞ is the joint distribution of the above random variables, conditionally independent given P, and each having conditional distribution P. For example, see Ewens (1972), Watterson (1974a,b;1976) and Kingman (1977, 1982). This structure will cover the N-coalescent models as well as their extended variants, when the subpopulations are among haploid (as in some fungi), diploid (in mamals), haploid or tetraploid (larger groups as in many plants etc.), or in general polyploid populations. It is worth noting here, as Kingman (1982) has pointed out, that the m-coalescent can be taken as a robust description of the genealogy of a sample of size m<<M from a neutral haploid population of large size M.

With the above preliminary remarks, to elucidate the concepts of exchangeability and partial exchangeability, we give two simple examples as below.

Example 1. Consider n matched pairs $\{(X_i,Y_i),\ i = 1,2,\dots,n\}$ with a bivariate distribution F(x,y). If we let X as 'control' and Y as the treatment response, then $\Omega(H) = \{F:F(x,y)=F(y,x)\}$ is equivalent to the complex null hypothesis that there is no treatment effect. The extension to the p-response is obvious. Such situationscommonly occur in medicine, genetics and biology besides other areas.

Example 2. In a situation let the random variable X_j indicate the observable jth evolutionary characteristic. If the values of one of the variates $X_1,\dots,X_n$ is measured on a randomly chosen member of a population from time to time with a view to detecting whether the population's characteristics are changing with time a natural null hypothesis for the generic data point

$$\underset{\sim}{Z} = (X_{11},\dots,X_{1k_1};\ \dots;X_{n1},\dots,X_{nk_n}) = (Z_1;\dots;Z_n) \tag{3}$$

is that its distribution is symmetric for all permutations in the group: $\Sigma = S_{k_1} \times S_{k_2} \times \dots \times S_{k_n}$. This situation has indeed many practical applications.

3. SELECTION SCHEME DISTRIBUTION THEORY

A large class of models have been proposed for selectivly neutral mutation in the infinite alleles case, in which mutation always gives rise to a completely new allele. Since different biological situations are best fitted by different models, naturally these models differ in reproductive structure and in population mechanism. However, in one respect all of these models are similar, that is if the population size is allowed to become large while the expected overall mutation rate remains fixed, then the stationary distribution of the allelic composition of a sample of a fixed size tends to the so-called Ewens sampling formula.

In essence Ewens selection scheme is the infinite version of the classical Wright-Fisher model. This formula is relevant to other situations exhibiting recurrent mutation and selective neutrality in large populations. Let N be the total population size, n a random sample from this population. Let a_j denote alleles with j representatives (j = 1,2,...,n) with $a_j \geq 0$ and $\sum_{j=1}^{n} ja_j = n$. If λ is the mutation rate per generation per individual, then as $N\to\infty$ with $\theta = 2\lambda N$ fixed, one has

$$P_N\{a_1,\ldots,a_n\} \to n!\ \theta^{k-1}\Big[\prod_{j=1}^{n}(a_j!j^{a_j})(\theta+1)(\theta+2)\ldots(\theta+n-1)\Big]^{-1} \tag{4}$$

where $k = \sum_{j=1}^{n} a_j$ is the different alleles found in the sample. The above limiting result was stated by Ewens (1972) and later proved by Karlin and McGregor (1972).

In a paper Watterson (1976) replaced an infinite alleles model by one with, for convenience say, (k+1) alleles, in which an allele mutates to any one of the other k alleles with equal probability - notice the underlying concept of exchangeability here. The diffusion approximation gives the stationary distribution of the proportions $(p_1,\ldots,p_{k+1})$ of the alleles by the Dirichlet distribution with probability density given by

$$f(p_1,\ldots,p_k) = \frac{\Gamma(\alpha_1+\ldots+\alpha_{k+1})}{\Gamma(\alpha_1)\ldots\Gamma(\alpha_{k+1})}\ p_1^{\alpha_1-1}\ldots p_k^{\alpha_k-1}\ (1-\sum_{j=1}^{k} p_j)^{\alpha_{k+1}-1} \tag{5}$$

where $0 < p_i$, $i = 1,2,\ldots,k$, $\sum_{i=1}^{k} p_i < 1$, and $\alpha_1 = \alpha_2 = \ldots = \alpha_{k+1} = \frac{\theta}{k}$.

The expression (5) has no non-degenerate limit as $k\to\infty$, see Stewart (1976). However, as Kingman (1975) shows, if we order p_j's : $p_{(1)} \geq p_{(2)} \geq \ldots \geq k_{(k+1)}$; the $\{p_{(j)}\}$ converge in joint distribution, with limit depending only on θ. This he called the Poisson-Dirichlet limit, and it can be regarded as a probability distribution on the space

$$S = \{(u_1,\ldots,u_k,\ldots) : u_1 \geq u_2 \geq \ldots \geq 0;\ \sum_{j=1}^{\infty} u_j = 1\} \tag{6}$$

Kingman (1977) has shown that a sequence of populations has the

Ewens sampling property if and only if it has the Poisson-Dirichlet limit. Since in practice many population structures are composed of several homogeneous subpopulations, by using partial exchangeability we give below an extension of Ewens, Watterson and Kingman results.

4. AN EXTENSION

In Watterson's model consider a nonhomogeneous population with q homogeneous subpopulations. In jth subpopulation let each 'element' mutate to any one of the other k_j elements with equal probability p_j. The size of this population is assumed to be (k_j+1). Thus, for $j = 1,2,\ldots,q$, we replace in Watterson's model k by k_j; θ by $\theta_j = 2\lambda_j k_j$ (fixed as $k_j\to\infty$); for each subpopulation $(\alpha_1,\ldots,\alpha_k)$ by $(\alpha_{j1},\ldots,\alpha_{jk_j})$, and $(p_1,\ldots,p_k)$ by $(p_{j1},\ldots,p_{jk_j})$ with the same restrictions on α's and p's as before but now within each subpopulation.

By using the partial exchangeability argument it turns out that there exists a mixing probability measure $\mu_{\underset{\sim}{p}}$ such that the expression (5) now becomes:

$$\int\Big[\prod_{j=1}^{q} f_j(p_{j1},\ldots,p_{jk_j})\Big]d\mu_{\underset{\sim}{p}} \quad . \tag{7}$$

The above integral has again no non-degenerate limit as $k_j\to\infty$ for $j = 1,2,\ldots,q$. However, if we order p_{ji}'s in each subpopulation as: $p_{(j1)} \geq p_{(j2)} \geq \cdots \geq p_{(jk_j)} \geq 0$; then under some regularity conditions $\{p_{(ji)}\}$ converge in joint distribution, the limit depending only on $\underset{\sim}{\theta} = (\theta_1,\ldots,\theta_q)$. This limiting distribution can be thought of as the Poisson-Dirichlet-mixture $PD(\underset{\sim}{\theta},\mu^*_{\underset{\sim}{\theta}})$ limit. That is,

$$\lim_{\substack{k_j\to\infty\\ 1\leq j\leq q}} \int\Big[\prod_{j=1}^{q} g_j(p_{(j1)},\ldots,p_{(jk_j)})\Big]d\mu_{\underset{\sim}{p}(\cdot)} =$$

$$= \int\Big[\prod_{j=1}^{q} PD(\theta_j)\Big]d\mu^*_{\underset{\sim}{\theta}} \equiv PD(\underset{\sim}{\theta},\mu^*_{\underset{\sim}{\theta}}).$$

The Poisson-Dirichlet-mixture limit $PD(\theta,\mu^*_{\underset{\sim}{\theta}})$ can be viewed as a probability distribution on the product space: $\times_j S_j$, for jth subpopulation S_j is defined as in the expression (6). The detailed arguments for the above development are somewhat intricate and involved - though similar to Kingman's proof excepting an addition of mixing measure through partial exchangeability.

In summary, then we have: a sequence of non-homogeneous populations having homogeneous subpopulations has the modified Ewens-Watterson-Kingman structure if and only if it has the Poisson-Dirichlet-mixture limit.

Remark: Note the sampling formulae of previous two sections have robustness property.

REFERENCES

1. Ahmad, R. (1982). On the structure and applications of restricted exchangeability. In: Exchangeability in Probability and Statistics (G. Koch and F. Spizzichino; eds.), 157-164. North-Holland Pub. Co. Amsterdam.
2. Ahmad, R. and Peterson, M.M. (1978). Restricted permutation symmetry and hypotheses-generating groups in statistics. In: Trans. 8th Prague Conf. on Information Theory, Statistical Decision Functions, Random Processes, Vol. A, 71-82. Academia, Czechoslovak Academy of Sciences, Prague.
3. De Finetti, B. (1931). Funzione caratteristica di un fenomeno aleatorio. Atti della R. Acad. Naz. dei Lincei, Ser 6, Mem. Sci. Fis. Mat. Nat., 251-299.
4. Ewens, W.J. (1972). The sampling theory of selectively neutral alleles. Theor. Pop. Biol., 3, 87-112.
5. Haag, J. (1924). Sur une problème général de probabilités et ses diverses applications. In: Proc. International Congress of Mathematicians, Tronto, 1924 (Tronto, 1928), 659-674.
6. Karlin, S. and McGregor, J.L. (1967). The number of mutant forms maintained in a population. In: 5th Berkeley Symp. on Mathematical Statistics and Probability, Vol. IV, 415-438.
7. Karlin, S. and McGregor, J.L. (1972). Addendum to a paper of W. Ewens. Theor. Pop. Biol., 3, 113-116.
8. Kingman, J.F.C. (1975). Random discrete distributions. J. Roy. Statist. Soc. B, 37, 1-22.
9. Kingman, J.F.C. (1977). The population structure associated with the Ewens sampling formula. Theor. Pop. Biol., 11, 274-283.
10. Kingman, J.F.C. (1982). Exchangeability and evolution of large populations. In: Exchangeability in Probability and Statistics (G. Koch and F. Spizzichino; eds.), 97-112. North-Holland Pub. Co. Amsterdam.
11. Koch, G. and Spizzichino, F. (1982). Exchangeability in Probability and Statistics (eds.). North-Holland Pub. Co. Amsterdam.
12. Stewart, F.M. (1976). Variability in the amount of heterozygosity maintained by neutral mutations. Theor. Pop. Biol., 9, 188-201.

13. Watterson, G.A. (1974a). The sampling theory of selectively neutral alleles. Adv. Appl. Prob., 6, 463-488.

14. Watterson, G.A. (1974b). Models for logarithmic series abundance distributions. Theor. Pop. Biol., 6, 217-250.

15. Watterson, G.A. (1976). The stationary distribution of the infinitely-many alleles diffusion model. J. Appl. Prob., 13, 639-651.

CYCLIC CHEMICAL SYSTEMS : AN ASYMPTOTIC STABILITY CRITERION

Edoardo Beretta and Claudio Lazzari

Istituto di Biomatematica, Facoltà di Scienze, Università di Urbino , Via A.Saffi,1 , I-61029 , Italy.

Definitions and Nomenclature.

In previous papers (Beretta et al,1979;Beretta and Lazzari,1983) we obtained asymptotic stability criteria concerning positive equilibria of chemical networks with mass action kinetics.

The evolution equations of a chemical network with " r " pseudoreactions and " n " internal species are:

$$\dot{\underline{c}} = \underline{f}(\underline{c}) \quad ; \quad \underline{f}(\underline{c}) := \underline{\underline{\nu}}\, \underline{v}(\underline{c}) \quad , \quad \underline{c} \in \overline{R^n_+} \quad . \tag{1}$$

$\underline{\underline{\nu}} = (\nu_{ij})$ is the " n x r " stoichiometric matrix; $\underline{v} = \underline{v}(\underline{c})$, $\underline{v} \in R^r$ is the velocity whose j-th component gives the net velocity $v_j(\underline{c})$ of the j-th reaction according to the mass action kinetics; $\underline{c} = \underline{c}(t)$ is the chemical composition of the network at the time $t \in [t_o, +\infty[$ .For a more detailed definition of a chemical network see Clarke (1980).Let $\underline{c}^o \in R^n_+$ be a positive equilibrium of (1).The equations (1), linearized around $\underline{c}^o$, are

$$\dot{\underline{x}} = \underline{\underline{Q}}\, \underline{x} \quad , \quad \underline{x} \in R^n \tag{2}$$

where $\underline{\underline{Q}}$ is the Jacobian matrix of (1) evaluated at $\underline{c}^o$ .Let $\mathcal{A}_1,..,\mathcal{A}_n$ the internal species and

$$\nu_{ij} := p_{ij} - r_{ij} \quad , \; i = 1,..,n \quad , \; j = 1,..,r \tag{3}$$

the net stoichiometry of the species $\mathcal{A}_i$ in the j-th pseudoreaction: r_{ij} is the stoichiometry of $\mathcal{A}_i$ as reagent and p_{ij} is the stoichiometry of $\mathcal{A}_i$ as product (r_{ij}, p_{ij} are nonnegative integers).

We associate a knot graph with the Jacobian matrix of a chemical network by the following rules:

(1) with each pseudoreaction, say the j-th, we associate two interactants

$$\mathfrak{I}^R_j := \{ \mathcal{A}_i : r_{ij} \neq 0 \} \quad , \quad \mathfrak{I}^P_j := \{ \mathcal{A}_i : p_{ij} \neq 0 \} \quad . \tag{4}$$

If for all " i " we have $r_{ij} = 0$ (or $p_{ij} = 0$), we obtain the empty interactant represented by the symbol $\square$.

(2) Every set of connected interactants (two interactants are connected if they have at least one common internal species) is complexed in one knot.

(3) With each pair of interactants ($\mathfrak{J}_j^R$, $\mathfrak{J}_j^P$) we associate one arc connecting the two knots (not necessarily distinct) to which $\mathfrak{J}_j^R$ and $\mathfrak{J}_j^P$ belong;no arc is associated with the pair ($\mathfrak{J}_j^{R,P}$, $\square$).

We proved (Beretta and Lazzari,1983) the following theorem:

Theorem 1. If the knot graph associated with the chemical network is a tree or it contains a balanced cycle,then : (i) the Jacobian matrix is D-symmetrizable and has real non-positive eigenvalues; (ii) rank $\underline{\underline{Q}}$ = rank $\underline{\underline{\nu}}$; (iii) any positive equilibrium $\underline{c}^o$ is locally stable in R^n and locally asymptotically stable in the linear manifold $\underline{c}^o + S$, S being the stoichiometric subspace defined by:

$$S := \mathrm{Im}\ \underline{\underline{\nu}} \quad , \quad s := \dim S = \mathrm{rank}\ \underline{\underline{\nu}} \quad . \tag{5}$$

The theses (i)-(iii) of theor.1 hold true even when the knot graph contains several balanced cycles at a given equilibrium $\underline{c}^o \in R^n_+$.

Let $K_1,\ldots,K_\delta$ the knots of a cycle of length " δ" and let ϕ_j^o ,ψ_j^o , j = 1,..,δ the forward and reverse velocities (the net velocity of a chemical reaction,say the j-th,is $v_j(\underline{c}) = \phi_j(\underline{c}) - \psi_j(\underline{c})$) of the cycle at a given positive equilibrium $\underline{c}^o$.The " unbalancement " of the cycle is the number

$$u = \prod_{j=1}^{\delta} (\phi_j^o / \psi_j^o) \quad . \tag{6}$$

It is always possible to direct an unbalanced cycle in such a way that $u > 1$.When $u = 1$ the cycle is said " balanced " and theor.1 applies.A cycle is balanced at $\underline{c}^o$if, as a particular case,$\underline{c}^o$ is detailed balanced,that is, $\phi_j^o = \psi_j^o$ for all j.

The aim of the next section is to derive an asymptotic stability criterion for the equilibria of a particular,but important,class of chemical networks whose knot graph contains one unbalanced cycle only.For example,most of the enzyme reactions belong to this class.

Asymptotic stability of unbalanced cycles.

By (1) the Jacobian matrix is $\underline{\underline{Q}} = \underline{\underline{\nu}}\,\underline{\underline{V}}$,where

$$\underline{\underline{V}} = (v_{ij})_{i = 1,..,r;\ j = 1,..,n} \tag{7}$$

$$v_{ij} = (\partial v_i(\underline{c}) / \partial c_j \big|_{\underline{c}^o}) \quad . \tag{8}$$

Let $\underline{\nu}_\rho$ the ρ-th column of $\underline{\underline{\nu}}$ and $\underline{v}_\rho$ the ρ-th row of $\underline{\underline{V}}$, $\rho = 1,..,r$. Then

$$\underline{\underline{Q}} = \sum_{\rho=1}^{r} \underline{\underline{A}}_\rho \tag{9}$$

where

$$\underline{\underline{A}}_\rho := \underline{\nu}_\rho \underline{v}_\rho \tag{10}$$

is the n x n matrix concerning the contribution of the ρ-th reaction to $\underline{\underline{Q}}$. Each matrix $\underline{\underline{A}}_\rho$ has real eigenvalues, one negative and equal to $\mathrm{tr}\, \underline{\underline{A}}_\rho$, the others equal to zero. According to (9),(10), $\underline{\underline{Q}}$ may be written as

$$\underline{\underline{Q}} = \underline{\underline{Q}}_t + \underline{\underline{A}}_\delta \quad , \quad \underline{\underline{A}}_\delta = \underline{\nu}_\delta \underline{v}_\delta \quad . \tag{11}$$

$\underline{\underline{Q}}_t$ is the Jacobian matrix of the tree graph and $\underline{\underline{A}}_\delta$ is the one of the closure branch which gives rise to the cycle of length δ. Obviously, the closure branch can be arbitrarily chosen among those of the cycle. By the theses (i),(ii) of theor.1, a linear non-singular transformation

$$\underline{\underline{P}} = \underline{\underline{D}}^{1/2} \underline{\underline{T}} \tag{12}$$

exists such that

$$\underline{\underline{Q}}_t^* = \underline{\underline{P}}^{-1} \underline{\underline{Q}}_t \underline{\underline{P}} = \mathrm{diag}\,(\lambda_1,..,\lambda_s,0,..,0) \; ; \quad \underline{\underline{A}}_\delta^* = \underline{\underline{P}}^{-1} \underline{\underline{A}}_\delta \underline{\underline{P}} \; . \tag{13}$$

In (12) $\underline{\underline{D}}$ is a diagonal positive matrix and $\underline{\underline{T}}$ an orthogonal one. λ_i, $i = 1,..,s$ are the real negative eigenvalues of $\underline{\underline{Q}}_t$ ordered so that $|\lambda_1| \geq |\lambda_2| \geq \dots \geq |\lambda_s| > 0$.
Let $\Lambda = \mathrm{tr}\, \underline{\underline{A}}_\delta$, $\Lambda < 0$, the unique non-zero eigenvalue of $\underline{\underline{A}}_\delta$. By the change of variables

$$\underline{z} := \underline{\underline{P}}^{-1} \underline{x} \tag{14}$$

the linear part of (1) reads

$$\dot{\underline{z}} = \mathrm{diag}\,(\lambda_1,..,\lambda_s,0,..,0)\, \underline{z} + \underline{\underline{A}}_\delta^* \underline{z} \quad . \tag{15}$$

Then, we can prove :

<u>Theorem 2</u>. Let the knot graph of the chemical network contain one cycle only of length δ for which: (a) $\phi_j^\circ > \psi_j^\circ$, $j = 1,..,\delta$; (b) the stoichiometry of any reaction of the cycle is linearly dependent on the others; then any equilibrium $\underline{c}^\circ$ such that

$$u^{1/2} < 1 + \frac{|\lambda_s|}{|\Lambda|} \tag{16}$$

is locally asymptotically stable in $\underline{c}^\circ + S$.

<u>Proof</u>. The stoichiometric subspace (Beretta and Lazzari, 1983) is

$$S = \mathrm{Im}\, \underline{\underline{\nu}} = \mathrm{span}\,\{\, \underline{q}_j : \underline{\underline{Q}}_t \underline{q}_j = \lambda_j \underline{q}_j \, , \, \lambda_j < 0 \,\} \; .$$

Thanks to hypothesis (b), $\underline{\nu}_\delta \in S$ and we can write

$$\underline{\nu}_\delta = \sum_{j=1}^{s} \beta_j \underline{q}_j \quad . \tag{17}$$

Let $\{\underline{e}_1,\dots,\underline{e}_n\}$ the standard base of R^n and $\underline{v}^*_\delta = \underline{v}_\delta \underline{\underline{P}}$ .According to (14), $\underline{e}_j = \underline{\underline{P}}^{-1} \underline{q}_j$, $j = 1,\dots,n$. Then, by (11), (13) and (17)

$$\underline{\underline{A}}^*_\delta \underline{z} = \sigma \underline{\beta} \tag{18}$$

with $\sigma \in R$, $\underline{\beta} \in R^n$ defined as

$$\sigma = \underline{v}^*_\delta \underline{z} = \sum_{k=1}^{n} v^*_{\delta k} z_k \quad ; \quad \underline{\beta} = \sum_{j=1}^{s} \beta_j \underline{e}_j \quad . \tag{19}$$

Substituting (18), (19) in (15) the cycle is described by

$$\begin{aligned} \dot{z}_j &= \lambda_j z_j + \beta_j \sigma \quad , \ j = 1,\dots,s \ , \\ \dot{z}_j &= 0 \quad , \ j = s+1,\dots,n \quad . \end{aligned} \tag{20}$$

We study the stability properties of $\underline{z} = \underline{0}$ in R^n relatively to the invariant subspace $S^* = \{ \underline{z} \in R^n : z_j = 0 \ , \ j = s+1,\dots,n \}$,obtained from S by $\underline{\underline{P}}^{-1}$. In S^* the linearized evolution equations (20) become :

$$\dot{z}_j = \lambda_j z_j + \beta_j \sigma \quad , \ j = 1,\dots,s \ , \text{with} \quad \sigma = \sum_{k=1}^{s} v^*_{\delta k} z_k \quad .$$

This is a linear system with a linear control " σ " which weights the effect of the closure branch on the asymptotically stable part due to the tree graph. The control theory (Minorsky, 1969) suggests the Liapunov function :

$$V = \frac{1}{2} \sum_{j=1}^{s} \eta_j z_j^2 + \frac{1}{2} \mu \sigma^2 \quad , \ \eta_j, \mu \in R_+ \quad . \tag{21}$$

Then,

$$\dot{V} = - \sum_{j=1}^{s} \eta_j |\lambda_j| z_j^2 + 2 \sigma \sum_{j=1}^{s} \gamma_j z_j - \mu (|\Lambda| + |\bar{\lambda}|) \sigma^2 \quad , \tag{22}$$

where $\gamma_j = \frac{1}{2} (\eta_j \beta_j + \mu \varepsilon_j v^*_{\delta j})$, $\lambda_j = \bar{\lambda} + \varepsilon_j$ $\bar{\lambda} \in R_-$, for $j = 1,\dots,s$.

Furthermore, to derive (22), we used $\Lambda = \mathrm{tr}\, \underline{\underline{A}}_\delta = \sum_{j=1}^{s} \beta_j v^*_{\delta j}$. In (22) we put $\dot{V} = W(\underline{z},\sigma)$, where $W(\underline{z},\sigma)$ is the quadratic form $W(\underline{z},\sigma) = (\underline{z},\sigma)^T \underline{\underline{\Gamma}} (\underline{z},\sigma)$ of the s+1 variables z_1, $\dots, z_s$, σ . $\underline{\underline{\Gamma}}$ is the real symmetric matrix

$$\underline{\underline{\Gamma}} = \begin{pmatrix} -\eta_1|\lambda_1| & & & & \gamma_1 \\ & \ddots & & 0 & \vdots \\ 0 & & \ddots & & \vdots \\ & & & -\eta_s|\lambda_s| & \gamma_s \\ \gamma_1 & \cdots & \cdots & \gamma_s & -\mu(|\Lambda|+|\bar{\lambda}|) \end{pmatrix} \quad . \tag{23}$$

By a theorem on the inertia of Hermitian matrices (Barnett,1971), $\underline{\underline{\Gamma}}$ is a negative definite matrix if :

$$- \mu(|\Lambda|+|\bar{\lambda}|) + \sum_{j=1}^{s} \frac{\gamma_j^2}{\eta_j |\lambda_j|} < 0 \tag{24}$$

Therefore,(24) is a sufficient condition for the negative definiteness of $\dot{V}$.Using the explicit expressions of γ_j and choosing $\eta_j = |\lambda_j|$, $j = 1,..,s$,after some algebra,it follows from (24) :

$$f\,|\underline{\beta}|\,|\underline{v}^*_\delta| < |\Lambda| + |\bar{\lambda}| \tag{25}$$

where $f = \max_\rho |\varepsilon_\rho/\lambda_\rho|$.We can set $|\bar{\lambda}| = |\lambda_s|$ and therefore it results $f< 1$.On the left of (25),by $|\cdot|$ we mean the Euclidean vector norm.Taking as subordinate matrix norm $\|\underline{\underline{A}}\| = (\max\lambda(\underline{\underline{A}}^T\underline{\underline{A}}))^{1/2}$,it follows :

$$|\underline{\beta}|\,|\underline{v}^*_\delta| = \|\underline{\underline{A}}^*_\delta\| \quad . \tag{26}$$

By the hypothesis (a) we can prove that

$$\|\underline{\underline{A}}^*_\delta\| \leq u^{1/2}\,|\Lambda| \quad , \tag{27}$$

where u ($u >1$) is the unbalancement of the cycle.It follows from (25),(26),(27) that,if

$$u^{1/2}\,|\Lambda| < |\Lambda|+|\lambda_s|$$

then (25) holds,and the asymptotic stability of the equilibrium is assured by the Liapunov function (21).This completes the proof.

To prove theor.2,we set $|\bar{\lambda}| = |\lambda_s|$ in (25) to obtain the sufficient condition (16). We observe that another choice of $|\bar{\lambda}|$ should lead to a better sufficient condition than (16).

Conclusions.

In the knot graph theory,theorems 1 and 2 offer a complete view of the local asymptotic stability criteria for positive equilibria of chemical networks with mass action Kinetics.About the validity of theorem 2,consider an open enzyme reaction system whose knot graph is a cycle of length δ ($\delta \geq 2$) ;then,the enzyme velocity v°,evaluated at any equilibrium $\underline{c}^\circ$,and the unbalancement u are related by

$$u = V_M / (V_M - v^\circ)$$

where V_M is the asymptotic maximum value of the enzyme velocity.From biochemical literature,it is known that an enzyme reaction works at a velocity $v^\circ << V_M$,that is $u \sim 1$.

Therefore,for this class of enzyme reactions theorem 2 appears as a valid tool to ascertain the asymptotic stability.

REFERENCES

Barnett,S. : Matrices in control theory. London,Van Nostrand Reinhold Company,1971.

Beretta,E.,Lazzari,C. : Asymptotic stability criteria for chemical networks.Submitted for publication to J.Math.Biol.,(1983).

Beretta,E.,Vetrano,F.,Solimano,F.,Lazzari,C. : Some results about nonlinear chemical systems represented by trees and cycles.Bull.Math.Biol.,41,641-664,(1979).

Clarke,B.L. : Stability of complex reaction networks. Adv.Chem.Phys.,Vol.XLIII, ,1-215,(1980).

Minorsky,N. : Theory of nonlinear control systems. New York,McGraw-Hill Book Company,1969.

ON EXISTENCE, UNIQUENESS AND ATTRACTIVITY OF STATIONARY SOLUTIONS TO SOME QUASILINEAR PARABOLIC SYSTEMS

R. Dal Passo
Istituto per le Applicazioni del Calcolo "M. Picone", CNR, Roma
Viale Policlinico 137, I-00161 ROMA

P. de Mottoni
Istituto di Matematica Applicata Università dell'Aquila
I-67040 POGGIO di ROIO (L'Aquila)

The purpose of the present note is twofold, namely to present some results (obtained by the authors in [3]) on existence, uniqueness and attractivity of stationary solutions of certain quasilinear systems, and to show how such results apply to some models of mathematical biology.

Consider the elliptic quasilinear problem formally defined by

$$(1)\qquad \begin{cases} \Delta\phi_i(u_1,u_2) + f_i(u_1,u_2) = 0 & \text{in } \Omega \quad i=1,2 \\ u_i = 0 & \text{in } \partial\Omega \quad i=1,2 \end{cases} ;$$

(H.1)
- i) Ω is an open bounded domain in $\mathbb{R}^n$ with $C^{2,\nu}$-boundary, $\partial\Omega$
- ii) f_i $(i=1,2)$ are real, locally Lipschitz continuous functions defined on $\mathbb{R}^+\times\mathbb{R}^+$; $f_i(0,0) = 0$ $(i=1,2)$,
- iii) ϕ_i $(i=1,2)$ are real, continuous functions on $\mathbb{R}^+\times\mathbb{R}^+$, vanishing at $(0,0)$.

Solutions to (1) will always be understood in the weak sense (cf. [1]). Along with (1); we shall consider the associated parabolic problem

$$(2)\qquad \begin{cases} \Delta\phi_i(u_1,u_2) + f_i(u_1,u_2) = \partial_t u_i & \text{in } \Omega\times(0,\infty),\ i=1,2 \\ u_i = 0 & \text{in } \partial\Omega\times(0,\infty),\ i=1,2 \\ u_i = u_{io} & \text{in } \Omega\times\{0\},\ i=1,2 . \end{cases}$$

The following (alternative) hypotheses (H.2.1) and (H.2.2) will play a central role:

(H.2.1) Φ_i depends on u_i only $(i=1,2)$; both Φ_i's satisfy:

(3) $\Phi \in C(\mathbb{R}^+)$, $\Phi(0) = 0$, Φ is increasing and either Φ or Φ^{-1} is locally Lipschitz continuous,

(H.2.2) $\left\{\begin{array}{l}\text{Both } \phi_i\text{'s are continuously differentiable, and} \\ \det \dfrac{\partial(\phi_1,\phi_2)}{\partial(u_1,u_2)} \neq 0 \text{ in } \mathbb{R}^+\times\mathbb{R}^+ \text{ ; moreover, calling } (\Phi_1,\Phi_2) \\ \text{the inverse map of } (\phi_1,\phi_2)\text{, the functions} \\ (U_1,U_2) \longrightarrow F_i(U_1,U_2) := f_i(\Phi_1(U_1,U_2),\Phi_2(U_1,U_2)) \; (i=1,2) \\ \text{are locally Lipschitz continuous on } \overline{\mathbb{R}}^+\times\overline{\mathbb{R}}^+ \text{ .}\end{array}\right.$

General Existence Results.

Existence results were established in [3] by combining monotonicity and topological methods: they basically require the knowledge of a couple of lower-upper solutions: for details, the reader is referred to Theorem 1.1 and 1.3 of [3].

Uniqueness Results.

To prove uniqueness of non-negative (non trivial solutions), use was made, in [3], of an extension of the concavity concept introduced by Krasnosel'skiĭ [4]. The main result can be stated as follows.

Assume (H.1) and either (H.2.1) or (H.2.2); assume moreover

(G.1)

i) (a-priori bound) There are two positive constants $\hat{U}_1$, $\hat{U}_2$ such that $F_1(U_1,\hat{U}_2) \leq 0$ for all $U_1 \geq \hat{U}_1$, and $F_2(\hat{U}_1,U_2) \leq 0$ for all $U_2 \geq \hat{U}_2$.

ii) (quasi-monotonicity) F_1 is nondecreasing in U_2, F_2 is nondecreasing in U_1 in S; where $S := [0,\hat{U}_1]\times[0,\hat{U}_2]$.

iii) The following quantities :

$$\liminf_{\Delta U_1\to 0} \frac{F_1(U_1+\Delta U_1,U_2)-F_1(U_1,U_2)}{\Delta U_1}, \text{ and } \liminf_{\Delta U_2\to 0} \frac{F_2(U_1,U_2+\Delta U_2)-F_2(U_1,U_2)}{\Delta U_2}$$

are bounded below in S .

and

(G.2) $F_i(tU_1,tU_2) \geq tF_i(U_1,U_2)$, i=1,2, for all $t \in (0,1)$, $(U_1,U_2)\in S$.

Then (1) has at most one non negative solution such that $\phi_1(u_1,u_2)$, $\phi_2(u_1,u_2) \in \mathcal{S} := \{(U_1,U_2)\in L^\infty(\Omega)\oplus L^\infty(\Omega);\ 0 \leq U_1(x)\leq \hat{U}_1,\ 0 \leq U_2(x) \leq \hat{U}_2\}$, having both components non zero.

Existence of Non trivial Solutions.

An important point is, of course, ensuring that the such unique solution does exist. To this end, a sufficient condition (cf. [3], Theor. 2.6) is the existence of a positive eigenvector W with negative eigenvalue λ to the problem

$$(\Delta \mathbb{I} - \omega \mathbb{I})W + \hat{G}_\omega(\varepsilon,\varepsilon)W + \lambda W = 0 \quad \text{in } \Omega$$
$$W = 0 \quad \text{in } \partial\Omega \quad ,$$

for some $\varepsilon \geq 0$, where $\hat{G}_\omega(U_1,U_2)$ is the 2x2 matrix defined by

$$[\hat{G}_\omega(U_1,U_2)]_{ij} = [G_\omega(U_1\delta_{j1},U_2\delta_{j2})]_{ij}$$

where in turn

$$[G_\omega(U_1,U_2)]_{ij} = \frac{1}{2}\,\frac{F_i(U_1,U_2)+\omega U_i}{U_j} \quad ,$$

ω being chosen "large" enough.

Attractivity Relative to Time Evolution.

Finally, in [3] "global" attractivity results for solutions to (1) (viewed as stationary solutions of (2)) have been proved in case of diagonal diffusion (i.e., assumptions (H.2.1)) and quasi-monotonicity of the f_i's, i.e. $\frac{\partial f_1}{\partial u_2} \geq 0$, $\frac{\partial f_2}{\partial u_1} \geq 0$ (or the opposite inequalities), (see Theorem 3.3 in [3]).

The results synthetized above will be applied to some models which generalize those suggested by V.Capasso and co-workers for describing the spread of certain infectious deseases [2]. For semilinear positive-feedback systems, see [5].

Such models are based on a two-dimensional differential system, describing the time course of a bacterial population in sea waters (density: u_1) and a (human) infective population in a urban community (density: u_2). The generalizations we are going to consider include nonlinear diffusion effects in the bacterial population, more precisely: in example 1, we shall consider "fast" diffusion of bacteria; in example 2, "slow" diffusion of bacteria, and in example 3, we shall study a case of crossed diffusion, in which the bacterial diffusion is enhanced by the presence of human population.

Example 1.

Consider the elliptic system

$$(4)\qquad \begin{cases} \Delta u_1^\alpha - a_{11}u_1 + a_{12}u_2 = 0 & \\ d\Delta u_2 + g(u_1) - a_{22}u_2 = 0 & \text{in } \Omega \\ u_1 = du_2 = 0 & \text{in } \partial\Omega \end{cases},$$

where

$$(\text{E}.0)\qquad \begin{cases} \text{i) } g \text{ is continuous and increasing on } \mathbb{R}^+; \\ \text{ii) } g(x)=h(x)x \text{ on } \mathbb{R}^+ \text{ with } h\in C^1(\mathbb{R}^+), \text{ h decreasing and such that} \\ \text{iii) } h(\hat{u}_1)= \dfrac{a_{11}a_{22}}{a_{12}} \text{ for some } \hat{u}_1 > 0; \\ \text{iv) } a_{11}, a_{12}, a_{22} > 0;\ d \geq 0\ ; \\ \text{v) } \Omega : \text{open bounded domain in } \mathbb{R}^n \text{ with } C^{2,\nu}\text{-boundary } \partial\Omega . \end{cases}$$

Here we assume

$$(\text{E}.1)\qquad \alpha\in(0,1).$$

Performing the change of variables $U_1 = u_1^\alpha$, $U_2 = u_2$, we re-write (4) as

$$(5)\qquad \begin{cases} \Delta U_1 - a_{11}U_1^{1/\alpha} + a_{12}U_2 = 0 & \\ d\Delta U_2 + h(U_1^{1/\alpha})U_1^{1/\alpha} - a_{22}U_2 = 0 & \text{in } \Omega \\ U_1 = U_2 = 0 & \text{in } \partial\Omega . \end{cases}$$

Putting $\hat{U}_1 = \hat{u}_1^\alpha$, $\hat{U}_2 = \dfrac{a_{11}}{a_{22}}\hat{U}_1$, and recalling the notations introduced previously, it can be easily verified that assumption (G.1) is fulfilled. To verify (G.2), observe first that, since $\alpha\in(0,1)$,

$$-a_{11}t^{1/\alpha}U_1^{1/\alpha} + a_{12}tU_2 \geq -a_{11}tU_1^{1/\alpha} + a_{12}tU_2$$

is obviously verified whenever $0 \leq U_1 \leq \hat{U}_1$, $0 \leq U_2 \leq \hat{U}_2$, $0 \leq t \leq 1$.

Thus, it remains to check that

$$h(t^{1/\alpha}U_1^{1/\alpha})t^{1/\alpha}U_1^{1/\alpha} - a_{22}tU_2 \geq th(U_1^{1/\alpha})U_1^{1/\alpha} - a_{22}tU_2$$

for t, U_1, U_2 as above. But this amounts to verifying the non-increasing character of $h(U_1^{1/\alpha})U_1^{1/\alpha-1}$, which results in turn from the fact that $(1-\alpha)h(u_1) + u_1h'(u_1) < 0$ in $(0,\hat{U}_1]$. Using the uniqueness theorem, we then conclude that (4) admits at most one non trivial non-negative solution with (u_1^α,u_2) having values in $[0,\hat{U}_1]\times[0,\hat{U}_2]$.

To ensure its existence, consider the linear eigenvalue problem

$$(6)\qquad \begin{cases} \Delta U_1 - \omega U_1 + (\omega - a_{11}\varepsilon^{1/\alpha - 1})U_1 + a_{12}U_2 + \lambda U_1 = 0 & \\ & \text{in } \Omega \\ d\Delta U_2 - \omega U_2 + h(\varepsilon^{1/\alpha})\varepsilon^{1/\alpha - 1}U_1 + (\omega - a_{22})U_2 + \lambda U_2 = 0 & \\ U_1 = U_2 = 0 & \text{in } \partial\Omega \ . \end{cases}$$

Let us look for an eigenfunction of the form $(\sigma_1\Phi_o, \sigma_2\Phi_o)$, $\sigma_i > 0$ $(i=1,2)$, Φ_o being the positive eigenfunction of Δ on Ω : $\Delta\Phi_o + \lambda_o\Phi_o = 0$ in Ω , $\Phi_o = 0$ on $\partial\Omega$, $\max\Phi_o = 1$. Substituting in (6) we see that, to have a nontrivial solution, λ is determinated by

$$\det\begin{pmatrix} -\lambda_o + \lambda - a_{11}\varepsilon^{1/\alpha - 1} & a_{12} \\ h(\varepsilon^{1/\alpha})\varepsilon^{1/\alpha - 1} & -\lambda_o d + \lambda - a_{22} \end{pmatrix} = 0 \ .$$

According to the previously recalled result, we see, after some calculation, that $\lambda < 0$ if

$$(7)\qquad d\,\lambda_o^2 + a_{22}\lambda_o < a_{12}h(\varepsilon^{1/\alpha})\varepsilon^{1/\alpha - 1} \ .$$

Thus we can conclude by saying that problem (4) with (E.0), (E.1) has a unique non trivial non-negative solution if (7) holds for some $\varepsilon > 0$; such solution is ("globally") attractive relative to the associated evolution system because of the general arguments recalled above.

Example 2.

Consider again system (4), with assumptions (E.0), but suppose now

$$(E.2)\qquad \alpha > 1 \ .$$

We shall distinguish the cases $d = 0$ and $d > 0$. In case $d = 0$ a uniqueness result is obtained in a straightforward way : indeed by simple algebra, we obtain from (4) a single equation for u_1 :

$$\Delta u_1^\alpha - a_{11}u_1 + \frac{a_{12}}{a_{22}} g(u_1) = 0 \quad \text{in} \quad \Omega,\ u_1 = 0 \quad \text{in } \partial\Omega \ ,$$

or, equivalently, putting $u_1^\alpha = U_1$:

$$(8)\qquad \Delta U_1 - a_{11}U_1^{1/\alpha} + \frac{a_{12}}{a_{22}} g(U_1^{1/\alpha}) = 0 \quad \text{in} \quad \Omega,\ U_1 = 0 \quad \text{in} \quad \partial\Omega.$$

A sufficient condition for uniqueness is therefore

$$(1-\alpha)(-a_{11} + \frac{a_{12}}{a_{22}} h(U_1^{1/\alpha})) + \frac{a_{12}}{a_{22}} U_1^{1/\alpha} h'(U_1^{1/\alpha}) < 0 \quad \text{in} \quad (0,\hat{U}_1] \ .$$

Since $\alpha > 1$, and $h' < 0$, we see that (E.0 - iii) ensures the uniqueness of the non-negative non trivial solution: on the other hand from (6) it is easily seen that the same assumption ensures as well the existence of such solution. Concluding, if $d = 0$, (E.2), (E.0) ensure the existence of a unique non trivial stationary solution. Its ("global") attractivity relative to the associated evolution problem results from the general results mentioned above. In case $d > 0$, our methods for proving uniqueness fail. We can, nevertheless, guarantee the existence of non trivial solutions by looking at lower-solutions of the form $(\sigma_1\Phi_o,\sigma_2\Phi_o)$, Φ_o as above. After some calculation, it turns out that such quantity is indeed a lower-solution for $0<\sigma_1<\varrho_1$, $0<\sigma_2<\varrho_2$ if there are positive constants ϱ_1,ϱ_2 such that $\varrho_2>\frac{a_{11}}{a_{12}}\varrho_1$, $h(\varrho_1)>(a_{22}+\lambda_o d)\frac{a_{11}}{a_{12}}$.

Example 3.

Consider the system

$$(9)\qquad \begin{cases} \Delta(u_1(1+\beta u_2)) - a_{11}u_1 + a_{12}u_2 = 0,\ g(u_1) - a_{22}u_2 = 0 & \text{in } \Omega \\ u_1 = u_2 = 0 & \text{in } \partial\Omega \end{cases},$$

where β is a real positive constant, and the remaining quantities are as in (E.0). Then (9) is equivalent to the elliptic equation

$$\Delta(u_1(1+\frac{\beta}{a_{22}}g(u_1))) - a_{11}u_1 + \frac{a_{12}}{a_{22}}g(u_1) = 0 \quad \text{in } \Omega,\ u_1 = 0 \text{ in } \partial\Omega .$$

Let $\Phi(u_1) := u_1(1+\frac{\beta}{a_{22}}g(u_1))$ and perform the change of variable $U_1 = \Phi(u_1)$ (note that $\Phi\in C^1(\mathbb{R}^+)$ and is increasing). We obtain

$$\Delta U_1 - a_{11}\Phi^{-1}(U_1) + \frac{a_{12}}{a_{22}}g(\Phi^{-1}(U_1)) = 0 \quad \text{in } \Omega \ ,\ U_1 = 0 \text{ in } \partial\Omega ,$$

whose analysis can be carried out with the usual tools, as in Example 2 (d=0) above. The uniqueness is ensured if

$$(10)\qquad (\frac{a_{12}}{a_{22}}h(\Phi^{-1}(U_1)) - a_{11})(\frac{\Phi^{-1}(U_1)}{U_1})' + \frac{a_{12}}{a_{22}}h'(\Phi^{-1}(U_1))(\Phi^{-1}(U_1))'\frac{\Phi^{-1}(U_1)}{U_1} < 0$$

in $[0,\hat{U}_1]$, where $\hat{U}_1 = \Phi(\hat{u}_1)$. Recalling that $\Phi^{-1\prime} > 0$, $h^{-1}(\Phi^{-1}(\cdot)) < 0$ and observing that $(\frac{\Phi^{-1}(U_1)}{U_1})' < 0$, we see that (10) holds as a consequence of (E.0-iii) - an assumption which, as easily checked, also guarantees the existence of a non trivial solution. Concluding, problem (9) (with (E.0)) admits a unique non-negative non trivial solution.

References

[1] D.G.Aronson,M.G.Crandall,L.A.Peletier: Stabilization of solutions of a degenerate nonlinear diffusion problem, Nonlinear Anal. TMA,6,(1982) 1001-1022.

[2] V.Capasso,S.L.Paveri-Fontana: A mathematical model for the 1973 cholera epidemic in the european mediterranean region, Rev. Epidém.et Santé Publ.27,(1979) 121-132.

[3] R.Dal Passo,P.de Mottoni: Some existence, uniqueness and stability results for a class of semilinear degenerate elliptic systems,Boll. U.M.I. SezC. "Anal.Funz.e Appl."1984.

[4] M.A.Krasnosel'skiĭ: Positive solutions of operator equations, P.Noordhoff Groningen,The Netherlands,(1964).

[5] R.H.Martin, Asymptotic stability analysis and critical points for nonlinear quasimonotone systems, J.Differential Equations 30,(1978) 391-423.

ASYMPTOTIC ESTIMATES FOR PRINCIPAL EIGENVALUES

Gabriella Del Grosso, Federico Marchetti

Dipartimento di Matematica
Istituto Matematico "G. Castelnuovo"

Città Universitaria - I-00100 ROMA, Italy

In the present paper some results recently obtained by the authors on the asymptotic behaviour of the principal eigenvalue of certain operators are summarized and discussed.

1. Introduction

The following situation is frequently encountered in applied mathematics.

A particle diffuses in a region G until it is trapped when it hits the boundary ∂G and one is interested in "computing" this (random) hitting time.

Typically ∂G will be the union of the boundaries of a number of small "holes" that have been delated from a simple connected manifold M to build G.

In mathematical physics M could be $\mathbf{R}^3$ or a bounded domain in $\mathbf{R}^3$ with Neumann boundary conditions, representing a (thermically or electrically) conducting medium and the holes would represent insulating portions [1], [2].

In biology similar models are encountered e.g. when studying the motion of ligands until their capture by receptors located on the surface of lymphocyte cells [3], [4], [5].

In each case one would like to compute the distribution of the hitting time of the diffusion modelling the motion of the particle. In many applications this distribution is tacitly assumed to be exponential, so that the mean hitting time is used to compute the forward rate of the process (free particle $\rightarrow$ bound particle), treated as obeying a law of mass action (cfr. [3], [4]). In other words one considers a continuous-time Markov chain with two states ("free" and "bound") in the limit of the law of large numbers [6], [7], [8]. This type of approximation is justified in the limit of very small trapping regions in compact M. In this limit, as is intuitive, the distribution of the hitting time is approximately exponential with parameter given by the principal eigenvalue of the generator of the stopped diffusion. On the other hand the situation when M is not compact is much less clearly understood (cfr. [9], [10]).

This discussion justifies the interest in giving estimates for the principal eigenvalue of elliptic 2nd order operators. It should also be mentioned that the problem is classical in geometry, where deletion of small "holes" is one of the basic tools in the surgery of manifolds (cfr. [11], [12]).

We conclude by formally stating the mathematical problem.

Let M be an n-dimensional Riemannian manifold and L a 2nd order elliptic differential operator, usually the Laplace-Beltrami operator. Let O_i M $i=1,2,\ldots$ be points uniformly far apart and B_i^ε be a geodesic ball of radius ε centered in O_i (ε is small enough so that no two balls overlap). Define $M \; = M \bigcup_i B_i^\varepsilon$ and consider L with homogeneous boundary conditions on ∂M_ε, whose principal eigenvalue we denote by λ_ε. The problem is to find an asymptotic estimate for λ_ε as $\varepsilon \downarrow 0$.

2. The Case of a Model

The problem posed in section 1 is rather hard in its full generality. In order to tackle the problem it is thus necessary to start with useful special cases.

Definition: A Riemannian compact n-manifold M is a compact model if there exists p,q M and global polar coordinates around p in M {q}, (r,θ), such that the metric can be written as

$$ds^2 = dr^2 + \Psi^2(r)d\theta^2 \qquad 0 < r < R$$

where

$$\Psi(0) = \Psi(R) = 0$$

$$\Psi(r) > 0 \qquad 0 < r < R$$

$$\Psi'(r) \to 1 \qquad r \to 0$$

$$\Psi'(r) \to -1 \qquad r \to R$$

and $d\theta^2$ is a metric on the (n-1)-dimensional unit sphere.

Remark: this definition is an adaptation to the compact case of a notion introduced in [14] and could be formulated in intrinsic terms. However the phrasing used emphasizes the relevant part as far as the following computations go.

In this metric the Laplace-Beltrami operator separates in a radial part. Assuming now that there is a single hole of radius ε centered in p, the eigenvalue problem becomes

$$Lu + \lambda u = \frac{1}{\Psi(r)}\frac{d}{dr}\Psi(r)\frac{du}{dr} + \lambda u = 0 \qquad (2)$$

$$u(\varepsilon) = 0$$

In this case we have

Theorem 1 ([15])

Let M be a compact model with respect to p, $M_\varepsilon = \{(r,\theta): R \geq r \geq \varepsilon\}$ and λ_ε the principal eigenvalue of L in M_ε with homogeneous Dirichlet boundary conditions. Then the following estimate holds as $\varepsilon \downarrow 0$:

$$\lambda_\varepsilon = \frac{\mathrm{vol}(S^{n-1})}{\mathrm{vol}(M)}\phi_n(\varepsilon) + o(\phi_n(\varepsilon)) \qquad (3)$$

where

$$\phi_n(x) = \begin{cases} (\log \frac{1}{x})^{-1} & n = 2 \\ (n-2)x^{n-2} & n > 2 \end{cases}$$

Proof (sketch): one proves an upper and lower bound on λ_ε and their ratio turns out to be $1 + o(\phi_n(\varepsilon))$.

The upper bound is found on the classical Rayleigh-Rietz principle

$$\lambda_\varepsilon = \underset{\substack{f \in L^2(M;\, \Psi dr d\theta) \cap D(L) \\ \|f\|_2 = 1}}{f}(df, df)$$

by choosing a suitable test function,

$$f(r,\theta) = 1 - \frac{\phi_n(r)}{\phi_n(\varepsilon)}$$

The lower bound is found by an upper estimate of the bounded solution of

$$Lv = -1 \qquad (4)$$

$$v(\partial M) =$$

which is known to be equal to the mean exit time, $E_r \tau_\varepsilon$. From this one deduces a bound on λ_ε because of the following general lemma.

Lemma [13]

Let M be a smooth paracompact n-dimensional manifold M and L a smooth 2nd order elliptic differential operator governing a regular diffusion. Let $M' \subset M$ be a domain with smooth boundary and τ the first exit time from M'. Denote

$$-\lambda_0 := \limsup_{t\to\infty} t^{-1} \log P_x(\tau > t)$$

the "principal eigenvalue" of L with zero boundary conditions, on $\partial M'$ (cfr. [15]). Then

$$\lambda_0^{-1} \leq \sup_{x\, M'} E_x \tau \tag{5}$$

(where $\lambda_0 = 0$ if the r.h.s. is infinite).

Theorem 1 can be partly generalized to a general manifold with one hole of radius ε. The estimates turn out to be

$$c_2 \phi_n(\varepsilon) \leq \lambda_\varepsilon \leq c_1 \phi_n(\varepsilon) \qquad \varepsilon \downarrow 0$$

for some positive constants c_i.

The idea behind the proof is to reason as in theorem 1. However since (4) cannot be solved explicitly, one can estimate $E_x \tau$ by comparing the actual process between two suitable elementary return processes on a small neighbourhood of the hole. For details we refer again to [13].

3. Manifolds with Many Holes

The solution of sec 2 is made possible by the symmetry of the operator and the boundary condition. When more than one hole is present, even special symmetries in the operator will not be enough, except in very special situations, such as a medal with two holes centered in p and q.

In [16] a method has been developed that can be helpful in such general cases. This has been applied to two specific situations.

First of all note that the diffusion generated on M by L can be constructed, as a solution to the martingale problem, in a locally finite atlas by adapting the gluing techniques used in [17].

The details of this construction, given a locally finite open atlas with precompact charts U_i for M are the following.

Consider the compactification M* of M (if M was compact to start with, we simply add an isolated point $*$) and the measurable space $\Omega \equiv \{x(\cdot)\in C(0,\infty);\ M^*): x(s) = *, x(t) = *, t \geq s\}$ with its Borel sigma-field. Further let us assume that the stopped martingale problem for L is well-posed for any initial point in each chart: U_i.

Let us define

$$m(x) = \min\{i: x\in U_i \text{ and } \operatorname{dist}(x, \partial U_i) \geq \operatorname{dist}(x, \partial U_j) \forall j\}$$

Remark: this choice is simply one of different possible markovian choices.

Define the sequence

$$T_0 = 0 \quad T_k = \inf\{t > T_{k-1}: x(t) \notin U_{m(x(T_{k-1}))}\}$$

and let $Q^i_{s,x}$ be the unique solution starting from (s,x) to the stopped martingale problem for L on U_i. By a simple adaptation of the aforementioned theorem it is easy to prove the following

Theorem

There is a unique measure P_x on $(\Omega$ such that $P_x((x(0) = x) = 1$, P_x solves the martingale problem for L on M and with the following properties

(i) $P_x((x(t)) = * \; ; \; t \geq Z) = 1$

(ii) $P_x|_{B_Z}$ is the inductive limit of measures p_x^k

(iii) $P_x^k = P_x^{k-1}$ on $B_{T_{K-1}}$

(iv) $P_x^k(\cdot|B_{T_{k-1}}) = Q^{i_k}_{T_{k-1},x(T_{k-1})}(\cdot)$

where $Z = \sup T_i$ is the explosion time and $i_k = m(x(T_{k-1}))$.

Note that finite stopping time S will satisfy $S = S \cap Z$. Thus for any a.s. finite stopping time we have the following

Corollary

Let T be a Markov time for $x(\cdot)$. Then

$$E^{P_x} T \quad \lim_{k\to\infty} E^{P_x} (T \cap T_k) = \lim_{k\to\infty} E^{P_x^k} (T \cap T_k) \tag{6}$$

This corollary is useful because the rhs is, at least in principle, computable in an explicit way from local problems on each U_i, enabling a localization of problem (4) and a subsequent passage to the limit.

To get a feeling of the type of formula one can get in this way, consider any of the following two cases:

Case 1: M is compact, there are two disks B_i and two charts U_i $i = 1,2$;

Case 2: For general M and any number of disks each chart contains exactly one disk and, in each chart, we can choose coordinates such that in each deleted chart L and the traces of the boundaries of the neighbouring charts are all equal.

Remark: these cases avoid the additional burden of keeping track of the random index of the chart each particular path will be visiting: in case 1 each path visits alternatively the only two charts available, in case 2 we can identify the problem as a diffusion in a bounded domain in R^n with an elementary return boundary condition without holding time. It is intuitive that this last case provides a more or less rough approximation for the general case, see also [16] for the use of this kind of approximation.

With these provisions we can prove the following

Theorem 2

Let u_i be the solution of

$$\begin{aligned} Lu_i &= -1 \qquad \text{in } U_i \setminus B_i \\ u_i|_{\partial B_i} &= 0 \\ u_i|_{\partial U_i} &= 0 \end{aligned} \tag{7}$$

and assume there exist kernels K^i such that the solution of

$$\begin{aligned} Lv &= 0 \qquad \text{in } U_i \setminus B_i \\ v|_{\partial B_i} &= 0 \\ v|_{\partial U_i} &= f \end{aligned} \tag{8}$$

are given by

$$v(x) = \int_{\partial U_i} K^i(x,y) f(y) ds_i(y) \tag{9}$$

Then the following expansion holds

$$E^{P_x} T = u_{m(x)}(x) + \sum_{n=2}^{\infty} \int_{\partial U_{m(n)}} ds_{m(x)}(y_1) K^{m(x)}(x,y_1)$$

$$\int_{\partial U_{m(x(T_1))}} ds_{m(x(T_1))}(y_2)\, K^{m(x(T_1))}(y_1,y_2) \ldots$$

$$\int_{\partial U_{m(x(T_{n-2}))}} ds_{m(x(T_{n-2}))}(y_n) K^{m(x(T_{n-2}))}\, y(y_{n-1},y_n)$$

$$u_{m(x(T_{n-1}))}(y_n)$$

$$(y_0 = x) \tag{10}$$

As an applicated consider, as an example of case 1, M=S with two holes centered in points at angular distance α. Using (5) a straightforward estimate of (10) yields

$$\lambda_0 \geq \frac{\log \operatorname{tg} \alpha/2}{\log \cos \alpha/2} (\log \frac{1}{\varepsilon})^{-1} + o((\log \frac{1}{\varepsilon})^{-1}) \tag{11}$$

As an example of case 2 consider $M = R^2$ with countably many holes whose center lie on a regular lattice of equilateral triangles of side a. Following the same procedure an estimate is

$$\lambda_0 \geq \frac{4 \log \sqrt{2 - \sqrt{3}}}{a^2} (\log \frac{1}{\varepsilon})^{-1} + o((\log \frac{1}{\varepsilon})^{-1}) \tag{12}$$

A check via the Rayleigh-Rietz principle suggests that these estimates are not asymptotically sharp. In the case of the sphere (11) could be improved by a finer estimate of (10).

The situation appears to be more delicate, as the non compactness of M does not guarantee that (5) is asymptotically sharp (cfr. [9], [10]).

References

[1] J. Rauch, M. Taylor, "Potential and Scattering Theory on Wildly Perturbed Domains", J. Func. Analysis 18 (1975) 27-59.

[2] G. Papanicolau, S.R.S. Varadhan "Diffusions in Regions with Many Small Holes" in B. Grigelionis (ed.) "Stochastic Differential Systems". Lecture Notes in Control and Information Sciences No. 25, Springer-Verlag, Berling, New York.

[3] H.C. Berg, E.M. Purcell "Physics of Chemoreception". Biophys. J. 20 (1977) 193-219.

[4] C. De Lisi, "The Biophysics of Ligand Receptor Interactions". Quart. Rev. Biophys. 13, 2.(1980) 201-230.

[5] G. Del Grosso, A. Gerardi, F. Marchetti, "A Diffusion Model for Patch Formation on Cellular Surfaces". Appl. Math. Optimiz. 7 (1981) 125-135.

[6] T.G. Kurtz, "Approximation of Population Processes". S.I.A.M., Philadelphia 1981.

[7] L. Arnold, M. Theodosopulu, "Deterministic Limit of the Stochastic Model of Chemical Reactions with Diffusion". Adv. Appl. Prob. 1980.

[8] C. De Lisi, G. Del Grosso, F. Marchetti, "A Theory of Measurement Error and its Implications for Spatial and Temporal Gradient Sensing During Chemotaxis". Cell. Biophys. (to appear).

[9] F. Marchetti, "Asymptotic Exponentiality of Exit Times" (to appear)
[10] E. Orlandi, "An Integral Method for Solving Certain Asymptotic Problems" (to appear)
[11] M. Schiffer, D.C. Spencer, "Functionals of Finite Riemannian Manifolds", Princeton University Press, Princeton 1954.
[12] C.A. Swanson, "Asymptotic Variational Formulae for Eigenvalues", Canad. Math. Bull. 6 (1963) 15-25.
[13] G. Del Grosso, F. Marchetti, "Asymptotic Estimates for the Principal Eigenvalue of the Laplacian in a Geodesic Ball, Appl. Math. Optimiz. (to appear)
[14] R.E. Greene, M. Wu, "Function Theory on Manifolds Which Possess a Pole". Lecture Notes in Mathematics No. 699, Springer-Verlag, Berlin, Heidelberg, New York 1979.
[15] M.D. Dousker, S.R.S. Varadhan, "On the Principal Eigenvalue of Elliptic Second Order Operators" in Proc. Intern. Symp. SDE Kyoto 1976, Wiley, New York 1978, p. 41-48.
[16] G. Del Grosso, F. Marchetti, "Principal Eigenvalues and Exit Times for Diffusions on Manifolds" (to appear)
[17] D.W. Stroock, S.R.S. Varadhan, "Multidimensional Diffusion Processes", Springer-Verlag, Berling, Heidelberg, New York 1979.

REMARKS ON THE CONNECTION BETWEEN POSITIVE FEEDBACK AND INSTABILITY IN REGARD TO MODEL SELECTION IN POPULATION BIOLOGY

J. Eisenfeld

Department of Mathematics
The University of Texas at Arlington
Arlington, Texas 76019

Abstract. The connection between positive feedback and instability has been widely observed in network theory. In this paper several points are made regarding this connection from the point of view of developing a criteria for determining which models from a class of possible models are unstable or admit an exchange of stabilities. The criteria may be used as a preliminary test of model design.

Introduction. Mathematical models in population biology are usually developed from a qualitative scheme of species interactions. This scheme, formalized by Odum [1], may be described as follows. To each ordered pair of species (i,j) one associates a sign +,-, or 0, depending on whether the population of species i is increased, decreased, or is uneffected by the presence of species j. The sign structure of the interactions may be described by a diagram in which an arrow is drawn from (vertex) j to i if the sign of the interaction (i,j) is not zero. The arrow is labeled + or - according to whether the sign of (i,j) is + or -. In some instances, e.g., in immunology[2], the system is so large and complex that it is difficult to ascertain from the available data the signs of some of the interactions. In such cases one must select from an (often large) number of possibilities. A criteria is presented here that could considerably narrow the range of possible qualitative schemes. Applications of this criteria to control circuits arising in immunology will appear elsewhere[3].

Having ascertained the qualitative scheme, one then proceeds to the formulation of the system of rate equations, which is assumed here to have the general form $dx/dt = F(x)$. The steady states are the solutions of $F(x) = 0$. Many types of phenomena are explained in terms of stable steady states. For instance, memory may be considered as a transition to a higher stable steady state [2]. Consider a hypothetical steady state, s. The linearized equations about s are $dz/dt = Jz$, where J is the Jacobian of F at s and z(t) represents a small perturbation from s. In order to maintain consistency with the model diagram one constructs the equations in such a manner that the (signed) graph of J coincides with the diagram. In other words, the sign of the Jacobian element J(i,j) should be the same as that of the interaction (i,j). Moreover, to insure stability of the steady state, J is required to be stable, i.e., all of its eigenvalues should have negative real parts. This brings up the following interesting question. What are the necessary and sufficient conditions on a graph, G, such that there exists at least one stable matrix whose graph is G?

The above question is answered, at least in part by the theory of qualitative stability [4] (also see May's discussion [5]). This theory gives necessary and sufficient conditions on a graph, G, such that every matrix having G as its graph is stable. These conditions are quite restrictive; for instance, cycles of length greater than two are prohibited. Moreover, considering that the diagram is based on biological properties which are independent of a particular steady state, the qualitative-stability conditions would require stability at every steady state, which is unlikely when there are more than one steady states.

In this paper the problem of the existence of stable steady states is approached from the negative side of the issue by obtaining sufficient conditions on a graph such that all matrices having that graph are not stable. The practical value of such an approach would be to rule out all qualitative schemes which are not consistent with the premise that there should exist at least one stable steady state. The sufficient conditions are given an intuitive interpretation and several additional remarks are made leading to criteria on the sign structure of species interactions to permit both stable and unstable steady states.

It should be noted that the literature on chemical and biological applications of graph theory is enormous[6]. The work of Clarke[7], which attempts to represent the Routh-Hurwitz stability criteria graphically for chemical networks, is particularly aligned in spirit to this paper.

2. Main results and discussion. Given a graph, G, recall that a cycle in G is a sequence of edges, (r,i), (q,r),...,(j,k),(i,j), where the vertices, i,r,...,k,j, are distinct. The length of the cycle is the number of edges. The sign of the cycle is negative if the number of negative edges is odd; otherwise the sign is positive. A generalized cycle, hereafter called a g-cycle, is a non-empty set of cycles such that any pair in the set are disjoint in the sense that they do not share a vertex in common. The length of a g-cycle is the combined length of the cycles in the set. The sign of a g-cycle is positive if the number of positive cycles is odd; otherwise the sign is negative. With this terminology in hand, one may now state the following result.

Theorem 1. Let J be an nxn matrix with graph, G. A sufficient condition for J to be unstable (i.e., to have an eigenvalue with a nonnegative real part) is that there is at least one strong component H of G and at least one integer k, which is $\leq$ the number of vertices in H, such that H does not contain at least one negative g-cycle of length k. Moreover, if there is at least one positive g-cycle of length k in H then J has an eigenvalue with a positive real part.

Remark 1. Theorem 1, proved in Section 3, is explained intuitively as follows. First, observe that a positive g-cycle, without the presence of a negative g-cycle to offset it, causes instability by virtue of an eigenvalue with a positive real part. This establishes a positive g-cycle as a destabilizing agent and a negative g-cycle a stabilizing agent. The connection between a positive cycle (also called a positive feedback loop and a deviation-amplifier) with instability has been discussed by several authors (e.g., [5,8-9]). In the case of two species A and B, each having a positive effect on the other (mutualism) the connection is clear. A perturbation of A above its steady-state value causes an increase in B, which, in turn, further amplifies A. If the initial disturbance decreases the population of A then A helps B less thereby allowing B to decrease which, in turn, helps A less allowing A to decrease further. A similar argument applies to a cycle with more than two species, e.g., A helps B, B suppresses C, C helps D, and D suppresses A. An increase in A causes an increase in B, which causes a decrease in C, which causes a decrease in D, which causes an increase in A, so that the net effect is to amplify the disturbance. Similar reasoning explains the connection between a negative cycle and perturbation-suppression. These amplification and suppression effects are more difficult to comprehend in the presence of more than one cycles. One might consider that a pair of positive cycles might cause amplifications in opposite directions thereby cancelling each other. This may explain why a g-cycle with an even number of positive cycles (its sign is negative) is construed in the theorem as a deviation -suppressor. Still, these simple-minded intuitive arguments leave much to be explained and they can be misleading (Remark 2).

Remark 2. Further points are to be discussed with the aid of the matrices whose sign patterns are:

$$J1 = \begin{pmatrix} - & + & 0 \\ + & 0 & - \\ 0 & + & 0 \end{pmatrix}, \quad J2 = \begin{pmatrix} - & + & + \\ + & 0 & - \\ 0 & + & 0 \end{pmatrix}, \quad J3 = \begin{pmatrix} - & 0 & + \\ + & 0 & - \\ 0 & + & 0 \end{pmatrix}.$$

It is easily seen by drawing the graph G1 of J1 that G1 is strongly connected and has negative g-cycles of lengths 1, 2 and 3. However, application of the Routh-Hurwitz shows immediately that J1 has an eigenvalue with a positive real part irrespective of the magnitudes of its elements. Therefore, the sufficiency condition of Theorem 1 is not sharp. Another point emerges by consideration of the graph G2 of J2. The cycles of G2 are those of G1 plus an additional positive cycle of length 3. Based on the intuitive argument given above one would expect the instability property of J1 to carry over to J2 but J2 can be stable. In fact, setting the magnitudes of the (2,3) and the (3,2) elements of J2 equal to v and the magnitudes of the remaining nonzero elements equal to u (keeping the signs fixed) yields a stable matrix when $v > u > 0$. Thus, in some cases, a positive cycle may have a stabilizing effect.

Remark 3. It should be kept in mind that a matrix with some eigenvalues having zero real parts and the rest having negative real parts is considered unstable even though there are examples of asymptotically stable states having Jacobians with this property. However, such a steady state would be structurally unstable (see the discussion by May [5]).

Remark 4. Several results on the stability properties of a matrix based only on the structure of its graph may be summarized. Following [4], two matrices are said to be sign-equivalent if they have the same sign structure, or equivalently, the same graph. It is interesting to classify graphs based on the stability properties of their associated sign-equivalent class of matrices. The graph is classified as S, (for stable) or U (for unstable), or E (for exchange of stabilities) depending on whether the A matrices in the associated sign-equivalence class are all stable, all unstable, or both stable and unstable matrices are admitted. From the results in [4] it is easy to show that a graph is in S iff :

(i) There are no cycles which are either positive or of length > 2.

(ii) There is at least one loop (a cycle of unit length) in each strong component.

(iii) There is a g-cycle that contains all the vertices.

A graph is in U if the conditions of Theorem 1 are satisfied but not conversely. The graphs in E are seemingly the most interesting for modeling complex systems. Only graphs in E permit possible Hopf bifurcations. To determine whether a graph is in E the following remarks may be helpful. Consider two graphs, G1 and G2, and suppose that G1 reduces to G2 upon removing a subset of its edges. Then if G2 admits stable matrices so does G1. This observation derives from the fact that the eigenvalues of a matrix depend continuously on the elements. For instance, a graph not satisfying condition (i) (above) and having a negative loop at each vertex is in E as is any graph, with 3 vertices, which contains the edges in the graph of J3 (above).

3. Proof of Theorem 1. Given a real nxn matrix J, with graph G and characteristic polynomial $p(\lambda)$, let ck denote the coefficient of the (n-k)th power of $p(\lambda)$. It is well known and easy to prove (either from the Routh-Hurwitz criteria or from the representation of ck in terms of the roots of $p(\lambda)$) that all the ck are positive (resp. nonnegative) if all the roots have negative (resp. nonpositive) real parts. In other words, if there is at least one ck ≥ 0 then J is unstable, and the strict inequality implies that J has an eigenvalue with a positive real part.

With each cycle σ in G, with edges (r,i),(q,r),...,(j,k),(i,j), write the product of the matrix elements associated with these edges as $h(\sigma) = J(r,i) J(q,r)\ldots J(j,k)J(i,j)$. Corresponding to each g-cycle in G, $\sigma = \sigma 1 \sigma 2 \ldots \sigma v$, where v is the number of cycles forming σ, set $W(\sigma) = (-1)\exp(v)h(\sigma 1)h(\sigma 2)\ldots h(\sigma v)$. It can be shown that ck is the sum of all W(σ) taken over the set of all g-cycles of length k, and moreover the sign of any W(σ) is the negative of the sign σ (see[7]). It follows that if there are no negative g-cycles of length k then ck $\leq$ 0 with the existence of positive g-cycles implying strict inequality.

In the case of a graph with more than one strong components, the set of eigenvalues of the graph is the union of the sets of eigenvalues of the strong components[9]. The desired conclusion is obtained by applying the relationship between cycles and coefficients and between coefficients and eigenvalues to each strong component.

REFERENCES

1. Odum, E.P., Fundamentals of Ecology, W.B. Saunders, London, 1953.

2. Hiernaux, J., Some remarks on the stability of the Idiotypic Network, Immunochem. 14(1977), 733-739.

3. Eisenfeld, J., and DeLisi, C., On conditions for qualitative instability of regulatory circuits with applications to immunological control circuits, in Biomedical Applications of Mathematics and Computers, J. Eisenfeld and C. DeLisi (Ed.s) North-Holland, New York, to appear.

4. Quirk, J.P., and Rupport, R.M., Qualitative Economics and the Stability of Equilibrium, Rev. Econ. Studies, 32(1965), 311-326.

5. May, R.M., Stability and Complexity in Ecosystems, Princeton University Press, Princeton, New Jersey, 1973.

6. King, R.B., Proceedings of the Symposium on Chemical Applications of Topology and Graph Theory, to appear.

7. Clarke, B.L., Graph theoretic approach to the stability analysis of steady state chemical reaction networks, J. Chem. Phys., 60(1974) 1482-1492.

8. Maruyama, M., The Second Cybernetics: Deviation amplifying mutual causal processes, Amer. Scientist, 51(1963) ,164-179.

9. Roberts, F.S., Discrete Mathematical Models, Prentice Hall, Englewood Cliffs, New Jersey, 1976.

Acknowledgment. The author expresses his gratitude to the Laboratory of Mathematical Biology, National Cancer Institute, National Institutes of Health, Bethesda, Maryland, for their support of this work while he was on leave there.

LYAPUNOV METHODS FOR A WIDE CLASS OF STOCHASTIC MODELS IN BIOLOGY

L. Ferrante(*), G. Koch(**)

Abstract: In this paper we show how Lyapunov functions may be used to investigate qualitative properties (recurrence, existence of invariant sets, etc.) for a class of stochastic diffusion models which describe interaction between two populations.

1. Introduction

In this work we show how Lyapunov functions with essentially the same structure may be built to investigate recurrence properties of a large class of two dimensional stochastic models. This class may be obtained as an extension of the popular Lotka-Volterra model and received much attention both in the deterministic and stochastic versions (see for example [1-8]), since it is able to describe interacting populations, chemical reactions, genetic evolution, and many other phenomena in the life sciences. The aim of the paper is to show that if one is ready to be satisfied with sufficient conditions (the only ones Lyapunov methods are expected to yield) then this technique features a high degree of flexibility. Indeed, in applicative fields, such as population biology, it usually happens that a model is intended to represent a phenomenon for which only a rough description is available [9]. Then Lyapunov-type techniques allow to get results based only on some general properties of the drift and diffusion coefficient. Moreover, quite often these coefficients do not satisfy global Lipschitz conditions which are usually assumed to guarantee uniqueness of the solution of the stochastic differential equations. In this context, it is remarkable that the use of Lyapunov functions allow to overcome this problem and, at the same time, to determine physically meaningful invariant subsets for the solution trajectories.

Let $x(t)$, $t \in R^+$, be a homogeneous diffusion process with values in R^n, which is the solution of the stochastic differential equation:

$$dx = \alpha(x)dt + \beta(x)dw \tag{1.1}$$

where $\beta(x)$ is the vector of drift components, $\alpha(x)$ is the matrix of diffusion components and w is a p-dimensional Wiener process with independent components.

The infinitesimal generator L of $x(t)$, as well as its recurrence properties and invariant measures, are defined in the usual way (see [10]). We shall repeatedly exploit the following well known theorems due to Has'minskij [11]:

Theorem 1.1. Let $K_1 \subset K_2 \subset \dots$ be a sequence of compact regions such that $\cup_j K_j = G$, G open in R^n. Let α, β be lipschitzian and bounded in each K_j. Let V be a $C_2(G)$ function such that:

a) $V \geq 0$, $LV \leq cV$, $c > 0$ in G

b) $\inf\{V(x): x \in G/K_j\} \to +\infty$, $j \to +\infty$

Let $P(x(t_o) \in G) = 1$, and $x(t_o)$ independent of $w(t) - w(t_o)$, $\forall t \geq t_o$. Then equation (2.1) admits a unique solution, $\forall t \geq t_o$, which is a (regular) homogeneous diffusion process and $P(x(t) \in G, \forall t \geq t_o) = 1$.

Theorem 1.2. Let U be bounded with regular boundary. Let x be a regular process, solution of (1.1) with α, β lipschitzian on compact sets. Let L be strictly uniformly elliptic in a neighbourhood $U_1 \supseteq \bar{U}$ with regular boundary. Let V be a $C_2(U^c)$ function such that:

$V \geq 0$, $LV \leq -d$, $d > 0$ in U^c

Then the process x is positively recurrent with respect to U_1 and U_1^c. Moreover x admits a unique invariant probability measure μ which is ergodic.

(*) Istituto di Medicina sperimentale e clinica, University of Ancona, Italy.

(**) Dipertimento di Matematica, Istituto Matematico "G. Castelnuovo", University of Roma, Italy.

2. **A general class of models.**

Let us consider the following class of models in R^2:

$$dx_1=x_1(-b_1(x_1)-b_2(x_2))dt+x_1(a_1(x)dw_1+a_2(x)dw_2)$$
$$dx_2=x_2(+b_3(x_1)-b_4(x_2))dt+x_2(a_3(x)dw_1+a_4(x)dw_2) \tag{2.1}$$

where w_1, w_2 are Wiener processes (which are taken independent with no loss of generality). Both drift and diffusion coefficients vanish on the coordinate axes; this is a necessary condition to guarantee that the closure $\bar{Q}$ of the open positive quadrant $Q=\{x: x_1>0, x_2>0\}$ is an invariant set for the solution of (2.1). Thus x_1, x_2 may have the physical meaning of (non negative) densities of two interacting species. All coefficients a_i, b_i are assumed to be Lipschitz continuous on compact sets in $\bar{Q}$. Moreover, it is assumed that there exists a $\bar{x}\in Q$ such that:

$$+b_1(\bar{x}_1)+b_2(\bar{x}_2)=0$$
$$+b_3(\bar{x}_1)-b_4(\bar{x}_2)=0 \tag{2.2}$$

Therefore the related deterministic model:

$$dx_1=x_1(-b_1(x_1)-b_2(x_2))dt$$
$$dx_2=x_2(+b_3(x_1)-b_4(x_2))dt \tag{2.3}$$

has (at least) an equilibrium point $\bar{x}$ in Q. Furthermore, all b_i's are taken as strictly increasing functions of their arguments. On one side, this guarantees that $\bar{x}$ is a (locally) asymptotically stable equilibrium point for (2.3). On the other side this is not a severe restriction since it simply amounts to say that each species incorporates a self saturation effect (crowding, lack of food, etc.), while the interaction between them has an increasing effect on x_2 ("predator") and a decreasing effect on x_1 ("prey"). An example which does not share this structure is discussed in [10]. With no loss of generality we take $b_2(\bar{x}_2)>0$, $b_3(\bar{x}_1)>0$. Furthermore, with slight restriction b_2 and b_3 are assumed to be $C_1(Q)$ and such that $x_1b_3'(x_1)$, $x_2b_2'(x_2)$ be continuous in $\bar{Q}$. The diffusion coefficients a_i account for the type of randomness (whether external or internal) one thinks of (see the introduction and [6]). In any case, they are assumed to be continuous in $\bar{Q}$, and such that L be strictly uniformly elliptic on compact sets in Q.

We now show how qualitative properties of (2.1) may be investigated by using just one properly selected Lyapunov function V:

$$V(x) = c + \int_{\bar{x}_1}^{x_1} b_3(z)/z\,dz - b_3(\bar{x}_1)\ln x_1 + \int_{\bar{x}_2}^{x_2} b_2(z)/z\,dz - b_2(\bar{x}_2)\ln x_2 \tag{2.4}$$

where c is a constant to be selected later.

<u>Proposition 2.1</u>: Let us consider the model (2.1) with the above mentioned assumptions. If there exist $M>0$, $0<\delta<1$ such that:

$$\frac{A_1(x)x_1b'_3(x_1)}{(b_3(x_1)-b_3(\bar{x}_1))(b_1(x_1)-b_1(\bar{x}_1))} \le \delta,\ x_1>M \tag{2.5}$$

$$\frac{A_2(x)x_2b'_2(x_2)}{(b_2(x_2)-b_2(\bar{x}_2))(b_4(x_2)-b_4(\bar{x}_2))} \le \delta,\ x_2>M \tag{2.6}$$

where $A_i(x)= 1/2(a_{2i-1}^2(x)+a_{2i}^2(x))$, $i=1,2$, then exists $k>0$ such that:

$$A_1(x)\le k,\quad x_1<M$$
$$A_2(x)\le k,\quad x_2<M$$

implies

$$LV\le -d \quad \text{in } Q-U$$

for a $d>0$ and U compact in Q.

<u>Proof.</u> The proposition is proved in [10] by a direct approach.

<u>Remark 1</u>. With reference to the case in which b_2, b_3 are polynomials; the above assumptions essentially amount to say that the diffusion coefficients grow at most linearly within a finite distance M from the axes, while when both x_1, x_2 diverge

they are allowed to grow at most as $x_1[b_1(x_1)]^{1/2}$, $x_2[b_4(x_2)]^{1/2}$. This condition is indeed a slight restriction in the case of external randomness, while it is usually satisfied in the internal randomness case [6].

Remark 2. Once this proposition is proved, condition a) of Th. 1.1 is fulfilled as soon as we choose in (2.4) $c=\max\{\sup LV, 0\}$. The condition b) is fulfilled as soon as we choose $\{K_j\}$ to be a sequence of compact growing regions such that $\cup_j K_j=Q$. Thus conditions of Th. 1.1 and Th. 1.2 are satisfied, and we conclude that for every $x(t_0)\in Q$, the process x is the unique (up to equivalence) solution of the system (2.1), Q is an invariant set for it, and x is positively recurrent with respect to Q. Moreover it admits a unique invariant ergodic probability measure.

When the initial condition is $x_1(0)=0$, $x_2(0)=\hat{x}_2>0$, both drift and diffusion terms of the first equation in the system (2.1) are zero: this implies the positive x_2-axis to be invariant for the process and the system reduces to the equation:

$$dx_2 = x_2(b_3(0) - b_4(x_2))dt + x_2(a_3(0,x_2)dw_1 + a_4(0,x_2)dw_2) \tag{2.7}$$

Proposition 3.2: If the initial condition of the system (2.1) belongs to the positive x_2-axis and

$$b_3(\theta) - b_4(0) - A_2(0,0) > 0 \tag{2.8}$$

then, the existence of $M>0, 0<\delta<1$ such that:

$$\frac{A_2(0,x_2)x_2b_2'(x_2)}{(b_2(x_2)-b_2(\bar{x}_2))(b_4(x_2)-b_4(\bar{x}_2))} \leq \delta, \quad x_2>M \tag{2.9}$$

implies that the process which moves on the positive x_2-axis is positive recurrent and admits a unique invariant probability measure which is ergodic.

Proof. Again by direct check (see [10]).

Similar results may obviously be found if the initial condition is on the x_1-axis.

As well known [12,13] condition (2.8) is necessary to guarantee existence of an invariant measure; indeed if inequality is reversed in (2.8) the process (2.1) along the positive x_2-axis turns out to be transient. For a non-Lyapunov proof of this fact, see [10]. Notice that in (2.9), $\bar{x}_2$ may be substituted by any equilibrium point along the x_2-axis. The existence of at least one such point is implied by (2.2) and (2.8).

Clearly, sufficient conditions for recurrence along each axis do not involve the behaviour along the same axis of the diffusion coefficient of the other variable. In [12] a conjecture is formulated to infer recurrence in Q from the behaviour of all coefficients along the two axes.

3. Some particular cases.

We consider now an extension of the well-known mathematical model for a prey-predator biological system in which the limited growth takes into account various causes referred to as "social phenomena" [11]:

$$\begin{aligned} dx_1 &= x_1(k_1 - b_{11}x_1^{n_1} - b_{12}x_2^{n_2})dt + x_1(a_1(x)dw_1 + a_2(x)dw_2) \\ dx_2 &= x_2(k_2 + b_{21}x_1^{n_1} - b_{22}x_2^{n_2})dt + x_2(a_3(x)dw_1 + a_4(x)dw_2) \end{aligned} \tag{3.1}$$

with $b_{ij}>0$, $n_i>0$, $i,j=1, 2$ and $k_i>0$, $i=1,2$ or alternatively with $k_1>0>k_2$, in which case we also assume $k_1/b_{11} > -k_2/b_{21}$. The coefficients $a_i(x)$, $i=1,\dots,4$, are assumed to be lipschitzian on compact sets and such that $A_i(x)\neq 0$, $i=1,2$, in $\bar{Q}$. Model (3.1) is easily verified to satisfy all the assumptions introduced for the model (2.1) and the function (2.4) is now:

$$V = \sum_{i=1}^{2} B_i \left(\frac{x_i^{n_i}}{\bar{x}_i^{n_i}} - \ln \frac{x_i^{n_i}}{\bar{x}_i^{n_i}} - 1\right) \tag{3.2}$$

where B_1, B_2 are suitable constants.

Under the further hypothesis of constant coefficients a_i, the model was already considered in [8]. Conditions (2.5), (2.6) become in this case:

$$\frac{A_i(x)n_i x_i^{n_i}}{b_{ii}(x_i^{n_i} - \bar{x}_i^{n_i})^2} \leq \delta, \text{for } x_i > M$$

Thus, proposition 2.1 guarantees uniqueness and regularity of the solution x(t) of the system (4.3), as well as existence and uniqueness of invariant measure, as soon as the A_i's are bound by a sufficiently small quantity near the axis, and grow at most as $x_i^{n_i}$ for $x_i \to +\infty$.

If we assume $A_i(x)=A_i$, i=1,2, constant, then sufficient conditions to apply Proposition 2.1 are

$$\begin{aligned} A_2 &< -\frac{(n_1-1)^2 b_{21}}{4b_{11}\, b_{12}\bar{x}_2^{n_2}} A_1^2 - \frac{n_1 b_{21} \bar{x}_1^{n_1}}{b_{12}\bar{x}_2^{n_2}} A_1 + b_{22}\bar{x}_2^{n_2} \\ A_1 &< -\frac{(n_2-1)^2 b_{12}}{4b_{22} b_{21} \bar{x}_1^{n_1}} A_2^2 - \frac{n_2 b_{12} \bar{x}_2^{n_2}}{b_{21}\bar{x}_2^{n_2}} A_2 + b_{11}\bar{x}_1^{n_1} \end{aligned} \qquad (3.3)$$

Inequalities (3.3) define a domain bounded by two parabolas, which degenerate in two straight lines for $n_1=n_2=1$. In [8], necessary and sufficient conditions on A_i, i=1,2 are found, which of course are weaker than (3.3), by exploiting a detailed analysis of suitably defined stopping times. These conditions can be shown (see [10]) to be equivalent to the conjecture of [12], which then, in this case, is demonstrated to be valid. Stability along x_2-axis can be investigated by mean of Proposition 2.2.

In the model (4.1) we assume that "social phenomena" in the i-th population affect in the same way (except possibly for the sign) the growth rate of both populations. In other instances this might be substituted by the assumption that "social phenomena" in the $x_1(x_2)$ population are more effective in the interaction with $x_2(x_1)$ than in the growth of $x_1(x_2)$ itself.

For example, in [14], the following model is proposed:

$$\begin{aligned} \dot{x}_1 &= x_1(k_1 - b_{11}x_1 - b_{12}x_2^2) \\ \dot{x}_2 &= x_2(k_2 + b_{21}x_1^2 - b_{22}x_2) \end{aligned} \qquad (3.4)$$

where $b_{ij} > 0$ and $k_i > 0$, i,j=1,2 (in which case we also assume $k_1/b_{12} > (k_2/b_{22})^2$) or alternatively $k_1 > 0 > k_2$ (in which case we assume $k_2/b_{12} > -(k_1/b_{11})^2$).

A stochastic version of (3.4) can be obtained by the usual procedure [6]. In this case equation (2.4) becomes:

$$V(x) = \frac{b_{21}}{2}(x_1^2 - \bar{x}_1^2) - b_{21}\bar{x}_1^2 \ln\frac{x_1}{\bar{x}_1} + \frac{b_{12}}{2}(x_2^2 - \bar{x}_2^2) - b_{12}\bar{x}_2^2 \ln\frac{x_2}{\bar{x}_2} \qquad (3.5)$$

and stochastic stability in guaranteed by Prop. 2.1 if the A_i's are bound by a sufficiently small quantity near the axis and grow at most linearly as $x_i \to \infty$.

REFERENCES

(1) A. Rescigno, I.W. Richardson, 'The Deterministic Theory of Population Dynamics' in: R. Rosen (ed) 'Foundation of mathematical biology' Vol. 3° Academic Press, New York, (1973).

(2) R.M. May, 'Stability and complexity in model ecosystems'. Princeton University Press, (1973).

(3) D. Ludwig, 'Stochastic population theories'. Lecture Notes in Biomathematics,

v. 3, Springer Verlag, (1974).

(4) L. Arnold and R. Lefever (eds.), 'Stochastic non linear systems in Physics Chemistry and Biology', Springer Series in Synergetics (Berlin-Heidelberg-New York, 1981).

(5) J.F. Crow, M. Kimura, 'An introduction to population genetics theory'. Harper and Row, (1970).

(6) G. Kōch, 'Stochastic models in Biology', Systems analysis - modeling-simulation, to appear.

(7) M. Barra; G. Del Grosso, A. Gerardi, G. Koch, F. Marchetti, 'Some basic properties of stochastic population models'. Lecture Notes in Biomathematics, v.32 155-164, Springer Verlag, (1978).

(8) H. Kesten, Y. Ogura, 'Recurrence properties of Lotka-Volterra models with random fluctuations'. J. Math. Soc. Japan 33, 335-366, (1981).

(9) R.J. Gratton, D.R. Appleton, M.K. Alwiswasy, 'The measurement of tumor growth rates'. In: Biomathematics and Cell Kinetics, A.J. Valleron, P.D.M. McDonald eds., Elsevier, (1978).

(10) L. Ferrante, G. Koch, 'An application of Lyapunov techniques to stochastic population models'. Rep., Dept of Mathematics, Ist. "G. Castelnuovo" University of Rome, (1983).

(11) Has'miskij, Stability of Systems of Differential Equations under Stochastic Perturbations of their parameters.

(12) M. Turelli, J.H. Gillespie, 'Conditions for the Existence of Stationary Densities for Some Two-Dimensional Diffusion Process with Applications in Population Biology', Theor. Pop. Biol. 17, 167-189, (1980).

(13) Y.V. Prohorov, Y.A. Rozanov 'Probability theory', Springer-Verlag, New York, (1969).

(14) G.E. Hutchinson, 'A note on the theory on competition between two social species", Ecology 28, 319-321 (1947).

EXISTENCE AND UNIQUENESS FOR A NON LINEAR CAUCHY PROBLEM OF HYPERBOLIC TYPE

Lucia GASTALDI - Franco TOMARELLI
Dipartimento di Matematica dell'Università
27100 PAVIA - ITALY

In this note we collect some results (see [3], [4]) concerning a Cauchy problem for a non linear and non local partial differential equation arising in a model of the contracting mechanism of the cardiac muscle. We refer to a model that has been introduced by physiologists and mathematicians of the University of Pavia, starting from the sliding filament theory of Huxley (see [1], [2], [5], [6]).

In the mathematical formulation, the unknown $u(x,t)$ represents the density of chemical links between myosin and actin in a sarcomere; $\gamma(t)$, $f(x)$, $g(x)$, $\phi(x)$ are given functions, which measure respectively the activation, the aptitude to form the links at the distance x, the aptitude to destroy the links and the density of the links at the initial time.

We have the following mathematical formulation.

Let T, γ, f, g, ϕ, Q and q be given, such that

(1) $T,q,Q \in \mathbb{R}$, $\quad 0 < q < Q \quad T > 0$

(2) $\gamma \in C^0([0,T])$, $\quad f,\ g,\ \phi \in C^{1,\alpha}(\mathbb{R})$, $\quad 0 < \alpha < 1$

(3) ϕ and f have compact support

(4) γ, f, g, ϕ are nonnegative functions

(5) $0 \leq \phi(x) \leq 1,\ x \in \mathbb{R}; \qquad -1 < \int_{\mathbb{R}} x\phi(x)dx < Q$

(6) $\lim_{|x|\to+\infty} g(x) = +\infty.$

PROBLEM 1 Find $u(x,t)$ in $C^{0,1}(\mathbb{R}\times[0,T])$ such that

(7) $u(x,t)$ has compact support in $\mathbb{R} \quad \forall\, t$

(8)
$$\frac{\partial}{\partial t}u(x,t) + \frac{d}{dt}\ln\frac{q-(\int_{\mathbb{R}} xu(x,t)dx - Q + q)^+}{(1+\int_{\mathbb{R}} xu(x,t)dx)} \cdot \frac{\partial}{\partial x}u(x,t) =$$
$$= \gamma(t)f(x) - (\gamma(t)f(x)+g(x))u(x,t) \qquad x \in \mathbb{R},\ \text{a.e. in } [0,T]$$

(9) $- 1 < \int_{\mathbb{R}} xu(x,t)dx < Q \qquad \forall\, t \in [0,T]$

(10) $u(x,0) = \phi(x) \qquad x \in \mathbb{R}.$

In (8), $(.)^+$ is the positive part, and the derivatives must be intended in the distributional sense.

The equation (8) takes two different experimental situations into account:

a) the so-called <u>isometric contraction</u>, characterized by the inequality

(11) $$\int_{\mathbb{R}} xu(x,t)dx \leq Q-q$$

this means that the length of the sarcomere remains constant;

b) the so-called <u>isotonic contraction</u>, characterized by the inequality

(12) $$\int_{\mathbb{R}} xu(x,t)dx \geq Q-q$$

this occurs when the force acting on the sarcomere is constant.

Actually the contraction of a cardiac muscle is a sequence of both isometric and isotonic contractions.

The subset of $[0,T]$ where the solution is isotonic is an unknown too. Hence problem 1 may be considered as a free-boundary problem.

In [2], the existence and the uniqueness of the solution of the only isometric problem have been proved; in [7] Torelli studied the existence of the solution of problem 1 with null initial density.

The main question concerning problem 1 is the uniqueness of the solution: since the nonlinearity in the equation (8) is discontinuous, it is not possible to apply the usual techniques in order to get the local uniqueness near the free boundary. We report here some results concerning problem 1 (for the details we refer to [3] and [4]).

<u>**THEOREM 1**</u>**. Under the assumptions (1)-(5), problem 1 has at least one solution $u(x,t)$ and any solution satisfies**

(13) $$0 \leq u(x,t) \leq 1 \qquad \forall\ (x,t) \in \mathbb{R} \times [0,T].$$

<u>**THEOREM 2**</u> **If γ, f, g, ϕ, T, Q and q satisfy (1)-(5) and γ is piecewise analitic, then problem 1 has only one solution.**

We give an outline of the proofs.

We set:

(14) $$H(x,t) = \gamma(t)f(x) + g(x)$$

For every v in $C^0([0,T])$ we define

(15) $$U_v(x,t) = \phi(x+v(0)-v(t)) \exp(-\int_0^t H(x+v(\tau)-v(t)\tau)d\tau) + $$
$$+ \int_0^t \gamma(s)f(x+v(s)-v(t)) \exp(-\int_s^t H(x+v(\tau)-v(t),\tau)d\tau)\, ds$$

(16) $$\Xi_v(t) = \int_{\mathbb{R}} x[\gamma(t)f(x)(1-U_v(x,t)) - g(x)U_v(x,t)]\, dx$$

(17) $$z_0 = \ln \frac{q - (\int_{\mathbb{R}} x\phi(x)dx - Q + q)^+}{q(1 + \int_{\mathbb{R}} x\phi(x)dx)}$$

(18) $$R(y) = \ln \frac{q - (y+q-Q)^+}{q(1+y)} \qquad -1 < y < Q$$

(19) $$\zeta(s) = -1 + \frac{q + \exp(-s) - (\exp(-s)-1-Q+q)^+}{q \exp(s) + 1} \qquad s \in \mathbb{R}$$

We have that

(20) $\zeta(R(y)) = y,\ \forall\, y \in (-1,Q);\qquad R(\zeta(s)) = s,\ \forall\, s \in \mathbb{R}$

The functions R and ζ are locally Lipschitz continuous. So they have classical derivatives almost everywhere in their domains of definition.

We set

$$\zeta'(s) = \frac{[1-\mathcal{H}(\exp(-s)-1-Q+q)]\,(q+\exp(-s))+[q+\exp(-s)-(\exp(-s)-1-Q+q)^+]\,q\exp(s)}{(\,q \exp(s) + 1\,)^2} \qquad \forall\, s \in \mathbb{R}$$

where $\mathcal{H}(s)$ is the Heaviside function, i.e.

$\mathcal{H}(s) = 0$ if $s \leq 0$, $\mathcal{H}(s) = 1$ if $s>0$.

We remark that $\zeta'(s)$ is defined for every $s \in \mathbb{R}$ and that it is equal almost everywhere to the derivative of ζ in the sense of distributions ($\frac{d}{ds}\zeta(s) = \zeta'(s)$ a.e. in $\mathbb{R}$).

In order to prove theorems 1 and 2 it is useful to introduce the following problem .

PROBLEM 2. Given γ, f, g, ϕ, T, Q and q satisfying (1)-(5), find $z \in C^{0,1}([0,T])$ s.t.

(21) $z(0) = z_0$

(22) $$\frac{d}{dt} z(t) = \frac{\Xi_z(t)}{\zeta'(z(t)) - \int_{\mathbb{R}} U_z(x,t)\, dx} \qquad \text{a.e. in } [0,T]$$

The problems 1 and 2 are equivalent in the sense that: if z(t) solves problem 2, then $U_z(x,t)$ given by (15) is a solution of problem 1, and, if u(x,t) solves problem 1, then $z(t) = R(\int_{\mathbb{R}} xu(x,t)dx)$ is a solution of problem 2.

By smoothing suitably the positive part, we approximate problems 1 and 2 by problems 1_n and 2_n respectively. By contraddiction we get an a priori estimate of $\max |z_n(t)|$, where z_n is a solution of problem 2_n. Then we obtain the existence of a solution of problem 2_n and of problem 1_n, by applying the following fixed point theorem.

<u>THEOREM</u> (Browder-Potter). Let X be a normed space, E a non-empty convex and closed subset of X, and F a map from $[0,1]xE$ in X, such that: a) F is continuous; b) $F([0,1]xE)$ is a relatively compact subset of X; c) $F(\lambda,v)\neq v$ $\forall v \in \partial E$ $\forall \lambda \in [0,1]$; d) $F(\{0\}x\partial E) \subset E$. Then there is $w \in E$ such that $w = F(1,w)$.

Getting the limit, as $n \to +\infty$, of suitable subsequences $\{u_{n_k}\}$, $\{z_{n_k}\}$ of solutions of problems 1_{n_k} and 2_{n_k} we obtain a solution of problem 1 and 2.

Let us consider now the question of the uniqueness of the solution. We have already noticed that the subset of $[0,T]$, where the contraction is isometric (or respectively isotonic) is an unknown of the problem. Anyway in every interval $[t_0,t_1]$ contained in such a region the solution of problem 1 satisfies the unilateral bound (11) (or respectively (12)) and a problem, the so-called isometric-problem (or respectively isotonic-problem), which is analogous to problem 1 with the obvious modification of the equation (8).
Both the isometric and isotonic problems have at most one solution. Therefore, if we know that in a fixed interval $[t_0,t_1]$ the contraction is isotonic (resp. isometric), that is, it satisfies (11) (resp. (12)), then we get the local uniqueness. But, in general, we cannot know whether $\int_{\mathbb{R}} xu(x,t)dx$ is less or greater than the critical value Q-q in any interval.
We notice that the uniqueness can be studied locally. So it is enough to prove this property for any admissible Cauchy datum ϕ in a neighbourhood of the origin.
If $\int_{\mathbb{R}} x\phi(x)dx$ is different from Q-q the solution is forced to be isometric (or isotonic) in same neighbourhood by a continuity argument.
Let us consider now the more interesting case, corresponding to $\int_{\mathbb{R}} x\,\phi(x)dx = Q-q$. In order to get the local uniqueness in this case, we have to prove that neither $\int xu(x,t)dx$ may oscillate infinitely

many times around the critical value, nor the solution may bifurcate. In the case of critical Cauchy datum, for any pair of related solutions u and z (respectively of problems 1 and 2), we show that $u(x+z(t)-z_0,t)$ and $z(t)$ have an asymptotic expansion near $t=0$, which depends only on the data. This fact entails the local uniqueness.

More precisely, let us assume that γ, f, g, ϕ, Q and q satisfy (1)-(5), γ is analitic and $\int_{\mathbb{R}} x\ \phi(x)dx = Q-q$.

We define

$$(23)\begin{cases} A_0(x) = \phi(x) \\ A_n(x) = f(x)\left[\gamma^{(n-1)}(0) - \sum_{j=0}^{n-1} \binom{n-1}{j}\gamma^{(j)}(0)A_{n-j-1}(x)\right]-g(x)A_{n-1}(x) \\ \qquad \forall\ n \geq 1 \end{cases}$$

Then, only two cases are possible:

I) If there is $N \in \mathbb{N}$ such that

$$\int_{\mathbb{R}} xA_n(x)dx = 0 \qquad n = 1, 2, \dots, N-1 \quad \text{and} \quad \int_{\mathbb{R}} xA_N(x)dx \neq 0$$

then

$$(24) \qquad z(t) - z_0 = o(t^{N-1})$$

$$(25) \qquad u(x+z(t)-z_0,t) = \sum_{n=0}^{N-1} A_n(x)\frac{t^n}{n!} = o_x(t^{N-1}).$$

II) If $\int_{\mathbb{R}} xA_n(x)dx = 0 \qquad \forall\ n \in \mathbb{N}$

then (24) and (25) hold for every $N \in \mathbb{N}$.

Here $o_x(t^n)$ is an infinitesimal of greater order than t^n uniformly in x.

In the first case, substituting the expansion of $u(x+z(t)-z_0,t)$ in (22), we get that $\left(\frac{d}{dt}\int_{\mathbb{R}} xu(x,t)dx\right)\Big|_{t=0}$ is different from zero, and has a fixed sign in a right neighbourhood of $t=0$ whatever u and z are. Then u enters in only one region either isometric or isotonic.

In case II we prove that there is a positive constant τ such that the series $\sum_{n=0}^{\infty} A_n(x)\ \frac{t^n}{n!}$ converges absolutely and uniformly in $\mathbb{R}\times[0,\tau]$ and that every solution of problem 2 is identically equal to z_0. Therefore the solution of problem 1 is unique.

REFERENCES

[1] Capelo A., Comincioli V., Minelli R., Poggesi C., Reggiani C., Ricciardi L., "Study and parameters identification of a rheological model for excised quiescent cardiac muscle", J. Biomechanics 14, 1-11, (1981).

[2] Comincioli V., Torelli A., "Mathematical aspects of the cross-bridge mechanism in muscle contraction", to appear on J. of Non-linear Analysis, Theory, Method and Applications.

[3] Gastaldi L., Tomarelli F., "A non linear hyperbolic Cauchy problem arising in the dynamic of cardiac muscle", to appear on Pubbl. N. 340 I.A.N. of C.N.R., Pavia (1983).

[4] Gastaldi L., Tomarelli F., "A uniqueness result for a nonlinear hyperbolic equation" to appear on Annali di Matematica Pura e Applicata.

[5] Huxley A.F., "Muscle structure and theories of contraction", Progr. Biophys. 7, 255-318, (1957).

[6] Minelli R., Comincioli V., Poggesi C., Reggiani C., Ricciardi L., "Mathematical models for isolated resting and active cardiac muscle", Progetto HUSPI 6, 105-130, (1981).

[7] Torelli A., "A non linear hyperbolic equation related to the dynamic of cardiac muscle", to appear on Rend. Sem. Lisbona (1982).

CATASTROPHE THEORY IN BIOLOGY

P. T. Saunders,
Department of Mathematics,
Queen Elizabeth College,
University of London, UK.

In view of the long and fruitful partnership between mathematics and physics, it was only natural that the first applications of mathematics in biology should take theoretical physics as a model. Indeed, Lotka (1924) entitled his pioneering work *Elements of Physical* (not, as in the 1956 reprint, *Mathematical) Biology,* claiming as his intention the 'application of physical principles and methods in the contemplation of biological systems'.

Now this was the obvious way to begin, and the physics-led approach has certainly contributed a great deal to our understanding of biology. By itself, however, it is not enough. It does not provide us with the means of tackling many essentially biological (as distinct from biophysical or biochemical) problems. Above all, it is not well adapted to the study of the highly complex systems which are the chief subject matter of biology and the other 'soft' sciences.

An important step towards putting this right has been the appearance of catastrophe theory (Thom, 1972). This has caused a considerable furor, but one that has concerned neither the mathematical foundations of the theory nor its applications to physics. The controversy has been about the way it has been used in biology, which suggests that the real point at issue is the way mathematics in general is to be employed in biological studies. Hence the aim of this article is not just to give some examples of catastrophe theory in biology (and so dispel the common misconception that no such examples exist) but also to touch on some wider issues as well. We begin by considering four different applications, different not only in their subject matter but also in the way in which the theory is used. While the descriptions are intended to give an adequate idea of how catastrophe theory enters into the problem, they are of necessity only outlines, and the reader is strongly urged to consult the original articles for more detail.

One of the earliest applications is Zeeman's demonstration that when a frontier forms in a previously undifferentiated region, it is unlikely to appear first in its final position. Instead, it will move as it forms.

The basic idea is quite simple. Consider a region of tissue, and, as is usual in such studies, suppose that the state of a cell is determined by the concentration, x say, of some substance which we may call a morphogen. Two variables are clearly of importance in specifying the value of x; these are s, the coordinate of the cell in a direction orthogonal to the frontier, and t, the time. We also make the standard assumption that the cells are at all times at or near biochemical equilibrium. Then the observation that at a certain time t_0 the variation of x with s is continuous but that at a later time t_1 it is discontinuous leads to the inference that a cusp catastrophe is involved, with x as the state variable and s and t as control variables.

The next step is the crucial one. Zeeman argues that genericity (i.e. repeatability of experiments) demands that the axis of the cusp in the control space cannot be taken parallel to the t-axis. This leads to the prediction of motion, and to a good deal of controversy as well. It is possible to defend Zeeman's original statement that all frontiers formed in this way *must* move, but it is more satisfactory to interpret the result simply as saying that it is likely that they will (Zeeman, 1978; see also Saunders, 1980). As this is contrary to what we would expect, it is a non-trivial prediction. What is more, there is evidence that it actually happens. Zeeman (1978) cites examples of experiments suggested by his work which did confirm the idea that a forming frontier is likely to move. And, as he points out, the result enables us to account for certain phenomena in development without the need for hypothetical diffusing morphogens.

It is also interesting to compare the above with a recent model proposed by Lewis *et al* (1977) for frontier formation. They suggest the following equation for the production of a gene product g:

$$\frac{dg}{dt} = k_1 S + \frac{k_2 g^2}{k_3 + g^2} - k_4 g$$

The idea is that the signal substance, S, initiates the production of g from zero concentration. Lewis and his co-workers found that for values of S between 0 and a threshold value S_c the product g will stabilize at a relatively low value, but that above S_c the only possible equilibrium concentration is much higher. There will therefore be a discontinuity in g, i.e. a boundary.

This has a familiar ring to anyone accustomed to catastrophe theory. Even the condition for equilibrium, $dg/dt = 0$, leads to a cubic equation for g, as one would expect. We note, however, that there is only a single control variable, S, and that it might be worth considering what significance if any the missing one has.

What is more important for the present discussion is that the model possesses an interesting feature not mentioned by Lewis *et al*. They suggest that the necessary gradient in S could be established by diffusion. If this is so, then at each point in the tissue the value of S will rise from zero to its final value as diffusion progresses. Now the threshold value S_c will obviously be reached at different times at different locations. Consequently, unless the production of g is somehow inhibited until the gradient has been set up, or the time scale for the g reaction is very much longer than that for the diffusion of S, the boundary between high and low equilibrium values of g will move as it forms. Catastrophe theory made it possible to predict this without postulating a particular model, and by the same token, the prediction is more robust because it does not stand or fall with in the chosen model.

In ecology it is generally assumed that the specific growth rate of a predator is roughly proportional to the number of prey available. Dent *et al* (1976) found, however, that in their microbial predator-prey system the predator's specific growth rate, λ, would remain constant for quite long periods of time, during which the prey density, H, could change by a factor of as much as a hundred. It would then change abruptly to a different value, which again would be maintained for a long time despite large variation in H. This observation of sudden jumps led Bazin & Saunders (1978, 1979) to interpret the system in terms of a catastrophe with λ as the state variable. They made the usual assumption that H was a control variable, but because it was clear that a cusp was the simplest catastrophe that could fit the data a second control variable was required, and they chose the time, t.

With this choice of variables, a plot of H against t becomes a diagram of the control space with the control trajectory already drawn. By marking on it the points at which the sudden changes in λ occur, it is possible to sketch the bifurcation set (Fig. 1). The picture is, however, unsatisfactory, because even allowing for the uncertainty in the shape of the bifurcation set it is clear that the jumps occur as the trajectory enters the cusp, not as it leaves it. It is possible to account for such behaviour, but only by postulating a mechanism which seems much more complicated than one would expect.

There was, however, some indication that this particular predator, the amoeba *Dictyostelium discoideum*, might be responding not to H but to the prey:predator ratio, H/P. So Bazin & Saunders repeated the procedure described above, but with H/P replac-

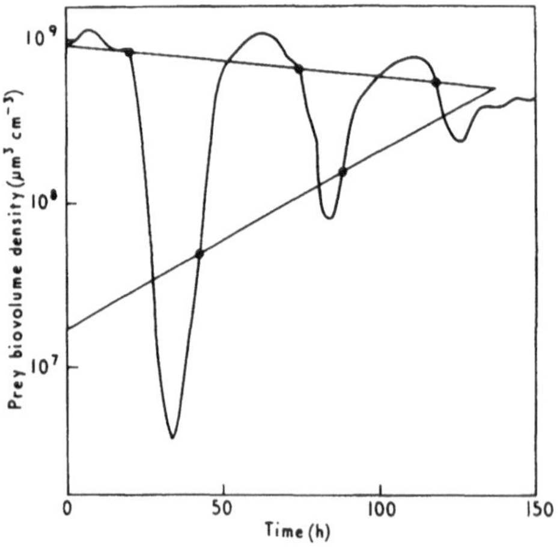

Fig. 1

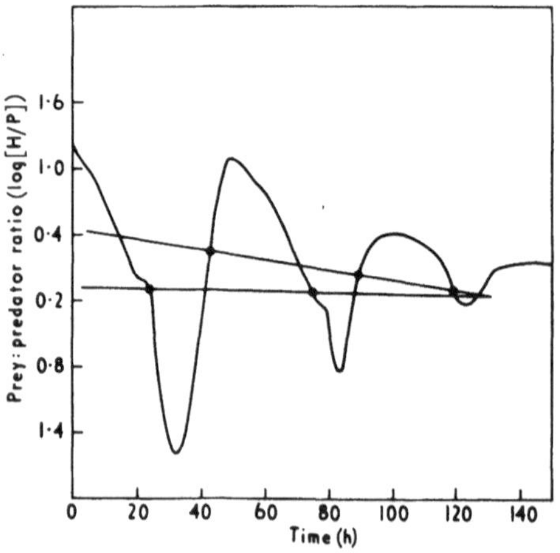

Fig. 2

ing H as a control variable (Fig. 2). They found that the configuration of jumps was now consistent with the perfect delay convention. This implies that a comparatively simple mechanism can be involved. Accordingly they drew the inference that it is indeed H/P and not H to which the amoebae respond.

Now it is known that the prey organism in this system secretes folic acid, and that this is an attractant for *D. discoideum.* The folic acid concentration could thus serve as a measure of H, but it can only serve as a measure of H/P if the amoebae themselves degrade it. In that case the folic acid concentration will increase with H and decrease with P, as required. Bazin & Saunders (1978) therefore predicted that these amoebae degrade folic acid. Independently, and almost simultaneously, Pan & Wurster (1978) reported that they do, commenting that they were unable to suggest a reason for this.

Thus catastrophe theory led to a firm (and, as it happens, confirmed) prediction. And while no mechanism has so far been proposed for the switch in λ, it should be much easier to make progress in that direction now that we know what the correct control variables are. This illustrates one of the advantages of catastrophe theory: it can allow us to solve problems in stages, and not have to come up with the solution in one fell swoop.

Seif (1979) has applied catastrophe theory to the study of thyrotropic response in humans. He begins from the observation that patients with primary hyperthyroidism who have been subjected to treatment show a bimodal distribution of responsiveness to protirelin, whereas the responsiveness of those with primary hypothyroidism is unimodal. He shows that if the state of the patient is defined by two variables, which he denotes by $H(o)$ and P and which represent hormone concentrations, then the points in the $H(o)$-P plane corresponding to bimodal responsiveness occupy a cusp-shaped region. This suggests the existence of a cusp catastrophe, with $H(o)$ and P as control variables and responsiveness, R, as the state variable. The model, based only on the observation of bimodality, also predicts that a patient cannot pass directly from hyperthyroidism to euthyroidism, but must first be guided to a hypothyroid state, and this is in accord with what is observed.

Seif also proposes a mechanism for the system. Catastrophe theory is not directly involved in the construction of the model but it does contribute to the analysis. By curve fitting, Seif obtains a transformation relating the variables $(R,H(o),P)$ to the (x,u,v) of the canonical cusp. In terms of the observed variables the equation of the equilibrium surface turns out to be

$$(a \log R - y_0)^3 + c(H(o) - H_0)(a \log R - y_0) + (P - P_0) = 0. \qquad (1)$$

Here, a, c, y_0, H_0, P are all parameters whose values are estimated empirically.

Now in analyzing his model of the system, Seif finds that the steady state solutions are of the form

$$\{\gamma+[(1+j+x_0)+(B-\kappa)u]^n\}u = n\rho[(k_3j - k_2x_0) - (k_3\kappa+k_2B)u][(1+j+x_0)+(B-\kappa)u]^{n-1} \quad (2)$$

Comparison of these two equations shows that n, which represents the number of secretory granules bound to a membrane, should be taken equal to 2, and the other quantities in (2), all of which are defined in terms of details of the model, can then be related to the phenomenologically determined quantities of (1).

This demonstrates how catastrophe theory can be used in otherwise conventional modelling. As usual, we write down the equations which we consider best describe the underlying processes, and we try to show that these predict the observed macroscopic behaviour. Unfortunately, if the system is at all complex, we may be unable to make any progress without a number of simplifying assumptions. Even a computer simulation will generally require good estimates of parameters whose values we do not know and cannot deduce save by exhaustive trial and error.

Catastrophe theory can assist by allowing us to use our observations directly to suggest the sorts of equations we are looking for, putting us in a position similar to the student who finds it much easier to solve an examination question if the answer is given on the paper so that he can work backwards from it. We do not have to know in advance what approximations we may make. Instead, we can determine which ones we must make if the model is to be consistent with the observations. This may lead to a greater understanding of the working of the system, by connecting certain internal features of it to the observed phenomena in ways which would otherwise be difficult to discover. Alternatively, we may be able to convince ourselves that no reasonable set of subsidiary hypotheses can save our model and that we may therefore reject it at once and not waste any more time on it.

The equations that are generally accepted to provide a good model of the nerve impulse were written down originally by Hodgkin & Huxley (1952) and have been studied extensively since. They permit steady state solutions which can bifurcate into periodic ones; this property makes them suitable for their purpose and also attracts the attention of mathematicians interested in bifurcations and catastrophes.

Recently Salgado Labouriau (1983) has applied Golubitsky & Langford's (1980) classification of degenerate Hopf bifurcations to the Hodgkin-Huxley equations. The existence of two organizing centres for previously known bifurcations was taken as suggesting that there should be a hidden, more degenerate organizing centre. This was found, and Salgado Labouriau was then able both to supply a topological explanation for the bifurcation diagrams that were already known and also to demonstrate the existence of others.

It is well known that the Hopf bifurcation is not an elementary catastrophe, and some see this as indicating that it must lie outside the scope of the theory.

Golubitsky & Langford, however, proved the Hopf theorem by catastrophe theoretic means. This enabled them immediately to extend it to the previously excluded degenerate cases, and it is on this extension that Salgado Labouriau's work depends.

This is a mathematical example of an approach which is of much more general applicability. Whenever we encounter a system whose behaviour suggests in some way the existence of a catastrophe, it is worth trying to express it in terms of catastrophe theory. At first sight this may seem no more than an exercise, and indeed there is no guarantee that it will not end up as just that, but experience shows that it can lead to real progress. Note, however, that it requires the mathematician to be in much closer contact with his subject matter than when he employs more conventional techniques; the point at which catastrophe theory makes its contribution may, as here, be in the mathematical analysis, but it may also be at a much earlier stage.

Some workers feel uncomfortable with catastrophe theory because they suspect it of trying to get something for nothing, and of claiming to break a sort of theoreticians' first law by getting more information out in the form of conclusions than was put in as data and hypotheses. In fact this is not the case. A complete model of a system contains a very great deal of information, and so requires a great deal of input. By comparison, the answer to one or two specific questions is a minute amount of information. It may be the most interesting information, but that is another issue altogether.

Statisticians have a criterion called 'efficiency' which they use to help them choose between two possible estimators of the same parameter. An efficient estimator is one that gives the desired accuracy on the basis of as few data as possible. It seems appropriate to extend the definition to mathematical techniques in general. We define an efficient technique as one that tells us what we want to know with the minimum of data and theoretical background as input. It generally does this by concentrating on the particular question that is asked and saying as little as possible about other matters.

In this sense, complete models are very inefficient, although they are highly desirable if we have enough information at our disposal to be able to afford the luxury. Catastrophe theoretic methods, in contrast, are generally very efficient. Put in this way, it becomes clear that the two approaches are trying to accomplish quite different things. If we insist on judging them on the same criteria (other than the ultimate one of whether we have learned anything from our analysis that we did not know before) it is hardly surprising that there will be controversy.

As applied mathematics extends its range far beyond physics, we are going to find ourselves more and more dealing with systems which are too complex to admit of modelling in the usual sense of the word. We shall have to engage far more in what Pielou (1981) calls 'investigating', the step by step elucidation of the properties of a system. This will require the development of new techniques which are efficient, in the sense we have just defined. Part of this is a task for pure mathematicians, and progress is already well under way. But there is also much to be done in learning how to use mathematics in this new context. This is the real challenge that catastrophe theory presents to applied mathematicians.

References:

Bazin, M.J. & Saunders, P.T. (1978). Determination of critical variables in a microbial predator-prey system. *Nature* (*London*), 275, 52-54.

Bazin, M.J. & Saunders, P.T. (1979). An application of catastrophe theory to the study of a switch in *Dictyostelium discoideum*. In *Kinetic Logic* (R. Thomas, ed.) Berlin: Springer.

Dent, V.E., Bazin, M.J. & Saunders, P.T. (1976). Behaviour of *Dictyostelium discoideum* amoebae and *Escherichia coli* grown together in chemostat culture. *Arch. Microbiol.*, 109, 187-194.

Golubitsky, M. & Langford, W.F. (1980). Classification and unfoldings of degenerate Hopf bifurcations. Preprint, Warwick University.

Hodgkin, A.L. & Huxley, A.F. (1952). A quantitative description of membrane current and its application to conduction and excitation in nerve. *J. Physiol.*, 117, 500-544.

Lewis, J., Slack, J.M. & Wolpert, L. (1977). Thresholds in development. *J. theor. Biol.*, 65, 579-590.

Lotka, A.J. (1924). *Elements of Physical Biology*. Baltimore: Williams & Wilkins.

Pan, P. & Wurster, B. (1978). Inactivation of the chemoattractant folic acid by cellular slime moulds and identification of the reaction product. *J. Bacteriol.*, 136, 955-959.

Pielou, E.C. (1981). The usefulness of ecological models: A stock-taking. *Q. Rev. Biol.*, 56, 17-31.

Salgado Labouriau, I. (1983). Degenerate Hopf bifurcation and nerve impulse. Preprint, Warwick University.

Saunders, P.T. (1980). *An Introduction to Catastrophe Theory*. Cambridge: Cambridge University Press.

Seif, F.J. (1979). Cusp bifurcation in pituitary thyrotropin secretion. In *Structural Stability in Physics* (W. Güttinger & H. Eikemeier, eds). Berlin: Springer.

Thom, R. (1972). *Stabilité Structurelle et Morphogénèse*. Reading: Benjamin.

Zeeman, E.C. (1974). Primary and secondary waves in developmental biology. In *Some Mathematical Questions in Biology* VIII (S.A. Levin, ed.). Providence: American Mathematical Society.

Zeeman, E.C. (1978). A dialogue between a mathematician and a biologist. *Biosciences Commun.*, 4, 225-240.

S-System Analysis of Biological Systems

Eberhard O. Voit and Michael A. Savageau

Zoologisches Institut der Universität zu Köln
Weyertal 119, 5000 Köln 41
Bundesrepublik Deutschland

Department of Microbiology and Immunology
The University of Michigan
6643 Medical Science II
Ann Arbor, MI
48109 U.S.A.

Introduction

For a long time, the major goal of biological and medical research was a more and more detailed analysis of the elementary parts of an organism. Now, in some of the simpler organisms like Escherichia coli, a very large number of proteins (Bloch et al., 1980) and many other compounds are known and the question arises how these elements work and interact in an integrated system. On a theoretical level, the first approach in answering this question has been a linear systems analysis that is based on the multitude of simple but powerful methods of linear mathematics. However, biological and medical phenomena often are quite nonlinear, and in most cases their underlying processes cannot simply be studied separately and then superimposed again, because these processes may influence each other. That is, a biological system as a whole frequently reacts differently than the sum of all its components. Such a system is called "cooperative" (if you like Latin) or "synergistic" (if you prefer Greek).

In our paper, we present a nonlinear formalism that mimics essential features of synergistic systems and that is yet simple enough to treat large numbers of mechanisms mathematically.

Derivation of the S-System

Let us assume that a synergistic system consists of n components X_i that interact with each other. Each interaction is assumed to be describable by a rational function. If we collect all interactions that increase the component X_i and those that decrease X_i, the change of X_i in time can be written as the difference of two positive rational functions (Savageau 1976, Savageau and Voit 1982):

$$\text{(1)} \qquad \dot{X}_i = R_i^+ (X_1,\ldots,X_n) - R_i^- (X_1,\ldots,X_n) \qquad (i=1,\ldots,n)$$

We approximate these rational functions by forming their Taylor series in a logarithmic space and retaining only the constant and linear terms. This procedure yields products of power functions in the corresponding Cartesian space:

$$\text{(2)} \qquad R_i^+ \simeq \alpha_i \prod_{j=1}^{n} X_j^{g_{ij}} \qquad R_i^- \simeq \beta_i \prod_{j=1}^{n} X_j^{h_{ij}}$$

$$\text{(3)} \qquad \dot{X}_i = \alpha_i \prod_{j=1}^{n} X_j^{g_{ij}} - \beta_i \prod_{j=1}^{n} X_j^{h_{ij}}$$

The time derivative of X_i is equal to a difference of two power functions that contain all variables X_j raised to certain powers. The parameters α_i and β_i are positive real constants, the doubly indexed parameters g_{ij} and h_{ij} may be positive or negative. We call this set of nonlinear differential equations an "S-system".

In general, S-systems are a valid approximation to the original equations within a local region surrounding the point about which the Taylor series expansion was performed. Consequently, the original equations and the S-system approximation will be dynamically equivalent for "small" perturbations about this nominal operating value.

Special Cases of the S-System

The S-system was originally derived as an approximation to the underlying rational-function nonlinearities of synergistic systems (hence, the term "S-system"). Nevertheless, many biological and non-biological laws and equations can be represented exactly as S-systems (Savageau 1980, Savageau 1982, Voit and Savageau 1983). Famous non-biological examples are Bessel functions, probability distribution functions and elliptical integrals and such physical laws as those describing electrical circuits, gravitation, unforced vibrations, and cooling and dilution processes. Biological examples include all well-known growth laws, the kinetics of chemical reactions, the spread of epidemics, including the Kermack-McKendrick model (in Busenberg 1983), Lotka's and Volterra's equations on predator-prey interactions and more intricate models on phage-host dynamics (e.g. in Bremermann 1983).

In order to show how such laws can be formulated exactly as S-systems, we have chosen the phenomenon of interspecific competition as an example from ecology. Two species are said to compete with one another, when they seek in the same community the same resources, such as food and space which are in short supply, or when they interact in a way that affects their growth and survival (Smith 1974). Growth or decrease of each of the two populations depends on the proliferation rate and the carrying capacity, i.e. the upper growth limit, that are typical for a species in its environment, but they also depend on size and growth of the other population. The system is usually described by a pair of differential equations (Czihak et al 1976):

$$\text{(4)}\qquad \begin{aligned} \dot{N}_1 &= r_1 N_1 - \frac{r_1}{k_1} N_1^{\,2} - \frac{r_1}{k_1} c_1 N_1 N_2 \\ \dot{N}_2 &= r_2 N_2 - \frac{r_2}{k_2} N_2^{\,2} - \frac{r_2}{k_2} c_2 N_2 N_1 \end{aligned}$$

The change in the number N_i of organisms of species i follows a logistic equation diminished by an "interaction term". r_i is the proliferation rate, k_i the carrying capacity of species i, c_1 and c_2 are called competitive coefficients for one species in relation to the other.

This pair of equations can be readily transformed into an S-system by defining new additional variables N_3 and N_4:

$$\text{(5)}\qquad N_3 = k_1 - N_1 \qquad\qquad N_4 = k_2 - N_2$$

N_1 and N_2 remain unchanged. With

$$\text{(6)}\qquad \alpha_i = r_i/k_i \qquad \beta_i = c_i \cdot \alpha_i \qquad\qquad (i=1,2)$$

the system (4) reads

$$\text{(7)}\qquad \begin{aligned} \dot{N}_1 &= \alpha_1 N_1 N_3 - \beta_1 N_1 N_2 \qquad & \dot{N}_3 &= \beta_1 N_1 N_2 - \alpha_1 N_1 N_3 \\ \dot{N}_2 &= \alpha_2 N_2 N_4 - \beta_2 N_1 N_2 \qquad & \dot{N}_4 &= \beta_2 N_1 N_2 - \alpha_2 N_2 N_4 \end{aligned}$$

Both methods of representation, the original and the S-system equivalent, use the same number of "parameters". The form (Eq.4) uses six coefficients and two initial conditions for the differential equations, whereas the S-system contains four initial conditions and four coefficients. They are in fact transformations of each other. It should be emphazised that these two representations are thus dynamically equivalent; they have the same steady states and the same dynamical properties.

The representation of well-known equations as S-systems is of academic interest but also of practical advantage. The academic progress lies in the fact that equations that showed no relation to each other before are now classified and categorized, like all the famous growth laws that were shown to represent a few parameter combinations in the continuous parameter space of an S-system. The practical advantage of one formalism that includes many examples as special cases is that the analysis of all phenomena describable in this formalism can be standardized. So, instead of writing new computer routines for each new example, one program can be written to solve, to analyze, and to illustrate the different aspects of the S-system.

Analysis of a Synergistic System

We would now like to demonstrate how the S-system formalism can be used to analyze synergistic systems. We will illustrate the different parts of the analysis (cf. Voit and Savageau 1982 b) with an example from biotechnology, the production of ethanol by yeast (cf. Voit and Savageau 1982 a).

The first step of an S-system analysis is the model design. All components of the synergistic system that could possibly be relevant have to be included in the S-system as variables X_i. Also, all possible interactions among the components must be considered. If, in fact, some of the included components or interactions are not relevant for the system behavior, it will become evident during the analysis. For illustration, an arrow diagram can be drawn that contains components and interactions. The graph for our example is shown in Figure 1.

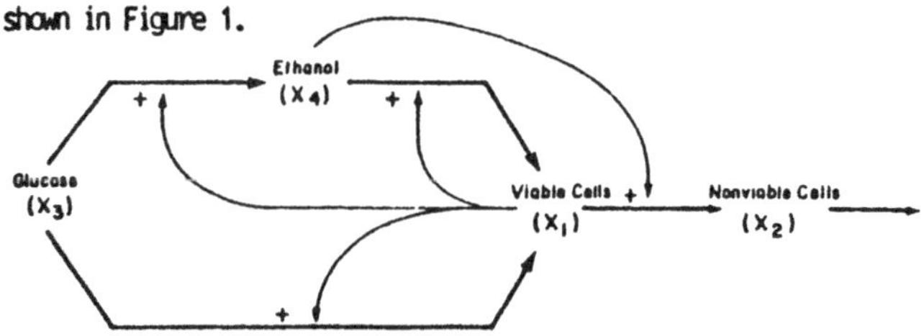

Figure 1: Arrow diagram describing the production of alcohol by yeast (Voit and Savageau 1982 a). See text for details.

The thick arrows show the flow of material in the system, the thin arrows heading from a component X_i to the center of a thick arrow from X_j to X_k signify that the flow from X_j to X_k is modified by the component X_i. Our example only contains activations as indicated by the plus signs at the thin arrowheads. Inhibitory effects would be represented as minus signs.

In our example, viable cells (X_1) use glucose (X_3) and perhaps ethanol (X_4) as substrate and they activate the transformation of glucose into alcohol. The viable cells eventually die as represented by the thick arrow from X_1 to X_2, a process that is amplified by alcohol. Finally, the nonviable cells may disappear by lysis.

The corresponding S-system is derived from the arrow diagram. The change in each component X_i is approximated by a difference of products of power functions as shown earlier (cf. Eq.(3)). These products contain all components that contribute to the change in X_i. The S-system in our example is shown in Eq.(8).

$$\begin{aligned}
\dot{X}_1 &= \alpha_1 X_1^{g_{11}} X_3^{g_{13}} X_4^{g_{14}} - \beta_1 X_1^{h_{11}} X_4^{h_{14}} \\
\dot{X}_2 &= \beta_1 X_1^{h_{11}} X_4^{h_{14}} - \beta_2 X_2^{h_{22}} \\
\dot{X}_3 &= - \beta_3 X_1^{h_{31}} X_3^{h_{33}} \\
\dot{X}_4 &= \alpha_4 X_1^{g_{41}} X_3^{g_{43}} - \beta_4 X_1^{h_{41}} X_4^{h_{44}}
\end{aligned} \tag{8}$$

As an example, let us study the first equation which describes the change in the number of viable cells. The cells proliferate autocatalytically, therefore, X_1 is present in the α-term, which represents all increasing influences. The proliferation depends on the substrate glucose, hence, X_3 is also included in the α-term. Because it could be that ethanol is used as an additional substrate, X_4 is included a. well. The degradation of X_1 is represented by the β-term. It contains the concentration of viable cells X_1 and the role ethanol plays for the death of cells. The other equations are constructed analogously.

The next step of the S-system analysis is the determination of values for the α and β parameters and for the doubly indexed parameters g_{ij} and h_{ij}. If experimental data are being analyzed, one approach for finding the optimal set of parameter values is to recall the geometrical meaning of the derivative as a slope. If for all time points the data and the measured slopes are plugged in the S-system, an algebraic system is obtained from which the desired parameter values can be estimated by algebraic or statistical methods.

When all parameter values are specified, the S-system is uniquely determined except for the initial conditions. These can be subject to a least-sqare fitting procedure or they can be estimated from the data with a method described elsewhere (Voit and Savageau 1982 b).

The S-system consists of simultaneous, nonlinear, first-order differential equations. An analytical solution was found for a class of S-systems that contains the well-known growth laws and probability distribution functions (Voit and Savageau 1984). However, since elliptical integrals can be transformed into S-systems, and since it is known that these integrals have no solution in terms of elementary mathematical functions, it is proven that S-systems in general cannot be solved analytically. In cases without an analytical solution, the algorithms by Runge and Kutta and by Gear (in Voit and Savageau 1982 b) proved efficient in solving S-systems numerically.

In our example of ethanol production by yeast, we analyzed data by Lee and Wang (1982). The optimal fit to their data is shown in Figure 2.

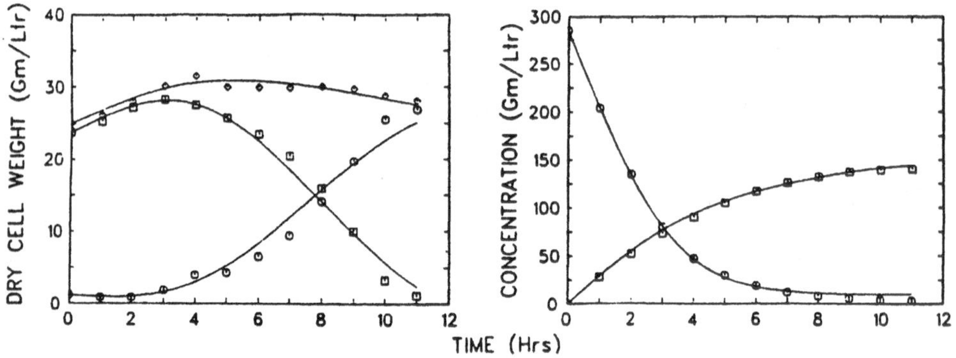

Figure 2: Optimal fit (Voit and Savageau 1982a) to the data of Lee and Wang (1982). Left panel: ▫ viable cells, o: nonviable cells, ◇: total dry weight; right panel: o : glucose, ▫ ethanol; abscissa: time (hrs). See text for further information.

The left panel in Fig. 2 shows how the model mimics the increase and decrease in viable cells and the accumulation of dead cells. The upper curve represents the total cell mass. Its decrease made it necessary to include a mechanism like lysis. The right panel shows the utilization of glucose and the production of ethanol. We compared different S-system descriptions, including and excluding lysis of nonviable cells and including and excluding alcohol catabolism by the viable cells. Statistical tests objectively showed that considering alcohol catabolism did not improve the fit significantly whereas the inclusion of lysis was worth the additional parameters.

Summary

The S-system formalism is a tool for analyzing nonlinear synergistic systems. It is derived from the very general assumption that the system nonlinearities follow rational functions. Because the approximation uses Taylor's theorem, it is guaranteed to be an accurate representation at least over a certain range of values. This range is considerably increased compared to a linear analysis, since the nonlinearities are approximated in a logarithmic space. Many experimental regulatory systems from biochemistry and genetics are validly described by the S-system formalism. Well-known biological and physical laws and mathematical functions can be written as S-systems.

The generality of the S-system approach to modeling biological systems makes it worthwhile to write a comprehensive computer program that reduces most parts of the analysis to a routine process. It allows the analyst to focus on model design and interpretation of results. These two parts require creative thinking and cannot yet be done by computer.

References

Bloch, P.L., Phillips, T.A., and Neidhardt, F.C.: Protein identifications on O'Farrell two-dimensional gels: Locations of 81 Escherichia coli proteins. J. Bacteriology 141,3, 1409-1420 (1980)

Bremermann,H.J.: Parasites at the origin of life. J.Math.Biol. 16, 165-180 (1983)

Busenberg,S.N., and Travis, C.C.: Epidemic models with spatial spread due to population migration. J.Math.Biol. 16, 181-198 (1983)

Czihak, G., Langer, H., and Ziegler, H.: Biologie. Springer-Verlag Heidelberg, New York (1976), pp.711-712.

Lee, S.S., and Wang, H.Y.: Repeated fed-batch rapid fermentation using yeast cells and an activated carbon extraction system. Biotech.Bioengr.Symp. 12, 221-231 (1982)

Savageau, M.A.: Biochemical Systems Analysis: A Study of Function and Design in Molecular Biology. Addison-Wesley, Reading, MA (1976)

Savageau, M.A.: Growth of complex systems can be related to the properties of their underlying determinants. Proc.Natl.Acad.Sci.U.S.A. 76,11, 5413-5417 (1979 a)

Savageau, M.A.: Allometric morphogenesis of complex systems: Derivation of the basic equation from first principles. Proc.Natl.Acad.Sci.U.S.A. 76,12, 6023-6025 (1979 b)

Savageau, M.A.: Growth equations: A general equation and a survey of special cases. Math. Biosci. 48, 267-278 (1980)

Savageau, M.A.: A suprasystem of probability distributions. Biometrical J. 24,1-8 (1982)

Savageau, M.A., and Voit, E.O.: Power-law approach to modeling biological systems, I. Theory. J.Ferment. Technol. 60,3, 221-228 (1982)

Smith, R.L.: Ecology and Field Biology. Harper & Row, Publ., New York, Evanston, San Francisco, London (1974)

Voit, E.O., and Savageau, M.A.: Power-law approach to modeling biological systems, II. Application to ethanol production. J.Ferment. Technol. 60,3, 229-232 (1982 a)

Voit, E.O., and Savageau, M.A.: Power-law approach to modeling biological systems, III. Methods of analysis. J.Ferment. Technol. 60,3, 233-241 (1982 b)

Voit, E.O., and Savageau, M.A.: Analytical solutions to a generalized growth equation. J. Math. Analysis and Appl. (in press 1984)

721/7.5-11.10.

Biomathematics

Managing Editor: **S. A. Levin**

Volume 9
W. J. Ewens

Mathematical Population Genetics

1979. 4 figures, 17 tables. XII, 325 pages.
ISBN-13: 978-3-540-15200-2

This graduate level monograph considers the mathematical theory of population genetics, emphasizing aspects relevant to evolutionary studies. It contains a definitive and comprehensive discussion of relevant areas with references to the essential literature. The sound presentation and excellent exposition make this book a standard for population geneticists interested in the mathematical foundations of their subject as well as for mathematicians involved with genetic ecolutionary processes.

Volume 10
A. Okubo

Diffusion and Ecological Problems: Mathematical Models

1980. 114 figures, 6 tables. XIII, 254 pages.
ISBN-13: 978-3-540-15200-2

This is the first comprehensive book on mathematical models of diffusion in an ecological context. Directed towards applied mathematicians, physicists and biologists, it gives a sound, biologically oriented treatment of the mathematics and physics of diffusion.

Volume 11
B. G. Mirkin, S. N. Rodin

Graphs and Genes

Translated from the Russian by H. L. Beus
1984. 46 figures. XIV, 197 pages. ISBN 3-540-12657-0

Contents: Graphs in the analysis of gene structure. - Graphs in the analysis of gene semantics. - Graphs in the analysis of gene evolution. - Epilogue: Cryptographic problems in genetics. - Appendix: Some notions about graphs. - References. - Index of genetics terms. - Index of mathematical terms.

Springer-Verlag
Berlin
Heidelberg
New York
Tokyo

Lecture Notes in Biomathematics

Vol. 1: P. Waltman, Deterministic Threshold Models in the Theory of Epidemics. V, 101 pages. 1974.

Vol. 2: Mathematical Problems in Biology, Victoria Conference 1973. Edited by P. van den Driessche. VI, 280 pages. 1974.

Vol. 3: D. Ludwig, Stochastic Population Theories. VI, 108 pages. 1974.

Vol. 4: Physics and Mathematics of the Nervous System. Edited by M. Conrad, W. Güttinger, and M. Dal Cin. XI, 584 pages. 1974.

Vol. 5: Mathematical Analysis of Decision Problems in Ecology. Proceedings 1973. Edited by A. Charnes and W. R. Lynn. VIII, 421 pages. 1975.

Vol. 6: H. T. Banks, Modeling and Control in the Biomedical Sciences. V. 114 pages. 1975.

Vol. 7: M. C. Mackey, Ion Transport through Biological Membranes, An Integrated Theoretical Approach. IX, 240 pages. 1975.

Vol. 8: C. DeLisi, Antigen Antibody Interactions. IV, 142 pages. 1976.

Vol. 9: N. Dubin, A Stochastic Model for Immunological Feedback in Carcinogenesis: Analysis and Approximations. XIII, 163 pages. 1976.

Vol. 10: J. J. Tyson, The Belousov-Zhabotinskii Reaktion. IX, 128 pages. 1976.

Vol. 11: Mathematical Models in Medicine. Workshop 1976. Edited by J. Berger, W. Bühler, R. Repges, and P. Tautu. XII, 281 pages. 1976.

Vol. 12: A. V. Holden, Models of the Stochastic Activity of Neurones. VII, 368 pages. 1976.

Vol. 13: Mathematical Models in Biological Discovery. Edited by D. L. Solomon and C. Walter. VI, 240 pages. 1977.

Vol. 14: L. M. Ricciardi, Diffusion Processes and Related Topics in Biology. VI, 200 pages. 1977.

Vol. 15: Th. Nagylaki, Selection in One- and Two-Locus Systems. VIII, 208 pages. 1977.

Vol. 16: G. Sampath, S. K. Srinivasan, Stochastic Models for Spike Trains of Single Neurons. VIII, 188 pages. 1977.

Vol. 17: T. Maruyama, Stochastic Problems in Population Genetics. VIII, 245 pages. 1977.

Vol. 18: Mathematics and the Life Sciences. Proceedings 1975. Edited by D. E. Matthews. VII, 385 pages. 1977.

Vol. 19: Measuring Selection in Natural Populations. Edited by F. B. Christiansen and T. M. Fenchel. XXXI, 564 pages. 1977.

Vol. 20: J. M. Cushing, Integrodifferential Equations and Delay Models in Population Dynamics. VI, 196 pages. 1977.

Vol. 21: Theoretical Approaches to Complex Systems. Proceedings 1977. Edited by R. Heim and G. Palm. VI, 244 pages. 1978.

Vol. 22: F. M. Scudo and J. R. Ziegler, The Golden Age of Theoretical Ecology: 1923–1940. XII, 490 pages. 1978.

Vol. 23: Geometrical Probability and Biological Structures: Buffon's 200th Anniversary. Proceedings 1977. Edited by R. E. Miles and J. Serra. XII, 338 pages. 1978.

Vol. 24: F. L. Bookstein, The Measurement of Biological Shape and Shape Change. VIII, 191 pages. 1978.

Vol. 25: P. Yodzis, Competition for Space and the Structure of Ecological Communities. VI, 191 pages. 1978.

Vol. 26: M. B Katz, Questions of Uniqueness and Resolution in Reconstruction from Projections. IX, 175 pages. 1978.

Vol. 27: N. MacDonald, Time Lags in Biological [illegible] pages. 1978.

Vol. 28: P. C. Fife, Mathematical Aspects of Reactin[illegible] Systems. IV, 185 pages. 1979.

Vol. 29: Kinetic Logic – A Boolean Approach to [illegible] Complex Regulatory Systems. Proceedings, 1977. Edited by R. Thomas. XIII, 507 pages. 1979.

Vol. 30: M. Eisen, Mathematical Models in Cell Biology and Cancer Chemotherapy. IX, 431 pages. 1979.

Vol. 31: E. Akin, The Geometry of Population Genetics. IV, 205 pages. 1979.

Vol. 32: Systems Theory in Immunology. Proceedings, 1978. Edited by G. Bruni et al. XI, 273 pages. 1979.

Vol. 33: Mathematical Modelling in Biology and Ecology. Proceedings, 1979. Edited by W. M. Getz. VIII, 355 pages. 1980.

Vol. 34: R. Collins, T. J. van der Werff, Mathematical Models of the Dynamics of the Human Eye. VII, 99 pages. 1980.

Vol. 35: U. an der Heiden, Analysis of Neural Networks. X, 159 pages. 1980.

Vol. 36: A. Wörz-Busekros, Algebras in Genetics. VI, 237 pages. 1980.

Vol. 37: T. Ohta, Evolution and Variation of Multigene Families. VIII, 131 pages. 1980.

Vol. 38: Biological Growth and Spread: Mathematical Theories and Applications. Proceedings, 1979. Edited by W. Jäger, H. Rost and P. Tautu. XI, 511 pages. 1980.

Vol. 39: Vito Volterra Symposium on Mathematical Models in Biology. Proceedings, 1979. Edited by C. Barigozzi. VI, 417 pages. 1980.

Vol. 40: Renewable Resource Management. Proceedings, 1980. Edited by T. Vincent and J. Skowronski. XII, 236 pages. 1981.

Vol. 41: Modèles Mathématiques en Biologie. Proceedings, 1978. Edited by C. Chevalet and A. Micali. XIV, 219 pages. 1981.

Vol. 42: G. W. Swan, Optimization of Human Cancer Radiotherapy. VIII, 282 pages. 1981.

Vol. 43: Mathematical Modeling on the Hearing Process. Proceedings, 1980. Edited by M. H. Holmes and L. A. Rubenfeld. V, 104 pages. 1981.

Vol. 44: Recognition of Pattern and Form. Proceedings, 1979. Edited by D. G. Albrecht. III, 226 pages. 1982.

Vol. 45: Competition and Cooperation in Neutral Nets. Proceedings, 1982. Edited by S. Amari and M. A. Arbib. XIV, 441 pages. 1982.

Vol. 46: E. Walter, Identifiability of State Space Models with applications to transformation systems. VIII, 202 pages. 1982.

Vol. 47: E. Frehland, Stochastic Transport Processes in Discrete Biological Systems. VIII, 169 pages. 1982.

Vol. 48: Tracer Kinetics and Physiologic Modeling. Proceedings, 1983. Edited by R. M. Lambrecht and A. Rescigno. VIII, 509 pages. 1983.

Vol. 49: Rhythms in Biology and Other Fields of Application. Proceedings, 1981. Edited by M. Cosnard, J. Demongeot and A. Le Breton. VII, 400 pages. 1983.

Vol. 50: D. H. Anderson, Compartmental Modeling and Tracer Kinetics. VII, 302 pages. 1983.

Vol. 51: Oscillations in Mathematical Biology. Proceedings, 1982. Edited by J. P. E. Hodgson. VI, 196 pages. 1983.

Vol. 52: Population Biology. Proceedings, 1982. Edited by H. I. Freedman and C. Strobeck. XVII, 440 pages. 1983.

Vol. 53: Evolutionary Dynamics of Genetic Diversity. Proceedings, 1983. Edited by G. S. Mani. VII, 312 pages. 1984.

Vol. 54: Mathematical Ecology. Proceedings, 1982. Edited by S. A. Levin and T. G. Hall[illegible] 513 pages. 1984.

[illegible] n Space and Time. Proceedings, [illegible]). Murray. VIII, 405 pages. 1984.

[illegible], Gonorrhea Transmission Dynam- [illegible]984.

[illegible] Biology and Medicine. Proceedings, 1983. Edited by V. Capasso, E. Grosso and S. L. Paveri-Fontana. XVIII, 524 pages. 1985.